SOLAR AND SPACE PHYSICS

A Science for a Technological Society

Committee on a Decadal Strategy for Solar and Space Physics (Heliophysics)

Space Studies Board

Aeronautics and Space Engineering Board

Division on Engineering and Physical Sciences

NATIONAL RESEARCH COUNCIL
OF THE NATIONAL ACADEMIES

THE NATIONAL ACADEMIES PRESS
Washington, D.C.
www.nap.edu

THE NATIONAL ACADEMIES PRESS **500 Fifth Street, NW** **Washington, DC 20001**

NOTICE: The project that is the subject of this report was approved by the Governing Board of the National Research Council, whose members are drawn from the councils of the National Academy of Sciences, the National Academy of Engineering, and the Institute of Medicine. The members of the committee responsible for the report were chosen for their special competences and with regard for appropriate balance.

This study is based on work supported by Contract NNH06CE15B between the National Academy of Sciences and the National Aeronautics and Space Administration and Grant AGS-1050550 between the National Academy of Sciences and the National Science Foundation. Any opinions, findings, conclusions, or recommendations expressed in this publication are those of the authors and do not necessarily reflect the views of the agencies that provided support for the project.

International Standard Book Number-3: 978-0-309-16428-3
International Standard Book Number-0: 0-309-16428-1
Library of Congress Control Number: 2013940083

Cover: Insets, from the top: Solar Dynamics Observatory, full-disk multiwavelength extreme ultraviolet image of the Sun (NASA/GSFC/AIA); depiction of a coronal mass ejection and Earthís magnetosphere (Steele Hill/NASA); Advanced Modular Incoherent Scatter Radars at Resolute Bay (Craig Heinselman, EISCAT); ultraviolet image of the southern polar region of Saturn with its aurora (here shown inverted) from the Hubble Space Telescope (NASA/ESA/J. Clarke, Boston University/Z. Levay, STScI); depiction of the regions of the heliosphere (NASA); and bow shock around the star LL Orionis (NASA, ESA, and the Hubble Heritage Team (STScI/AURA). Background images: View from the International Space Station of the Midwestern United States and the aurora borealis (NASA).

Copies of this report are available free of charge from:

Space Studies Board
National Research Council
500 Fifth Street, NW
Washington, DC 20001

Additional copies of this report are available from the National Academies Press, 500 Fifth Street, NW, Keck 360, Washington, DC 20001; (800) 624-242 or (202) 334-3133; http://www.nap.edu.

Printed in the United States of America

OTHER RECENT REPORTS OF THE SPACE STUDIES BOARD AND THE AERONAUTICS AND SPACE ENGINEERING BOARD

Assessment of a Plan for U.S. Participation in Euclid (Board on Physics and Astronomy [BPA] with Space Studies Board [SSB], 2012)
Assessment of Planetary Protection Requirements for Spacecraft Missions to Icy Solar System Bodies (SSB, 2012)
Continuing Kepler's Quest: Assessing Air Force Space Command's Astrodynamics Standards (Aeronautics and Space Engineering Board [ASEB], 2012)
Earth Science and Applications from Space: A Midterm Assessment of NASA's Implementation of the Decadal Survey (SSB, 2012)
The Effects of Solar Variability on Earth's Climate: A Workshop Report (SSB, 2012)
NASA Space Technology Roadmaps and Priorities: Restoring NASA's Technological Edge and Paving the Way for a New Era in Space (ASEB, 2012)
NASA's Strategic Direction and the Need for a National Consensus (Division on Engineering and Physical Sciences, 2012)
Recapturing NASA's Aeronautics Flight Research Capabilities (ASEB, 2012)
Report of the Panel on Implementing Recommendations from the New Worlds, New Horizons Decadal Survey (BPA and SSB, 2012)
Reusable Booster System: Review and Assessment (ASEB, 2012)
Technical Evaluation of the NASA Model for Cancer Risk to Astronauts Due to Space Radiation (SSB, 2012)

Assessment of Impediments to Interagency Collaboration on Space and Earth Science Missions (SSB, 2011)
An Interim Report on NASA's Draft Space Technology Roadmaps (ASEB, 2011)
Limiting Future Collision Risk to Spacecraft: An Assessment of NASA's Meteoroid and Orbital Debris Programs (ASEB, 2011)
Panel Reports—New Worlds, New Horizons in Astronomy and Astrophysics (BPA and SSB, 2011)
Preparing for the High Frontier—the Role and Training of NASA Astronauts in the Post-Space Shuttle Era (ASEB, 2011)
Recapturing a Future for Space Exploration: Life and Physical Sciences Research for a New Era (SSB with ASEB, 2011)
Sharing the Adventure with the Public—The Value and Excitement of "Grand Questions" of Space Science and Exploration: Summary of a Workshop (SSB, 2011)
Summary of the Workshop to Identify Gaps and Possible Directions for NASA's Meteoroid and Orbital Debris Programs (ASEB, 2011)
Vision and Voyages for Planetary Science in the Decade 2013-2022 (SSB, 2011)

Advancing Aeronautical Safety: A Review of NASA's Aviation Safety-Related Research Programs (ASEB, 2010)
Capabilities for the Future: An Assessment of NASA Laboratories for Basic Research (Laboratory Assessments Board with SSB and ASEB, 2010)
Controlling Cost Growth of NASA Earth and Space Science Missions (SSB, 2010)
Defending Planet Earth: Near-Earth-Object Surveys and Hazard Mitigation Strategies (SSB with ASEB, 2010)
An Enabling Foundation for NASA's Space and Earth Science Missions (SSB, 2010)
Life and Physical Sciences Research for a New Era of Space Exploration: An Interim Report (SSB with ASEB, 2010)
New Worlds, New Horizons in Astronomy and Astrophysics (BPA and SSB, 2010)
Revitalizing NASA's Suborbital Program: Advancing Science, Driving Innovation, and Developing a Workforce (SSB, 2010)

Limited copies of SSB reports are available free of charge from

Space Studies Board
National Research Council
The Keck Center of the National Academies
500 Fifth Street, NW, Washington, DC 20001
(202) 334-4777/ssb@nas.edu
www.nationalacademies.org/ssb/ssb.html

COMMITTEE ON A DECADAL STRATEGY FOR SOLAR AND SPACE PHYSICS (HELIOPHYSICS)

DANIEL N. BAKER, University of Colorado, Boulder, *Chair*
THOMAS H. ZURBUCHEN, University of Michigan, *Vice Chair*
BRIAN J. ANDERSON, Johns Hopkins University, Applied Physics Laboratory
STEVEN J. BATTEL, Battel Engineering
JAMES F. DRAKE, JR., University of Maryland, College Park
LENNARD A. FISK, University of Michigan
MARVIN A. GELLER, Stony Brook University
SARAH GIBSON, National Center for Atmospheric Research
MICHAEL HESSE, NASA Goddard Space Flight Center
J. TODD HOEKSEMA,* Stanford University
MARY K. HUDSON,* Dartmouth College
DAVID L. HYSELL, Cornell University
THOMAS J. IMMEL, University of California, Berkeley
JUSTIN KASPER, Harvard-Smithsonian Center for Astrophysics
JUDITH L. LEAN, Naval Research Laboratory
RAMON E. LOPEZ, University of Texas, Arlington
HOWARD J. SINGER, NOAA Space Weather Prediction Center
HARLAN E. SPENCE, University of New Hampshire
EDWARD C. STONE, California Institute of Technology

*An asterisk indicates additional service on the survey's Solar Probe Plus Study Group, which was chaired by Louis J. Lanzerotti, New Jersey Institute of Technology.

PANEL ON ATMOSPHERE-IONOSPHERE-MAGNETOSPHERE INTERACTIONS

JEFFREY M. FORBES, University of Colorado, Boulder, *Chair*
JAMES H. CLEMMONS, Aerospace Corporation, *Vice Chair*
ODILE de la BEAUJARDIERE, Air Force Research Laboratory
JOHN V. EVANS, COMSAT Corporation (retired)
RODERICK A. HEELIS,* University of Texas, Dallas
THOMAS J. IMMEL, University of California, Berkeley
JANET U. KOZYRA, University of Michigan
WILLIAM LOTKO, Dartmouth College
GANG LU, High Altitude Observatory
KRISTINA A. LYNCH, Dartmouth College
JENS OBERHEIDE, Clemson University
LARRY J. PAXTON, Johns Hopkins University, Applied Physics Laboratory
ROBERT F. PFAFF, NASA Goddard Space Flight Center
JOSHUA SEMETER, Boston University
JEFFREY P. THAYER, University of Colorado, Boulder

PANEL ON SOLAR WIND–MAGNETOSPHERE INTERACTIONS

MICHELLE F. THOMSEN, Los Alamos National Laboratory, *Chair*
MICHAEL WILTBERGER, National Center for Atmospheric Research, *Vice Chair*
JOSEPH BOROVSKY, Los Alamos National Laboratory
JOSEPH F. FENNELL, Aerospace Corporation
JERRY GOLDSTEIN, Southwest Research Institute
JANET C. GREEN, National Oceanic and Atmospheric Administration
DONALD A. GURNETT, University of Iowa
LYNN M. KISTLER, University of New Hampshire
MICHAEL W. LIEMOHN, University of Michigan
ROBYN MILLAN, Dartmouth College
DONALD G. MITCHELL, Johns Hopkins University, Applied Physics Laboratory
TAI D. PHAN, University of California, Berkeley
MICHAEL SHAY, University of Delaware
HARLAN E. SPENCE, University of New Hampshire
RICHARD M. THORNE, University of California, Los Angeles

*An asterisk indicates additional service on the survey's Solar Probe Plus Study Group, which was chaired by Louis J. Lanzerotti, New Jersey Institute of Technology.

PANEL ON SOLAR AND HELIOSPHERIC PHYSICS

RICHARD A. MEWALDT, California Institute of Technology, *Chair*
SPIRO K. ANTIOCHOS,* NASA Goddard Space Flight Center, *Vice Chair*
TIMOTHY S. BASTIAN, National Radio Astronomy Observatory
JOE GIACALONE, University of Arizona
GEORGE M. GLOECKLER,* University of Michigan and University of Maryland (emeritus professor)
JOHN W. HARVEY,* National Solar Observatory
RUSSELL A. HOWARD, Naval Research Laboratory
JUSTIN KASPER, Harvard-Smithsonian Center for Astrophysics
ROBERT P. LIN,† University of California, Berkeley
GLENN M. MASON, Johns Hopkins University, Applied Physics Laboratory
EBERHARD MOEBIUS, University of New Hampshire
MERAV OPHER, Boston University
JESPER SCHOU, Stanford University
NATHAN A. SCHWADRON, Boston University
AMY R. WINEBARGER, NASA Marshall Space Flight Center
DANIEL WINTERHALTER, Jet Propulsion Laboratory
THOMAS N. WOODS, University of Colorado, Boulder

STAFF

ARTHUR A. CHARO, Senior Program Officer, Space Studies Board, *Study Director*
ABIGAIL A. SHEFFER, Associate Program Officer, Space Studies Board

MAUREEN MELLODY, Program Officer, Aeronautics and Space Engineering Board
LEWIS B. GROSWALD, Research Associate, Space Studies Board
CATHERINE A. GRUBER, Editor, Space Studies Board
DANIELLE PISKORZ, Lloyd V. Berkner Space Policy Intern
LINDA M. WALKER, Senior Program Assistant, Space Studies Board
TERRI BAKER, Senior Program Assistant, Space Studies Board (until April 2012)
BRUNO SÁNCHEZ-ANDRADE NUÑO, National Academies Christine Mirzayan Science and Technology Policy Fellow
HEATHER D. SMITH, National Academies Christine Mirzayan Science and Technology Policy Fellow

MICHAEL H. MOLONEY, Director, Space Studies Board, and Director, Aeronautics and Space Engineering Board

*An asterisk indicates additional service on the survey's Solar Probe Plus Study Group, which was chaired by Louis J. Lanzerotti, New Jersey Institute of Technology.
†Dr. Lin died on November 17, 2012.

SPACE STUDIES BOARD

[†]Dr. Brill died on March 27, 2013.
[‡]Dr. Lin died on November 17, 2012.

AERONAUTICS AND SPACE ENGINEERING BOARD

Preface

Strategic planning activities within NASA's Science Mission Directorate (SMD) and several National Science Foundation (NSF) divisions draw heavily on reports issued by the National Research Council (NRC), particularly those from the Space Studies Board (SSB). Principal among these SSB inputs is identification of priority science and missions and facilities in the decadal science strategy surveys. The first true decadal strategy for the field of solar and space physics, *The Sun to the Earth—and Beyond: A Decadal Research Strategy in Solar and Space Physics,* was published in 2003. That comprehensive study reviewed relevant research and applications activities, listed the key science questions, and recommended specific spacecraft missions and ground-based facilities and programs for the period 2003-2012. Supplemented by several subsequent SSB studies—for example, *A Performance Assessment of NASA's Heliophysics Program* (2009); *Distributed Arrays of Small Instruments for Solar-Terrestrial Research: Report of a Workshop* (2006); *Plasma Physics of the Local Cosmos* (2004); *Exploration of the Outer Heliosphere and the Local Interstellar Medium: A Workshop Report* (2004); and *The Role of Solar and Space Physics in NASA's Space Exploration Initiative* (2004)—the 2003 survey report provided key guidance for the SMD's solar and space physics (called *heliophysics* at NASA) programs and NSF's related atmospheric and geosciences programs during the first decade of the 21st century.

The successful initiation of many of the missions and programs recommended in the preceding studies, combined with important discoveries by a variety of ground- and space-based research activities, demonstrated the need for a second decadal survey of solar and space physics. Thus, in March 2010, Edward J. Weiler, NASA's associate administrator for the SMD, requested that a new decadal strategy survey be initiated (Appendix A). The request was seconded by the leadership of NSF's Division of Atmospheric and Geospace Sciences. Specific tasks outlined in the request included the following:

- An overview of the science and a broad survey of the current state of knowledge in the field, including a discussion of the relationship between space- and ground-based science research and its connection to other scientific areas;
- Determination of the most compelling science challenges that have arisen from recent advances and accomplishments;

- Identification—having considered scientific value, urgency, cost category and risk, and technical readiness—of the highest-priority science targets for the interval 2013-2022, recommending science objectives and measurement requirements for each target rather than specific mission or project design/implementation concepts; and
- Development of an integrated research strategy that will present means to address these targets.

In response to this request, the NRC appointed the 19-member Committee on a Decadal Strategy for Solar and Space Physics (Heliophysics) and 86 additional experts organized into three discipline panels—the Panel on Atmosphere-Ionosphere-Magnetosphere Interactions, the Panel on Solar Wind-Magnetosphere Interactions, and the Panel on Solar and Heliospheric Physics—and five informal working groups. The discipline panels were charged with the task of defining the current state of research in their discipline and determining priorities for scientific investigations in those areas. One member of each panel also served on the survey committee as a designated liaison. The working groups—Theory, Modeling, and Data Exploitation; Explorers, Suborbital, and Other Platforms; Innovations: Technology, Instruments, and Data Systems; Research to Operations/Operations to Research; and Education and Workforce—were similarly charged to assess the state of the field in these cross-cutting areas and to determine areas of critical needs. These working groups were composed approximately half by members of the survey committee and panels and half by volunteer consultants. The members of the working groups are listed in the section that follows this preface. In total, the survey committee, discipline panels, and working groups involved 105 members. This effort has been supported by NASA and NSF.

Work on the decadal survey began in August 2010 with preparations for the first meeting of the survey committee in September 2010. The survey committee met to consider the survey charge; to hear from NASA and NSF regarding their expectations for the survey, as well as from the National Oceanic and Atmospheric Administration and representatives from previous NRC decadal survey committees; and to determine the set of tasks for the discipline panels and working groups. The survey committee held a total of six meetings (September 1-3, 2010, and in 2011, February 1-3, April 12-14, June 14-16, August 29-21, and November 16-18).

The discipline panels each convened first in November 2010 and held a total of three meetings:

- Panel on Atmosphere-Ionosphere-Magnetosphere Interactions—November 15-17, 2010, January 12-14, 2011, and June 1-3, 2011;
- Panel on Solar Wind-Magnetosphere Interactions—November 17-19, 2010, January 18-20, 2011, and June 20-21, 2011;
- Panel on Solar and Heliospheric Physics—November 29-December 1, 2010, January 10-12, 2011, and May 25-27, 2011.

The working groups held one meeting each, with the exception of the Research to Operations/Operations to Research group, which held both a town hall meeting and a follow-up meeting. Each of the five working groups reported its progress during the survey committee meetings and by writing a white paper report to summarize its findings.

One significant difference from the 2003 decadal survey of solar and space physics was the decision to contract with the Aerospace Corporation to perform an independent cost and technical evaluation (CATE) of notional missions. This effort was made to increase the cost realism of notional missions and to facilitate cost comparisons between missions. The survey committee and panels reviewed 288 mission concept white papers, which were submitted in response to an invitation to the research community.[1] The panels mapped concepts against their prioritization of science targets and also considered factors such as technical readiness, scientific impact on particular disciplines, and, in some cases, operational utility. Then the survey committee selected 12 concepts for further study and CATE assessment. At the end of that process, six concepts were chosen for consideration, leading to the survey committee's recommendations of priorities. The details of this process are described in Appendix E.

In June 2011, NASA requested a clarification of the survey's scope to explicitly consider the Solar Probe Plus (SPP) mission. Previously, NASA had instructed the committee to assume that SPP and Solar Orbiter, along with missions that were in advanced development, were part of the baseline program and should not be subject to review or prioritization.[2] This new request specified that the SPP mission was not to be reprioritized, but that the survey committee should comment on the scientific rationale for the mission in the context of scientific developments since the publication of the 2003 decadal survey. Also, the survey committee was asked to provide appropriate programmatic or cost triggers as part of the anticipated decision rules to guide NASA in the event of major technical, cost, or programmatic changes during the development of SPP.

To address the change in the survey's scope, the NRC formed a study group specifically tasked to address these questions for the Solar Probe Plus mission. The group was chosen to minimize conflicts of interest while maintaining relevant scientific and technical expertise. The Solar Probe Plus Study Group was led by Louis J. Lanzerotti, New Jersey Institute of Technology, who served as an unpaid consultant to the survey. The remainder of the study group was made up of members of the survey committee and the discipline panels. The study group held one meeting in August 2011 and presented its conclusions in an internal report to the survey committee in October 2011.

The three discipline panels cast their scientific prioritization in the form of discipline goals and priorities, from which they derived more detailed scientific "imperatives" and, finally, implementation scenarios or reference mission concepts. *It is important to recognize that panel-specific priorities and recommendations (imperatives) are not equivalent to report recommendations—these can be offered only by the decadal survey committee.* The panels' various scientific inputs, assessments, and priorities for new ground- and space-based initiatives were integrated by the survey committee and an overall prioritization derived. The survey committee's prioritization of new spacecraft initiatives was heavily influenced by the CATE assessment provided by the Aerospace Corporation.

Chapter 1 of this report discusses the committee's key science goals for obtaining the necessary scientific knowledge for a society dependent on space, and it outlines the basic strategy that underpins the survey committee's recommendations for the next decade. Chapter 2 provides a retrospective view of recent notable successes in the field and exemplary achievements during the past decade and the research goals for the several subdisciplines of solar and space physics for the next decade. Chapters 4, 5, and 6 discuss specific implementation plans for a balanced program for NSF and NASA. Finally, Chapter 7 provides the survey committee's vision for a space weather and space climate program for the nation that could provide the new, integrated capabilities needed to serve the needs of a society ever more reliant on space. Chapter 3 provides an overview of the societal relevance of the field of solar and space physics and addresses the

[1] The survey's website, http://sites.nationalacademies.org/SSB/CurrentProjects/SSB_056864, includes links to the survey's request for information and to the submissions that were received in response. Appendix I also lists the responses received.

[2] As per its charge from NASA, the survey committee did not "grandfather" missions or mission concepts from the previous decadal survey unless they were already in advanced development. In some cases, a concept from the previous survey was reproposed for consideration by the present survey committee (e.g., MagCon); in others, elements of an earlier concept were incorporated into a new proposal.

current state of efforts to develop a capability to predict harmful space weather events. The recommendations of the survey committee are presented in the Summary and in Chapter 4. These recommendations were informed in part by the detailed analyses provided by the survey's three discipline study panels, whose work is summarized in Chapters 8, 9, and 10.

The work of the survey could not have been accomplished without the important help given by individuals too numerous to list, from a variety of public and private organizations, who made presentations at committee meetings, hosted outreach seminars and town meetings,[3] drafted white papers, and participated in mission studies. Finally, the committee acknowledges the exceptionally important contributions made by the following individuals at the Aerospace Corporation: Randy Persinger, Robert C. Kellogg, Robert J. Kinsey, Mark J. Barrera, David A. Bearden, Debra L. Emmons, Robert E. Bitten, and Matthew J. Hart.

[3]During the initiation of the decadal survey, town hall meetings and outreach events were held at the University of California, Los Angeles; the University of California, Berkeley; the University of Maryland, College Park; the University of New Hampshire; the University of Texas, Dallas; the University of Michigan; the National Center for Atmospheric Research; the National Astronomy and Ionosphere Center (Arecibo Observatory); Southwest Research Institute; and at the National Science Foundation's Upper Atmosphere Facilities Fall 2010 meeting in Roanoke, Va.

Acknowledgment of Members of the Decadal Survey Working Groups

The Committee on a Decadal Strategy for Solar and Space Physics (Heliophysics) acknowledges with gratitude the contributions of the members of the five working groups that assessed the state of the field and areas of critical need in the cross-cutting areas listed below:

- *Theory, Modeling, and Data Exploitation:* James F. Drake, Jr., University of Maryland, *Co-Lead*; Jon A. Linker, Predictive Science, Inc., *Co-Lead*; William Daughton, Los Alamos National Laboratory; Jan Egedal, Massachusetts Institute of Technology; Joe Giacalone, University of Arizona; Joseph Huba, Naval Research Laboratory; Janet Kozyra, University of Michigan; Dana Longcope, Montana State University; Mark Miesch, High Altitude Observatory; Vahe Petrosian, Stanford University; Tai D. Phan, University of California, Berkeley; Merav Opher, Boston University; Marco Velli, Jet Propulsion Laboratory; and Michael Wiltberger, National Center for Atmospheric Research.
- *Explorers, Suborbital, and Other Platforms:* Kristina A. Lynch, Dartmouth College, *Co-Lead*; Brian J. Anderson, John Hopkins University, Applied Physics Laboratory, *Co-Lead*; Gregory D. Earle, University of Texas, Dallas; Frank Hill, National Solar Observatory; David L. Hysell, Cornell University; Larry Kepko, NASA Goddard Space Flight Center; David Klumpar, Montana State University; Robert P. Lin, University of California, Berkeley; Anthony J. Mannucci, Jet Propulsion Laboratory; Robyn Millan, Dartmouth College; Robert F. Pfaff, NASA Goddard Space Flight Center; Harlan E. Spence, University of New Hampshire; Michael Thompson, University Corporation for Atmospheric Research; and Allan Weatherwax, Siena College.
- *Innovations: Technology, Instruments, and Data Systems:* Andrew B. Christensen, Dixie State College; *Co-Lead*; J. Todd Hoeksema, Stanford University, *Co-Lead*; Mihir Desai, Southwest Research Institute; Keith Goetz, University of Minnesota; Joseph Gurman, NASA Goddard Space Flight Center; Rod Heelis, University of Texas, Dallas; Gordon Hurford, University of California, Berkeley; Neal Hurlburt, Lockheed Martin; Jeff Kuhn, University of Hawaii; Ralph McNutt, Johns Hopkins University, Applied Physics Laboratory; Steve Mende, University of California, Berkeley; Eberhard Moebius, University of New Hampshire; Danny Morrison, Johns Hopkins University, Applied Physics Laboratory; Charles Swenson, Utah State University; and Daniel Winterhalter, Jet Propulsion Laboratory.

- *Research to Operations/Operations to Research:* Michael Hesse, NASA Goddard Space Flight Center, *Co-Lead*; Ronald E. Turner, Analytic Services, Inc. (ANSER), *Co-Lead*; John Allen, NASA Headquarters; Odile de la Beaujardiere, Air Force Research Laboratory; Joe Fennell, The Aerospace Corporation; Tamas Gombosi, University of Michigan; Kelly Hand, Air Force Space Command; Russ Howard, Naval Research Laboratory; Louis J. Lanzerotti, New Jersey Institute of Technology; Scott Pugh, Department of Homeland Security; Geoff Reeves, Los Alamos National Laboratory; Pete Riley, Predictive Science, Inc.; Al Ronn, Northrop Grumman Co.; Robert W. Schunk, Utah State University; Howard Singer, National Oceanic and Atmospheric Administration, Space Weather Prediction Center; and Kent Tobiska, Space Environment Technologies.
- *Education and Workforce:* Cherilynn Morrow, Georgia State University, *Co-Lead*; Mark Moldwin, University of Michigan, *Co-Lead*; Bryan Mendez, University of California, Berkeley; James Drake, University of Maryland; Nicholas Gross, Boston University; Michael Liemohn, University of Michigan; Ramon Lopez, University of Texas, Arlington; Joshua Semeter, Boston University; Allan Weatherwax, Siena College; and Amy Winebarger, NASA Marshall Space Flight Center.

Acknowledgment of Reviewers

This report has been reviewed in draft form by individuals chosen for their diverse perspectives and technical expertise, in accordance with procedures approved by the Report Review Committee of the National Research Council (NRC). The purpose of this independent review is to provide candid and critical comments that will assist the institution in making its published report as sound as possible and to ensure that the report meets institutional standards for objectivity, evidence, and responsiveness to the study charge. The review comments and draft manuscript remain confidential to protect the integrity of the deliberative process. We wish to thank the following individuals for their review of this report:

Vassilis Angelopoulos, University of California, Berkeley,
Roger D. Blandford, Stanford University,
John C. Foster, Massachusetts Institute of Technology,
Timothy J. Fuller-Rowell, University of Colorado,
William C. Gibson, Southwest Research Institute,
John T. Gosling, University of Colorado, Boulder,
J. Randy Jokipii, University of Arizona,
Charles F. Kennel, Scripps Institution of Oceanography, University of California, San Diego,
David Y. Kusnierkiewicz, Johns Hopkins University, Applied Physics Laboratory,
J. Patrick Looney, Brookhaven National Laboratory,
Janet G. Luhmann, University of California, Berkeley,
William H. Matthaeus, University of Delaware,
Atsuhiro Nishida, Institute of Space and Astronautical Science (emeritus professor),
Patricia H. Reiff, Rice University,
Arthur D. Richmond, National Center for Atmospheric Research,
Karel Schrijver, Lockheed Martin Advanced Technology Center,
A. Thomas Young, Lockheed Martin Corporation (retired), and
Gary P. Zank, University of California, Riverside.

Although the reviewers listed above have provided many constructive comments and suggestions, they were not asked to endorse the conclusions or recommendations, nor did they see the final draft of the report before its release. The review of this report was overseen by Martha P. Haynes, Cornell University, and George A. Paulikas, The Aerospace Corporation (retired). Appointed by the NRC, they were responsible for making certain that an independent examination of this report was carried out in accordance with institutional procedures and that all review comments were carefully considered. Responsibility for the final content of this report rests entirely with the authoring committee and the institution.

Contents

PART II: REPORTS TO THE SURVEY COMMITTEE FROM THE DISCIPLINE PANELS

APPENDIXES

This report is dedicated to the memory of
ROBERT P. LIN (1942-2012),
a pioneering space scientist and a beloved colleague.

Summary

From the interior of the Sun to the upper atmosphere and near-space environment of Earth, and outward to a region far beyond Pluto where the Sun's influence wanes, advances during the past decade in space physics and solar physics—the disciplines NASA refers to as heliophysics—have yielded spectacular insights into the phenomena that affect our home in space. This report, from the National Research Council's (NRC's) Committee on a Decadal Strategy for Solar and Space Physics (Heliophysics), is the second NRC decadal survey in heliophysics. Building on the research accomplishments realized over the past decade, the report presents a program of basic and applied research for the period 2013-2022 that will improve scientific understanding of the mechanisms that drive the Sun's activity and the fundamental physical processes underlying near-Earth plasma dynamics; determine the physical interactions of Earth's atmospheric layers in the context of the connected Sun-Earth system; and greatly enhance the capability to provide realistic and specific forecasts of Earth's space environment that will better serve the needs of society. Although the recommended program is directed primarily to NASA (Science Mission Directorate–Heliophysics Division) and the National Science Foundation (NSF) (Directorate for Geosciences–Atmospheric and Geospace Sciences) for action, the report also recommends actions by other federal agencies, especially the National Oceanic and Atmospheric Administration (NOAA) and those parts of NOAA charged with the day-to-day (operational) forecast of space weather. In addition to the recommendations included in this summary, related recommendations are presented in the main text of Part I of this report.

RECENT PROGRESS: SIGNIFICANT ADVANCES FROM THE PAST DECADE

As summarized in Chapter 2 and discussed in greater detail in Chapters 8-10, the disciplines of solar and space physics have made remarkable advances over the past decade—many of which have come from the implementation of the program recommended in the 2003 solar and space physics decadal survey.[1] Listed below are some of the highlights from an exciting decade of discovery:

- New insights, gained from novel observations and advances in theory, modeling, and computation, into the variability of the mechanisms that generate the Sun's magnetic field, and into the structure of that field;
- A new understanding of the unexpectedly deep minimum in solar activity;
- Significant progress in understanding the origin and evolution of the solar wind;
- Striking advances in understanding both explosive solar flares and the coronal mass ejections that drive space weather;
- Groundbreaking discoveries about the surprising nature of the boundary between the heliosphere—the immense magnetic bubble containing our solar system—and the surrounding interstellar medium;
- New imaging methods that permit researchers to directly observe space weather-driven changes in the particles and magnetic fields surrounding Earth;
- Significantly deeper knowledge of the numerous processes involved in the acceleration and loss of particles in Earth's radiation belts;
- Major advances in understanding the structure, dynamics, and linkages in other planetary magnetospheres, especially those of Mercury, Jupiter, and Saturn;
- New understanding of how oxygen from Earth's own atmosphere contributes to space storms;
- The surprising discovery that conditions in near-Earth space are linked strongly to the terrestrial weather and climate below;
- Evidence of a long-term decline in the density of Earth's upper atmosphere, indicative of planetary change; and
- New understanding of the temporal and spatial scales involved in magnetospheric-atmospheric coupling in Earth's aurora.

It is noteworthy that some of the most surprising discoveries of the past decade have come from comparatively small missions that were tightly cost-constrained, competitively selected, and principal-investigator (PI)-led—recommendations in the present decadal survey reflect this insight.

Enabled by advances in scientific understanding as well as fruitful interagency partnerships, the capabilities of models that predict space weather impacts on Earth have also made rapid gains over the past decade. Reflecting these advances and a society increasingly vulnerable to the adverse effects of space weather, the number of users of space weather services has also grown rapidly. Indeed, a growing community has come to depend on constant and immediate access to space weather information (see Chapter 7).

KEY SCIENCE GOALS FOR THE NEXT DECADE

The significant achievements of the past decade have set the stage for transformative advances in solar and space physics for the coming decade. Reports from the survey's three interdisciplinary study panels (Chapters 8-10) enumerate the highest-priority scientific opportunities and challenges for the com-

[1] National Research Council, *The Sun to the Earth—and Beyond: A Decadal Research Strategy in Solar and Space Physics,* The National Academies Press, Washington, D.C., 2003; and National Research Council, *The Sun to the Earth—and Beyond: Panel Reports,* The National Academies Press, Washington, D.C., 2003.

ing decade; collectively, they inform the survey's four key science goals, each of which is considered of equal priority:

Key Science Goal 1. Determine the origins of the Sun's activity and predict the variations in the space environment.

Key Science Goal 2. Determine the dynamics and coupling of Earth's magnetosphere, ionosphere, and atmosphere and their response to solar and terrestrial inputs.

Key Science Goal 3. Determine the interaction of the Sun with the solar system and the interstellar medium.

Key Science Goal 4. Discover and characterize fundamental processes that occur both within the heliosphere and throughout the universe.

GUIDING PRINCIPLES AND PROGRAMMATIC CHALLENGES

To achieve these four key science goals, the survey committee recommends adherence to the following principles (Chapter 1):

- To make transformational scientific progress, the Sun, Earth, and heliosphere must be studied as a coupled system;
- To understand the coupled system requires that each subdiscipline be able to make measurable advances in achieving its highest-priority science goals; and
- Success across the entire field requires that the various elements of solar and space physics research programs—the enabling foundation comprising theory, modeling, data analysis, innovation, and education, as well as ground-based facilities and small-, medium-, and large-class space missions—be deployed with careful attention both to the mix of assets and to the schedule (cadence) that optimizes their utility over time.

The survey committee's recommendations reflect these principles while also taking into account issues of cost, schedule, and complexity. The committee also recognizes a number of challenges that could impede achievement of the recommended program: the assumed budget might not be realized or missions could experience cost growth; the necessary activities have to be coordinated across multiple agencies; and the availability of appropriately sized and affordable space launch vehicles, particularly medium-class launch vehicles, is limited.

RECOMMENDATIONS—RESEARCH AND APPLICATIONS

The survey committee's recommendations are listed in Tables S.1 and S.2; a more complete discussion of the research recommendations—the primary focus of this survey—is found in Chapter 4, along with a discussion of the applications recommendations, while Chapter 7 presents the committee's vision, premised on the availability of additional funds, of an expanded program in space weather and space climatology. The committee's recommendations are prioritized and integrated across agencies to form an effective set of programs consistent with fiscal and other constraints. An explicit cost appraisal for each NASA research recommendation is incorporated into the budget for the overall program (Chapter 6); however, for NSF programs, only a general discussion of expected costs is provided (Chapter 5).

TABLE S.1 Summary of Top-Level Decadal Survey Research Recommendations

Priority	Recommendation	NASA	NSF	Other
0.0	Complete the current program	X	X	
1.0	Implement the DRIVE initiative Small satellites; midscale NSF projects; vigorous ATST and synoptic program support; science centers and grant programs; instrument development	X	X	X
2.0	Accelerate and expand the Heliophysics Explorer program Enable MIDEX line and Missions of Opportunity	X		
3.0	Restructure STP as a moderate-scale, PI-led line	X		
3.1	Implement an IMAP-like mission	X		
3.2	Implement a DYNAMIC-like mission	X		
3.3	Implement a MEDICI-like mission	X		
4.0	Implement a large LWS GDC-like mission	X		

TABLE S.2 Summary of Top-Level Decadal Survey Applications Recommendations

Priority	Recommendation	NASA	NSF	Other
1.0	Recharter the National Space Weather Program	X	X	X
2.0	Work in a multiagency partnership to achieve continuity of solar and solar wind observations	X	X	X
2.1	Continue solar wind observations from L1 (DSCOVR, IMAP)	X		X
2.2	Continue space-based coronagraph and solar magnetic field measurements	X		X
2.3	Evaluate new observations, platforms, and locations	X	X	X
2.4	Establish a space weather research program at NOAA to effectively transition research to operations			X
2.5	Develop and maintain distinct funding lines for basic space physics research and for space weather specification and forecasting	X	X	X

Research Recommendations

Baseline Priority for NASA and NSF: Complete the Current Program

The survey committee's recommended program for NSF and NASA assumes continued support in the near term for the key existing program elements that constitute the Heliophysics Systems Observatory (HSO) and successful implementation of programs in advanced stages of development.

NASA's existing heliophysics flight missions and NSF's ground-based facilities form a network of observing platforms that operate simultaneously to investigate the solar system. This array can be thought of as a single observatory—the Heliophysics Systems Observatory (HSO) (see Figure 1.2). The evolving HSO lies at the heart of the field of solar and space physics and provides a rich source of observations that can be used to address increasingly interdisciplinary and long-term scientific questions. Missions now under development will expand the HSO and drive scientific discovery. For NASA, these include the following:

- The Radiation Belt Storm Probes (RBSP; Living With a Star (LWS) program, 2012 launch[2]) and related Balloon Array for RBSP Relativistic Electron Losses (BARREL; first launch 2013) will determine the mechanisms that control the energy, intensity, spatial distribution, and time variability of Earth's radiation belts.

[2]Following its launch on August 30, 2012, RBSP was renamed the Van Allen Probes.

• The Interface Region Imaging Spectrograph (IRIS; Explorer program, 2013 launch) will deliver pioneering observations of chromospheric dynamics just above the solar surface to help determine their role in the origin of the fluxes of heat and mass into the corona and wind.
• The Magnetospheric Multiscale mission (MMS; Solar-Terrestrial Probes (STP) program, 2014 launch) will address the physics of magnetic reconnection at the previously inaccessible tiny scale where reconnection is triggered.

Compelling missions that are not yet in advanced stages of development but are part of a baseline program whose continuation NASA asked the survey committee to assume include the following:[3]

• Solar Orbiter (European Space Agency-NASA partnership, 2017 launch) will investigate links between the solar surface, corona, and inner heliosphere from as close as 62 solar radii.
• Solar Probe Plus (SPP; LWS program, 2018 launch) will make mankind's first visit to the solar corona to discover how the corona is heated, how the solar wind is accelerated, and how the Sun accelerates particles to high energy.

The powerful fleet of space missions that explore our local cosmos will be significantly strengthened with the addition of these missions. However, their implementation as well as the rest of the baseline program will consume nearly all of the resources anticipated to be available for new starts within NASA's Heliophysics Division through the midpoint of the overall survey period, 2013-2022.

For NSF, the previous decade witnessed the initial deployment in Alaska of the Advanced Modular Incoherent Scatter Radar (AMISR), a mobile facility used to study the upper atmosphere and to observe space weather events, and the initial development of the Advanced Technology Solar Telescope (ATST), a 4-meter-aperture optical solar telescope—by far the largest in the world—that will provide the most highly resolved measurements ever obtained of the Sun's plasma and magnetic field. These new NSF facilities join a broad range of existing ground-based assets that provide an essential global synoptic perspective and complement space-based measurements of the solar and space physics system. With adequate science and operations support, they will enable frontier research even as they add to the long-term record necessary for analyzing space climate over solar cycles.

R1.0 Implement the DRIVE Initiative

The survey committee recommends implementation of a new, integrated, multiagency initiative (DRIVE—Diversify, Realize, Integrate, Venture, Educate) that will develop more fully and employ more effectively the many experimental and theoretical assets at NASA, NSF, and other agencies.

The DRIVE initiative encompasses specific, cost-effective augmentations to NASA and NSF heliophysics programs. Its implementation will bring existing "enabling" programs to full fruition and will provide new opportunities to realize scientific discoveries from existing data, build more comprehensive models, make theoretical breakthroughs, and innovate. With this in mind, the committee has as its first priority for both NASA and NSF—after completion of the current program—the implementation of an integrated, multiagency initiative comprising the following components:

[3] In accordance with its statement of task, the survey committee did not reprioritize any NASA mission that was in formulation or advanced development. In addition, the study charge specified that Solar Orbiter and Solar Probe Plus would not be included in any prioritization of future mission opportunities.

- **D**iversify observing platforms with microsatellites and midscale ground-based assets.
- **R**ealize scientific potential by sufficiently funding operations and data analysis.
- **I**ntegrate observing platforms and strengthen ties between agency disciplines.
- **V**enture forward with science centers and instrument and technology development.
- **E**ducate, empower, and inspire the next generation of space researchers.

The five DRIVE components are defined in Chapter 4, with specific and actionable recommendations offered for each element. Implementation of the NASA portion of the DRIVE initiative would require an augmentation to existing program lines equivalent to approximately $33 million in current (2013) dollars (see Chapter 6).[4] The cost and implementation of the NSF portion of DRIVE are described in Chapter 5. Although the recommendations for NSF within the DRIVE initiative are not prioritized, the survey committee calls attention to two in particular:

The National Science Foundation should:

- **Provide funding sufficient for essential synoptic observations and for efficient and scientifically productive operation of the Advanced Technology Solar Telescope, which provides a revolutionary new window on the solar magnetic atmosphere.**
- **Create a new, competitively selected mid-scale project funding line in order to enable midscale projects and instrumentation for large projects.** There are a number of compelling candidates for a midscale facilities line, including the Frequency-Agile Solar Radiotelescope (FASR), the Coronal Solar Magnetism Observatory (COSMO), and several other projects exemplifying the kind of creative approaches necessary to fill gaps in observational capabilities and to move the survey's integrated science plan forward.

R2.0 Accelerate and Expand the Heliophysics Explorer Program

The survey committee recommends that NASA accelerate and expand the Heliophysics Explorer program. Augmenting the current program by $70 million per year, in fiscal year 2012 dollars, will restore the option of Mid-size Explorer (MIDEX) missions and allow them to be offered alternately with Small Explorer (SMEX) missions every 2 to 3 years. As part of the augmented Explorer program, NASA should support regular selections of Missions of Opportunity.

The Explorer program's strength lies in its ability to respond rapidly to new concepts and developments in science, as well as in the program's synergistic relationship with larger-class strategic missions.[5] The Explorer mission line has proven to be an outstanding success, delivering—cost-effectively—science results of great consequence. The committee recommends increased support of the Explorer program to enable significant scientific advances in solar and space physics. As discussed in Chapter 4, the committee believes that the proper cadence for Heliophysics Explorers is one mission every 2 to 3 years. The committee's recommended augmentation of the Explorer program would facilitate this cadence and would also allow selection of both small- and medium-class Explorers. Historically, MIDEX missions offered an opportunity to resolve many of the highest-level science questions, but they have not been feasible with the current Explorer budget.

---

[4]The survey committee assumed inflation at 2.7 percent in program costs, the same as the percentage used by NASA for new starts.

[5]National Research Council, *Solar and Space Physics and Its Role in Space Exploration*, The National Academies Press, Washington, D.C., 2003, p. 36.

Regular selections of Missions of Opportunity will also allow the research community to respond quickly and to leverage limited resources with interagency, international, and commercial flight partnerships. For relatively modest investments, such opportunities can potentially address high-priority science aims identified in this survey.

R3.0 Restructure Solar-Terrestrial Probes as a Moderate-Scale, PI-Led Line

The survey committee recommends that NASA's Solar-Terrestrial Probes program be restructured as a moderate-scale, competed, principal-investigator-led (PI-led) mission line that is cost-capped at $520 million per mission in fiscal year 2012 dollars including full life-cycle costs.

NASA's Planetary Science Division has demonstrated success in implementing mid-size missions as competed, cost-capped, PI-led investigations via the Discovery and New Frontiers programs. These are managed in a manner similar to Explorers and have a superior cost-performance history relative to that of larger flagship missions. The committee concluded that STP missions should be managed likewise, with the PI empowered to make scientific and mission design trade-offs necessary to remain within the cost cap (Chapter 4). With larger-class LWS missions and smaller-class Explorers and Missions of Opportunity, this new approach will lead to a more balanced and effective overall NASA Heliophysics Division mission portfolio that is implemented at a higher cadence and provides the vitality needed to accomplish the breadth of the survey's science goals. The eventual recommended minimum cadence of STP missions is one every 4 years.

Although the new STP program would involve moderate missions being chosen competitively, the survey committee recommends that their science targets be ordered as follows so as to systematically advance understanding of the full coupled solar-terrestrial system:

R3.1 The first new STP science target is to understand the outer heliosphere and its interaction with the interstellar medium, as illustrated by the reference mission[6] Interstellar Mapping and Acceleration Probe (IMAP; Chapter 4). Implementing IMAP as the first of the STP investigations will ensure coordination with NASA Voyager missions. The mission implementation also requires measurements of the critical solar wind inputs to the terrestrial system.

R3.2 The second STP science target is to provide a comprehensive understanding of the variability in space weather driven by lower-atmosphere weather on Earth. This target is illustrated by the reference mission Dynamical Neutral Atmosphere-Ionosphere Coupling (DYNAMIC; Chapter 4).

R3.3 The third STP science target is to determine how the magnetosphere-ionosphere-thermosphere system is coupled and how it responds to solar and magnetospheric forcing. This target is illustrated by the reference mission Magnetosphere Energetics, Dynamics, and Ionospheric Coupling Investigation (MEDICI; Chapter 4).

The rationale for all the selections and for their ordering is detailed in Chapter 4.

---

[6] In this report, the committee uses the terms "reference mission" and "science target" interchangeably, given that the mission concepts were developed specifically to assess the cost of addressing particular high-priority science investigations. The concepts presented in this report underwent an independent cost and technical analysis by the Aerospace Corporation, and they have been given names for convenience; however, the actual recommendation from the committee is to address the science priorities enumerated in the reference mission concept.

Living With a Star

Certain landmark scientific problems are of such scope and complexity that they can be addressed only with major missions. In the survey committee's plan, major heliophysics missions would be implemented within NASA's LWS program; the survey committee recommends that they continue to be managed and executed by NASA centers. Other integral thematic elements besides the flight program are essential to the LWS science and technology program: the unique LWS research, technology, strategic capabilities, and education programs remain of great value.

R4.0 Implement a large Living With a Star mission to study the ionosphere-thermosphere-mesosphere system in an integrated fashion.

The survey committee recommends that, following the launch of RBSP and SPP, the next LWS science target focus on how Earth's atmosphere absorbs solar wind energy. The recommended reference mission is Geospace Dynamics Constellation (GDC).

As detailed in Chapter 4, the GDC reference mission would provide crucial scientific measurements of the extreme variability of conditions in near-Earth space. Within anticipated budgets, the completion of the baseline LWS program, which includes the launch of two major missions—RBSP in 2012 and SPP in 2018—does not allow for the launch of a subsequent major mission in heliophysics until 2024, 6 years after SPP. This establishes what the survey committee regards as the absolute minimum cadence needed for major missions.

Applications Recommendations: Enabling Effective Space Weather and Climatology Capabilities

Multiple agencies of the federal government have vital interests related to space weather, and efforts to coordinate these agencies' activities are seen in the National Space Weather Program (NSWP). Nonetheless, the survey committee concluded that additional approaches are needed to develop the capabilities outlined in the 2010 National Space Policy document and envisioned in the 2010 NSWP plan.[7] Chapter 7 presents the committee's vision for a renewed national commitment to a comprehensive program in space weather and climatology (SWaC). Enabling an effective SWaC capability will require action across multiple agencies and an integrated program that builds on the strengths of individual agencies.

A1.0 Recharter the National Space Weather Program

As part of a plan to develop and coordinate a comprehensive program in space weather and climatology, the survey committee recommends that the National Space Weather Program be rechartered under the auspices of the National Science and Technology Council. With the active participation of the Office of Science and Technology Policy and the Office of Management and Budget, the program should build on current agency efforts, leverage the new capabilities and knowledge that will arise from implementation of the programs recommended in this report, and develop additional capabilities, on the ground and in space, that are specifically tailored to space weather monitoring and prediction.

---

[7] *National Space Policy of the United States of America,* June 28, 2010, available at http://www.whitehouse.gov/sites/default/files/national_space_policy_6-28-10.pdf. Committee for Space Weather, Office of the Federal Coordinator for Meteorological Services and Supporting Research, *National Space Weather Program Strategic Plan,* FCM-P30-2010, August 17, 2010, available at http://www.ofcm.gov/nswp-sp/fcm-p30.htm.

A2.0 Work in a multiagency partnership to achieve continuity of solar and solar wind observations.

The survey committee recommends that NASA, NOAA, and the Department of Defense work in partnership to plan for continuity of solar and solar wind observations beyond the lifetimes of ACE, SOHO, STEREO, and SDO. In particular:

A2.1 Solar wind measurements from L1 should be continued, because they are essential for space weather operations and research. The DSCOVR L1 monitor and IMAP STP mission are recommended for the near term, but plans should be made to ensure that measurements from L1 continue uninterrupted into the future.

A2.2 Space-based coronagraph and solar magnetic field measurements should likewise be continued.

Further, the survey committee concluded that a national, multifaceted program of both observations and modeling is needed to transition research into operations more effectively by fully leveraging expertise from different agencies, universities, and industry and by avoiding duplication of effort. This effort should include determining the operationally optimal set of observations and modeling tools and how best to effect that transition. With these objectives in mind:

A2.3 The space weather community should evaluate new observations, platforms, and locations that have the potential to provide improved space weather services. In addition, the utility of employing newly emerging information dissemination systems for space weather alerts should be assessed.

A2.4 NOAA should establish a space weather research program to effectively transition research to operations.

A2.5 Distinct funding lines for basic space physics research and for space weather specification and forecasting should be developed and maintained.

Implementation of a program to advance space weather and climatology will require funding well above what the survey committee assumes will be available to support its research-related recommendations to NASA (see Table S.1). The committee emphasizes that implementation of an initiative in space weather and climatology should proceed only if it does not impinge on the development and timely execution of the recommended research program.

RECOMMENDED PROGRAM, DECISION RULES, AND AUGMENTATION PRIORITIES FOR NASA

Recommended Program

The committee's recommended program for NASA's Heliophysics Division is shown in Figure S.1. As detailed in Chapter 6, the plan restores the medium-class Explorers and, together with small-class Explorer missions and Missions of Opportunity, achieves the recommended minimum mission cadence. The plan also begins the DRIVE initiative as early in the decade as budgets allow, with full implementation achieved by mid-decade. However, funding constraints affect the restoration and recommended rebalance of heliophysics program elements such that full realization of the survey committee's strategy is not possible until after 2017 (see Figure S.1).

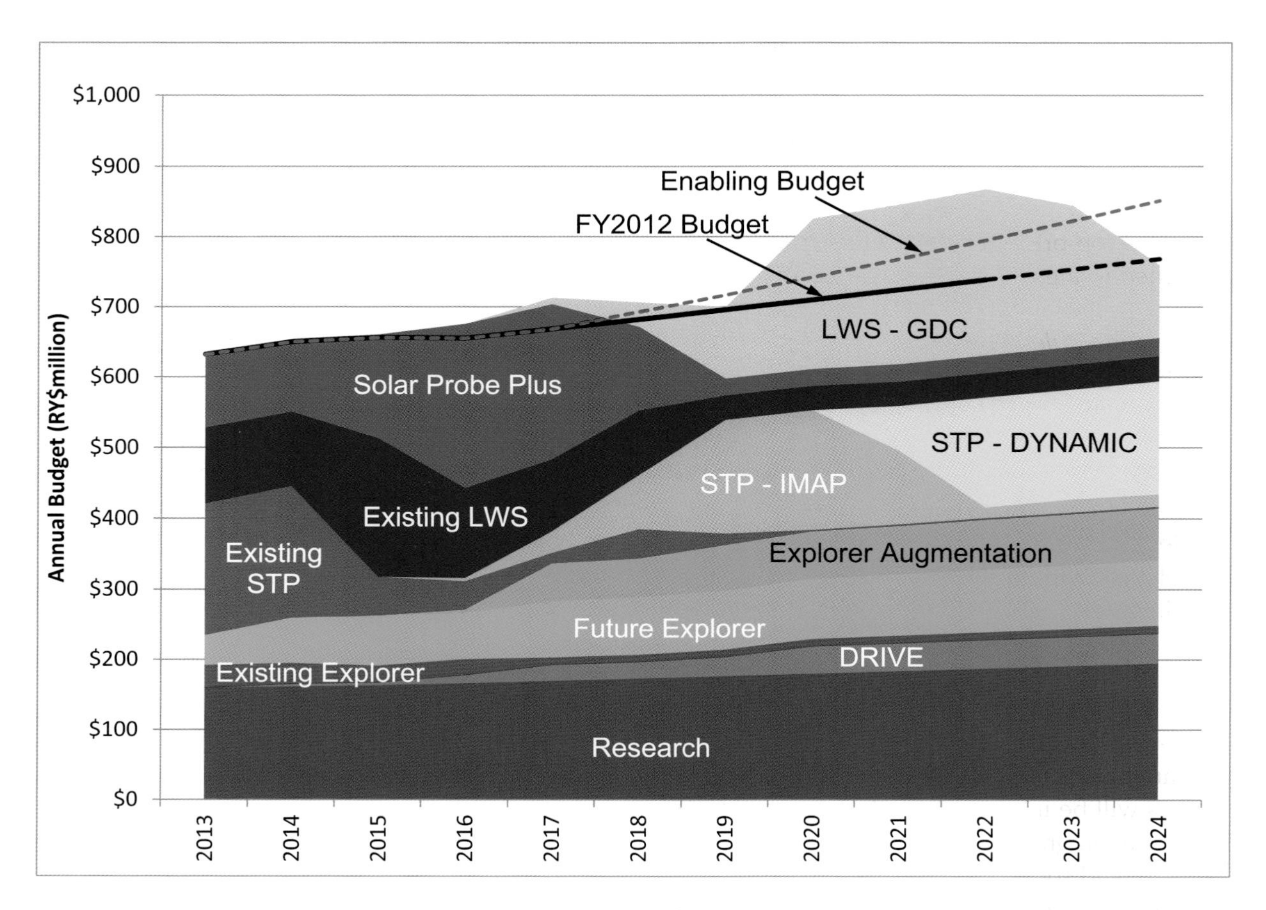

FIGURE S.1 Heliophysics budget and program plan by year and category from 2013 to 2024. The solid black line indicates the funding level from 2013 to 2022 provided to the survey committee by NASA as the baseline for budget planning, and the dashed black line extrapolates the budget forward to 2024. After 2017 the amount increases with a nominal 2 percent inflationary factor. Through 2016 the program content is tightly constrained by budgetary limits and fully committed for executing existing program elements. The red dashed "Enabling Budget" line includes a modest increase from the baseline budget starting in 2017, allowing implementation of the survey-recommended program at a more efficient cadence that better meets scientific and societal needs and improves optimization of the mix of small and large missions. From 2017 to 2024 the Enabling Budget grows at 1.5 percent above inflation. (Note that the 2024 Enabling Budget is equivalent to growth at a rate just 0.50 percent above inflation from 2009.) Geospace Dynamics Constellation, the next large mission of the LWS program after Solar Probe Plus, rises above the baseline curve in order to achieve a more efficient spending profile, as well as to achieve deployment in time for the next solar maximum in 2024. NOTE: LWS refers to missions in the Living With a Star line, and STP refers to missions in the Solar-Terrestrial Probes line.

Decision Rules to Ensure That Balanced Progress Is Maintained

The recommended program for NASA cost-effectively addresses key science objectives. However, the survey committee recognizes that the already tightly constrained program could face further budgetary challenges. For example, with launch planned in 2018, the Solar Probe Plus project has not yet entered the implementation phase when expenditures are highest.[8] Significant cost growth in this very important, but technically challenging, mission beyond the current cap has the potential to disrupt the overall NASA heliophysics program.

To guide the allocation of reduced resources, the committee recommends the following decision rules intended to provide flexibility and efficiency if funding is less than anticipated, or should some other disruptive event occur. These rules, discussed in greater depth in Chapter 6, would help maintain progress toward the top-priority, system-wide science challenges identified in this survey. The decision rules should be applied in the order shown to minimize disruption of higher-priority program elements:

Decision Rule 1. Missions in the STP and LWS lines should be reduced in scope or delayed to accomplish higher priorities (Chapter 6 gives explicit triggers for review of Solar Probe Plus).

Decision Rule 2. If further reductions are needed, the recommended increase in the cadence of Explorer missions should be scaled back, with the current cadence maintained as the minimum.

Decision Rule 3. If still further reductions are needed, the DRIVE augmentation profile should be delayed, with the current level of support for elements in the NASA research line maintained as the minimum.

Augmentations to Increase Program Value

The committee notes that the resources assumed in crafting this decadal survey's recommended program are barely sufficient to make adequate progress in solar and space physics; with reduced resources, progress will be inadequate. It is also evident that with increased resources, the pace at which the nation pursues its program could be accelerated with a concomitant increase in the achievement of scientific discovery and societal value. The committee recommends the following augmentation priorities to aid in implementing a program under a more favorable budgetary environment:

Augmentation Priority 1. Given additional budget authority early in the decade, the implementation of the DRIVE initiative should be accelerated.

Augmentation Priority 2. With sufficient funds throughout the decade, the Explorer line should be further augmented to increase the cadence and funding available for missions, including Missions of Opportunity.

Augmentation Priority 3. Given further budget augmentation, the schedule of STP missions should advance to allow the third STP science target (MEDICI) to begin in this decade.

Augmentation Priority 4. The next LWS mission (GDC) should be implemented with an accelerated, more cost-effective funding profile.

---

[8]On January 31, 2012, Solar Probe Plus passed its agency-level confirmation review and entered what NASA refers to as mission definition or Phase B of its project life cycle.

TABLE S.3 Fulfilling the Key Science Goals of the Decadal Survey

Advances in Scientific Understanding and Observational Capabilities		Goals
Advances owing to implementation of the existing program	Twin Radiation Belt Storm Probes will observe Earth's radiation belts from separate locations, finally resolving the importance of temporal and spatial variability in the generation and loss of trapped radiation that threatens spacecraft.	2, 4
	The Magnetospheric Multiscale mission will provide the first high-resolution, three-dimensional measurements of magnetic reconnection in the magnetosphere by sampling small regions where magnetic field line topologies reform.	2, 4
	Solar Probe Plus will be the first spacecraft to enter the outer atmosphere of the Sun, repeatedly sampling solar coronal particles and fields to understand coronal heating, solar wind acceleration, and the formation and transport of energetic solar particles.	1, 4
	Solar Orbiter will provide the first high-latitude images and spectral observations of the Sun's magnetic field, flows, and seismic waves, relating changes seen in the corona to local measurements of the resulting solar wind.	1, 4
	The 4-meter Advanced Technology Solar Telescope will resolve structures as small as 20 km, measuring the dynamics of the magnetic field at the solar surface down to the fundamental density length scale and in the low corona.	1, 4
	The Heliophysics Systems Observatory will gather a broad range of ground- and space-based observations and advance increasingly interdisciplinary and long-term solar and space physics science objectives.	All
New starts on programs and missions to be implemented within the next decade	The DRIVE initiative will greatly strengthen researchers' ability to pursue innovative observational, theoretical, numerical, modeling, and technical advances.	All
	Solar and space physicists will accomplish high-payoff, timely science goals with a revitalized Explorer program, including leveraged Missions of Opportunity.	All
	The Interstellar Mapping and Acceleration Probe, in conjunction with the twin Voyager spacecraft, will resolve the interaction between the heliosphere—our home in space—and the interstellar medium.	2, 3, 4
	A new funding line for mid-size projects at the National Science Foundation will facilitate long-recommended ground-based projects, such as COSMO and FASR, by closing the funding gap between large and small programs.	All
New starts on missions to be launched early in the next decade	The Dynamical Neutral Atmosphere-Ionosphere Coupling mission's two identical orbiting observatories will clarify the complex variability and structure in near-Earth plasma driven by lower-atmosphere wave energy.	2, 4
	The Geospace Dynamics Constellation will provide the first simultaneous, multipoint observations of how the ionosphere-thermosphere system responds to, and regulates, magnetospheric forcing over local and global scales.	2, 4
Possible new start this decade given budget augmentation and/or cost reduction in other missions	The Magnetosphere Energetics, Dynamics, and Ionospheric Coupling Investigation will target complex, coupled, and interconnected multiscale behavior of the magnetosphere-ionosphere system by providing global, high-resolution, continuous three-dimensional images and multipoint in situ measurements of the ring current, plasmasphere, aurora, and ionospheric-thermospheric dynamics.	2, 4

EXPECTED BENEFITS OF THE RECOMMENDED PROGRAM

Implementation of the survey committee's recommended program will ensure that the United States maintains its leadership in solar and space physics and will, the committee believes, lead to significant—even transformative—advances in scientific understanding and observational capabilities (Table S.3). In turn, these advances will support critical national needs for information that can be used to anticipate, recognize, and mitigate space weather effects that threaten human life and the technological systems society depends on.

Part I

REPORT FROM THE DECADAL SURVEY COMMITTEE

1

Enabling Discovery in Solar and Space Physics

EARTH'S DYNAMIC SPACE ENVIRONMENT

We live on a planet whose orbit traverses the tenuous outer atmosphere of a variable magnetic star, the Sun. This stellar atmosphere is a rapidly flowing plasma—the solar wind—that envelops Earth as it rushes outward, creating a cavity in the galaxy that extends to some 140 astronomical units (AU).[1] There, the inward pressure from the interstellar medium balances the outward pressure of the solar plasma, forming the heliopause, the boundary of our home in the universe. Earth and the other planets of our solar system are embedded deep in this extended stellar atmosphere, or heliosphere (Figure 1.1), the domain of solar and space physics.

The energy Earth receives from the Sun determines Earth's environment. This energy, primarily visible light, but also including ultraviolet and X-ray radiation, establishes the temperature, structure, and composition of Earth's uppermost atmosphere and ionosphere. The Sun also has a corpuscular output—the magnetized solar wind and energetic particles—that expands throughout interplanetary space, interacting with Earth and affecting humans in numerous ways (see Chapter 3). Because these latter sources of the Sun's power are highly variable in location, intensity, and time, Earth's near-space environment is profoundly dynamic and hosts numerous phenomena that present hazards to spacecraft, humans in space, and society's ground-based technological infrastructure.

Solar and space physics research seeks to understand the history, evolution, and detailed workings of the Sun and to characterize and understand Earth's space environment, including its upper atmosphere, and its response to the periodic, but highly variable, forcing by the Sun. The Earth system and its parent star also provide an accessible cosmic laboratory for studies that can lead to understanding the environs of other planets, stars, and cosmic systems. The research elements of solar and space physics span solar electromagnetic and radiative processes, the generation of solar magnetic fields, the solar wind and interplanetary magnetic fields, and their evolution, development, and interaction with planets and moons that

[1] An astronomical unit (AU) is the mean distance between Earth and the Sun; it is approximately 150 million kilometers (km). For comparison, the distance from the Sun to the Pluto-Charon system is currently approximately 32 AU.

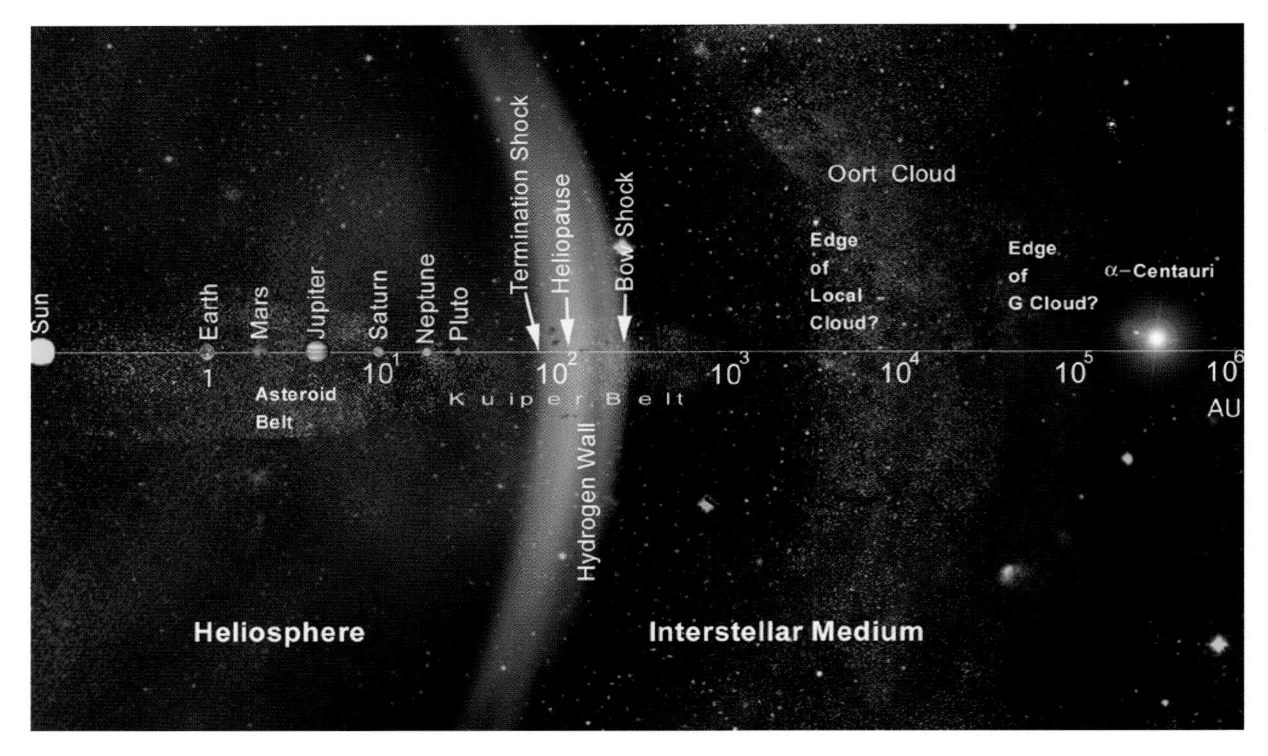

FIGURE 1.1 The solar system and its nearby galactic neighborhood are illustrated here on a logarithmic scale extending from <1 to 1 million AU. The Sun and its planets are shielded by a bubble of solar wind—the heliosphere—and the boundary between the solar wind and interstellar plasma is called the heliopause. Beyond this bubble is a largely unknown region—interstellar space. NOTE: The G cloud is a cloud of interstellar gas near the Local Interstellar Cloud in which the solar system is embedded. SOURCE: Image and text adapted from NASA and available at http://interstellar.jpl.nasa.gov/interstellar/probe/introduction/ scale.html. Also see, "Living in the Atmosphere of the Sun," at http://sunearthday.nasa.gov/2007/ locations/ttt_heliosphere_57.php.

have their own magnetospheres[2] and atmospheres. These magnetosphere-atmosphere systems are often strongly coupled, and they mediate the solar wind interaction with the planet (or moon) in ways unique to each body. Moreover, as human exploration extends farther into space—by means of both robotic probes and human spaceflight—and as society's technological infrastructure is linked increasingly to space-based assets and impacted by the dynamics of the space environment, the need to characterize, understand, and predict the dynamics of our environment in space becomes ever more pressing.

As a discipline, modern solar and space physics—now also referred to as heliophysics—can trace its origins back to the evening of January 31, 1958, when a Juno (Jupiter-C) rocket blasted into space, lofting the first U.S. artificial Earth satellite into orbit. This spacecraft, dubbed Explorer I, joined Sputnik II, a satellite that had been launched 2 months earlier by the Soviet Union. The Explorer I mission was truly groundbreaking because it carried a small scientific payload, prepared by a team of university researchers led by James A. Van Allen, that would make the first revolutionary discovery of the space age, namely, that

[2] Earth's magnetosphere is formed by the interaction of the solar wind and our planet's intrinsic magnetic field.

Earth is enshrouded in what later became known as the Van Allen belts, toroidal bands of extraordinarily high-energy, high-intensity radiation.[3]

The scope of studies in solar and space physics has since expanded to encompass the study of Earth's space environment, the solar wind and its interactions with other planets, and the Sun's role in creating and controlling the electrically charged plasma that fills the heliosphere. Progress in the field has critical impacts on society because we are increasingly dependent on a growing array of technologically advanced, but vulnerable, electronic devices in space. Because the Sun's output is highly variable in location, intensity, and time, Earth's near-space environment is a profoundly dynamic one and hosts numerous phenomena that present hazards to spacecraft, humans in space, and ground-based infrastructure on Earth (see Box 1.1, "Severe Space Weather Events—Understanding Societal and Economic Impacts").

Beyond understanding our local environment, space physics strives to understand how particles are accelerated to very high energies and how such particles subsequently move in magnetic fields around distant planets, distant stars, and—by extension—distant galaxies. Space physics provides the fundamental knowledge to prescribe how energy is transported and converted to form the remarkable tapestry of cosmic objects that have been observed. Studying the Earth system and its parent star provides the cosmic laboratory and the prototype that form the basis for understanding the environs of virtually all other planets, stars, and entire cosmic systems.

The programs, initiatives, and investments in the field that are outlined in this report are designed to make fundamental advances in current scientific knowledge of the governing processes of the space environment—from the interior of the Sun, to the atmosphere of Earth, to the local interstellar medium. These advances also enable predictive capabilities to be improved to the point that highly reliable forecasts can be made regarding the state of the space environment, particularly the disruptive space weather disturbances that threaten society and the economy, and their important technical infrastructure. The wealth of scientific insights that this recommended program will enable will also provide direct benefits to other scientific fields, including astrophysics, planetary science, and laboratory plasma physics. Many of the proposed activities will involve international collaboration and cooperation, thereby leveraging U.S. investments while simultaneously sustaining a U.S. leadership role in this science. Quite importantly, action on this decadal survey's recommendations will attract into the field the new talent that is required to ensure the continued vitality of solar and space physics.

FRAMING THE 2013-2022 DECADAL SURVEY

In this report, the Committee on a Decadal Strategy for Solar and Space Physics (Heliophysics) provides specific recommendations to its sponsors, NASA and the National Science Foundation (NSF), but its guidance is relevant to other federal departments and agencies, especially the National Oceanic and Atmospheric Administration (NOAA), the Department of Defense (DOD), the U.S. Geological Survey, and the Federal Emergency Management Agency. In developing its recommendations, the committee considered programs that vary widely in scale, ranging from what NASA's Heliophysics Division denotes as a flagship-class mission—one costing over $1 billion—to NSF grants programs that are smaller by some four orders of magnitude. In addition, the survey committee gave considerable attention to the cadence at which recommended programmatic elements should be repeated. For example, large spaceflight missions at NASA may be so complex and place such high demands on the community that they can be implemented only once or twice per decade. By contrast, smaller missions in the Explorer-class of spacecraft, or NSF ground-based facilities, can be developed and brought to scientific fruition on timescales of 3 to 5 years.

[3] Explorer II failed to reach orbit; data from Explorer III and Explorer I together provided the information that is credited with the discovery of the radiation belts.

BOX 1.1 SEVERE SPACE WEATHER EVENTS—UNDERSTANDING SOCIETAL AND ECONOMIC IMPACTS

The societal impact of space weather was dramatically demonstrated approximately a century before the launch of Explorer 1 when awe-inspiring auroral displays were seen over nearly the entire world on the night of August 28-29, 1859. In New York City, thousands watched "the heavens . . . arrayed in a drapery more gorgeous than they have been for years." The aurora witnessed that Sunday night, the *New York Times* told its readers," will be referred to hereafter among the events which occur but once or twice in a lifetime."[1] Even more spectacular displays occurred on September 2, 1859. For residents of Havana, Cuba, the sky that night "appeared stained with blood and in a state of general conflagration."[2] Earth had experienced a one-two punch from the Sun, the likes of which have not been recorded since. From August 28 through September 4, 1859, auroral displays of remarkable brilliance, color, and duration were observed around the world, as far south as Central America in the Northern Hemisphere and as far north as Santiago, Chile, in the Southern Hemisphere.

Even after daybreak, when the auroras were no longer visible, disturbances in Earth's magnetic field were so powerful that ground-level magnetic field monitoring sensors were driven off scale. Telegraph networks in many locations experienced major disruptions and outages. In several regions, operators disconnected their systems from the batteries and sent messages using only the current induced by the aurora. In fact, telegraphs were completely unusable for nearly 8 hours in most places around the world.

Humanity was just beginning to develop a dependence on high-tech systems in 1859. The telegraph was the technological wonder of its day. There were no high-power electrical lines crisscrossing the continents or sensitive satellites orbiting Earth, both of which are vulnerable to events of the sort that disrupted telegraph systems in the 19th century. There certainly was not yet a dependence on instantaneous communication and satellite remote imaging of Earth's surface. Now, in the early part of the 21st century, as the Sun is ramping up its activity in solar cycle 24, decision makers are asking: Has there been adequate preparation for severe space weather events, and what might be the consequences of worst-case events like that of the storm of 1859?[3]

To evaluate the nation's capabilities for forecasting and monitoring storms in space and for coping with their effects on Earth, the Space Studies Board of the National Research Council invited representatives of industry, academia, and the government to participate in a workshop in 2008 on the impacts of severe space weather on society and the economy. The workshop participants explored a number of issues, including the following:[4]

- The electric power, spacecraft, aviation, and Global Positioning System (GPS)-based industries are the main industries whose operations can be adversely affected by severe space weather. With increasing aware-

And data analysis, theory, and modeling programs for both NASA and NSF can make great strides even more quickly on timescales as short as 1 to 3 years.

Currently, the globally connected Sun-Earth system is studied by a multi-element system of solar and space physics observatories—the Heliophysics Systems Observatory (HSO; see Figure 1.2)—that are supported by NASA and NSF. Augmented by a constellation of missions operated by NOAA and DOD and by the implementation of the components recommended in this report, the system has remarkable potential to support simultaneous observing from distributed, strategically chosen vantage points. However, despite its evident strengths, much of the HSO's collective capabilities are somewhat fragile due to the aging of the current satellite fleet and ground-based facilities. Long-term, continuous observations of key parts of

ness and understanding of space weather's effects on vulnerable technological systems, these industries have adopted procedures and technologies designed to mitigate the impacts of space weather on their operations and customers.

- Relying on space weather forecasts and real-time data, power system operators modify the way the grid is operated during severe geomagnetic disturbances to protect against outages and equipment damage.
- Spacecraft manufacturers draw on current scientific knowledge of the space environment in an attempt to cost-effectively design and build commercial spacecraft that can operate 24/7 under severe space weather conditions. Spacecraft operators factor space weather conditions into decision making about whether to launch or to perform certain on-orbit operations.
- New signals and codes are being implemented in GPS satellites that are expected to help mitigate the effects of ionospheric disturbances on space-based navigation. Nonetheless, the Federal Aviation Administration will maintain a backup navigation system that is independent of the GPS.
- To preserve reliable communications during intense solar energetic particle events, airline companies re-route, at considerable expense, flights scheduled for polar routes. A secondary reason flights are diverted is to reduce the cumulative dose of radiation to which passengers and, especially crew, are exposed.
- Such measures notwithstanding, the potential for the space-weather-related disruption of critical technologies remains. Of particular concern is the vulnerability of the electric power grid on which the U.S. national infrastructure depends and which, despite the mitigation procedures adopted since 1989, could, in the event of an unusually strong geomagnetic storm, experience both widespread power outages and permanent equipment damage.

[1] *New York Times*, "The Aurora Borealis; The Brilliant Display on Sunday Night; Phenomena Connected with the Event; Mr. Meriam's Observations on the Aurora—E.M. Picks Up a Piece of the Aural Light; The Aurora as Seen Elsewhere—Remarkable Electrical Effects," August 30, 1859.

[2] M.A. Shea and D.F. Smart, Compendium of the eight articles on the "Carrington Event" attributed to or written by Elias Loomis in the *American Journal of Science,* 1859-1861, *Advances in Space Research* 38:313-385, 2006, p. 326.

[3] For example, the Federal Emergency Management Agency held an exercise at the NOAA Space Weather Prediction Center in Boulder, Colorado, to investigate the consequences of a worst-case scenario. See Jon Hamilton, "Solar Storms Could Be Earth's Next Katrina," NPR News, February 26, 2010, available at http://www.npr.org/templates/story/story.php?storyId=124125001.

[4] Adapted from National Research Council, *Severe Space Weather Events—Understanding Societal and Economic Impacts*, The National Academies Press, Washington, D.C., 2009.

the heliospheric system are particularly hard to maintain as the rising cost[4] of developing and launching individual observing elements limits the number of investigations that can be accomplished within available budget resources.

The diminished frequency of spacecraft launches is an example of a threat to the program recommended in this report. Its long-term detrimental consequences include reduced opportunities to retain

[4] Rising mission costs are the result of external factors such as the increased cost of launch vehicles over the past decade; they also reflect internal programmatic weaknesses. See National Research Council, *Controlling Cost Growth of NASA Earth and Space Science Missions,* The National Academies Press, Washington, D.C., 2010.

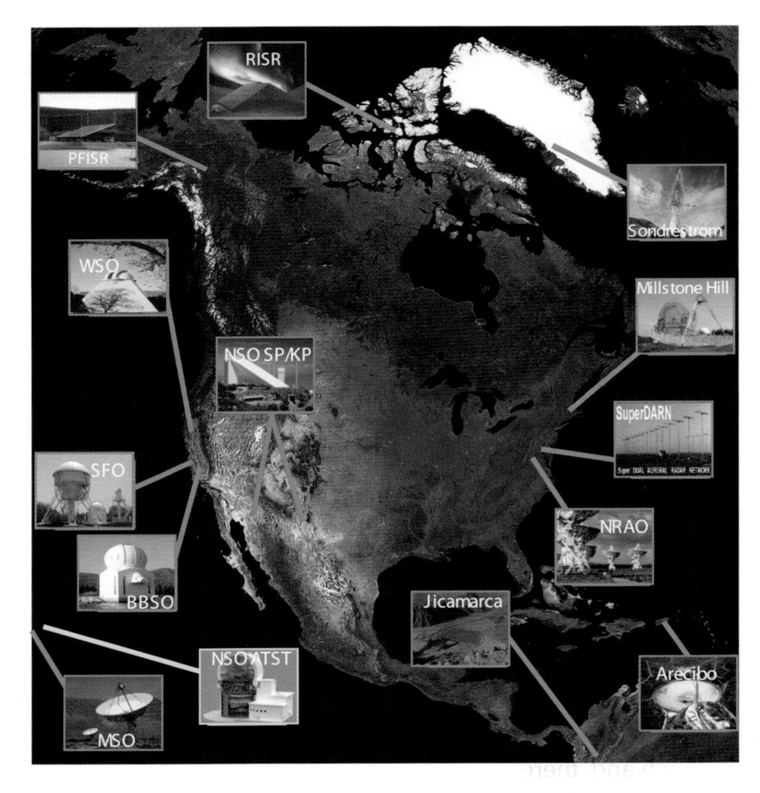

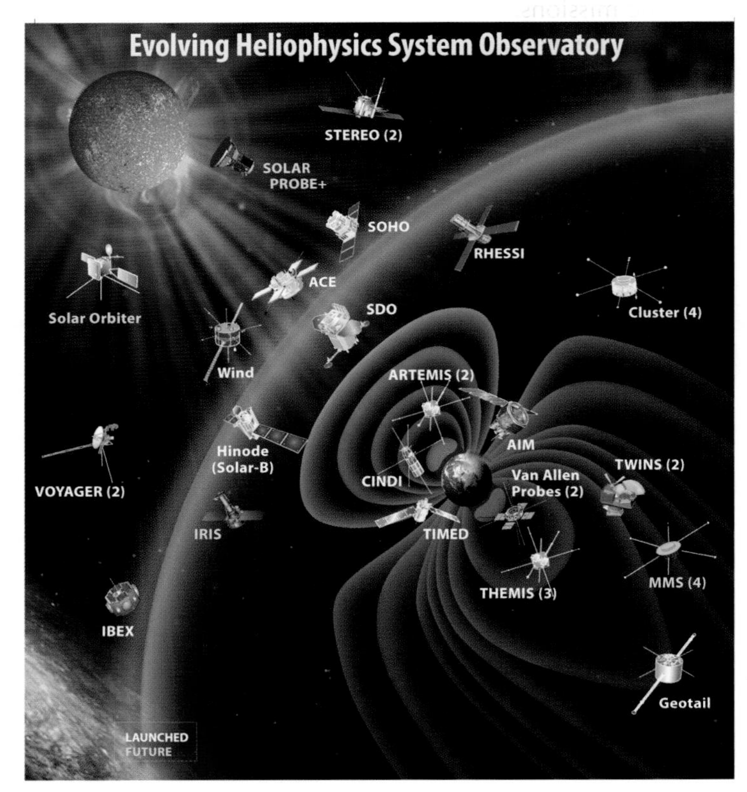

FIGURE 1.2 NSF and NASA's current and near-future (indicated by yellow text) program of solar and space physics missions, which form the Heliophysics Systems Observatory (HSO). Details about NSF facilities are available at http://www.nsf.gov/funding/pgm_summ.jsp?pims_id=12808&org=AGS&from=home and http://www.nsf.gov/div/index.jsp?div=AST; details about NASA missions are available at http://www.nasa.gov/topics/solarsystem/sunearthsystem/main/Missions_Heliophysics.html. SOURCE: *Top:* Courtesy of NSF. *Bottom:* Courtesy of NASA.

experienced flight hardware development personnel and the inability to attract fresh talent to the field. Another prominent threat is the lack of availability of a U.S. medium-class launch vehicle with a track record comparable to that of the Delta-II, which ended production as a result of the decline in orders from the U.S. Air Force. The limited availability of a medium-class space launch system places severe constraints on new missions, restricting them to smaller and lighter scientific payloads suitable for available launchers like the Pegasus or Taurus, or requiring the use of heavier-lift and much more expensive evolved expendable launch vehicles. New entrants to the field, most prominently Space Exploration Technologies' (SpaceX's) Falcon 9, which NASA has begun to use for resupply missions to the International Space Station, and Orbital Sciences'Antares (neé Taurus II) launch vehicle, which had its first test flight in April 2013, offer the potential for reduced launch costs and medium-lift capabilities. Both vehicles are still early in their development and utilization, and it remains to be seen if they can meet future needs for reliable, low-cost launch vehicles (see Box 1.2, "Access to Space," for further analysis of this issue).

An already-lean solar and space physics program is also threatened by the prospect of level or even declining budgets for the foreseeable future. The rising cost of executing space missions only exacerbates this problem,[5] and the resultant shortfalls affect programs and, indirectly, the "pipeline" of future engineers and scientists who choose to enter the field (see Appendix D). In the coming years, the solar and space physics enterprise will be challenged by demands to maintain and expand the breadth of its system-level observatory to meet the needs of a space-faring nation.

The International Traffic in Arms Regulations (ITAR), a set of U.S. government regulations that controls the export of spaceflight hardware, designs, or design and development information, is also a threat to progress. Although there are efforts within the government to streamline and rationalize the process for review and approval of such exports, ITAR remains an obstacle to international cooperation in space research and, thereby, an impediment to opportunities to enhance science returns and reduce costs in many scientific missions. Notably, problems with ITAR compliance extend even to collaborations with nations that are close allies of the United States.

In summary, although the survey committee found the solar and space physics community generally to be vibrant and more integrated among relevant government agencies than in prior eras, it also found significant weaknesses and threats to the continued health of the enterprise. Nevertheless, the survey committee concluded that if stakeholders (including agencies, the science community, and policy makers) vigorously exploit the community's capabilities, great opportunities for science and for society are still within reach. It is within this context that the survey committee makes its recommendations for programs and activities that will build on existing capabilities in an affordable and cost-effective manner and make fundamental contributions worthy of public investment.

KEY SCIENCE GOALS FOR A DECADE

Significant accomplishments in solar and space physics over the past decade have set the stage for transformative advances in the decade to come. Reports from the survey's three interdisciplinary study panels (Chapters 8-10) enumerate the scientific opportunities and priorities for the interval addressed by this decadal survey, 2013-2022; these provide detail and context for the survey committee's four key science goals.

---

[5]NRC, *Controlling Cost Growth*, 2010.

BOX 1.2 ACCESS TO SPACE

NASA procures launch services primarily through the NASA Launch Services (NLS) contract. The latest NLS contract, NLS II, was announced on September 16, 2010, and includes launch vehicle offerings from four vendors.[1] This contract was subsequently amended in 2012 through its "on-ramp" provision to add SpaceX's Falcon 9 and Orbital Sciences' Antares launch vehicles.[2,3] Prior to this, NASA added the United Launch Services' Delta II launch vehicle—long a workhorse for launching NASA science missions—back to its NLS II contract. However, that contract modification was limited to a maximum of five Delta II purchases, as opposed to the indefinite procurement nature of the other contracts.[4] Therefore, the Delta II is being phased out once again through attrition. Of the launch vehicles available through NLS II, four are in current production (the Pegasus, Taurus, Falcon 9, and Atlas V). The Atlas V and Falcon 9 meet or exceed Delta II-class capability, but prices for the Atlas V have increased dramatically. Pricing for the Falcon 9—though currently less costly—remains uncertain for the time being. Furthermore, Orbital Sciences' Taurus launch vehicle failed in three of its last four launch attempts (the 2001 QuikTOMS, 2009 Orbiting Carbon Observatory, and 2011 Glory missions—all NASA Earth science missions), resulting in a suspension of its use by NASA indefinitely. Currently, the Pegasus is the only small-class launch vehicle manifested for future launches, but even then it is currently slated to launch only one mission in the near future: the Interface Region Imaging Spectrograph (IRIS) mission in 2013.[5]

The challenges facing the launch vehicle industry that have played out over the past decade or more have also resulted in a lower launch cadence for NASA science missions. Decreasing launch rates exacerbate a tendency for missions in development to grow in size and complexity as longer development times, higher overall mission costs, and fewer overall missions increase community expectations for the few missions that do make it to space. This situation creates a feedback loop, with increasing costs driving increasing expectations and decreasing risk tolerance, which can further increase costs.

Although increased competition is projected for higher-performance launch vehicles, plans to develop alternative small-class launch vehicles appear less firm. SpaceX has ceased development of its Falcon 1/1e offerings in favor of its Falcon 9.[6] The publicly available launch manifest shows just one Falcon 1e launch in 2014, compared with 37 Falcon 9/F9 Dragon/Falcon heavy launches through 2017.[7]

The development of SpaceX's Falcon 9 and Orbital Sciences' Antares launch vehicles has garnered a great deal of attention in the Earth and space sciences community, especially since both meet or exceed Delta II capabilities. However, neither has the long historical record of the Delta II for the simple fact that there have not been many launches of the Falcon 9 and the first launch of Antares is not scheduled until 2013. The Falcon 9 was selected as the launch vehicle for the Jason-3 mission to be launched in 2014 for a price of approximately $82 million.[8] Nevertheless, its successes to date have been few in number and limited to the delivery of cargo to the International Space Station (in May and October 2012 and March 2013). The Falcon 9 has not yet placed a robotic spacecraft in Earth orbit, although it did successfully place its Dragon cargo capsule in orbit for a few hours before deorbiting it in a controlled reentry. While encouraged by these early achievements, the space sciences community can only be cautiously optimistic until new launch service providers have a more exten-

Key Science Goal 1

Key Science Goal 1. Determine the origins of the Sun's activity and predict the variations in the space environment. The Sun and its variability drive space weather on a wide variety of timescales. Research driven by this goal investigates the origin of this variability inside the Sun and how its influence penetrates to near-Earth space to interact with other, terrestrial drivers of change in Earth's space environment.

sive track record. The new providers promise significant cost savings; however, the high launch demand from government customers that would allow for such savings is by no means guaranteed, especially in the austere budget environment that is anticipated for at least the next several years, and despite 24 planned launches on SpaceX's manifest from commercial industry customers through 2017.

Noncommercial vehicles (e.g., Minotaur) exist that are capable of launching Delta II-class payloads, but the Commercial Space Act[9] precludes their use absent special dispensation (e.g., use of the Minotaur for a non-DOD payload requires a waiver from the Secretary of Defense). International launch vehicles (e.g., Ariane, H-II) are also widely available; however, international launch vehicles would require a partnership arrangement with a foreign agency and no exchange of funds.

The need for reliable and affordable access to space is by no means new or unique to NASA's Earth science program. The loss in 2009 and 2011 of two NASA Earth science missions due to launch vehicle failures, however, underscores the urgency of addressing the need.

NOTE: Unless indicated otherwise, the information supplied here was current as of Spring 2012.

[1] See NASA, "NASA Awards Launch Services Contracts," press release, September 16, 2010, available at http://www.nasa.gov/home/hqnews/2010/sep/C10-053_Launch_Services_Contract.html. NOTE: This contract was amended in May and June 2012, respectively, to add SpaceX's Falcon 9 and Orbital Sciences' Antares launch vehicles.

[2] See NASA, "NASA Modifies Launch Service Contract to Add Falcon 9 Rocket," press release, May 14, 2012, available at http://www.nasa.gov/home/hqnews/2012/may/HQ_C12-019_NLS_Falcon_9.html.

[3] See NASA, "NASA Adds Orbital's Antares to Launch Services II Contract," press release, June 26, 2012, available at http://www.nasa.gov/home/hqnews/2012/jun/HQ_C12-027_NLS_II_mod.html.

[4] See NASA, "NASA Modifies Launch Service Contract to Add Delta II Rocket," press release, September 30, 2011, available at http://www.nasa.gov/home/hqnews/2011/sep/HQ_C11-044_Delta_Ramp.html.

[5] See NASA, "NASA Launch Services Manifest," available at http://www.nasa.gov/pdf/315550main_NASA%20FPB%2007_24_12%20Manifest%20Release%2008_01_2012_508.pdf, accessed October 9, 2012. IRIS was launched in late June 2013.

[6] Production of the Falcon 1 was suspended in 2011; see G. Norris, "SpaceX Puts Falcon 1 on Ice," *Aviation Week*, September 28, 2011, available at http://www.aviationweek.com/Article.aspx?id=/article-xml/asd_09_28_2011_p01-01-375285.xml.

[7] See SpaceX, "Launch Manifest," available at http://www.spacex.com/launch_manifest.php, accessed October 9, 2012.

[8] See NASA, "NASA Selects Launch Services Contract for Jason-3 Mission," press release, July 16, 2012, available at http://www.nasa.gov/home/hqnews/2012/jul/HQ_C12-029_RSLP-20_Launch_Services.html.

[9] See Public Law 105-303, available at http://www.nasa.gov/offices/ogc/commercial/CommercialSpaceActof1998.html. To use a Minotaur, the NASA administrator must obtain approval from the secretary of defense and certify to Congress that use of a noncommercial launch vehicle will result in cost savings to the federal government, meet all mission requirements, and be consistent with international obligations of the United States.

SOURCE: Material presented here is drawn from National Research Council, *Earth Science and Applications from Space: A Midterm Assessment of NASA's Implementation of the Decadal Survey*, The National Academies Press, Washington, D.C., 2012.

Earth's star is far more complex and variable than it appears to the naked eye in the daytime sky. (See Figure 1.3.) Sunspots as large as several Earths dot the surface with intense magnetic fields thousands of times stronger than the average background field. Each active region on the Sun can grow, decay, and reorganize on timescales of minutes to months, and collectively the spots emerge in groups that define the roughly 11-year activity cycle and 22-year magnetic cycle of the Sun. The typical solar wind speed and solar irradiance vary in concert with the cycle. As often as three times per day during solar maximum, the

FIGURE 1.3 A series of active regions, lined up one after the other across the upper half of the Sun, twisted and interacted with each other over 4.5 days from September 28 to October 2, 2011. As seen in extreme ultraviolet light, the magnetically intense active regions sported coils of arcing loops. SOURCE: Courtesy of NASA Solar Dynamics Observatory Atmospheric Imaging Assembly.

Sun ejects billions of tons of material with embedded magnetic fields. These coronal mass ejections (CMEs), when aimed at Earth, can reach our planet in less than a day at speeds more than five times that of the background solar wind, the expanding solar atmosphere that fills the entire solar system and heliosphere. CMEs can cause large geomagnetic storms that affect terrestrial systems such as electric power grids. They are often associated with solar flares, the most intense explosions in the solar system.

Solar flares bathe Earth in excess radiation across the electromagnetic spectrum, but it is the intense X-ray and ultraviolet radiation that significantly affects Earth's dayside ionosphere and the communication and navigation systems that are vulnerable to the state of the ionosphere. Flares release enormous energy in minutes and can remain intense over many hours. While the largest flares occur about once per solar cycle, several hundred per solar cycle are categorized as strong to extreme. Solar particle events are yet another signature of the variable Sun. Relativistic-energy particles, created near the Sun and by shock waves propagating toward Earth, can arrive at Earth in as little as 15 minutes. Historically, the most intense events occur about once per solar cycle, whereas strong to extreme particle storms occur about 15 times per cycle. Extreme particle storms affect human and robotic activities in space; they can also disrupt airline operations over polar routes, and they have potentially important long-term health impacts on flight crews. Almost every aspect of the Sun is variable and affects life on Earth.

The most recent solar minimum (comprising the end of cycle 23) was longer and deeper than any in the past century. Cycle 24 started 13 years after cycle 23 and may be weaker than any during the space age. Yet, despite increasingly accurate measurements of the flows beneath the solar surface and more sophisticated models of the dynamo that drives the activity, scientists cannot confidently make a physical prediction of the emergence of a strong sunspot, let alone the level of activity that will be present at the end of the coming decade. The missing information will come, in part, from measurements of the hard-to-view solar poles. The deep, ponderous flows that carry patterns of magnetic flux to the poles regulate the seeding of the deep-seated dynamo that generates subsequent solar cycles.

While large-scale dynamics drive cyclic solar changes, the actual mechanisms of variability—brightness, heating, mass flow—depend also on the summation of the myriad interactions that take place on the smallest scales. Against that seething background of continual small-scale activity, global structures store immense energy on a vast scale. The build-up of magnetic stress can be modeled, but not observed directly. What triggers catastrophic energy release in a flare or CME remains a puzzle. Only by sensing the solar wind directly with a probe near the Sun will it be possible to distinguish what accelerates the ordinary wind and more energetic particles.

Key Science Goal 2

> ***Key Science Goal 2. Determine the dynamics and coupling of Earth's magnetosphere, ionosphere, and atmosphere and their response to solar and terrestrial inputs.*** The regions of Earth's space environment are coupled by interactions among neutral gas, electrically charged particles, and plasma waves occurring over a range of spatial and temporal scales. The transport of energy and momentum through this environment exhibits varying degrees of feedback and complexity, requiring research approaches that treat it as a coupled system.

The space environment of Earth is profoundly affected by the solar wind. When the entrained solar wind magnetic field encounters Earth's magnetic field where the two magnetic fields are pointing in opposite directions, they can annihilate through the process of magnetic reconnection. (See Figure 1.4.) Magnetic reconnection drives convection that carries energetic particles toward Earth where they are injected and trapped in orbits around Earth to form the outer radiation belt. Although the broad view of how reconnection takes place and drives convection in the magnetosphere is now well established, the underlying physics of magnetic reconnection in the magnetosphere is not yet understood well enough to predict when, where, and how fast this process will occur and how it contributes to the transport of mass, energy, and momentum. Understanding charged-particle acceleration, scattering, and loss, which control the intensification and depletion of the radiation belts, is therefore a priority of solar and space physics.

During magnetic storms intense ion upwelling from the ionosphere—the inner boundary of the magnetosphere—into the magnetosphere is so strong that it can alter magnetospheric dynamics by modifying magnetic reconnection both on the dayside and on the nightside. The ionosphere is also the site of fundamental plasma-neutral gas interactions that must be unraveled to understand the dynamics of the neutral atmosphere-ionosphere system. When magnetospheric currents are disrupted, it is the ionosphere that provides the alternate path for magnetospheric currents to flow, heating Earth's atmosphere.

Research and observations are clearly needed to understand energy transport, cooling, and structuring across this system. The intense energy input from the magnetosphere, reaching up to a terawatt or more, typically occurs in regions spanning less than 10 degrees in latitude, but during storms energy is redistributed throughout the polar regions and down to middle latitudes. Theory still cannot explain how the global thermosphere "inflates" several hours after the onset of high-latitude heating, nor have studies yet captured the subsequent cooling with the fidelity that is needed to predict changes in satellite orbits occurring during magnetic storms.

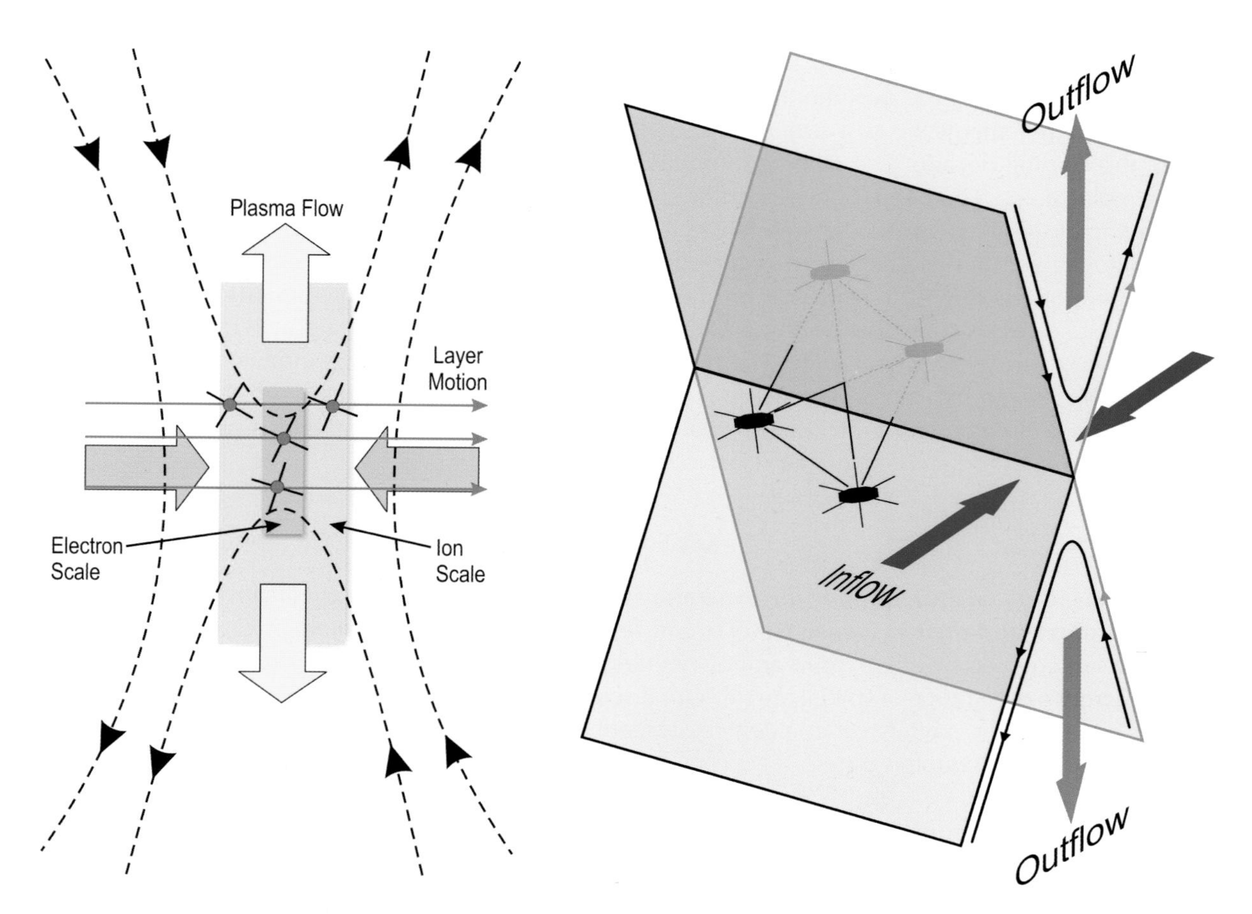

FIGURE 1.4 Schematic of magnetic reconnection fields, flows, and diffusion regions. The four-spacecraft Magnetospheric Multiscale (MMS) mission is targeted to resolve fundamental questions regarding the physics of magnetic reconnection. The nominal configuration of the MMS spacecraft during transit of magnetopause and magnetotail current layers is shown to scale with the key physical scale lengths. SOURCE: Courtesy of MMS-SMART Science Team, J.L. Burch, principal investigator, Southwest Research Institute.

Earth's upper atmosphere and ionosphere are a rich laboratory for the investigation of radiative processes and plasma-neutral coupling in the presence of a magnetic field. The behavior can be extraordinarily complex: plasma-neutral collisions and associated neutral winds drive turbulence that cascades to very small spatial scales and regularly disrupts communications. The primary mechanism through which energy and momentum are transferred from the lower atmosphere to the upper atmosphere and ionosphere is through the generation and propagation of waves. Although the presence and importance of waves are without dispute, the relevant coupling processes operating between the neutral atmosphere and ionosphere involve a host of multiscale dynamics that are not understood at present. Understanding the energy budget, dynamical behavior, and day-to-day variability in this region that is heavily influenced by wave energy from below and above presents a significant challenge to solar and space physics.

The release of greenhouse gases (e.g., carbon dioxide and methane) into the atmosphere is changing the global climate, warming the lower atmosphere but cooling the upper atmosphere. In the lower atmosphere the opacity of greenhouse gases traps energy by capturing the radiant infrared energy from Earth's surface and transferring it to thermal energy via collisions with other molecules. In the thermosphere, however, where intermolecular collisions are less frequent, greenhouse gases promote cooling by acquiring energy via collisions and then radiating this energy to space in the infrared.

Continued cooling of the thermosphere will alter atmosphere-ionosphere coupling, thereby altering global currents in the magnetosphere-ionosphere system and thus fundamentally altering magnetosphere-ionosphere coupling. These trends are only now apparent from the long-term set of satellite data and ground-based ionosonde networks, the latter of which show that the ionosphere is dropping to lower altitudes over time. This is a remarkable planetary change attributable, at least in part, to human society's modification of the atmosphere.

Key Science Goal 3

Key Science Goal 3. Determine the interaction of the Sun with the solar system and the interstellar medium. The outer regions of the heliosphere remain only sparsely explored.[6] Regions of high intrinsic interest, they are also the location where anomalous cosmic rays are generated. Anomalous and galactic cosmic rays penetrate into the inner solar system, posing a danger for humans and spacecraft traveling to locations beyond those afforded protection by Earth's magnetic field.

The supersonic flow of the solar wind transitions to a subsonic flow that merges with the local interstellar medium at a distance of about 100 AU from the Sun. The heliosheath lies beyond this "termination shock" and extends out to the heliopause, an as-yet-unexplored region that separates the domain of the Sun from the local interstellar medium. The two Voyager spacecraft, launched in 1977, crossed the termination shock in 2004 and 2007, and they are now exploring the heliosheath—a region of prodigious particle acceleration (see Figure 1.5). In terms of the total amount of energy placed into energetic particles, the heliosheath is a more copious accelerator than the Sun.

The heliosheath and the heliopause are the principal barriers against entry of galactic cosmic rays (GCRs) into the solar system, with the largest reduction in GCR intensity occurring in these regions. How this reduction occurs, how it depends on solar activity, and what is the maximum intensity at which GCRs can penetrate into the inner heliosphere are not known. GCRs generate isotopes in Earth's atmosphere that are preserved as an archive of historical solar activity—for example, carbon-14 abundance can be measured in tree rings. Understanding the production and formation of these unique records of past changes in solar activity is essential for interpreting changes in Earth's climate over the past millennium.

The Voyager spacecraft are currently at widely separated locations, and their measurements are limited by their 35-year-old instrumentation. Nevertheless, recent Voyager measurements have enabled discovery of unanticipated structures in the heliosheath. Measurements from the Interstellar Boundary Explorer (IBEX) have revealed that the interstellar magnetic field organizes the distribution of energetic ions. These unique measurements from the Voyagers as they cross the heliopause into the local interstellar medium, along with measurements from IBEX and the proposed IMAP mission, promise a new understanding of the important acceleration processes occurring in these unexplored regions. In particular, they will help

[6]The Voyager spacecraft continue to make in situ measurements of the outer heliosphere, and their entry into interstellar space sometime in the next decade will be an historic event. Remote sensing measurements from instruments on near-Earth spacecraft complement Voyager measurements and facilitate a multipronged attack on fundamental science questions.

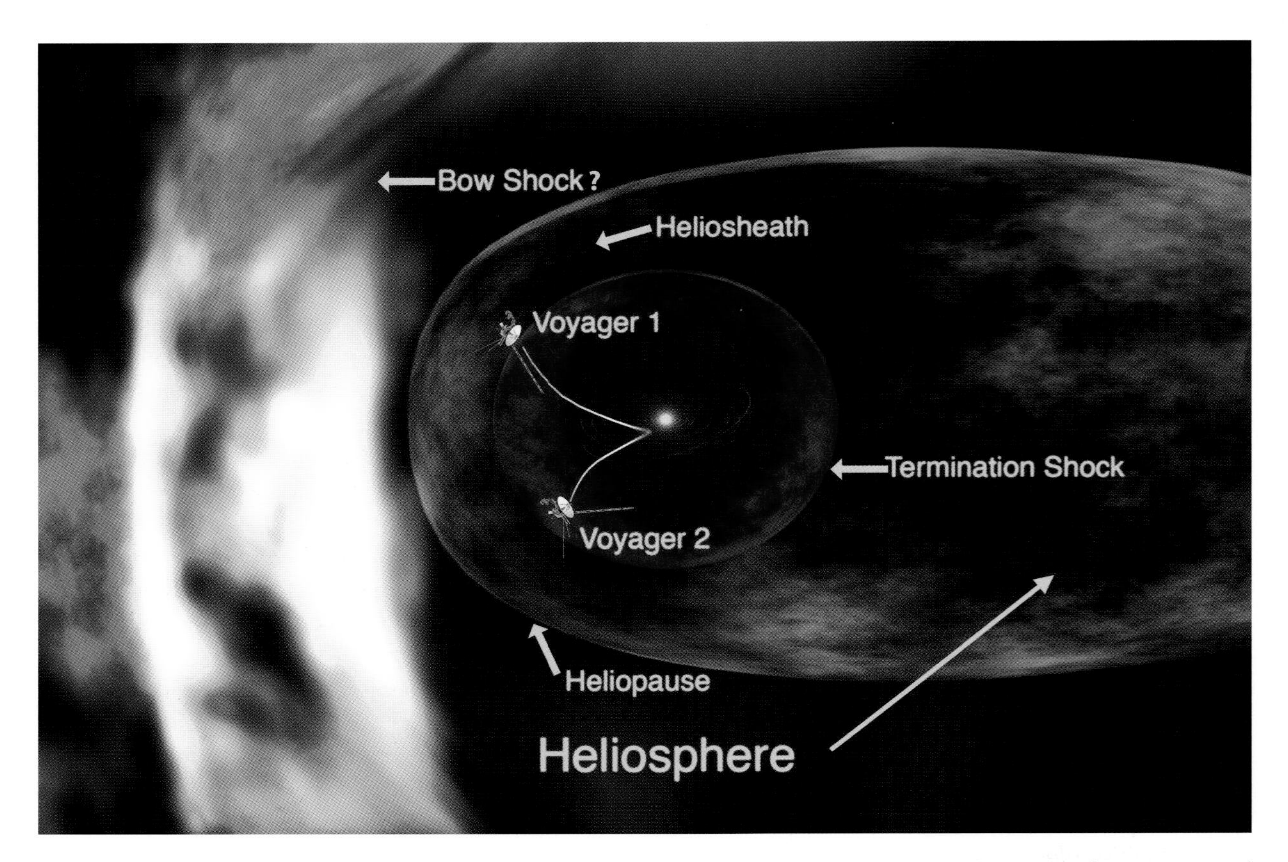

FIGURE 1-5 A schematic image of the regions of the heliosphere, including the approximate locations of Voyagers 1 and 2. Voyager 1 crossed the termination shock at 94 AU in December 2004, and Voyager 2 crossed at 84 AU in August 2007. For details, see http://science1.nasa.gov/missions/voyager/. SOURCE: Courtesy of NASA/JPL/Walt Feimer. NOTE: This figure was prepared several years ago when scientists expected Voyager to discover a bow shock. New results from the Interstellar Boundary Explorer (IBEX) have shown that instead of a shock, there is a softer interaction—a bow wave, similar to how water piles up ahead of a moving boat. For IBEX results, see D.J. McComas, D. Alexashov, M. Bzowski, H. Fahr, J. Heerikhuisen, V. Izmodenov, M.A. Lee, E. Moebius, N. Pogorelov, N.A. Schwadron, and G.P. Zank, The heliosphere's interstellar interaction: No bow shock, *Science* 336:1291, doi: 10.1126/science.1221054, May 2012.

---

researchers understand the mechanisms by which the solar system is protected against GCRs and how effective this protection is likely to be in the face of changing solar conditions.

Key Science Goal 4

> ***Key Science Goal 4. Discover and characterize fundamental processes that occur both within the heliosphere and throughout the universe.*** Advances in understanding of solar and space physics require the capability to characterize fundamental physical processes that govern how energy and matter are transported. Such understanding is also needed to improve the capability to predict space weather.

Underlying the extraordinarily complex and dynamic space environment are identifiable fundamental processes that can sometimes be explored as independent problems. These fundamental processes can also play a role in other astrophysical settings. In that sense, the Sun, the heliosphere, and Earth's magnetosphere and ionosphere serve as cosmic laboratories for studying universal plasma phenomena, with applications to laboratory plasma physics, fusion research, and plasma astrophysics.[7] Discoveries from these fields, of course, also contribute to the scientific progress in solar and space physics.

There are numerous examples of the universal processes that control the dynamics of the space environment:

- *Dynamos.* Turbulence in the convection zone of the solar interior twists and transports magnetic fields and ultimately determines the large-scale solar dipolar magnetic field, which reverses polarity on average every 11 years. Similar dynamos produce magnetic fields in stars, magnetars, and even in galaxies, black holes, and other compact objects. Earth's ionosphere exhibits several neutral wind dynamo processes that at high latitudes generate large-scale electric fields that affect the magnetosphere, and at low latitudes control the growth of plasma densities generated by solar radiation.
- *Solar and planetary winds.* The heating and subsequent outward expansion of the solar atmosphere create the solar wind. Produced by a variety of mechanisms, winds are features of essentially all stars. Winds from the poles of Earth—the polar wind—can similarly fill the magnetosphere with ionospheric plasma.
- *Magnetic reconnection.* Magnetic fields in regions of opposing field direction can annihilate each other to convert magnetic energy into high-speed flows, heated plasma, and energetic particles. This explosive release of energy drives flares on the Sun and other stars and possibly the magnetospheres of galactic accretion disks and astrophysical jets. Reconnection in Earth's magnetosphere leads to the erosion of Earth's protective magnetic shield during storms and is the driver of magnetospheric substorms.
- *Collisionless shocks.* Shock waves appear throughout the heliosphere where they facilitate the transition from supersonic to subsonic flow, heat the plasma, and act as accelerators of energetic particles. Shocks are widely observed in astrophysical systems in the form of supernova shocks, which are a predicted source of galactic cosmic rays, at the termination of astrophysical jets, and more generally during collisions and mergers of galaxies.
- *Turbulence.* Plasma turbulence is ubiquitous in the space environment and throughout the broader universe. It carries energy from the interior of the Sun to its surface and drives the solar dynamo. It is also one of the proposed mechanisms for heating the ambient corona and accelerating energetic particles in flares. Turbulence drives the transport of particles, and energy in the magnetosphere—the region of space dominated by Earth's magnetic field and radiation belt—heats electrons and ions in the auroral region and is ubiquitous in the charged upper layers of Earth's atmosphere (the ionosphere). The recognition that turbulence may facilitate accretion has transformed understanding of the environments of compact astrophysical objects and even of the mechanisms of star and planetary formation.
- *Plasma-neutral interactions.* The interaction of the ionized plasma in Earth's magnetosphere and neutral particles in the ionosphere/thermosphere lead to ionization, outflows into the magnetosphere, and the generation of neutral winds whose rich dynamics have only recently been appreciated. Similar interactions between neutral and ionized particles take place at the Sun and in the solar wind. In the broader universe, plasmas are often only partially ionized, so that plasma motions are often constrained by mass loading due to neutrals. The observed structuring of molecular clouds is believed to result from the ionization dynamics of radiation and plasma-neutral interaction.

---

[7]See National Research Council, *Plasma Physics of the Local Cosmos,* The National Academies Press, Washington, D.C., 2004.

OPTIMIZING A SCIENCE PROGRAM

With the above key science goals in mind, and recognizing that many of the outstanding problems in solar and space physics require an integrated observational approach, the survey committee's strategy was to construct a program that can achieve progress across the coupled domains that define the entire field. The strategy builds on existing and planned programs, optimally deploys recommended new assets at an appropriate cadence, and includes actions and decisions that should be taken in the event that the expected funding profile cannot be attained or, conversely, if it is augmented. The survey committee's overall strategy is summarized in the seven steps outlined in Box 1.3.

The Heliophysics Systems Observatory, made up of NASA's existing heliophysics flight missions and NSF's ground-based facilities, is an ever-rich source of observations that address increasingly interdisciplinary and long-term science objectives. The success of these exciting and productive activities at NASA and NSF is fundamentally important to long-term scientific progress in solar and space physics. With prudent management and careful cost containment, the survey committee concluded that completion of the ongoing program is precisely the right first step for the next decadal interval and as such represents the baseline priority.

In framing its recommendations for progress over the next decade, the survey committee identified five assets of the solar and space physics program as necessary to realize the full scientific potential of the field. They are the (cross-agency) enabling foundation, ground-based facilities, and small, moderate-scale, and large-scale (major) space missions. Making optimal use of these assets, which differ widely in cost and expected operating lifetime, requires careful attention to deployment cadence and overlap. Further details are presented below.

BOX 1.3 STEPS OF A STRATEGY FOR OPTIMIZING A SOLAR AND SPACE PHYSICS SCIENCE PROGRAM FOR 2013-2022

1. Identify the highest-priority science goals and major programmatic assets for each of the subdisciplines of solar and space physics.
2. Recognize that to make meaningful scientific progress, the Sun-heliosphere-Earth system must be studied as a coupled system.
3. Recognize that understanding the coupled Sun-heliosphere-Earth system requires that each subdiscipline must be able to make reasonable progress in achieving its highest-priority science goals.
4. Recognize that to make balanced progress across the subdisciplines, all the assets available to solar and space physics—the enabling foundation of theory, modeling and data analysis; ground-based facilities; and small, moderate, and large space missions—must be optimally deployed.
5. Recognize that each asset available to solar and space physics has a cadence at which it can most effectively be a successful contributor.
6. Construct a program that can achieve such balanced progress across the subdisciplines, optimally deploying all of the assets, each with a reasonable cadence.
7. Determine the actions and decisions that will be taken in the event that the expected funding profile cannot be attained or some other disruptive event occurs or, conversely, in the event that the budget is augmented or the cost to missions is reduced.

The Enabling Foundation

The enabling foundation for solar and space physics is common across its subdisciplines; it includes theory, modeling, and data analysis; innovative platforms and technologies; and education.[8] The survey committee found an immediate need to strengthen this foundation to garner maximum knowledge from data being collected from space and from the ground. Further benefits include the introduction of new innovative and cost-effective approaches to space exploration, and the generation and continued development of a world-leading science community.

This insight motivated the survey committee to amalgamate thrusts in these areas into a multiagency initiative, labeled DRIVE—Diversify, Realize, Integrate, Venture, and Educate. The elements of DRIVE are described in detail in Chapter 4, which also includes DRIVE-related recommendations, several of which are designed to strengthen the enabling foundation. The DRIVE initiative encompasses specific augmentations to existing enabling programs. Its implementation will provide opportunities to realize scientific discoveries from existing data, to build more comprehensive models, to make theoretical breakthroughs, and to innovate.

Ground-Based Facilities

NSF ground-based solar and space physics facilities, which are managed within two of its divisions, are key existing elements of the Heliophysics Systems Observatory. The Astronomy Division of the Mathematics and Physical Sciences Directorate manages the National Solar Observatory (NSO), with ongoing synoptic observations and the ATST under construction. It is also the home of the National Radio Astronomy Observatory, which includes solar observational capability. The Atmospheric and Geospace Sciences Division (AGS) of the Geosciences Directorate manages the National Center for Atmospheric Research (NCAR) and its High Altitude Observatory (HAO), which supports a broad range of research from the Sun to Earth and operates the Mauna Loa Solar Observatory. AGS also supports the SuperDARN coherent-scatter radar system and four large incoherent-scatter radar facilities including the Arecibo Radio Observatory, which remains the largest-aperture telescope in the world for astrophysical, planetary, and atmospheric studies. University-based observatories are also funded by both NSF and NASA.

When it begins operation in 2018, the 4-meter ATST will be, by far, the largest optical solar telescope in the world. It will provide revolutionary measurements of the solar magnetism that controls energetic mass eruptions and variability in the Sun's output. ATST measurements at the currently unreachable small size of density fluctuations will reveal how magnetic fields and mass interact at their fundamental size scales. ATST will also provide the first measurements of magnetic fields low in the corona where mass and energy are injected to form the heliosphere. It will be revolutionary in the capabilities it will provide to measure the dynamics of the magnetic field at the solar surface down to the fundamental density length scale. It will also be able to remotely sense coronal magnetic fields in locations where they have never been measured. For full realization of its potential, ATST requires adequate, sustained funding from NSF for operation, data analysis, development of advanced instrumentation, and research grant support for the ATST user community. DRIVE "Realize" includes a recommendation for a significant increase in funding allocation to properly operate and to realize the full and remarkable scientific potential of ATST.

Important ground-based research is also accomplished through midscale research projects that are larger in scope than typical single principal-investigator (PI)-led projects (like those funded by NSF's Major Research Instrumentation line) and smaller than facilities (such as those developed under NSF's Major

[8]See National Research Council, *An Enabling Foundation for NASA's Space and Earth Science Missions*, The National Academies Press, Washington, D.C., 2010.

Research Equipment and Facilities Construction line). The Advanced Modular Incoherent Scatter Radar (AMISR) is an example of a midscale project widely seen to have transformed research in ground-based solar and space physics.

Although different NSF directorates have programs to support unsolicited midscale projects at various levels, these may be overly prescriptive and uneven in their availability, and practical gaps in proposal opportunities and funding levels may be limiting the effectiveness of midscale research across the agency. It is unclear, for instance, how projects like the AMISR would be initiated and accomplished in the future, because no budget line is available that matches this scale of facility development. Mechanisms for the continued funding of management and operations at existing midscale facilities are also not entirely clear. In addition, there is a need for a means of funding midscale projects, many of which have been identified by this decadal survey committee as cost-effective additions of high priority to the overall program. These include the Frequency-Agile Solar Radiotelescope (FASR), the Coronal Solar Magnetism Observatory (COSMO), and several other projects exemplifying the kind of creative approaches that are necessary to fill gaps in observational capabilities and to move the survey's integrated science strategy forward. A mid-scale funding line will have a major impact on conducting science at ground-based facilities, and it would rejuvenate broadly utilized assets by taking advantage of innovations to address key science challenges. DRIVE "Diversify" includes a recommendation for a new, competitively selected midscale funding line at NSF.

Small Space Missions

Since the 2003 decadal survey,[9] a new experimental capability has emerged for very small spacecraft, which can act as stand-alone measurement platforms or can be integrated into a greater whole. These platforms are enabled by innovations in miniature, low-power, highly integrated electronics and nanoscale manufacturing techniques, and they provide potentially revolutionary approaches to experimental space science. For example, small, low-cost satellites may be deployed into regions where satellite lifetimes are short but where important, hitherto insufficiently characterized scientific linkages take place. The vulnerability of microsatellites to this radiation environment is also of interest from a space-weather viewpoint. NSF's CubeSats initiative promotes science done by very small satellites, and the enthusiastic response from the university community argues for a launch cadence greater than the current one per year. DRIVE "Diversify" includes a recommendation targeting the development of very-small-satellite flight opportunities as a key growth area for both NASA and NSF.

NASA's Explorer space projects have proven historically to be highly successful and cost-effective, providing insights into both the remotest parts of the universe and the detailed dynamics of Earth's local space environment. The 1997 Advanced Composition Explorer (ACE) now stands as a sentinel to measure, in situ, transient disturbances from the Sun and the energetic particles that are a danger to humans in space. RHESSI and the recently retired TRACE spacecraft study the dynamics of the solar corona, where large flares and potentially damaging solar storms originate. The relatively recently launched THEMIS constellation and the Aeronomy of Ice in the Mesosphere mission were both accomplished under the aegis of the Explorer program and are revolutionizing understanding of magnetospheric dynamics and global atmospheric changes, respectively. IBEX, which observes the heliospheric boundary from high Earth orbit, found a bright ribbon of energetic neutral atom emission whose origin is requiring reconsideration of fundamental concepts of the heliosphere and local interstellar medium interaction. The Interface Region Imaging Spectrograph (IRIS), the most recently selected Heliophysics Explorer, will explore the flow of

[9]National Research Council, *The Sun to the Earth—and Beyond: A Decadal Research Strategy in Solar and Space Physics*, The National Academies Press, Washington, D.C., 2003; and National Research Council, *The Sun to the Earth—and Beyond: Panel Reports*, The National Academies Press, Washington, D.C., 2003.

energy and plasma at the foundation of the Sun's atmosphere. In summary, Explorers are among the most competitive solicitations in NASA science, and they offer relatively frequent opportunities for all researchers to propose new and exciting ideas that are selected on the basis of science content, relationship to overall NASA strategic goals, and feasibility of execution.

The survey committee believes that an adequate cadence for Heliospheric Explorers is one mission every 2 to 3 years, a rate that was possible before the major reduction in 2005 in the Explorer program. The survey committee also notes that competition for the MIDEX class of Explorers, which historically has offered an opportunity to resolve the highest-level science questions, has not been possible under the current Explorer budget. Finally, the survey committee notes that the Explorer program is the home for Missions of Opportunity, which make it possible to achieve fundamental science at a fraction of the cost of stand-alone missions by hosting payloads through partnering with other agencies, nations, or commercial spaceflight providers. For example, the Solar-C mission now confirmed by Japan presents a future opportunity for the United States to provide instrumentation to a major foreign mission and in so doing to obtain a high science return for relatively low cost. Thus, an augmentation to the Explorer program and restoration of the MIDEX component of Explorers, described in detail in Chapter 4, are required to achieve the optimal cadance and to leverage resources with commercial, interagency, and international opportunities.

Moderate-Scale Space Missions

Many of the most important solar and space physics science questions cannot be addressed by Explorer-class missions. The survey committee has considered the most critical topics that can be realistically addressed by moderate-scale NASA missions and lists the community's top three priorities below (see Chapter 4 for details). To achieve an acceptable flight rate, the survey committee recommends that the Heliophysics Division's Solar-Terrestrial Probes (STP) program be reconfigured as a moderate-mission program modeled after the successful Discovery and New Frontiers programs of NASA's Planetary Science Division. Like these planetary missions, the new Solar-Terrestrial Probes should be led by a principal investigator, selected competitively, cost-capped, and managed in a manner similar to Explorers. Such programs exhibit superior cost-performance history compared with the more traditional mode for major missions (see Appendix E).

The survey committee believes that an adequate cadence for moderate-scale space missions is one every 4 years. In view of the expected budget constraints discussed below, it is evident that it will not be possible to achieve this cadence until the end of the decade. Although moderate missions are to be selected competitively, each moderate mission's science goal has to be defined in advance to achieve the strategic objective of balanced progress.

In Chapter 4, the survey committee presents the rationale for and priority order of science investigations that best achieve these objectives. In descending order for implementation they are as follows:

1. A mission to understand the interaction of the outer heliosphere with the interstellar medium—one that will be coordinated with NASA's Voyager mission and will also provide critical data on solar wind inputs to the terrestrial system. An illustrative example is the Interstellar Mapping and Acceleration Probe (IMAP).
2. A mission designed to substantially advance understanding of the variability in space weather driven by lower-atmosphere weather on Earth, illustrated by the Dynamical Neutral Atmosphere-Ionosphere Coupling (DYNAMIC) mission.
3. A mission that probes how the magnetosphere-ionosphere-thermosphere system is coupled and how it responds to solar and magnetospheric forcing, illustrated by the Magnetosphere Energetics, Dynamics,

and Ionospheric Coupling Investigation (MEDICI). The survey committee notes that MEDICI could not begin before 2024 absent a NASA Heliophysics Division budget augmentation or a reduction in the cost of other missions.

Each of the survey committee's recommended STP moderate-scale space missions was subjected to a cost and technical evaluation (CATE) process (Box 1.4), and projected cost assessments were found to be consistent with the proposed life-cycle cost of $520 million for a mission in this renewed STP line. However, these costs are possible only so long as the STP missions are executed as competitively selected, cost-capped, and PI-led missions, as recommended.

Major Space Missions

Certain very-high-priority science investigations are of such scope and complexity that they can be undertaken only with major-mission-based research. In the current decade, two such missions are already underway—the Magnetospheric Multiscale mission, which is scheduled for launch in 2014, and the Solar Probe Plus (SPP) mission, which is scheduled for launch in 2018. In its deliberations, the survey committee

BOX 1.4 SURVEY PRIORITIZATION AND THE CATE PROCESS

In September 2010, shortly after the present decadal survey commenced, the survey committee distributed widely to the solar and space physics community a request for information (RFI) for "a concept paper (e.g., about a mission or extended mission, observation, theory, or modeling activity) that promises to advance an existing or new scientific objective, contribute to fundamental understanding of the Sun-Earth system, and/or facilitate the connection between science and societal needs (e.g., improvements in space weather prediction)."[1]

Each submission was assigned to one or more of the survey's three study panels for review, and each submission was assigned to a specific reader who prepared a short presentation that was discussed by the panel. Concepts of particular interest were discussed in more detail at a subsequent session, and, at the end of a lengthy review process, panels developed a short list of concepts for consideration by the survey committee. In making their selections, panels mapped concepts against their prioritization of science targets and also considered factors such as technical readiness, scientific impact on a particular discipline or disciplines, and, in some cases, operational utility. It is important to note that the concepts that were eventually forwarded to the survey committee were in some cases an amalgam of more than one submission; they also drew on the expertise of panel members.

At a meeting of the decadal survey committee, 12 mission concepts were selected for further study by the Aerospace Corporation, which under contract to the National Research Council provided a preliminary cost and technical evaluation (CATE) of the concepts. At a subsequent meeting, the survey committee reviewed the results of the pre-CATE and selected six concepts for a more thorough analysis, including an evaluation of the options for descopes and other trade-offs that would affect estimated cost. Details of this process are explained in Appendix E.

The CATE process provided an independent analytical approach to realistically assessing the cost and risk related to recommended initiatives that are typically expected at an early stage of formulation. Each of the Solar-Terrestrial Probes and Living With a Star missions recommended in this report were chosen from among the six candidate missions that underwent the detailed CATE process.

[1] Further information about the RFI along with a compilation of submissions, which numbered nearly 300, is available on the decadal survey website at http://sites.nationalacademies.org/SSB/CurrentProjects/SSB_056864# White_Papers_and_Community_Input.

found a need for the development and launch of constellations of spacecraft, which are necessary to provide simultaneous measurements from broad regions of space and thereby separate spatial from temporal effects to reveal the true couplings between adjacent regions of space. By necessity, constellation missions require large investments, and their design, assembly, and execution are challenging. The expected budget profile for NASA's Heliophysics program is such that launch of the next major mission in heliophysics cannot be reasonably expected before 2024, or 6 years after SPP. This constraint—and not the absence of proposals to undertake compelling science investigations—defines a cadence for major missions.

In contrast to the STP program, which the survey committee recommends be community-based like the Explorers, major missions are appropriately undertaken by NASA centers. NASA's LWS program is the proper vehicle for large-class, center-led, major missions. In Chapter 4 the survey committee describes the next science target best addressed by the LWS program when the budget of the Heliophysics Division allows: a mission to understand how Earth's atmosphere absorbs solar wind energy, illustrated by the Geospace Dynamics Constellation (GDC).

IMPLEMENTATION STRATEGIES

The foundational assets described above form the cornerstones of the solar and space physics research program for the coming decade. Each is of exceptional importance; however, maximizing science return while operating in a highly constrained fiscal environment requires that they be prioritized and then implemented at an appropriate cadence. Further, a strategy is needed to address unforeseen technical or budgetary problems, such as budget problems attributable to the cost growth of individual program elements or unexpected changes in the overall (top-line) budget. Described below is the survey committee's approach to prioritizing program elements and formulating decision rules to address budget shortfalls.

Funding Priorities for NASA's Heliophysics Program

By employing the assets described above at an appropriate cadence, and using the decision rules described below, a viable program can be crafted that should allow solar and space physics to preserve the strategic goal of balanced progress even under less favorable budgetary circumstances. The DRIVE initiative has the fastest cadence (new competitions annually for many of its components), followed by a 2- to 3-year cadence for the Explorer program, a 4-year cadence for the moderate-scale mission STP program, and a 6-year cadence for the LWS major mission program. It follows that the first new NASA implementation priority is the augmentation required by the DRIVE initiative, followed by the augmentation for the Explorer program, the initiation of the STP moderate-scale mission program, and finally the continuation of the LWS major mission program. Chapter 6 provides a detailed discussion of the survey committee's proposed implementation of these program elements within the budget projected by NASA over the next 5 years and extrapolated by the committee for an additional 5 years.

Decision Rules

Decision rules are strategies to preserve an orderly and effective program for solar and space physics in the event that less funding than anticipated is available, or some other disruptive event occurs. As described in more detail in Chapter 6, the rules, with one exception, affect the year that mission and foundational activities commence and the cadence at which they repeat in the coming decade. The survey committee also provides decision rules to aid in implementing a program under a more favorable budget. In particular,

TABLE 1.1 Fulfilling the Key Science Goals of the Decadal Survey

Advances in Scientific Understanding and Observational Capabilities		Goals
Advances owing to implementation of the existing program	Twin Radiation Belt Storm Probes will observe Earth's radiation belts from separate locations, finally resolving the importance of temporal and spatial variability in the generation and loss of trapped radiation that threatens spacecraft.	2, 4
	The Magnetospheric Multiscale mission will provide the first high-resolution, three-dimensional measurements of magnetic reconnection in the magnetosphere by sampling small regions where magnetic field line topologies reform.	2, 4
	Solar Probe Plus will be the first spacecraft to enter the outer atmosphere of the Sun, repeatedly sampling solar coronal particles and fields to understand coronal heating, solar wind acceleration, and the formation and transport of energetic solar particles.	1, 4
	Solar Orbiter will provide the first high-latitude images and spectral observations of the Sun's magnetic field, flows, and seismic waves, relating changes seen in the corona to local measurements of the resulting solar wind.	1, 4
	The 4-meter Advanced Technology Solar Telescope will resolve structures as small as 20 km, measuring the dynamics of the magnetic field at the solar surface down to the fundamental density length scale and in the low corona.	1, 4
	The Heliophysics Systems Observatory will gather a broad range of ground- and space-based observations and advance increasingly interdisciplinary and long-term solar and space physics science objectives.	All
New starts on programs and missions to be implemented within the next decade	The DRIVE initiative will greatly strengthen researchers' ability to pursue innovative observational, theoretical, numerical, modeling, and technical advances.	All
	Solar and space physicists will accomplish high-payoff, timely science goals with a revitalized Explorer program, including leveraged Missions of Opportunity.	All
	The Interstellar Mapping and Acceleration Probe, in conjunction with the twin Voyager spacecraft, will resolve the interaction between the heliosphere—our home in space—and the interstellar medium.	2, 3, 4
	A new funding line for mid-size projects at the National Science Foundation will facilitate long-recommended ground-based projects, such as COSMO and FASR, by closing the funding gap between large and small programs.	All
New starts on missions to be launched early in the next decade	The Dynamical Neutral Atmosphere-Ionosphere Coupling mission's two identical orbiting observatories will clarify the complex variability and structure in near-Earth plasma driven by lower-atmosphere wave energy.	2, 4
	The Geospace Dynamics Constellation will provide the first simultaneous, multipoint observations of how the ionosphere-thermosphere system responds to, and regulates, magnetospheric forcing over local and global scales.	2, 4
Possible new start this decade given budget augmentation and/or cost reduction in other missions	The Magnetosphere Energetics, Dynamics, and Ionospheric Coupling Investigation will target complex, coupled, and interconnected multiscale behavior of the magnetosphere-ionosphere system by providing global, high-resolution, continuous three-dimensional images and multipoint in situ measurements of the ring current, plasmasphere, aurora, and ionospheric-thermospheric dynamics.	2, 4

by increasing the cadence of the recommended mission and foundational activities in DRIVE, Explorers, and the STP and LWS lines, a commensurate increase in program value can be obtained.

A DECADE OF TRANSFORMATIVE SCIENCE

Implementation over the coming decade of the multifaceted program recommended in this report will allow substantial progress to be achieved in meeting the survey's four key science goals. Indeed, as summarized in Table 1.1, the survey committee anticipates a decade of transformative advances in scientific understanding and observational capabilities, both space- and ground-based, upon implementation of the existing program and execution of the recommended program.

Throughout this report, the committee emphasizes the necessity of adopting a systems approach to achieve appropriately balanced progress in understanding an interconnected solar-heliospheric-terrestrial and planetary system. Of particular importance is the capability to advance solar and space physics science that is directly relevant to societal needs. However, the resources assumed in crafting this survey's recommended program are barely adequate to make required progress; with reduced resources, progress will be inadequate. It is also evident that with increased resources, the cadence of the assets by which the nation pursues the recommended program can be increased, with a concomitant increase in the pace of scientific discovery and in the value added for society.

2

Solar and Space Physics: Recent Discoveries, Future Frontiers

SCOPE AND RELEVANCE OF THE DISCIPLINE

To appreciate the complex structure and evolution of Earth's home in space, one need only look at the striking image of the extended solar atmosphere, the corona, taken during the July 11, 2010, solar eclipse (Figure 2.1, left panel). Turbulent convection below the Sun's visible surface is the engine that drives the extreme ultraviolet (EUV) and X-ray radiation and the solar wind. The solar magnetic field is churned and twisted by this subsurface convection and in turn produces the fine-scale structure of the solar corona. The right panel of Figure 2.1 shows magnetic lines of force from a physics-based prediction of coronal structure based on solar surface magnetic field measurements for the same event. The correspondence between the imaged corona and the simulated magnetic field structure is striking.

The corona is the source of both EUV radiation and the solar wind, an outward flowing plasma and entrained magnetic field with speeds in the range of 400 to 800 kilometers per second, or around a million miles per hour. Solar ultraviolet and X-ray radiation, for example from solar flares, reaches Earth directly in 8 minutes, where it is absorbed in the thermosphere, the uppermost portion of Earth's atmosphere. This photon energy heats the thermosphere and produces the electrically conductive ionosphere within the thermosphere. The ionosphere is linked both with the neutral atmosphere below through waves generated in the troposphere near Earth's surface that propagate upward through the atmosphere and with the magnetosphere above via electric currents and the flow of charged particles. In contrast with the EUV radiation, the solar wind does not impact Earth directly but instead encounters Earth's dipolar magnetic field, which deflects the solar wind and channels electric currents and energetic particles to the polar regions, shielding the middle and equatorial atmosphere.

Earth is therefore best understood not as orbiting the Sun in isolation through a vacuum, but as a physical system intimately linked to the highly variable solar atmosphere that engulfs the entire solar system. The magnetized solar atmosphere, solar wind, and Earth's magnetosphere, ionosphere, and atmosphere are connected through a chain of interactions that govern the state of our space environment. Furthermore, the Sun occasionally sends out powerful mass ejections, which are accompanied by shock waves that accelerate charged particles to very high speeds, up to nearly the speed of light. These disturbances in the

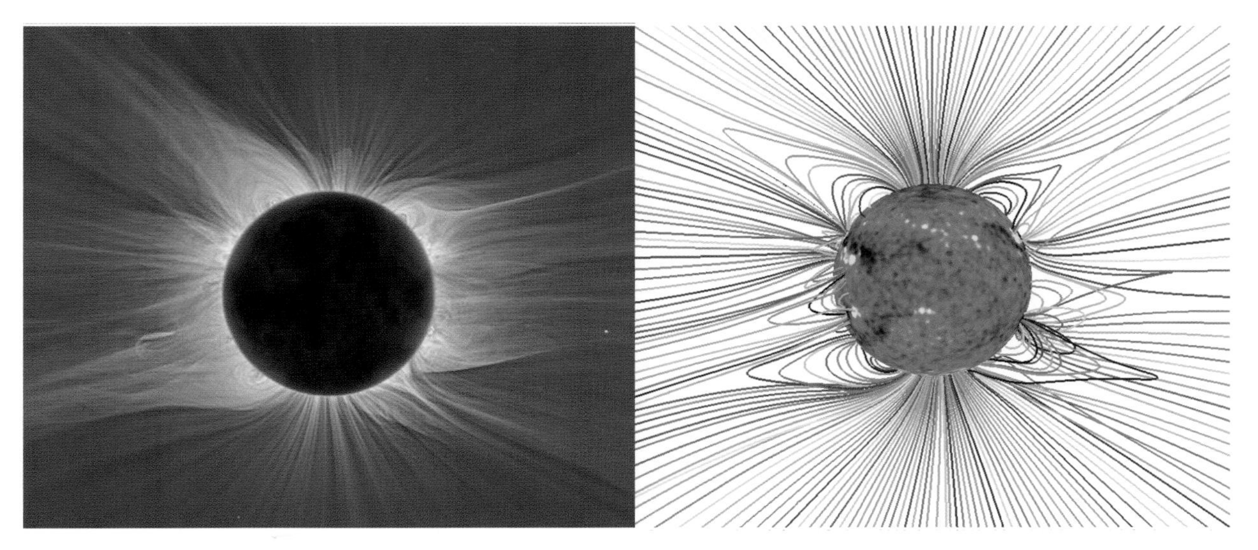

FIGURE 2.1 *Left:* White-light image of the solar corona out to 4 solar radii during the solar eclipse of July 11, 2010. *Right:* Predictive Science, Inc., prediction of the magnetic field during the July 11, 2010, eclipse, using observations of photospheric magnetic field and numerical simulation of magnetized fluid. SOURCE: *Left:* Courtesy of M. Druckmüller, M. Dietzel, S. Habbal, and V. Rušin; available at http://www.predsci.com/corona/jul10eclipse/jul10eclipse.html. *Right:* Courtesy of Predictive Science, Inc.

solar wind intensify the Van Allen radiation belts, drive the aurora and powerful electric currents on Earth, and violently churn the ionosphere and uppermost atmosphere.

There is a growing appreciation that solar systems are commonplace in the universe and that the physical processes active in Earth's heliosphere are universal. Deepening understanding of our own home in space therefore informs humanity's understanding of some of the most basic workings of the universe. As human exploration extends farther into space via robotic probes and human flight, and as society's technological infrastructure is increasingly linked to assets that are affected by the space environment, a deeper and fundamental understanding of these governing processes becomes ever more pressing (see Chapter 3).

The principles governing the Sun-Earth system include the physics of plasmas and of neutral and ionized atmospheres; atomic and molecular physics; radiative transport; and relativistic particle acceleration. The problems in solar and space physics are the basis of some of the most daunting challenges in these fields. For example, the physical regimes of plasmas in the heliosphere range from the highly collisional environment of the Sun's convective zone to nearly collision-free environments of the Sun's outer corona, as well as the interplanetary medium and planetary magnetospheres. In each regime, different theoretical approaches must be used to describe the system, and no single theoretical treatment applies throughout this vast range of regimes. Moreover, the dynamics of the system are governed by processes that span a broad range of spatial and temporal scales and are often the product of nonlinear or chaotic processes. For convenience, the broad discipline of solar and space physics, often referred to as heliophysics, is divided into the following three areas, each of which is described in much greater detail in Part II, Chapters 8 through 10, of this report:

- *Solar and heliospheric physics (SHP)*—which covers the physics of the outer regions of the Sun and the solar wind and its expansion through interplanetary space;
- *Solar wind-magnetosphere interactions (SWMI)*—which deals with the interaction of the solar wind with magnetized bodies (principally Earth and other planets) and the resulting dynamics of their magnetospheres and the associated coupling to the underlying ionosphere or planetary surface; and
- *Atmosphere-ionosphere-magnetosphere interactions (AIMI)*—which concerns the dynamics of planetary ionospheres owing to solar, magnetospheric, and atmospheric drivers and coupling.

The work of the three decadal survey panels representing these areas provided the basis for distillation of the key science challenges discussed in this chapter.

A DECADE OF HELIOPHYSICS DISCOVERY

The decade 2003-2012 was a time of significant progress in all areas of solar and space physics. Dramatic advances were made in establishing the relationships among solar activity, resulting interplanetary disturbances, the response of Earth's space environment, and the dynamics of the outer boundaries of our solar system with interstellar space. The links between the solar dynamo, convection, active regions, flares, coronal mass ejections (CMEs), and disturbances in the interplanetary medium are now identified. Researchers have identified candidate mechanisms that accelerate ions and electrons to relativistic energies in the inner heliosphere. They know the interplanetary conditions that drive geomagnetic activity and storms and have identified the dominant dynamic characteristics of the coupled magnetosphere-ionosphere-thermosphere system. Finally, they have now begun to explore the outermost reaches of the Sun's influence at the boundary between the heliosphere and interstellar space.

These developments occurred in coordination with advances in physics-based numerical simulations that provide the foundation for understanding phenomena in terms of underlying physical processes, yielding insights into the basic physics of the systems and attaining a measure of predictive capability. Researchers are now poised to answer questions concerning universal physical processes, advance understanding of the complex coupling and nonlinear dynamics of the heliosphere, and apply this understanding for mitigation of harmful impacts to Earth's technological infrastructures. To show how the recommendations of this report follow from the flow of scientific discovery, a selection of the most salient discoveries and advances are presented below.[1]

The Sun and Heliosphere

The Solar Dynamo and Activity

Over the past decade, the solar dynamo, which is the source of the Sun's magnetic field and the resultant dissipation that drives solar activity, continued as a high-priority focus of research. The results of this work also have important implications for understanding stellar dynamos. Solar activity reached normal levels in cycle 23, but the minimum between cycles 23 and 24 in 2008-2009 reached low levels not seen

[1] For a more complete discussion of ongoing missions and their contributions, see, for example, NASA, "Senior Review 2010 of the Mission Operations and Data Analysis Program for the Heliophysics Operating Missions," July 5, 2010, available at http://science.nasa.gov/media/medialibrary/2010/07/22/SeniorReview2010-MODAProgramPublic_V3.pdf. Also see, NASA, "Heliophysics: State of the Discipline," in *Heliophysics: The Solar and Space Physics of a New Era: Recommended Roadmap for Science and Technology 2009-2030*, 2009 Heliophysics Roadmap Team Report to the NASA Advisory Council Heliophysics Subcommittee, May 2009, available at http://sec.gsfc.nasa.gov/2009_Roadmap.pdf.

for nearly a century. Cycle 24 had been predicted to be more active than cycle 23, and the unexpected deep minimum focused attention on the need to improve understanding of the solar dynamo. Ground-based and SOHO space-based measurements prior to the activity minimum revealed unusually low magnetic flux near the poles of the Sun, and these low flux levels suggest that solar activity at the maximum of the current solar cycle will be low relative to that of recent past cycles.

Poleward meridional flows in the solar convective zone may be responsible for concentrating solar magnetic flux at the poles. Observations of this flow were made possible with great improvements in space-based (SOHO and SDO) and ground-based (GONG) helioseismic measurements of the solar interior. These observations have revealed changes in zonal and meridional flows consistent with the low polar flux and have shown that solar active regions exhibit subsurface helical flows whose strength is closely related to flare activity. Helioseismic observations are needed to firmly establish if scientists have indeed found a key to understanding the engine of solar activity.

The deep solar minimum in 2008-2009 provided an opportunity to study the heliosphere under conditions not present since the dawn of the space age and—enabled by STEREO—to study it for the first time in a truly global fashion: cosmic-ray fluxes near Earth reached the highest levels on record; reduced heating of Earth's upper atmosphere by solar ultraviolet radiation led to unprecedented low drag on satellites; and the radiation belts reached historically low levels of intensity. This enhanced galactic cosmic-ray flux was caused by reduced modulation in a historically weak solar wind with slower speeds, lower magnetic field, and historically low activity.

The extended solar minimum, prolonged period of low sunspot numbers, and record cosmic-ray intensity led to suggestions that the Sun might be entering an extended period of minimum activity such as that observed (in sunspot, ^{14}C and ^{10}Be data) during the Dalton minimum (1800-1820) or the Maunder minimum (1645-1715). Recent low activity was used to set a lower limit for total solar irradiance (TSI), a key factor in climate change. Measurements of TSI have consistently shown a cycle variation on the order of 0.1 percent but give conflicting results about its absolute value. These conflicts have recently been resolved, in favor of the lower values shown in Figure 3.1. It remains uncertain whether the TSI levels during the recent solar minimum are indicative of levels expected for a prolonged cessation of solar activity.

The past decade has seen spectacular advances in understanding of the structure of the solar magnetic field. Increases in processor speed and massively parallel computational techniques have enabled greater than 100-fold improvements in the spatial resolution of simulations. The resolution of observations has improved with data from the 0.5-m-aperture telescope of the Hinode satellite and through image processing techniques applied to ground-based data from the 1-m-class apertures and the new 1.6-m New Solar Telescope (NST). Researchers have now identified the main physical processes at work in sunspot penumbral filaments, bright umbral dots, bright faculae, and small-scale magnetic structures. Figure 2.2 is a side-by-side comparison of the results of a numerical simulation of a sunspot and an image from the NST; it shows astonishing correspondence in the background circulation pattern (granulation), the umbra fibrils, and the central spot.

Solar Wind Origins

New observations of the photosphere and lower corona have revealed significant information on the mechanisms of coronal heating, which is ultimately the driver of the solar wind. High-resolution chromospheric images from Hinode's Solar Optical Telescope unveiled relentless dynamics and contorted structures. A new type of spicule (a radial jet of plasma) was discovered that may play a critical role in transferring mass and energy to the corona. The narrowband EUV images from the SDO Atmospheric Imaging Assembly have revealed that coronal loops cannot be in a steady state as previously believed.

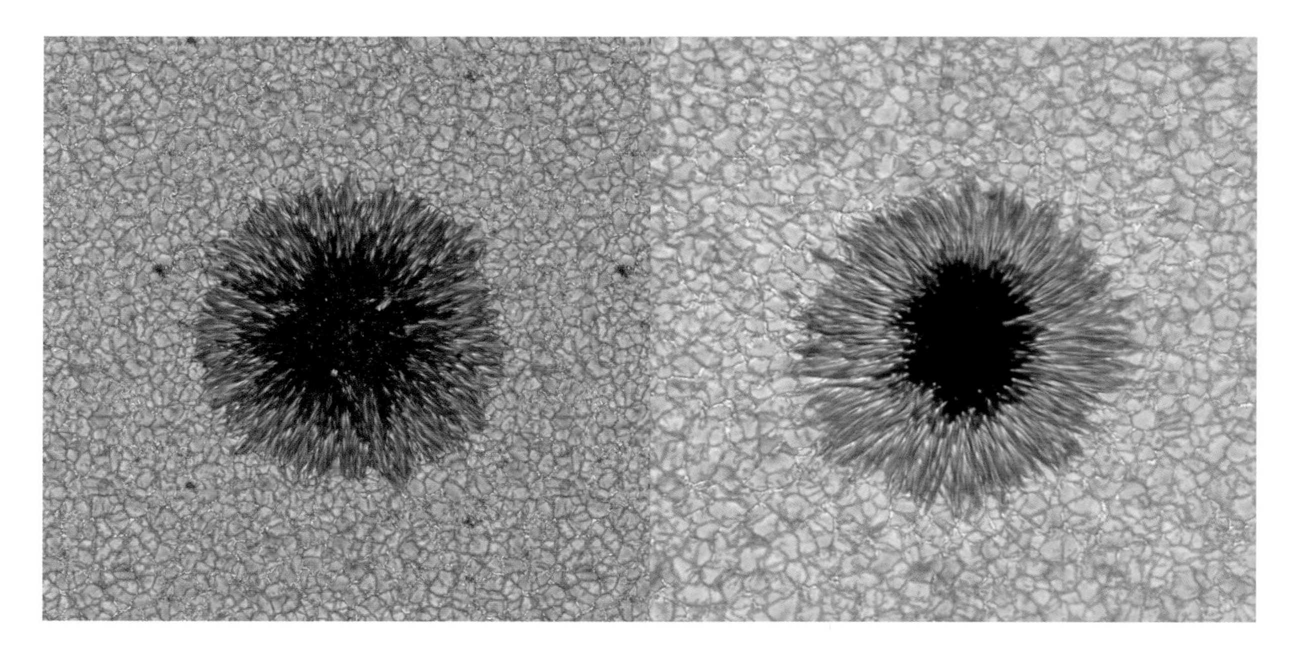

FIGURE 2.2 Numerical simulation of a sunspot (*left*) and a very-high-resolution image (*right*) from the New Solar Telescope at Big Bear Observatory, which is operated by the New Jersey Institute of Technology. Detailed comparisons have elucidated the physics of such solar features. SOURCE: *Left*, courtesy of M. Rempel, High Altitude Observatory. *Right*, courtesy of Big Bear Solar Observatory. For further information on the simulation, see M. Rempel, Numerical sunspot models: Robustness of photospheric velocity and magnetic field structure, *Astrophysical Journal* 750(1):62, 2012.

Furthermore, elemental fractionation signatures, identical to those in the quiet coronal loops, have been observed in slow solar wind.

The transition from the chromosphere to the solar wind is governed by the magnetic field of the corona. However, Hinode and SDO can measure the photospheric field but not the coronal magnetic field. Two advances of the past decade offer promise to fill this data gap: the first observations were made of the full chromospheric vector field on the disk, and the first maps were obtained of the coronal field above the solar limb using ground-based observations. Further advances in measuring the coronal field are crucial for understanding the origins of the solar wind and the driver of solar activity and its impact on Earth's space environment.

Significant progress was made toward achieving closure between theory/models and observations. The first semi-realistic global-scale three-dimensional magnetohydrodynamic (MHD) numerical simulations of the corona were performed with spatial resolution sufficient to enable comparison with modern observations (e.g., Figure 2.1). Modeling the chromosphere, however, remains a significant challenge, because in this region the classical description of the transport of energy begins to break down and dynamically important spatial scales may not be resolved. Three-dimensional numerical simulations cannot yet address all the physical ingredients on scales larger than a few granules or one supergranule, but many of these challenges can be overcome in the coming decade if these efforts are adequately supported.

Explosive Release of Magnetic Energy

Flares and CMEs are the dominant sources of the solar energetic particles (SEPs) that threaten human spaceflight. Significant progress was made in understanding how magnetic energy is explosively released in flares. RHESSI hard X-ray (HXR) imaging-spectroscopy measurements revealed that accelerated electrons often contain ~50 percent of the magnetic energy released in flares and indicate that energy-release/electron-acceleration is associated with magnetic reconnection. In large flares, HXR imaging of flare-accelerated ~30-MeV ions shows that these emissions originate from small foot points linked to magnetic loop structures rather than over an extended region, indicating that ion acceleration is also related to magnetic reconnection. The energy in >~1-MeV ions and that in >20-keV electrons appear comparable. Thus, understanding the remarkably efficient conversion of magnetic energy to particle energy flares is a significant challenge.

Major advances were also made in understanding photon energy release from flares. For the first time, flares were detected in TSI by the SORCE/TIM instrument showing that the total radiated energy and CME kinetic energy can be comparable. The SDO/EVE instrument discovered an EUV late phase in flares delayed many minutes from the X-ray peak. Global EUV observations by SDO/AIA and STEREO/EUVI revealed long-distance "sympathetic" interactions between magnetic fields in flares, eruptions, and CMEs likely owing to distortions of the coronal magnetic field.

The understanding of how CMEs and flares are produced and related has also progressed. CME velocity profiles below ~4 R_S are in sync with flare-HXR energy releases. The magnetic flux-rope structure of models of CMEs is consistent with the observations of many events. Furthermore, shocks produced by fast CMEs can be identified in coronagraph images, suggesting that scientists are close to pinning down the sources of SEPs. Achieving a predictive capability for SEP energy spectra and transport variability is a greater challenge.

Structure and Dynamics of the Solar Wind

Major progress was made over the past decade in understanding solar wind structure and dynamics, a key to understanding the Sun's influence on Earth's geospace environment. The conceptual picture from Ulysses and ACE was that the sources of the slow and fast solar wind were at low-latitude and high-latitude regions of the Sun, respectively. Fast, slow, and transient (associated with CMEs) solar wind can now be identified and distinguished by ionic composition signatures (Fe charge states, Fe/O, O^{7+}/O^{6+}), and so the origins of solar wind parcels can be directly identified from in situ observations. Coronal mass ejections interact with these solar wind streams, leading to dynamic fluid interactions and also particle acceleration through a variety of processes. Microstructure of the solar wind, presumably related to structures in the corona, may now be analyzed with the most powerful set of in situ observations, sometimes using several observational platforms. The cascade of turbulence to short spatial scales and its ultimate dissipation are the likely source of energy for heating the expanding solar wind. Observations and models have produced major advances on this topic. Temperature anisotropies with respect to the local magnetic field of solar wind H^+ and He^{2+} were shown to be limited by the mirror and firehose instabilities.[2] These observations constrain the possible mechanisms of solar wind heating. Scientists have also discovered that magnetic reconnection between adjacent domains of opposing magnetic fields is ubiquitous in the solar wind but appears to involve little particle acceleration near heliospheric reconnection sites—a surprise, given the

[2]See J.C. Kasper, A.J. Lazarus, and S.P. Gary, Wind/SWE observations of firehose constraint on solar wind proton temperature anisotropy, *Geophysical Research Letters* 29(17):20-1-20-4, 2002; B.A. Maruca, J.C. Kasper, and S.P. Gary, Instability-driven limits on helium temperature anisotropy in the solar wind: Observations and linear Vlasov analysis, *Astrophysical Journal* 748(2): 137, 2012.

efficiency of energetic particle production in flares. Unexpectedly, most of these reconnection sites have been found away from the heliospheric current sheet. The observations have also emphasized the importance of observations nearer to the Sun to enhance understanding of the roles of waves, wave turbulence, and reconnection physics in driving solar wind dynamics.

Solar Energetic Particles

New observations of solar energetic particles have yielded a number of surprises. Solar-cycle 23 produced 16 ground-level events in ground-based neutron monitors, which allowed researchers to establish that most large SEP events have a recent, preceding CME from the same active region. This finding indicates that the most intense events may involve the acceleration of particles in one or more flares that produce a seed population of energetic ions that can then reach very high energy through classical diffusive shock acceleration at the CME-driven shock. The measured enrichments by ACE of ^{3}He and Fe in many large SEP events are consistent with this picture. Continuing observations from STEREO, ACE, and other platforms as well as upcoming Solar Orbiter and Solar Probe Plus missions will provide key measurements in the source regions of these events and data on their spatial extent and evolution so that the complex dynamics of SEP acceleration and transport to the geospace environment can be unraveled.

Exploring the Heliosphere's Outer Limits

A series of groundbreaking discoveries were made as the Voyager spacecraft approached and crossed the termination shock (TS) and entered the heliosheath on their way to the heliopause, the outer boundary of the Sun's domain in the universe. These measurements and results from the Interstellar Boundary Explorer (IBEX) and Cassini have significantly altered understanding of how the solar system interacts with the interstellar medium and have also quantitatively confirmed a number of scientific predictions about the heliospheric boundary region. The TS, which is where the solar wind can no longer maintain its supersonic velocity as it pushes against the interstellar medium, had long been accepted as the driver of anomalous cosmic ray (ACR) acceleration, but when the two Voyager spacecraft crossed the TS, neither found evidence that the local TS is the source of ACRs. The source of the ACRs is now a subject of fierce scientific debate. In addition, consistent with earlier theoretical predictions, most of the supersonic-flow energy did not heat the ambient solar wind but likely went into supra-thermals (not measureable with the Voyager instruments). The most recent observations may indicate the presence of an unexpected transition region in which the outward solar wind flow stagnates.

Energetic neutral atom (ENA) maps by IBEX and Cassini show an unpredicted "ribbon" of emissions from the outer heliosphere, apparently ordered by the local interstellar magnetic field (Figure 2.3). The ribbon evolves on timescales as short as 6 months, demonstrating that the heliosphere/interstellar-medium interaction is highly dynamic. The role of the interstellar magnetic field in shaping the outer heliosphere is stronger than was expected prior to the recent influx of new data. Models based on these observations suggest that the local interstellar magnetic field provides most of the pressure in the local cloud. The unexpected results from Voyager, IBEX, and Cassini observations demonstrate how little is really understood about the interactions of stars with their interstellar environments.

Solar Wind-Magnetosphere Interactions

Advances in the physics of magnetospheres, their dynamics, and their coupling with the solar wind and ionospheres were made on a number of fronts. Global imaging and in situ observation networks revealed

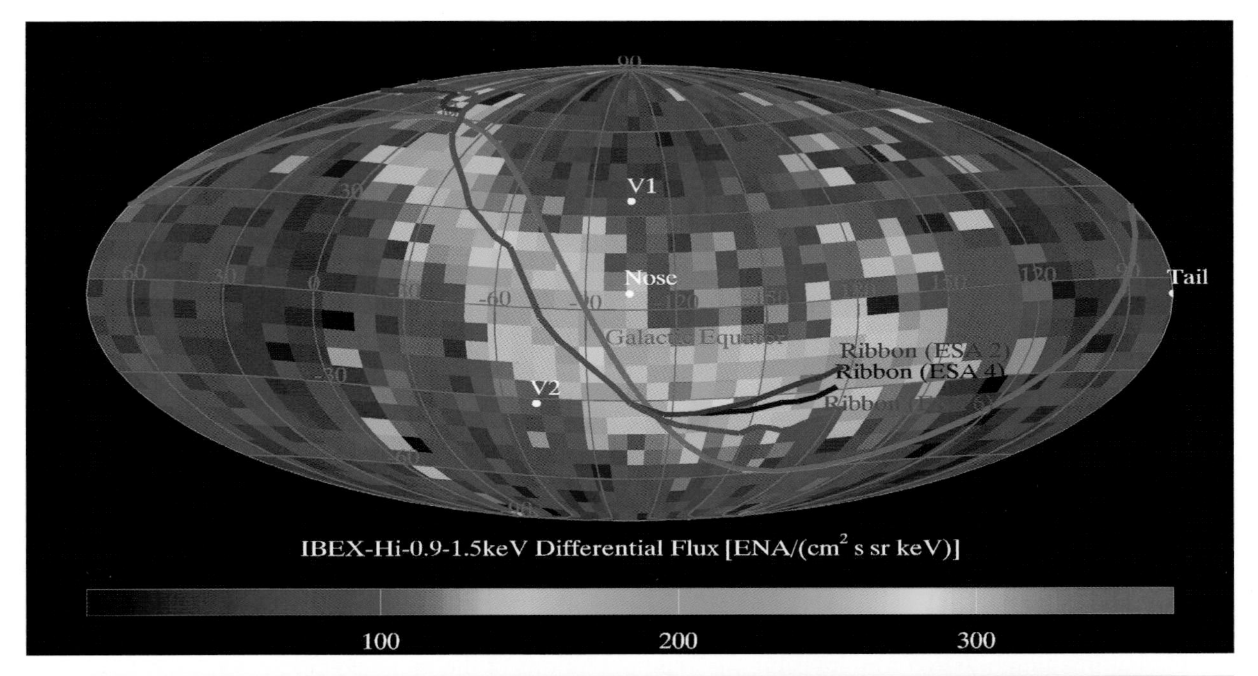

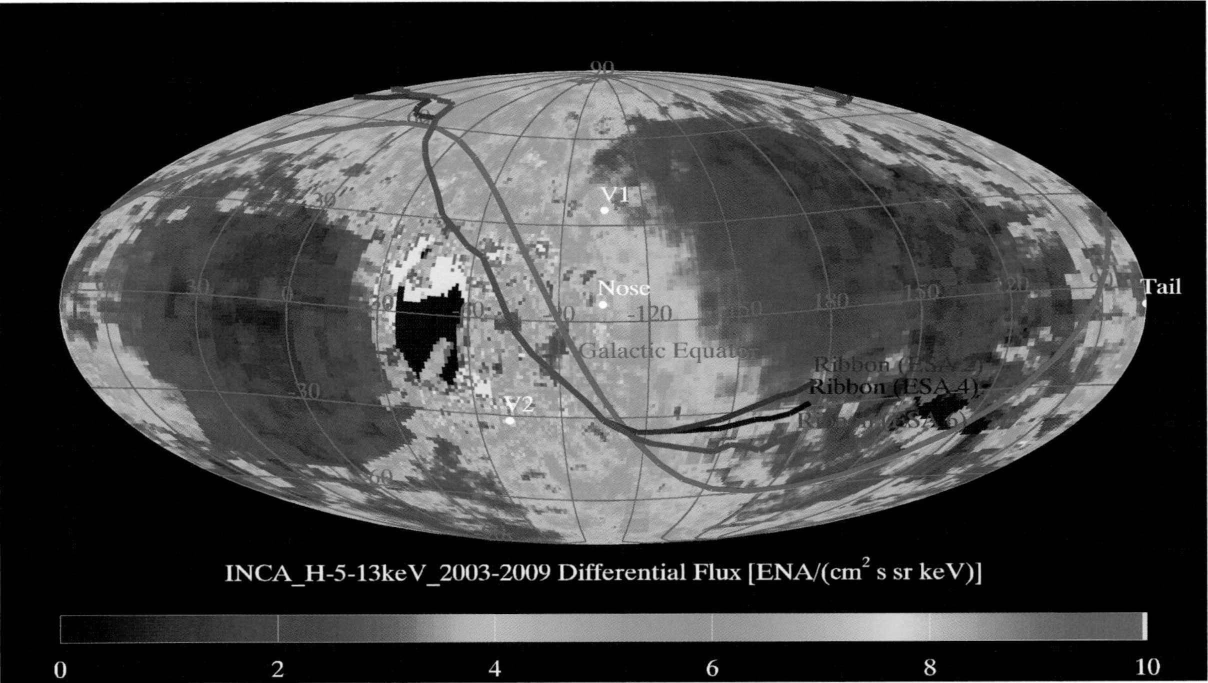

FIGURE 2.3 The unexpected ribbon seen in 0.9- to 1.5-keV energetic neutral atoms (ENAs) with IBEX and the 5- to 13-keV INCA belt. These maps depict integrated line-of-sight global maps of energetic neutrals. Previous models, based on ENA production in the heliosheath, predicted concentrated, uniform emission near the nose. None of the earlier models predicted the ribbon or belt. SOURCE: Courtesy of the Interstellar Boundary Explorer Mission Team.

unexpected dynamics associated with plasma convection, particle acceleration, and particle transport. Key advances were made on the underlying fundamental physical processes that govern the nonlinear dynamics of the system, including reconnection, wave-particle interactions, and turbulence. Observations and simulations of the dramatically different magnetospheres of Jupiter and Saturn provided key tests of current understanding and highlight the great variety of behavior exhibited by different systems.

These advances were enabled by combining a wide array of observations in concert with theory, laboratory plasma experiments, and revolutionary computational models. Critical observations were returned from instruments on suborbital rockets and balloons, and from an extensive ground-based network of radars, lidars, imagers, magnetometers, and riometers. Instrumental to these advances were spacecraft observations returned from new satellites launched during the decade or just before (e.g., Cluster, IMAGE, THEMIS, TWINS) as well as data returned from earlier missions and data collected by instruments flown on non-NASA satellites.

Global Dynamics

The global dynamics of the magnetosphere are controlled by the changing north-south component of the interplanetary magnetic field (IMF), which drives global circulation in the magnetosphere, as shown in Figure 2.4. Changes in the IMF and solar wind dynamic pressure produce storms, light up the aurora, and drive a host of other global responses.

Global imaging of heretofore invisible plasma populations of the magnetosphere was used to identify its large-scale response to this variable solar wind forcing. The plasmasphere, which is the region of cool-dense plasma that co-rotates with Earth, was imaged in the extreme ultraviolet. Observations revealed that strong storms strip off the outer part of the plasmasphere in plumes, which convect outward to the dayside magnetopause (Figure 2.5) and map to produce ionospheric density enhancements of the type shown in Figure 3.3.

The magnetospheric equatorial ring current is enhanced during geomagnetic storms, and it perturbs the strength of the magnetic field at Earth's surface. Understanding its dynamics is crucial for establishing a predictive capability of the response of geospace to storms. The injections of ring current ions were imaged for the first time, establishing their configuration and composition. Numerical models and global ENA imaging revealed that the ring current is highly asymmetric during the main phase of storms, which suggests a strong coupling with the ionosphere. The peak of the ring-current proton distribution during the main phase of magnetic storms was shown to occur consistently in the early morning and not in the afternoon as had been expected. This can happen only if the ionosphere feedback fundamentally alters the electric field that is responsible for magnetospheric convection.

Fundamental Physical Processes: Magnetic Reconnection and Wave-Particle Interactions

The understanding of fundamental physical processes that govern system-level dynamics advanced on a number of fronts over the past decade. Substantial progress was made in understanding how magnetic reconnection works. The first quantitative predictions of detailed magnetic and plasma flow signatures were spectacularly confirmed with in situ observations. Similarly, sophisticated kinetic simulations finally yielded a consistent understanding of signatures of the onset of magnetic reconnection in the tail.

Increased computing power has facilitated simulations of the essential physics and structure of the diffusion region, where magnetic field lines[3] reconnect and change their connectivity (Figure 1.4). It was

[3] Field lines are a convenient construct for understanding magnetic field connectivity and topology.

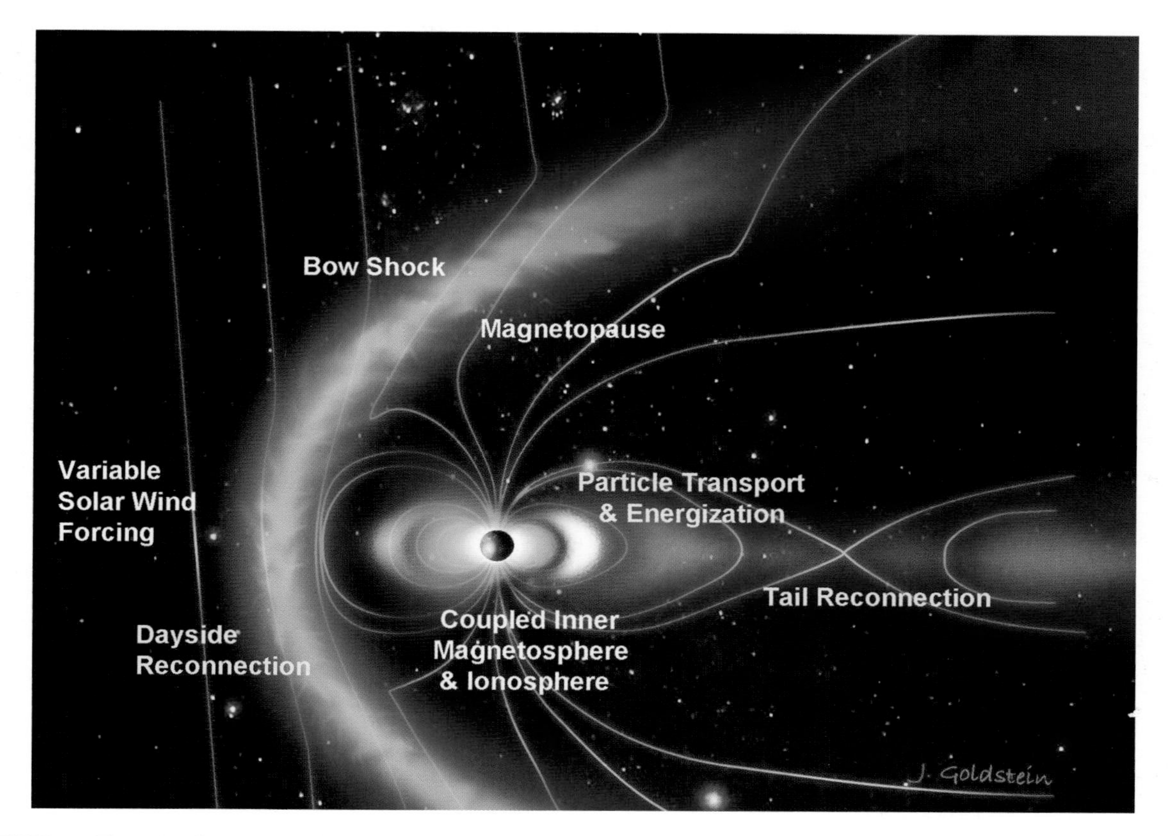

FIGURE 2.4 The critical processes that drive the magnetosphere. To achieve a full understanding of the complex, coupled, and dynamic magnetosphere, it is important to understand how global and mesoscale structures in the magnetosphere respond to variable solar wind forcing, and how plasmas and processes interact within the magnetosphere and at its outer and inner boundaries, by using a combination of imaging and in situ measurements. SOURCE: Courtesy of Jerry Goldstein, Southwest Research Institute.

shown that at the small spatial scales where reconnection occurs, the decoupling of ion and electron motion as a result of their very different mass plays a key role in facilitating the rapid rate of reconnection seen in the observations. Ions become demagnetized in a much larger region than did the electrons, which changes the forces that accelerate particles away from the x-line compared with the usual MHD description. These ideas led to predictions that facilitated the first direct detection of the ion diffusion region (where the ions decouple from the magnetic field) in the magnetosphere and in the laboratory, as well as glimpses of the much smaller electron diffusion region (where the electrons decouple from the magnetic field). The observations in the vicinity of the diffusion region revealed surprisingly that reconnection can accelerate electrons to hundreds of kiloelectronvolts, potentially providing a seed population for subsequent acceleration in the inner magnetosphere to form the electron radiation belts. Discoveries were also made regarding the triggering and modulation of reconnection. Prior to around 2000, computational resources had simulations limited to two spatial dimensions. New capabilities to perform fully three-dimensional simulations revealed that the added dimension facilitates the growth of plasma instabilities that may break up the diffusion region, making reconnection highly turbulent.

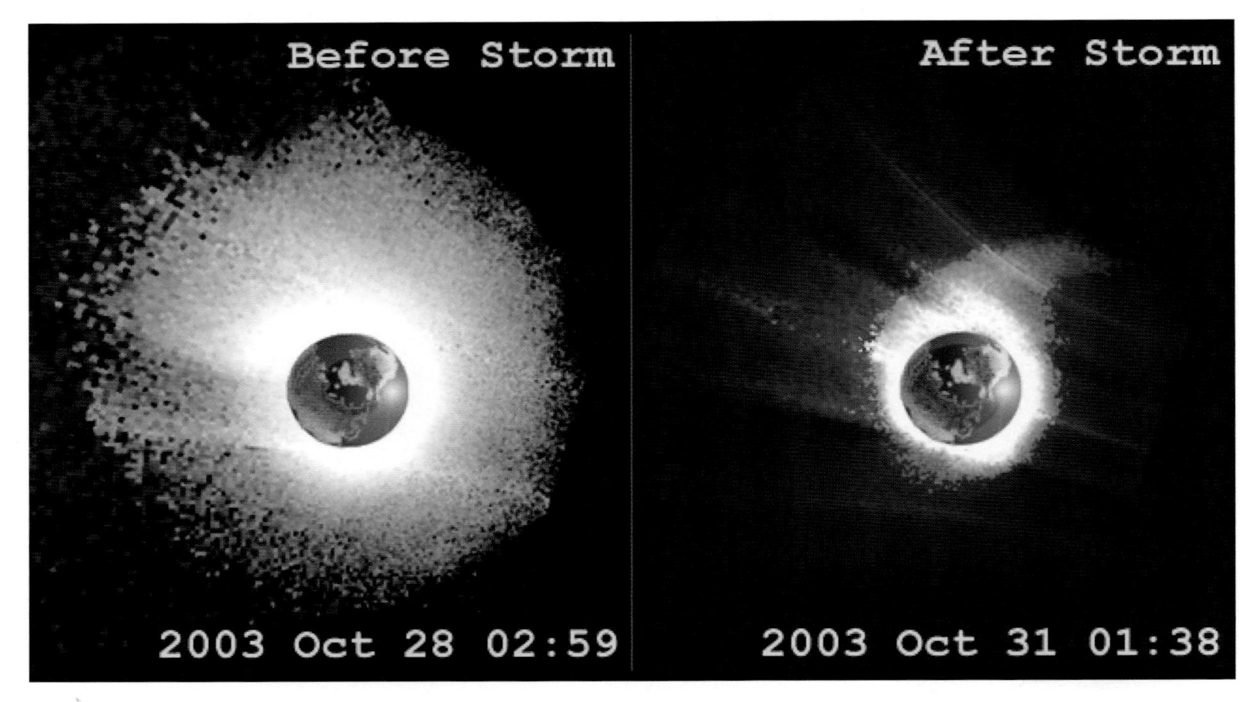

FIGURE 2.5 Measurements by the EUV instrument on the IMAGE satellite. EUV images before a storm and after a storm, when the plasmapause reaches its minimum radial extent due to erosion by enhanced convection. SOURCE: Reprinted from M.K. Hudson, B.T. Kress, H.-R. Mueller, J.A. Zastrow, and J. Bernard Blake, Relationship of the Van Allen radiation belts to solar wind drivers, *Journal of Atmospheric and Solar-Terrestrial Physics* 70(5):708-729, 2008, copyright 2008, with permission from Elsevier.

Observationally, reconnection seems to behave differently in different regions. Although reconnection in the magnetotail and in the magnetosheath, where multiple reconnection sites have been identified, appears to be transient and turbulent, it can, at other times, be quite steady in time and extended in space at the dayside magnetopause and in the solar wind. Reconnection in the magnetotail produces bursts of narrow channels of high-speed flow. Multi-spacecraft observations have revealed that these reconnection-generated flow channels initiate magnetospheric substorms and drive the Earth-ward convection in the magnetotail; however, further multi-spacecraft studies may be necessary to complete the pattern of global magnetospheric circulation predicted four decades ago. Finally, observational analyses will benefit greatly from the inclusion of reconnection scenarios more general than the standard X-point picture, including more general geometries identified in both theory and simulations.

Wave-particle interactions (WPIs) have been established as key drivers of particle energy gain and loss in the radiation belts (Figure 2.6). Plasma instability theory, global simulations that include WPI processes, and wave observations have demonstrated that the mixing of energetic and low-energy plasmas drives instabilities distributed throughout the ring current and radiation belt. Satellite observations of radiation-belt electrons demonstrate that local acceleration due to WPIs may at times dominate acceleration due to diffusive radial transport. Statistical analyses of satellite wave observations were used to quantify the rates of energization and scattering. The results have been incorporated into time-dependent models of the radiation belts and the ring current. Scientists now know that storm-time particle dynamics are the result

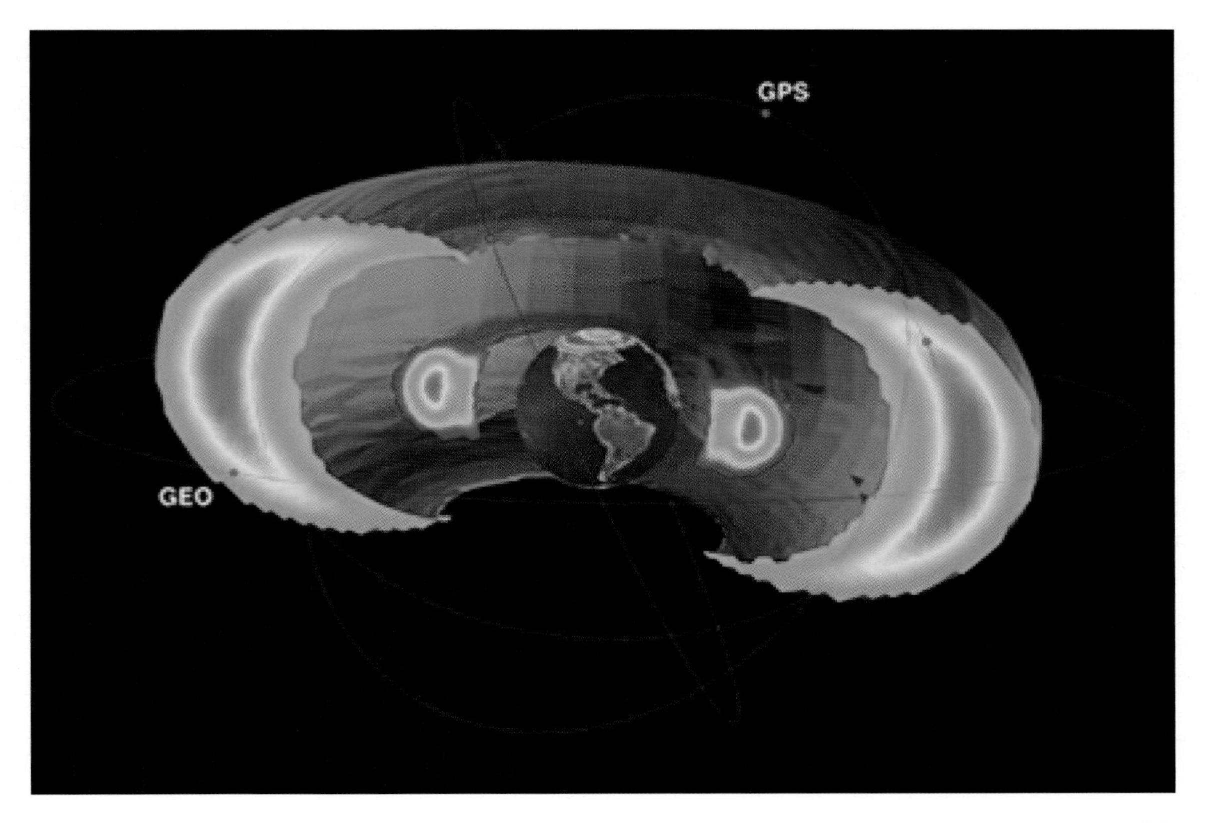

FIGURE 2.6 Model-generated image showing the two main radiation belts, the outer belt and the inner belt. The model was developed at the Air Force Research Laboratory. Colors in the radiation belts indicate relative number flux. The auroral zone colors reflect precipitation to the atmosphere. Shown here are representative orbits for three Global Positioning System and one geosynchronous spacecraft. SOURCE: Courtesy of R.V. Hilmer, Air Force Research Laboratory.

of a delicate balance between acceleration and loss of relativistic particles mediated by waves produced by local plasma instabilities.

Nonlinear Dynamics

In the past decade there has been tremendous improvement in understanding of how the magnetosphere responds to storm-time disturbances as a coherent system of coupled, mutually interacting plasmas. Imaging and global simulations have played a central role by providing quantitative contextual information that ties together single-point observations and allows assessment of the global behavior implied by local observations. These advances were coupled with continuous measurements of the solar wind and IMF and numerous in situ observations in space and ground-based and remote sensing networks to yield discoveries of characteristic global responses. Researchers now realize that there are multiple nonlinear dynamic linkages whose consequences for coupled magnetosphere-ionosphere behavior are revealed only when they are integrated together in the global system. As a result the system exhibits characteristic nonlinear, chaotic dynamics, or "emergent" behavior, that could never have been predicted without knowledge of the coupling physics.

The electrodynamic coupling between the magnetosphere and ionosphere modifies the simple dissipative response in dramatic ways. The interaction of the ring current with the ionosphere severely distorts inner magnetospheric convection, which feeds back on the ring current itself, skewing its peak toward dawn. Duskside flow channels arising from ionospheric coupling remain long past periods of peak solar wind driving. These studies shattered the notion that the inner magnetosphere is well shielded from the outer magnetosphere and is quiescent.

It is now established that storm-time acceleration and injection of the ring current depend on pre-storm loading of the magnetosphere from the solar wind. Spacecraft observations and numerical simulations reveal that solar-wind plasma entry into the magnetosphere is surprisingly efficient under "quiescent" conditions of a northward interplanetary magnetic field. This plasma in turn participates as a substantial element of the storm-time ring-current development when southward interplanetary magnetic fields couple with and energize the magnetosphere. The plasmasphere in turn controls whether ring current injections act to enhance or deplete energetic particle populations, overturning the decades-old idea of a passive, quiescent plasmasphere. The overlap of freshly injected, hot, ring current plasma with the dense plasmasphere produces local instabilities. The resulting waves scatter radiation belt particles, depleting the radiation belts. The predictions that cold dense plasma facilitates the acceleration of energetic electrons to relativistic energies were also confirmed by observations. Thus, it has been established that pre-injection dynamics are critical in establishing the state of the plasmasphere and governing the radiation belt storm response.

The ionosphere can also be a significant source of plasma for the magnetosphere over the past decade. The understanding of ionospheric ion outflow advanced significantly, and the conditions promoting extraction of ionospheric plasma to high altitudes and into the magnetosphere were established. Solar wind density and dynamic pressure increases were shown to lead to enhanced ionospheric outflow, but the greatest outflow rates were also closely correlated with the electromagnetic energy flux into the ionosphere. The energy flux from the solar wind yields intense ionospheric ion outflows supporting the theoretical predictions that this outflow requires a multistep process involving a combination of local heating by waves and electromagnetic forcing. It was also demonstrated that ionospheric outflow has dramatic consequences for the dynamic evolution of the magnetosphere. Outflows merge with plasmas of solar wind origin in the plasma sheet, creating a multi-species plasma that alters the dynamics of magnetic reconnection. Multi-fluid global simulations confirmed the major role that ionospheric outflow plays in the creation of periodic substorms or so-called sawtooth intervals.

The discoveries of preconditioning interactions and of efficient pathways for magnetosphere-ionosphere coupling, and the identification of the dynamics that emerge, provide the basis for a program of research to achieve a quantitative, predictive understanding of system behavior under extreme conditions.

Magnetospheres of Other Planets

The past decade saw many advances in understanding the structure, dynamics, and linkages in other planetary magnetospheres or systems with magnetospheric-like aspects. For the terrestrial planets, they range from insights on atmospheric loss at Mars and the identification of Venus lightning from high-altitude radio wave measurements to observations of magnetospheric dynamics at Mercury that reflect dramatically stronger solar wind-magnetosphere coupling than that at Earth. There have also been advances in theoretical understanding and observational tests of the impact of solar wind dynamic pressure variations on Jovian auroral emissions, and significant progress in understanding magnetospheric interactions with Jupiter's satellites, especially Io. ENA imaging demonstrated that an extensive torus of neutral gas from Europa has a significant impact on Jupiter's magnetosphere. Io-genic plasma is transported outward by flux-tube interchange processes on the dayside but by centrifugal instabilities and plasmoid ejection in the evening and at night. Intense bursts of energetic particles are accelerated in regions ~200 Jupiter radii down the tail on the dusk flank. These discoveries demonstrate the great range of physical processes that the Jovian system exhibits, presenting an enormous opportunity for advancing understanding of magnetospheric dynamics.

A major highlight of the decade came from the extensive measurements of Saturn's highly structured magnetosphere and satellite system by the Cassini spacecraft. Plumes of water gas and ice crystals emanate from rifts in the south polar region of Enceladus (Figure 2.7). Flux tube interchange in the middle magnetosphere followed by plasmoid release in the magnetotail was revealed as the primary transport mechanism for cold Enceladus plasma.

Solar wind pressure variations strongly modulate the activity in the outer magnetosphere, including Saturnian kilometric radio emission and the acceleration of energetic particles in Saturn's ring current. These results remain a challenge to explain and demonstrate the critical role the study of these other systems has in advancing magnetospheric physics.

Atmosphere-Ionosphere-Magnetosphere Interactions

A broad range of national, international, and multiagency programs facilitated major advances in the science of Earth's ionosphere and thermosphere and their interactions with the magnetosphere and the lower atmosphere. A major surprise is that the ionosphere-thermosphere system exhibits unexpected structuring during solar-quiet conditions. New Global Positioning System (GPS)-based assets from ground and space led to fundamental discoveries of dynamics of the global ionospheric density. Reactive feedback processes of thermospheric upwelling and intense ionospheric ion outflows were demonstrated to occur in new ways and were shown to have profound consequences for magnetospheric dynamics. The storm-time response of the system is now better characterized than ever before, and key gaps in understanding of the linkages between drivers and responses have been identified. Finally, tropospheric forcing from below was discovered to play a surprisingly strong role in the dynamics and structure of the ionosphere and thermosphere.

It is worth noting here the importance of international and cross-agency support that has made these scientific discoveries possible. For example, the COSMIC mission—a six-satellite joint U.S.-Taiwanese mission to improve understanding of both weather and space weather—carries instruments developed by JPL and the Naval Research Laboratory and was launched by the U.S. Air Force (USAF), and the data it collects are downloaded at NOAA and NASA facilities and processed at the NSF-supported National Center for Atmospheric Research (NCAR). The C/NOFS mission is another more recent example of scientifically productive cooperation between the USAF and NASA. Perhaps not surprisingly, several scientific discoveries involve physical processes that extend across the regions of interest of these agencies, and across nations.

Active Ionosphere During Solar Minimum

Gradual changes in solar activity, solar wind, solar EUV radiation, and Earth's magnetic field play a significant role in defining the long-term variation in the geospace environment. The most recent solar minimum produced a prolonged period of low solar EUV fluxes and corresponding heating rates. At the same time, the thermospheric densities dropped to anomalously low levels, lower than any observed in the past four solar cycles. No numerical model has yet been able to predict or reproduce the density observations, which are thought to have resulted from some combination of low solar and geomagnetic activity, cooling from increasing greenhouse gas concentrations, and possibly additional chemical or dynamical

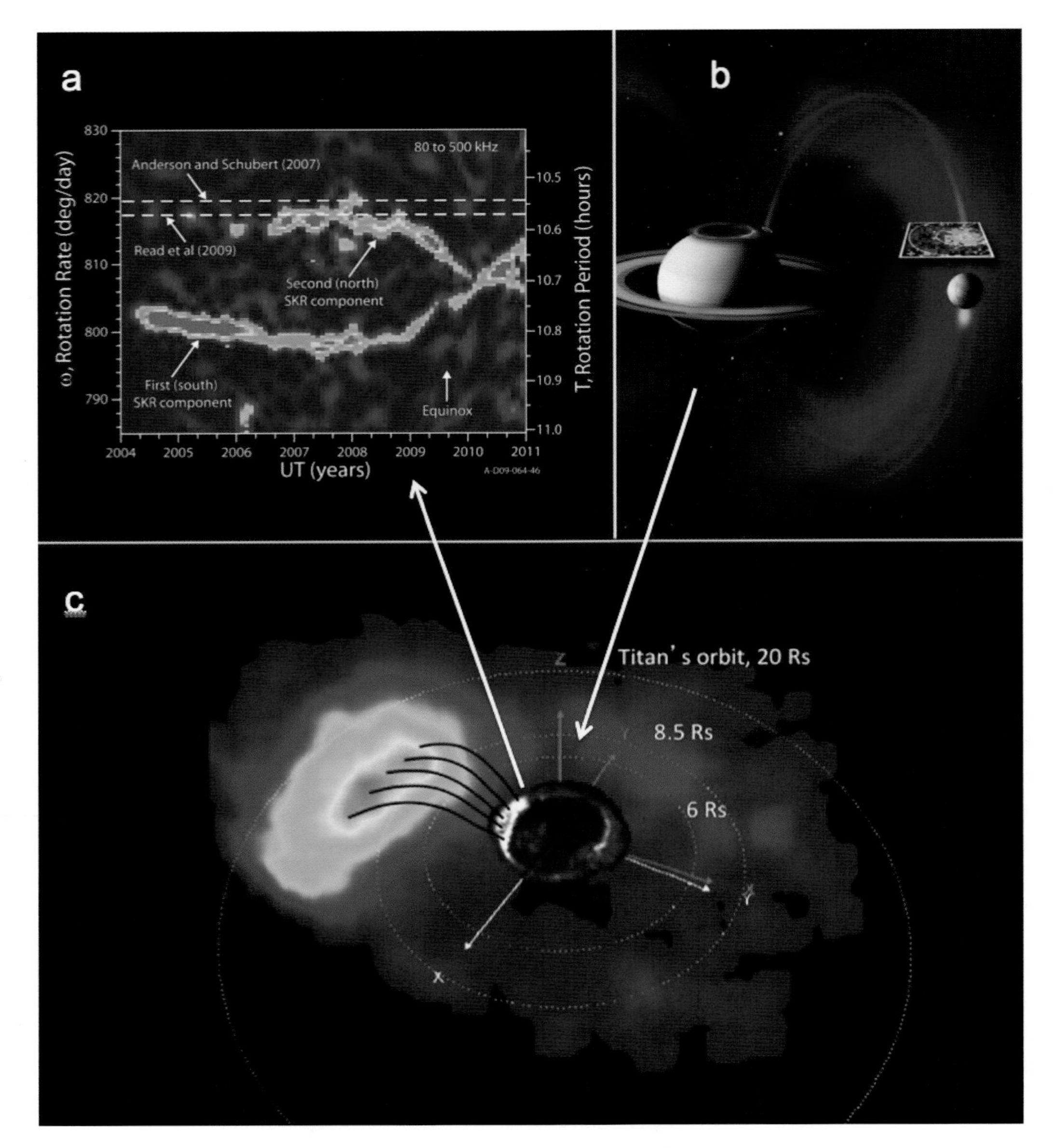

FIGURE 2.7 Saturn's Enceladus-dominated, rotating magnetosphere. (a) Saturnian kilometric radio (SKR) emission periodicities. (b) Enceladus, its geysers, resulting plasma, and connection to Saturn's ionosphere. (c) Return of energized plasma after tail plasmoid loss. These injections yield bright auroral displays in the same region as the SKR emissions. SOURCE: (a) D.A. Gurnett, J.B. Groene, A.M. Persoon, J.D. Menietti, S.-Y. Ye, W.S. Kurth, R.J. MacDowall, and A. Lecacheux, The reversal of the rotational modulation rates of the north and south components of Saturn kilometric radiation near equinox, *Geophysical Research Letters* 37:L24101, doi:10.1029/2010GL045796, 2010. Copyright 2010 American Geophysical Union. Reproduced by permission of American Geophysical Union. (b) JHUAPL/NASA/JPL/University of Colorado/Central Arizona College/SSI. (c) Adapted from D.G. Mitchell, S.M. Krimigis, C. Paranicas, P.C. Brandt, J.F. Carbary, E.C. Roelof, W.S. Kurth, D.A. Gurnett, J.T. Clarke, J.D. Nichols, J.-C. Gérard, et al., Recurrent energization of plasma in the midnight-to-dawn quadrant of Saturn's magnetosphere, and its relationship to auroral UV and radio emissions, *Planetary and Space Science* 57(14-15):1732-1742, doi:10.1016/j.pss.2009.04.002, 2009.

changes propagating upward from the atmosphere below. The extended epoch of low solar EUV and the reduced neutral densities at low-Earth-orbit altitudes led to the unexpectedly long mission life for the German Challenging Mini-Satellite Payload (CHAMP) satellite. But despite the extended solar quiet period, the ionosphere displayed a surprising array of dynamics, including complex density structures in the morning hours near local dawn that were documented by the USAF C/NOFS mission and NASA's CINDI experiment and other space- and ground-based assets. It is now known that a quiet Sun does not correspond to a calm, benign ionosphere and that deleterious impacts on navigation and communications occur under these conditions in unexpected ways.

Global Density Structures and Reactive Feedback

Global GPS maps of ionospheric density showed, for the first time, large-scale dense plumes of plasma extending from middle latitudes to the auroral zone at the onset of magnetic storms (see Figure 3.3). During such events, plasmaspheric imaging of He^+ ions by IMAGE showed corresponding structures in the inner magnetosphere, where plasma was sheared away from the plasmasphere and advected toward the magnetopause (see Figure 2.5). The plasmaspheric structure was never expected to appear in the ionosphere, and the discovery points to a process critical to enhancing auroral ion outflow during storms.

Localized structures in the neutral density were discovered by international geodesy programs. The CHAMP and NASA/German Gravity Recovery and Climate Experiment (GRACE) missions led to the discovery of localized neutral upwelling very near the poles associated with strong Joule heating that occurs during geomagnetically calm or moderate conditions. This result demonstrated the surprising range of conditions wherein neutral densities are sufficiently altered to modify the decay rates of satellites in low Earth orbit. Understanding of the generation of these localized densities is not yet mature enough to predict their occurrence.

Recent results from NASA's FAST and IMAGE satellites revealed intense outflows of ionospheric ions during storms. The solar wind-magnetosphere interaction on the dayside, that is, magnetopause reconnection, is a copious source of electromagnetic energy that propagates along the magnetic field into the ionosphere at high latitudes near noon. This energy is converted to heat and momentum through ion-neutral interactions and promotes resonant heating of O^+ that drives outflows. The O^+ flows upward and is carried into the magnetotail by the reconnection-convection cycle. The resultant large O^+ densities in the tail plasma sheet appear to change reconnection dynamics in the tail, leading to the ~3-hour planetary-scale (sawtooth) oscillations or quasi-periodic substorms in the magnetosphere. The influence of the O^+ outflow on global dynamics is only one of a number of instances in which nonlinear reactive feedback leads to nonlinear dynamics.

Storm Dynamics

Several of the geomagnetic storms driven by CMEs during solar cycle 23 were considered "great" storms that led to highly nonlinear dynamics. Ionosphere observations indicated the emergence of a daytime superfountain effect, lifting the ionosphere to new heights and increasing its total electron content by as much as 250 percent. Other extreme responses included very-large-amplitude traveling ionospheric disturbances and modifications in the equatorial plasma irregularities that impact communications.

During cycle 23, there were 89 great storms that drove the geomagnetic storm index Dst[4] below −100, but only one, associated with the extremely fast CME launched by the spectacular X17 flare of October

[4]The Dst (disturbance–storm time) index is used to define geomagnetic storms. Quiet times usually have a Dst of between +20 and −20 nanoteslas.

28, 2003, topped –400. Fortunately that same active region had rotated past Earth when the largest flare ever measured by spacecraft erupted on November 4 with an energy index of X28.

During these great storms, the atmosphere responded with dramatic changes in neutral composition, winds, temperature, and mass density. Thermosphere mass density at 400-km altitude increased by more than 400 percent and recovered to pre-storm levels exceptionally rapidly, indicating a strong overcooling mechanism. Although many of the responses of the atmosphere-ionosphere-magnetosphere (AIM) system to these storms have been documented, the mechanisms responsible for producing these effects are poorly understood—scientists have not been able to emulate these effects in simulations. In particular, scientists cannot yet predict the impacts of so-called superstorms, storms comparable in magnitude to the Carrington event of 1859 that had an astounding estimated Dst of –850.

Tropospheric Driving

One of the most exciting developments in recent years has been the realization that tropospheric weather and climate can strongly affect the upper atmosphere and ionosphere. Ultraviolet imaging of Earth by the NASA IMAGE and TIMED satellites in the 2000-2003 time frame provided an unprecedented new view of the equatorial ionosphere that revealed a large, longitudinal variation in density, with peaks over rainforests. In the same period, atmospheric models developed at NCAR were gaining new capability showing that atmospheric tides driven by tropospheric heat released in thunderstorms would propagate well above 100 km and potentially modify the ionosphere-thermosphere (IT) system. Large-scale changes in the structure of the ionosphere on seasonal timescales were also revealed, which also matched the seasonal changes in tropical weather conditions. Since its launch in 2006, the COSMIC mission has observed a number of ionospheric features that point to forcing from below: tidal influence on total electron content and the F region of the ionosphere; wave signatures in the ionosphere and plasmasphere; a geographically fixed (with the Weddell Sea) ionospheric anomaly; and complex structure in ionosphere F-region density potentially attributable to tropospheric storm systems. These results have been matched by extensive numerical modeling efforts (e.g., the Whole Atmosphere Community Climate Model; WACCM) focused on understanding how atmospheric waves and tides of tropospheric origin propagate through the lower and middle atmosphere, and with the upper-atmospheric general circulation models now also being driven by stratospheric lower-boundary forcing that mirrors the tropospheric inputs, or with input of fitted wave data at approximately 100 km, the boundary of space. Further, the signature of tropospheric forcing has subsequently been observed in upper-thermospheric composition and temperature.

These and other observations and model studies have unequivocally revealed that Earth's IT system owes a considerable amount of its longitudinal, local-time, seasonal-latitudinal, and day-to-day variability to atmospheric waves that begin near Earth's surface and propagate into the upper atmosphere. Current estimates indicate that waves propagating upward from the lower atmosphere contribute about as much to the energy transfer in the IT system as does forcing from above in the forms of solar EUV and UV radiation, precipitating particles, resistive heating, and winds driven by magnetospheric convection.

Thermospheric Climate Change

A systematic decrease by several percent per decade in thermosphere mass density is now evident in the record of satellite orbit decay measured since the beginning of the space age. An effect predicted in the 1980s, this change is thought to be largely in response to the increase in atmospheric CO_2, which, although it acts to trap infrared heat in the lower atmosphere, acts as a radiative cooler in the upper atmosphere. Thermospheric cooling is therefore an unambiguous signature of a human-influenced change in the upper atmosphere. Much work is now focused on understanding the impact of climate change on the thermosphere and ionosphere and on identifying signatures in the thermosphere that can be used to assist in monitoring and clarifying the sources and mechanisms of climate change.

KEY SCIENCE CHALLENGES

The advances over the past decade focus attention on the challenges that are most pressing, both scientifically and practically. Significantly, recent progress includes the greatest advances to date toward achieving the predictive capability needed to safeguard the global technological infrastructure. Distilled from the science goals presented by the survey's three interdisciplinary panels, the challenges identified below by the survey committee are major areas of ongoing inquiry that provided the context for development of the program of research advocated by this survey for the coming decade. Scientifically important in their own right, the frontiers of heliophysics are also important as a source of practical knowledge for maintaining the operability of the assets of our increasingly technological society. Despite the challenges in studying these systems, the experience of the past decade demonstrates that scientists and engineers in the field have achieved dramatic progress in advancing the state of knowledge, and progress in the areas outlined below can reasonably be expected to be equally impressive. Even though the committee anticipates that key components of these challenges will be resolved in the coming decade, some questions will undoubtedly remain open, and other challenges are likely to emerge.

Challenges Related to the Sun and Heliosphere

The magnetic field of the Sun is, directly or indirectly, the driver of much of the dynamics of the heliosphere. Thus, understanding how the solar magnetic field is generated—the dynamo problem—is a key challenge. Despite the complexity of the solar magnetic field at multiple scales, as is evident, for example, in Figure 2.1, the global solar magnetic field exhibits an approximately 22-year periodicity, with the polarity of the field reversing every 11 years. Various solar phenomena, of which number of sunspots is the most familiar, exhibit this same 11-year cycle. Indeed, evidence for decadal-scale periodicities has been found in the luminosities of other Sun-like stars. Scientists know that the twisting and amplification of seed magnetic fields in the Sun's convective zone, the outer one-third of the Sun, are the source of the solar magnetic field. The resulting solar dynamo has therefore become a prototype for understanding how magnetic fields are generated throughout the universe. Although researchers have shown that quasi-periodic reversals are a natural consequence of dynamo action, they have not yet established why the solar dynamo produces a field that reverses with a nearly regular 11-year period. Moreover, the deep and prolonged minimum in the present cycle, cycle 24, was totally unexpected, demonstrating that understanding of the solar magnetic field has not yet risen to the state of a predictive science. The following remains a primary challenge: **SHP-1. Understand how the Sun generates the quasi-cyclical magnetic field that extends throughout the heliosphere.**

The complex structure of the light emitted from the corona as shown in Figure 2.1 reflects the corresponding structure of the magnetic field since hot plasma making up the corona very quickly spreads out along lines of force of the magnetic field. Scientists can now calculate this complex magnetic structure with computational models using the ground-based measurements of the magnetic field at the Sun's visible surface, or photosphere. However, the time variation of the complex magnetic field in the Sun's tenuous outer atmosphere or corona, which often takes the form of explosive events, is not fully understood and remains at the frontier of heliophysics research. Probing the details of the solar magnetic field at multiple heights in its atmosphere, and at very high temporal and spatial resolution, is the goal of NSF's Advanced Technology Solar Telescope (ATST).

Active regions are locations where these explosive events are concentrated. There, magnetic energy is released in the form of ejected plasma, electromagnetic radiation, and heat that energize the local plasma. Figure 1.3 shows a series of active regions seen in EUV light from the Solar Dynamics Observatory (SDO) that form a chain across the upper half of the Sun. These arrays of loops emerge from the churning solar atmosphere below and are embedded in plasmas with temperatures of around 10^7 K. The photosphere, by comparison, is relatively cold at 6,000 K. The mechanisms that produce the hot corona of the Sun and other stars still defy definitive explanation, and determining how this occurs is a high-priority science goal of NASA's Solar Probe Plus (SPP) mission and also of the Solar Orbiter ESA/NASA joint mission.

How the corona is generated and what physical processes heat the coronal plasma and control its dynamics are not yet understood, thereby defining the second major challenge: **SHP-2. Determine how the Sun's magnetism creates its hot, dynamic atmosphere.**

An important result of recent research is the discovery of the critical role that magnetic reconnection plays in modulating the energy flux from the Sun. The turbulent flows of the Sun's surface twist and distort the coronal magnetic fields, thereby increasing their energy. The magnetic energy accumulates over days, weeks, or perhaps longer. When adjacent magnetic fields pointing in opposite directions become sufficiently strong, the magnetic fields explosively annihilate each other during magnetic reconnection (see Figure 1.3). The released magnetic energy drives high-speed flows, heats the local plasma, and contributes in complex ways to accelerating particles to relativistic energies, producing the intense bursts of energized particles that characterize solar flares. This process occurs almost continuously in the active regions in the corona (see Figure 1.3). As a result, the corona and heliosphere are filled with high-energy radiation, both particle and electromagnetic (UV, X rays, and gamma rays).

The strongest of these reconnection events propel CMEs into the solar wind, and the CMEs steepen into shocks that accelerate ions and electrons to high energy. Figure 2.8 shows a numerical simulation of a CME, illustrating the scale of the ejected field and plasma. When directed Earth-ward, CMEs generate large geomagnetic storms and intense energetic particle events in near-Earth space. The energetic particles from these shocks pose significant threats to human and robotic space exploration.[5]

The success of simulations in reproducing many of these observations testifies to the maturity of scientific understanding of these significant events. However, even though it is now possible to predict where on the Sun a CME will originate, it is not yet possible to predict CMEs' timing, speed, energy, or momentum, nor is there full scientific understanding of how a CME converts so much of its energy into particle radiation. The planned SPP and SO missions will provide crucial information related both to the reconnection process and to CME initiation. These issues present a third challenge: **SHP-3. Determine how magnetic energy is stored and explosively released and how the resultant disturbances propagate through the heliosphere.**

The heliopause, where the Sun's extended atmosphere ends and the galactic medium begins, is a region that is rich in unique and unexplored physics. It is also the boundary that, in part, controls the penetration of high-energy galactic cosmic rays into near-Earth space. Interstellar neutrals are crucial to the outer heliosphere because they stream into the heliosphere unimpeded by the heliospheric magnetic field and dump energy into the solar wind. They are the dominant energy source of the outer heliosphere. A revolution in understanding of the outer heliosphere is unfolding as the Voyager spacecraft provide the first in situ data from this region and NASA's Interstellar Boundary Explorer (IBEX) and Cassini missions use energetic neutral atoms to remotely sense processes occurring in the same region (see Figure 2.3).

During the next decade, the Voyager spacecraft are expected to exit Earth's heliosphere, entering interstellar space. For the first time, operating spacecraft will enter into our local galaxy and gather local

[5]National Research Council, *Space Radiation Hazards and the Vision for Space Exploration: Report of a Workshop*, The National Academies Press, Washington, D.C., 2006.

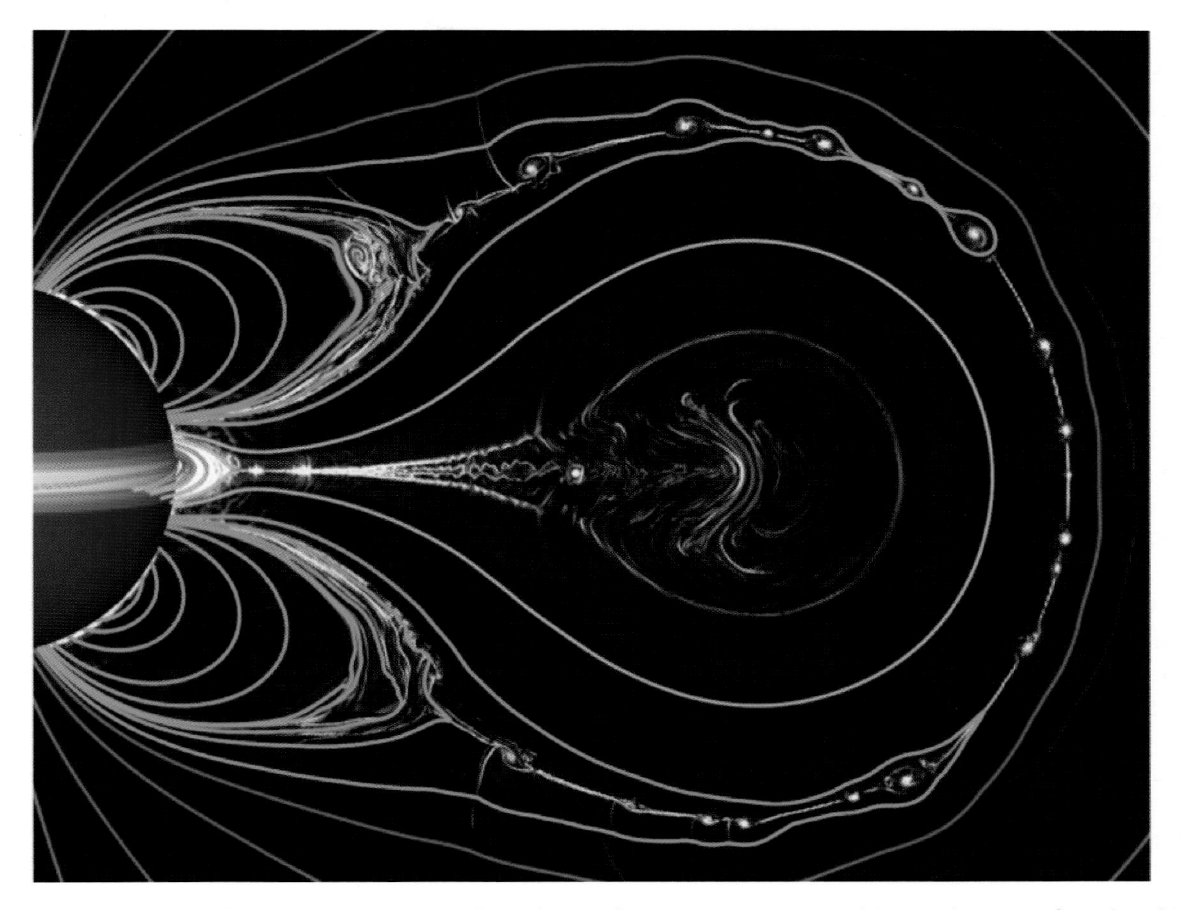

FIGURE 2.8 In this ultra-high-resolution numerical simulation of a reconnection-initiated CME and eruptive flare, the white contours indicate high current densities. Note the vertical flare current sheet below the erupting plasmoid. The plasmoid undergoes a sudden acceleration coincident with the onset of the flare (reconnection in this sheet). SOURCE: J.T. Karpen, S.K. Antiochos, and C.R. DeVore, The mechanisms for the onset and explosive eruption of coronal mass ejections and eruptive flares, *Astrophysical Journal* 760(1):81, 2012.

measurements from the interstellar medium—a truly historic event. The coming decade will therefore provide critical understanding of the heliospheric boundary regions and the processes that shape the interaction of the heliosphere with its local galactic medium. This motivates a fourth science challenge: **SHP-4. Discover how the Sun interacts with the local interstellar medium.**

Challenges Related to Solar Wind-Magnetosphere Interactions

While the broad view of how reconnection takes place and drives convection in the magnetosphere is now well established, the underlying physics of magnetic reconnection in the collisionless regime of the magnetosphere is not yet understood well enough to enable prediction of when, where, and how fast this process will occur and how it contributes to mass, energy, and momentum transport. NASA's Magnetospheric Multiscale Mission (MMS) is designed to carry out in situ measurements in the magnetosphere to establish the mechanisms that control how magnetic field lines reconnect. The results are expected to have profound implications for understanding reconnection within the heliosphere and in astrophysical settings throughout the universe. They are also highly relevant to understanding reconnection events in tokomak plasmas and in laboratory-based reconnection experiments. The centrality of reconnection in such diverse settings motivates the following primary challenge: **SWMI-1. Establish how magnetic reconnection is triggered and how it evolves to drive mass, momentum, and energy transport.**

Magnetic reconnection in the magnetotail drives convection that carries energetic particles toward Earth, where they are injected and trapped in orbits around Earth to form the extraterrestrial ring current, a region of relatively high energy ions and electrons that is most intense near the equator at distances of 3 to 7 R_E from Earth's center (see Figure 2.6). The outer radiation belt therefore overlaps the orbit radius of geostationary satellites (6.6 R_E) where the vast majority of communications and Earth-monitoring spacecraft reside. These satellites can be damaged by energetic radiation belt electrons whose flux is strongly enhanced during intense solar activity and the resultant storms in the magnetosphere. Understanding charged particle acceleration, scattering, and loss, which control the intensification and depletion of the radiation belts, is therefore a priority of solar and space physics.

The high variability of the radiation belts is evident in Figure 2.9, which shows a near-equatorial satellite view of energetic electron fluxes. The acceleration of particles in the radiation belts is believed to arise from a combination of compression as particles move from the weak magnetic field region in the distant magnetotail into the region of high magnetic field near Earth and the interaction with intense waves generated in the radiation belts themselves. NASA's Radiation Belts Storm Probes (RBSP; renamed the Van Allen Probes) mission is designed to determine the mechanisms that control the energy, intensity, spatial distribution, and time variability of the radiation belts. To understand the response of the magnetospheric system to driving by the solar wind, the following challenge must be addressed: **SWMI-2. Identify the mechanisms that control the production, loss, and energization of energetic particles in the magnetosphere.**

At around 100-km altitude, the atmosphere starts to transition from being neutrally dominant to being dominated by charged particles. The ionosphere, which is often thought to be the inner boundary of the magnetosphere, overlaps with the thermosphere. At these altitudes, the neutral density is about 1,000 times larger than the ion density, but the electromagnetic forces on the ions are significantly larger than the forces on the neutrals, so they become more and more important as the altitude increases. This region of the atmosphere is quite thick, being a couple of hundred kilometers in altitude, in comparison to the troposphere, which is only 10 km thick, but it pales in comparison to the vast space carved out by the magnetosphere, which extends out 10 to 100 Earth radii. If one were to calculate the whole mass of all of the particles in the magnetosphere, it would be about an order of magnitude less than the mass of the ionosphere, even though the ionosphere is so much smaller. This is because the density of the ionosphere is so much larger than the near-vacuum of the magnetosphere.

Magnetic field lines converge in the polar regions in the ionosphere. The magnetospheric convection cycle described above maps to middle and high latitudes in the ionosphere where the resulting flows transport and mix plasma and the more dense neutral gas. Ionospheric conductance facilitates field-aligned currents that produce resistance to the convection flows to the magnetosphere. The closure of these currents in the ionosphere drives neutral-gas winds and expels ions upward along the magnetic field.

During magnetic storms the intense upwelling of ions from the ionosphere into the magnetosphere is so strong that ionospheric O^+ can dominate the high-altitude ion pressures. This alters magnetospheric dynamics by modifying magnetic reconnection on both the dayside and the nightside. Figure 2.10 shows simulations of the magnetospheric response to changes in the IMF which, when O^+ outflow is properly included, results in the repeated onset of magnetic reconnection events that intensify the aurora and asso-

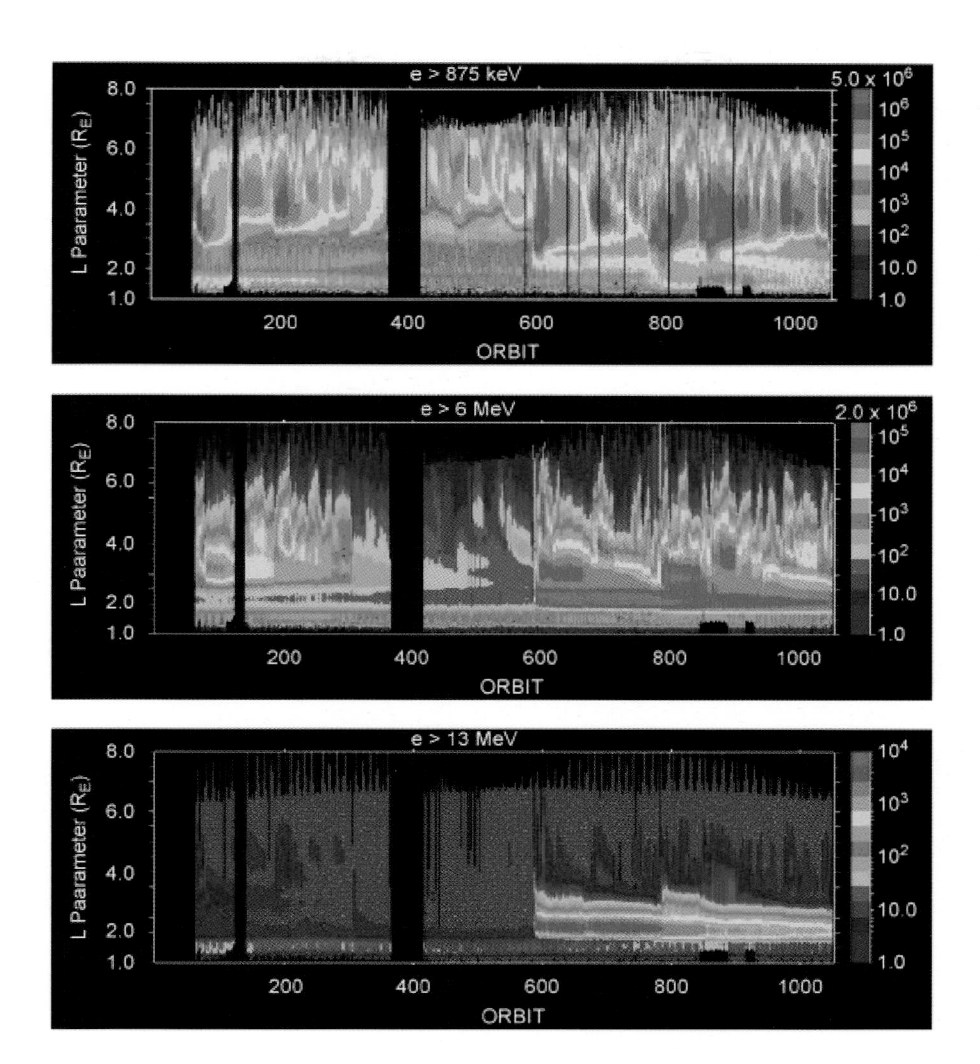

FIGURE 2.9 Energetic electron variability as measured during the 14-month Combined Release and Radiation Effects Satellite (CRRES) mission lifetime extending past geosynchronous orbit, 6.6 Earth radii (R_E) from Earth's center (22,000 miles above sea level), where spacecraft remain overhead as Earth rotates, a heavily populated orbit; data from July 1990 to October 1991, the maximum of solar cycle 22. A new radiation belt with energy greater than 13 million electron volts was created on a timescale of minutes in response to a strong interplanetary shock caused by a coronal mass ejection. This new radiation belt persisted until 1994. No solar wind observations preceding the arrival of the shock were available during this event due to absence of an L1 measurement (M. Blanc, J.L. Horwitz, J.B. Blake, I. Daglis, J.F. Lemaire, M.B. Moldwin, S. Orsini, R.M. Thorne, and R.A. Wolfe, Source and loss processes in the inner magnetosphere, *Space Science Review* 88(1-2):137-206, 1999). SOURCE: Reprinted from M.K. Hudson, B.T. Kress, H.-R. Mueller, J.A. Zastrow, and J.B. Blake, Relationship of the Van Allen radiation belts to solar wind drivers, *Journal of Atmospheric and Solar-Terrestrial Physics* 70(5):708-729, 2008, copyright 2008, with permission from Elsevier.

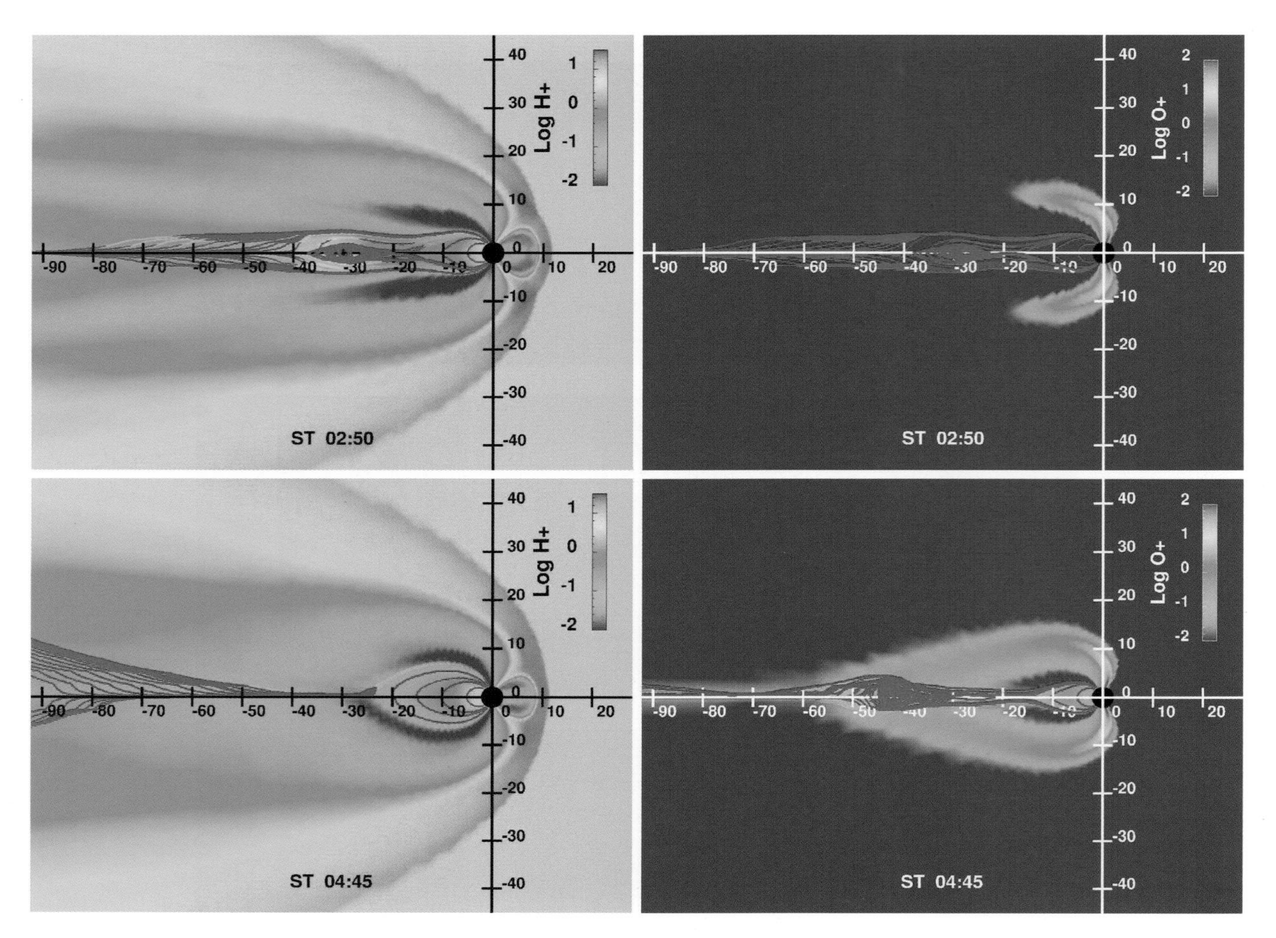

FIGURE 2.10 Multifluid MHD simulation results of substorm initiation without (left-hand panels) and with (right-hand panels) O+ outflow from the ionosphere. The colors indicate the densities of the two species in the simulations. The left-hand panels show only H+, the only species in the simulation, whereas the right-hand panels show the ionospheric O+, which is added to the H+. The red lines in each panel show magnetic field lines in the region of interest. In the upper panels, both simulations show a plasmoid release at 2 hours 50 minutes into the simulation, as indicated by the looped field lines beyond ~20 Earth radii. In both simulations, this plasmoid will depart rapidly downtail. In the lower panels (~2 hours later), the magnetosphere has stabilized in the simulation without O+, while the result with O+ shows a second plasmoid release in the region accessible to the O+. The addition of O+ as a distinct fluid with a significant contribution to the mass density makes the magnetosphere repetitively unstable. SOURCE: M. Wiltberger, W. Lotko, J.G. Lyon, P. Damiano, and V. Merkin, Influence of cusp O+ outflow on magnetotail dynamics in a multifluid MHD model of the magnetosphere, *Journal of Geophysical Research—Space Physics* 115:A00J05, 2010.

ciated ionospheric currents. These events inject plasma stored in the geomagnetic tail Earth-ward. This plasma acts as the seed population for the radiation belts and drives the plasma waves that are responsible for the scattering and loss of radiation belt electrons. In addition, storm-time ionospheric heating and convection produce large changes in the neutral and plasma densities that alter ionospheric conductances on a global scale.

Since the feedback of the ionosphere and thermosphere as a source of plasma and dissipation for the magnetosphere has such profound effects, the evolution of the ionosphere and magnetosphere must be studied as a globally coupled system. Thus, a key challenge is as follows: **SWMI-3. Determine how coupling and feedback between the magnetosphere, ionosphere, and thermosphere govern the dynamics of the coupled system in its response to the variable solar wind.**

Earth's magnetosphere is a prototype of a universal plasma system: an object with a global magnetic field that is subjected to an externally flowing plasma and forms a magnetosphere. Five other planets in Earth's solar system have magnetospheres: Mercury, Jupiter, Saturn, Uranus, and Neptune. Ganymede, one of Jupiter's satellites, also has its own tiny magnetosphere embedded within Jupiter's giant one. Although planetary systems exhibit analogous structures, the contrasting dynamics, boundary conditions, and magnetic fields make their detailed study of unique importance for testing theories and models.

Jupiter's moon Io, deep within the enormous Jovian magnetosphere, is a copious source of neutral gas, which, upon ionization, is a dominant drag force on the rapidly co-rotating magnetic field of the planet. Similarly, the moons of Saturn, particularly Titan and Enceladus, are major sources of plasma that affects the dynamics of Saturn's magnetosphere. A key enigma of the Saturnian system is the source of the regular, 10-hour 46-minute periodicity in Saturn's radio emissions, which differs from its rotation period by 6 minutes. This difference, discovered in data from the Cassini and Voyager spacecraft, remains unexplained. The magnetospheres of Uranus and Neptune are largely unexplored but present unique cases that will likely further challenge scientific understanding. Finally, the tiny magnetosphere of Mercury is an extreme example of a magnetospheric system because it possesses no ionosphere. In such a situation the coupling processes that operate are radically different. Thus, these other systems present a suite of vastly different configurations. The opportunity to test current theories and models on these widely varying systems motivates a fourth challenge: **SWMI-4. Critically advance the physical understanding of magnetospheres and their coupling to ionospheres and thermospheres by comparing models against observations from different magnetospheric systems.**

Challenges Related to Atmosphere-Ionosphere-Magnetosphere Interactions

Understanding ionosphere-thermosphere interactions is a major area of inquiry, especially during geomagnetic storms. The intense energy input from the magnetosphere, reaching up to terawatts, typically occurs in regions spanning less than 10 degrees in latitude but during storms is redistributed throughout the polar regions and down to middle latitudes over timescales from tens of minutes to hours. High-latitude heating (mainly below 200-km altitude) causes N_2-rich air to upwell. Strong winds driven by this heating transport N_2 equatorward. The mixing with ambient atomic oxygen produces dramatic changes in the ratio between atomic oxygen and molecular nitrogen. This global response was first discovered more than a decade ago, but researchers still cannot explain why it takes several hours for the global thermosphere to "inflate" after the high-latitude heating begins.

The ionospheric plasma also experiences major reconfigurations during storms as magnetospheric convection drives the mixing of low- and high-density regions of the ionosphere. Figure 3.3 shows an example of a plasma plume extending over thousands of kilometers that formed during the main phase of a geomagnetic storm. Redistributions of plasma by large-scale electric fields also occur in the middle and lower latitudes. At the onset of a storm, electric fields penetrate from the polar region and lift the equatorial ionosphere, depleting the equatorial density and producing anomalously high ionospheric densities on field lines that connect the high-altitude equator with ionospheric latitudes north and south of the equator. Convection in the polar regions also drives large-scale thermospheric winds that in turn carry ionospheric plasma across the polar regions to lower latitudes.

The storm response of the ionosphere and thermosphere produces structures over a wide range of time and spatial scales. To understand the storm-time behavior of this system, researchers must address the following science challenge: **AIMI-1. Understand how the ionosphere-thermosphere system responds to, and regulates, magnetospheric forcing over global, regional, and local scales.**

An important element of the dynamics of the IT system is the transfer of energy and momentum between the plasma and neutral components of the system and the role that electric and magnetic fields serve in accentuating and sometimes moderating this interchange. The pathways through which ions and neutrals interact are of course fundamental to space physics, given that they occur at all planets with atmospheres, at comets, and within the magnetospheres of Jupiter and Saturn. For example, in Earth's ionosphere at an altitude from 100 to 130 km the collisions between ions and electrons and neutrals enable current to flow across the local magnetic field, which facilitates closure of currents flowing along magnetic fields from the magnetosphere. The proper description of these cross-field currents requires the development of an accurate model of the plasma "conductivity," yet the dynamics of ionospheric conductivity are among the most poorly quantified parameters of the IT system. Earth's equatorial region is a rich laboratory for the investigation of plasma-neutral coupling in the presence of a magnetic field. The behavior can be extraordinarily complex: plasma-neutral collisions and associated neutral winds drive turbulence that cascades to very small spatial scales and regularly disrupts communications. The chemical interaction of a variety of ion species further complicates the dynamics.

A different suite of interactions occurs at middle latitudes. Spontaneous airglow emissions at 6,300 Å exhibit waves propagating to the south-west. They are thought to originate as neutral density waves at high latitudes which then interact with the mid-latitude ionosphere to create the structures, but their occurrence is curiously unrelated to levels of magnetic activity.

Thus, plasma-neutral coupling plays a critical role in ionospheric dynamics across the full range of latitudes. Researchers must therefore address the following challenge: **AIMI-2. Understand the plasma-neutral coupling processes that give rise to local, regional, and global-scale structures and dynamics in the AIM system.**

Numerous recent observations and simulations show that the IT system owes much of its longitudinal, local-time, seasonal, and even day-to-day variability to meteorological processes in the troposphere and stratosphere. The primary mechanism through which energy and momentum are transferred from the lower atmosphere to the upper atmosphere and ionosphere is through the generation and propagation of waves. The absorption of solar radiation (e.g., by tropospheric H_2O and stratospheric O_3) excites a spectrum of thermal tides. Figure 2.11 shows the spatial structure in daytime convective clouds that is believed to introduce longitudinal structure in the ionosphere, seen in Figure 2.11 in ultraviolet emissions. Surface topography and unstable shear flows excite planetary waves and gravity waves extending from planetary to very small (~tens to hundreds of kilometers) spatial scales and having periods from tens of days down to minutes. Convective tropospheric weather systems radiate additional thermal tides, gravity waves, and other classes of waves.

Those waves that propagate vertically grow exponentially with height into the more rarified atmosphere. Some of the waves spawn additional waves and turbulence. Figure 2.12 shows sodium layer observations revealing amazing wave structures at the base of the thermosphere, illustrating the rich spectrum of dynamics that occurs. Although the presence and the importance of waves are not in dispute, the relevant coupling processes operating between the neutral atmosphere and ionosphere involve a host of multiscale dynamics that are not understood at present. This leads to another major scientific challenge: **AIMI-3. Understand how forcing from the lower atmosphere via tidal, planetary, and gravity waves influences the ionosphere and thermosphere.**

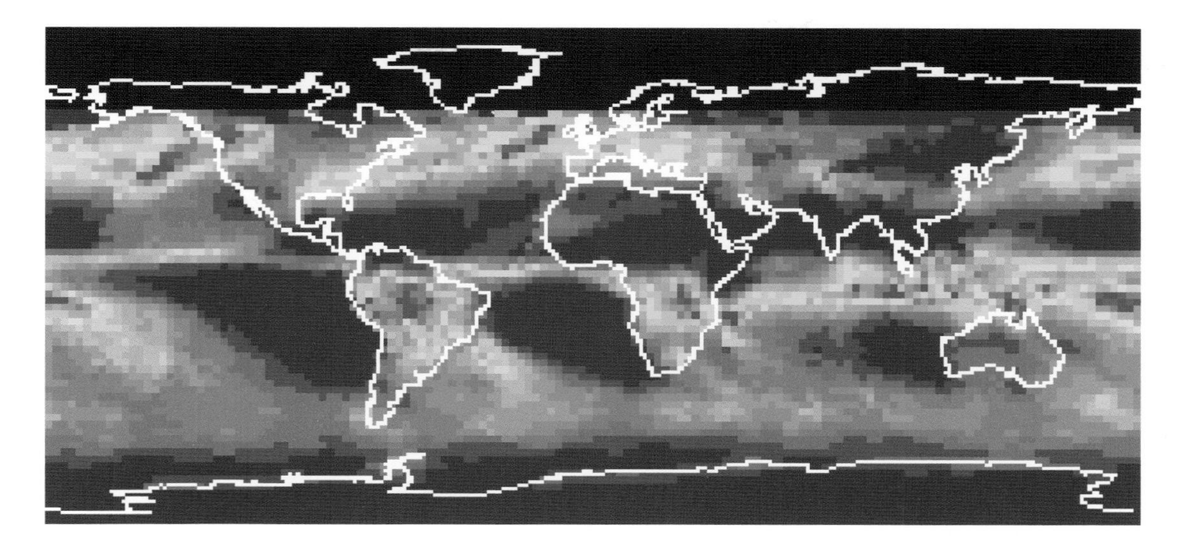

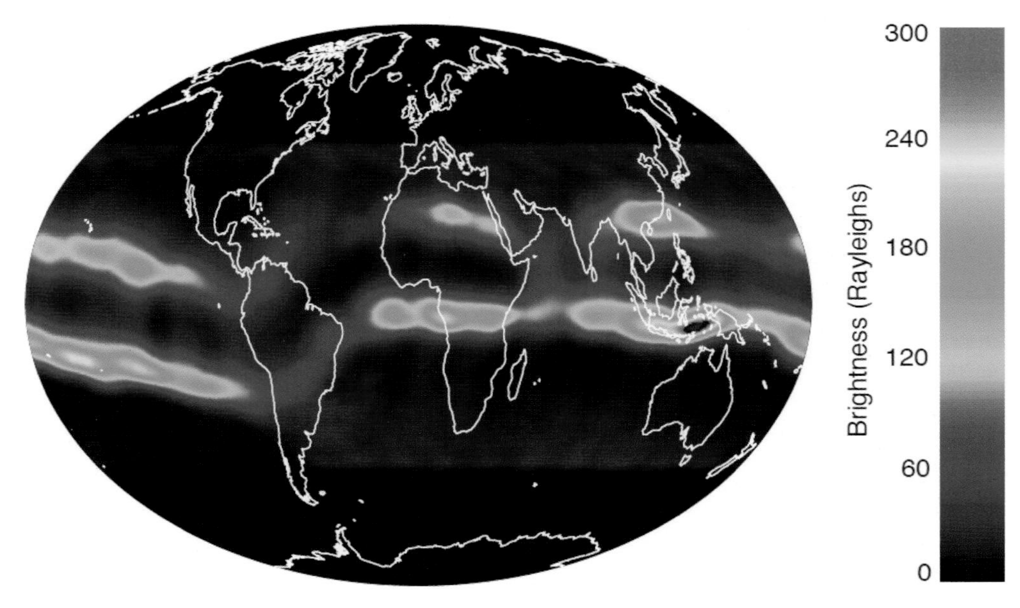

FIGURE 2.11 *Top:* Mean 1984-2009 January daytime convective cloud amount in percentage from ISCCP-D2. Blue indicates 10-15 percent, yellow/green indicates approximately 8 percent, and red indicates 0-4 percent. *Bottom:* Average ionospheric equatorial densities derived from TIMED GUVI observations of 135.6-nm OI emissions showing unexpected wave structure in ionospheric densities on the same longitude scales as the tropospheric pressure waves. The double-banded structure is due to the neutral wind dynamo at the magnetic equator which transports equatorial plasma north and south of the equator. SOURCE: *Top:* The International Satellite Cloud Climatology Project (ISCCP) D2 data/images (described in W.B. Rossow and R.A. Schiffer, Advances in understanding clouds from ISCCP, *Bulletin of the American Meteorological Society* 80:2261-2288, 1999) were obtained in January 2005 from the ISCCP website (available at http://isccp.giss.nasa.gov and maintained by the ISCCP research group at the NASA Goddard Institute for Space Studies, New York, N.Y.). *Bottom:* S.L. England, X. Zhang, T.J. Immel, J.M. Forbes, and R. Demajistre, The effect of non-migrating tides on the morphology of the equatorial ionospheric anomaly: Seasonal variability, *Earth, Planets and Space* 61:493-503.

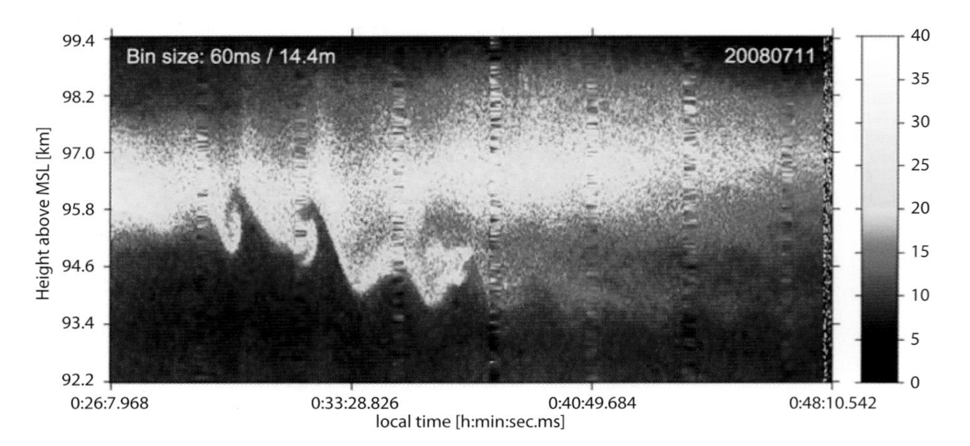

FIGURE 2.12 High-resolution sodium lidar observations of breaking gravity waves at the base of the thermosphere. A 6-meter, zenith-pointing telescope comprising a spinning mercury mirror was coupled to a sodium lidar system and revealed amazing detail in MLT instability structures, identified as Kelvin-Helmholtz billows evident at the base of the sodium layer, at a temporal resolution of 60 milliseconds and a spatial resolution of 15 meters. SOURCE: T. Pfrommer, P. Hickson, and C.-Y. She, A large-aperture sodium fluorescence lidar with very high resolution for mesopause dynamics and adaptive optics studies, *Geophysical Research Letters* 36:L15831, doi:10.1029/2009GL038802, 2009. Copyright 2009 American Geophysical Union. Reproduced by permission of American Geophysical Union.

The release of greenhouse gases (e.g., CO_2 and CH_4) into the atmosphere is changing Earth's surface climate by warming the lower atmosphere; these gases are also changing geospace climatology by cooling the upper atmosphere. In the lower atmosphere, the opacity of greenhouse gases to infrared radiation traps energy by capturing the radiant infrared energy from Earth's surface and transferring it to thermal energy via collisions with other molecules. In the thermosphere, however, where intermolecular collisions are less frequent, greenhouse gases promote cooling by acquiring energy via collisions and then radiating this energy to space in the infrared. This well-understood role of CO_2 as an effective radiator of energy in the upper atmosphere has produced a systematic decrease in thermospheric mass density by several percent per decade near the 400-km altitude. This systematic decrease follows from the record of satellite orbit decay measured since the beginning of the space age (Figure 2.13).

There are two other consequences of climate change for the ionosphere and thermosphere. First, changes in tropospheric weather patterns and atmospheric circulation may alter the occurrence of ionospheric instabilities triggered by tropospheric gravity waves propagating into the upper atmosphere. This change will affect the prevalence of the resulting ionospheric irregularities. Second, continued cooling of the thermosphere will reduce satellite drag, thereby increasing orbital debris lifetimes, and will lower the effective ionospheric conductivity. The latter change will alter global currents in the magnetosphere-ionosphere system and therefore fundamentally alter magnetosphere-ionosphere coupling. The survey committee, therefore, identifies the following science challenge: **AIMI-4. Determine and identify the causes for long-term (multi-decadal) changes in the AIM system.**

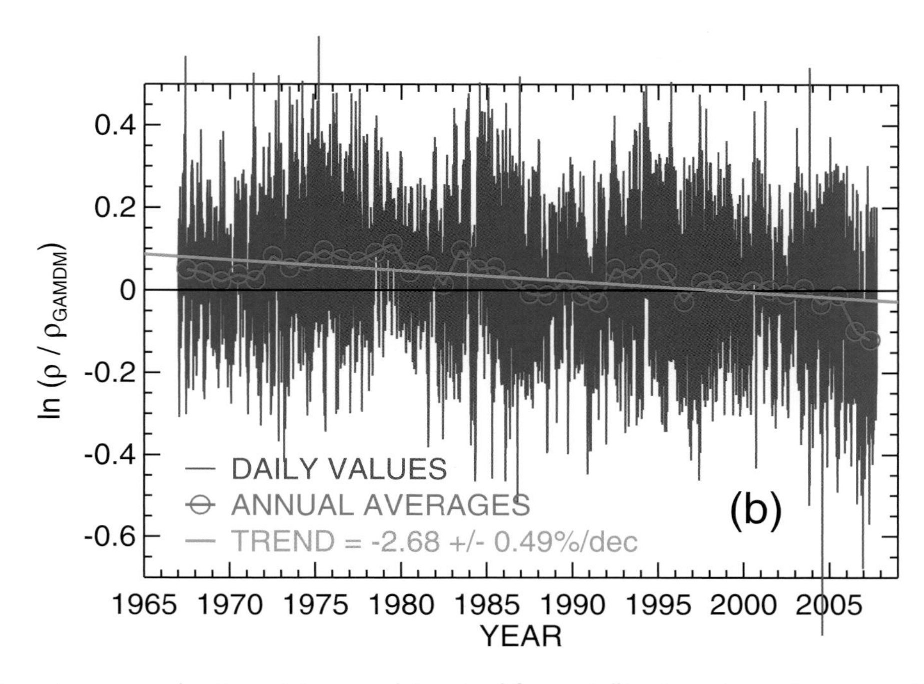

FIGURE 2.13 Long-term mass density variations as determined from satellite drag observations normalized to 400-km altitude demonstrating a consistent long-term cooling of the thermosphere consistent with increased CO_2, which at these altitudes cools the atmosphere by providing a mechanism to radiate energy at infrared wavelengths—the same property that traps heat lower in the atmosphere. SOURCE: J.T. Emmert, J.M. Picone, and R.R. Meier, Thermospheric global average density trends, 1967-2007, derived from orbits of 5000 near-Earth objects, *Geophysical Research Letters* 35:L05101, doi:10.1029/2007GL032809, 2008. Copyright 2008 American Geophysical Union. Reproduced by permission of American Geophysical Union.

RISING TO THE CHALLENGES OF THE COMING DECADE

Achievement of the survey committee's four key science goals (see Chapter 1) for the coming decade requires addressing the 12 science challenges, discussed above and also listed in Table 2.1, for the three subdisciplines of solar and space physics. In turn, addressing science challenges requires optimal use of existing assets, as well as initiation of new programs that will drive future discovery. Chapters 4, 5, and 6 outline the survey committee's recommendations for the upcoming decade and discuss how they may be implemented by NSF and NASA. The survey committee's recommendations were informed by a recognition that the interconnected nature of the science of solar and space physics requires a research effort that spans the entire front of science challenges. New missions, as described in Chapter 4, can be carefully chosen to address the most pressing of these science challenges. It will be evident, however, that in the foreseeable future nearly half of the science challenges are not targeted by any new heliophysics mission.

The survey committee views the Explorer line as a critical asset for broadening the field of inquiry to include questions not addressed by upcoming or recommended missions. In addition, the rich array of existing assets of NASA, NSF, NOAA, and DOD, as well as the use of non-science space platforms, also facilitates scientific discovery in solar and space physics provided that these assets are adequately supported and that research and analysis efforts are sustained. The central importance of L1 in situ observations of the

TABLE 2.1 Solar and Space Physics Decadal Science Challenges

The Sun and Heliosphere	
SHP-1	Understand how the Sun generates the quasi-cyclical magnetic field that extends throughout the heliosphere.
SHP-2	Determine how the Sun's magnetism creates its hot, dynamic atmosphere.
SHP-3	Determine how magnetic energy is stored and explosively released and how the resultant disturbances propagate through the heliosphere.
SHP-4	Discover how the Sun interacts with the local interstellar medium.
Solar Wind-Magnetosphere Interactions	
SWMI-1	Establish how magnetic reconnection is triggered and how it evolves to drive mass, momentum, and energy transport.
SWMI-2	Identify the mechanisms that control the production, loss, and energization of energetic particles in the magnetosphere.
SWMI-3	Determine how coupling and feedback between the magnetosphere, ionosphere, and thermosphere govern the dynamics of the coupled system in its response to the variable solar wind.
SWMI-4	Critically advance the physical understanding of magnetospheres and their coupling to ionospheres and thermospheres by comparing models against observations from different magnetospheric systems.
Atmosphere-Ionosphere-Magnetosphere Interactions	
AIMI-1	Understand how the ionosphere-thermosphere system responds to, and regulates, magnetospheric forcing over global, regional, and local scales.
AIMI-2	Understand the plasma-neutral coupling processes that give rise to local, regional, and global-scale structures and dynamics in the AIM system.
AIMI-3	Understand how forcing from the lower atmosphere via tidal, planetary, and gravity waves influences the ionosphere and thermosphere.
AIMI-4	Determine and identify the causes for long-term (multi-decadal) changes in the AIM system.

interplanetary medium in particular motivates the continuation of these observations to support a broad range of research in solar and space physics.

Finally, the nearly explosive growth in the ability to model complex phenomena in solar and space physics with realistic numerical simulations suggests that the field is on the cusp of greatly expanded predictive power and fundamental understanding. The advanced state of theory and simulation also provides a powerful opportunity to couple efforts in this area with observations, which will always remain limited in key aspects, to realize the full potential of the observations and their implications for understanding the underlying physical processes that they reflect. Reaching scientific closure and advancing predictive understanding therefore depend critically on robust support for theory and modeling across the spectrum of science challenges.

In summary, the program of solar and space physics research recommended in this report is specifically designed to make the most effective use of the nation's resources in a program that maximizes scientific advances and furthers understanding of the space weather threats to a society that is increasingly reliant on technologies that are vulnerable to solar and geospace activity.

3

Addressing Societal Needs

Societal aspirations for a secure, prosperous, and technologically sophisticated future are increasingly influenced by Earth's near-space environment. Central to the decadal survey's strategy is the intent to achieve scientific results that will be useful to society; this chapter briefly reviews societal needs for improved knowledge of solar and space physics phenomena related to the effects of space weather.[1]

IMPACTS OF EARTH'S NEAR-SPACE ENVIRONMENT

The Climate System

The Sun's photon energy is the main source of heat for the entire Earth system. Slight variations in this light with the Sun's 11-year activity cycle (Figure 3.1), and possibly on longer timescales, continuously affect Earth's climate and atmosphere, at times temporarily masking the effects of changing concentrations of greenhouse gases. Earth's atmosphere protects us from biologically damaging shorter-wavelength solar ultraviolet emissions. These emissions fluctuate by orders of magnitude more than variations in the total brightness, altering the ozone layer significantly. Solar-induced photochemical and dynamical changes in Earth's middle atmosphere may affect the climate at lower altitudes and in the upper atmosphere and ionosphere through dynamical coupling of changing wave structures.

Precipitating electrons from the aurora affect the atmospheric chemistry in the polar regions. During major magnetic storms, significant enhancements of nitric oxide concentrations occur in the auroral zone and can literally propagate downward from space into the stratosphere. Nitric oxide plays an important role in the chemistry of stratospheric ozone. The downward transport of auroral products is difficult to trace and varies with the general circulation of the polar atmosphere, but it is clear that major space storms can potentially modify stratospheric composition and reduce ozone densities for a period of time following their occurrence.

[1] See also Box 1.1, "Severe Space Weather Events—Understanding Societal and Economic Impacts."

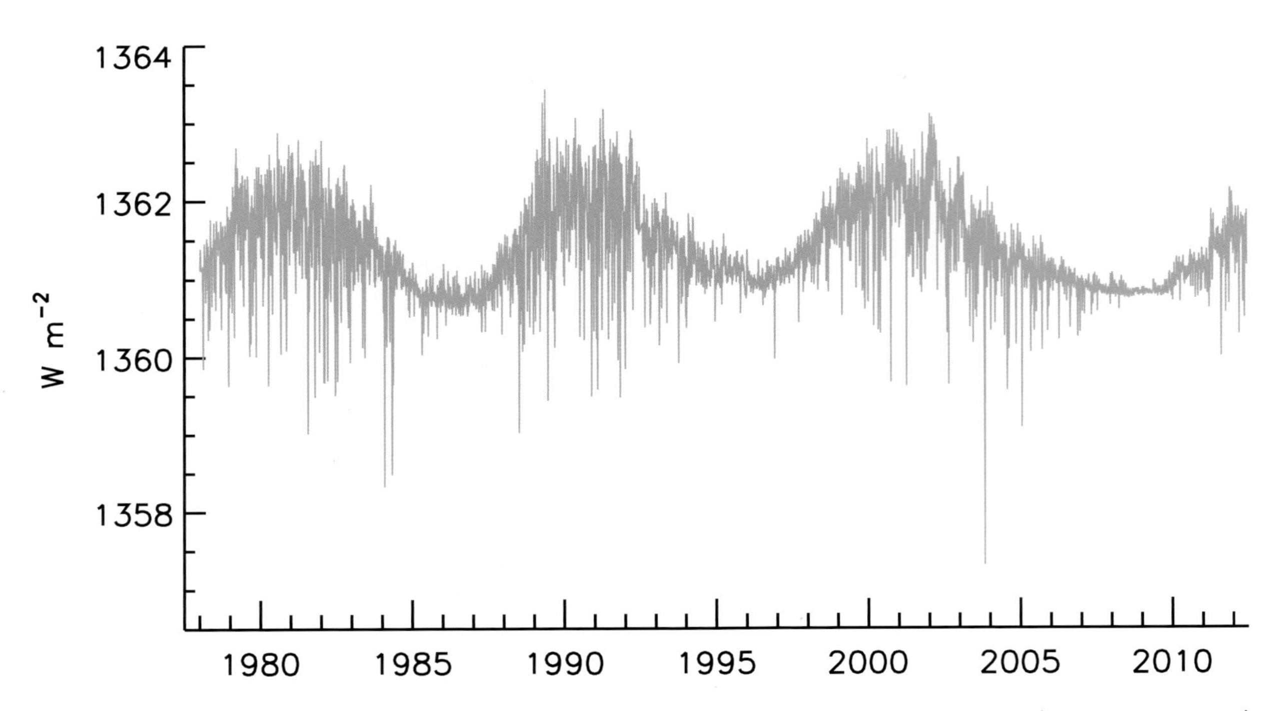

FIGURE 3.1 Total solar irradiance (TSI) observed over the past three solar cycles (since 1978), varying between 1,357.5 and 1,363.5 W/m^2. This composite time-series plot is based on the lower TSI level established with new laboratory calibrations of TSI instruments (see Figure 10.5 in Chapter 10). Differences between levels of irradiance during the solar minimum epochs (1986, 1996, 2008) are not significant because of instrument uncertainties. SOURCE: Replotted courtesy of Judith Lean, Naval Research Laboratory, after G. Kopp and J.L. Lean, A new, lower value of total solar irradiance: Evidence and climate significance, *Geophysical Research Letters* 38:L01706, doi:10.1029/2010GL045777, 2011.

Satellite Infrastructure

Satellites orbiting Earth support essential societal infrastructure and now form the basis for a total global economy in excess of $250 billion per year.[2] To inform our daily activities and decisions, we rely on weather predictions based on measurements from satellites. Satellites serve as communication relays and platforms for direct broadcasts of data and television signals. The nation's military protects U.S. strategic interests around the world through continuous surveillance from satellites and depends on satellites for global communication, continuous situational awareness, and geolocation related to national security. Although a relatively new technology, the use of signals from Global Positioning System (GPS) satellites is pervasive, facilitating everyday activities that range from navigation to financial transactions.

The magnetosphere is the domain of nearly all Earth-orbiting satellites, affecting those in low, medium, and geostationary orbits, as well as those in high-apogee orbits. It is a region filled with charged particles, including the intense radiation belts that vary continuously in response to changes in the solar wind and to the solar disturbances that strongly affect the space environment. (See Figure 3.2.) Charged particles affect space technology in a variety of ways: at their most benign they cause surface charging and discharging,

[2]See *Report on the Space Economy Symposium,* March 13, 2009, available at http://spaceeconomy.gmu.edu/ses2009/symposiumreport2009.pdf.

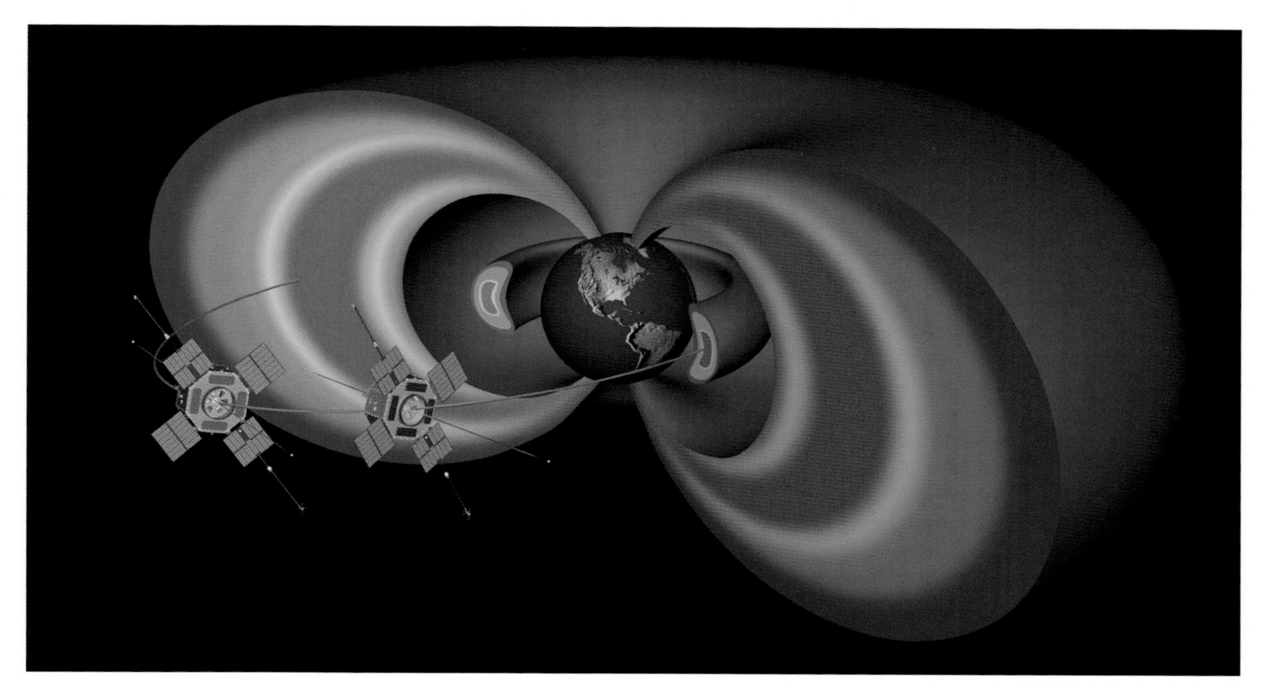

FIGURE 3.2 A diagram of the Van Allen radiation zone surrounding Earth. This cutaway image shows the weak inner zone, the "slot" region that is relatively devoid of trapped radiation, and the more intense and highly variable outer Van Allen belt. The two spacecraft of the Radiation Belt Storm Probes mission are shown schematically. SOURCE: Courtesy of NASA.

and at their most destructive they damage electronics components, including the temporary (single-event) upset of spacecraft commanding. Furthermore, upper atmospheric heating associated with the dynamics of the space environment can dramatically change drag effects on low-Earth-orbiting satellites, notably the International Space Station (ISS).

Ionospheric Variability, Communication, and Navigation

The specification and forecasting of ionospheric scintillation (i.e., radio propagation fluctuations due to plasma density irregularities) is a high priority for both civilian and military space operations. These scintillations are prevalent at low geographic latitudes as well as in the auroral regions, disrupting radio communications in critical geographical locations. They occur more frequently and extend to higher altitudes during times of high solar activity. Changes in ionospheric total electron content during geomagnetic storms compromise the performance of GPS technology vital for aviation and many other commercial and defense applications (see Figure 3.3). The participating electrons from the aurora also charge space systems, such as the ISS, generating the danger of arcing associated with discharges.

Radiation and Human Space Exploration

Humans venturing into space are vulnerable to damage caused by episodic radiation in the form of energetic particles from the Sun and from cosmic rays that constantly impinge on the solar system from

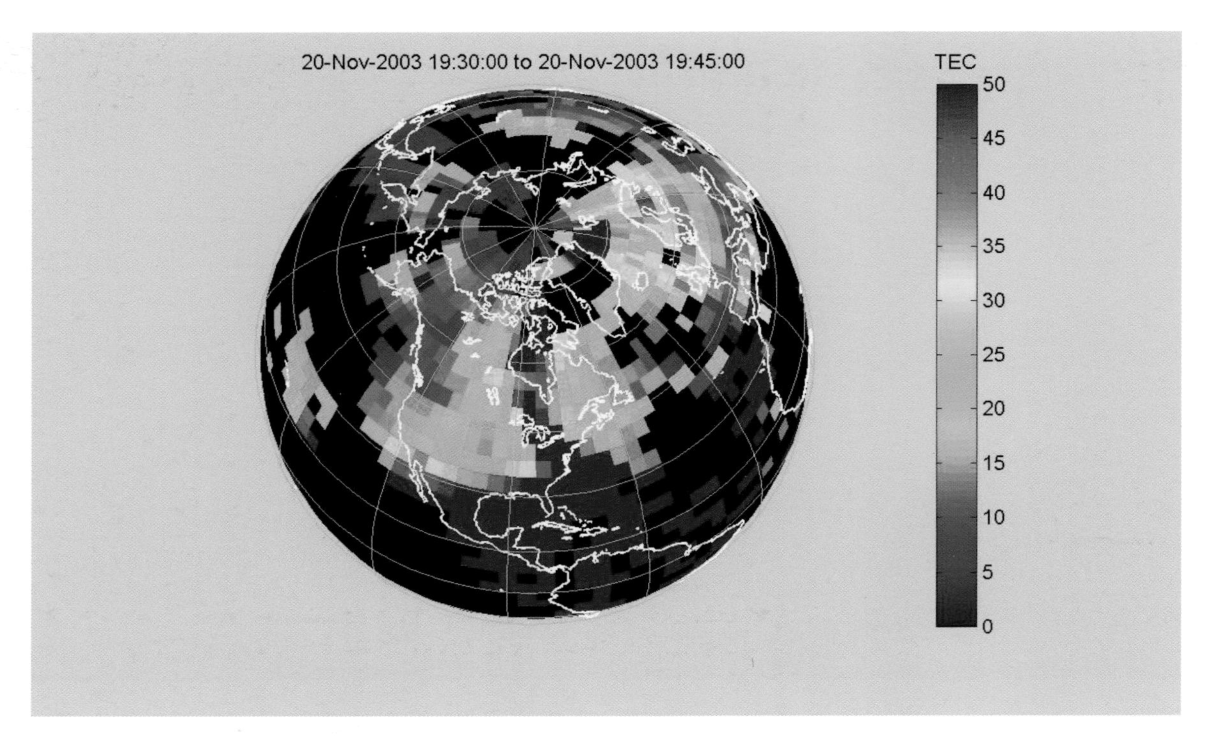

FIGURE 3.3 Total electron content (TEC), a measure of column-integrated electron density, derived from dual-frequency Global Positioning System (GPS) receivers for the November 20, 2003, geomagnetic storm. The tongue of enhanced electron density (red hue) is a signature of storm-time circulation and reflects the interplay of ionosphere-magnetosphere dynamics during active periods. Gradients in TEC (shown here as variations in hue) can cause GPS receivers to lose location lock, affecting many GPS-dependent systems. SOURCE: A. Coster and J. Foster, Space weather impacts of the subauroral polarization stream, *Radio Science Bulletin* 321:28-36, 2007; copyright 2007 Radio Science Press, Belgium, for the International Union of Radio Science (URSI); used with permission.

the galaxy. These dangers are well recognized[3] if not yet predictable: astronauts on the ISS must retreat to protected areas in the event of high-energy solar radiation. Space radiation from all sources will pose important hazards for space systems and astronauts on long-duration flights away from Earth's magnetic field.

Geomagnetic Effects on the Electric Power Grid

The electric power grid, the backbone of modern society, is particularly vulnerable to space environmental effects. As the geomagnetic field changes as a consequence of impacts from solar eruptions, large currents are generated and guided into the ionized layers of Earth's upper atmosphere, where they contribute substantially to the outward expansion of the outer atmosphere, causing satellite drag effects. Even more importantly, temporal changes in the currents induce voltages in the ground. The ground conductor of the power grid is connected to this source of voltage, and consequently currents other than those associ-

[3] National Research Council, *Managing Space Radiation Risk in the New Era of Space Exploration*, The National Academies Press, Washington, D.C., 2008.

ated with power transmission can flow on the grid. These geomagnetically induced currents (GICs) have the potential to overload transformers, causing, at a minimum, reductions in efficiency. Large-amplitude GICs can age, or even destroy, a transformer.

Owing to the critical place electric power has in the maintenance of society, a number of recent studies[4,5,6] have emphasized the need for further research with the objective to understand, and then prevent and mitigate, deleterious GIC-based hazards. Ground-based assets such as long-distance pipelines are also susceptible, responding to the strong currents induced by geomagnetic storms. Some of the many manifestations of disturbances from the Sun that now have the potential to disrupt society are illustrated in Figure 3.4.

Severe Solar Storms

Powerful solar flares (e.g., see Figure 3.5) and their accompanying ejections of mass and energetic particles occur episodically. In an extreme event in 1859, a large solar eruption triggered a geomagnetic storm that sparked fires in telegraph offices across the United States and triggered aurorae as far south as Central America. Such a powerful event directly striking Earth today could severely affect the power grid, destroying transformers and causing widespread outages.

Although during the space age a direct hit of this magnitude has not yet occurred, severe solar storms have nevertheless damaged spacecraft and power grids, producing, for example, widespread power outages in Quebec in 1989 and in South Africa in 2003. Very energetic solar particles (SEPs) that are accelerated in solar flares and Earth-ward-propagating interplanetary shocks penetrate along open magnetic field lines into Earth's polar ionosphere, where they degrade high-frequency communications over the poles. This interference forces the airline industry to reroute transpolar flights, at a significant cost in time and fuel. European flight crews on shorter high-latitude routes are categorized as "radiation workers" and are monitored by film badges because of their increased exposure to SEPs. Lacking the knowledge for predictive mitigation of severe solar storm impacts, operators currently simply assume (and hope) that the rarity of extreme events and the vastness of space will protect against the most deleterious consequences.

Although particularly intense events occur about once per solar cycle and strong to extreme particle storms can occur about 15 times per cycle, somewhat less intense geomagnetic storms occur even more often. Even during these weaker geomagnetic storms, large changes in ionospheric currents threaten transformers in long-distance east-west power lines in North America and northern Europe.

THE CHALLENGE OF PREDICTING SPACE WEATHER EVENTS

Science cannot now reliably predict, with sufficient warning, the disturbances from space that might threaten society at any particular time. The physical processes that control space weather differ in complex ways from those that control the weather of the neutral atmosphere of Earth. Within the entire system numerous phenomena have to be addressed on a wide range of physical scales—for example, the gas density can vary from 10^{19} in Earth's atmosphere to just a few particles per cubic centimeter in the solar wind, and the relevant scale lengths can vary from centimeters to astronomical units (AUs). The many different and complex interactions include electromagnetic forces that accelerate and control the flow of

[4] J. Kappenmann, Metatech Corporation, *Low-Frequency Protection Concepts for the Electric Power Grid: Geomagnetically Induced Current (GIC) and E3 HEMP Mitigation*, prepared for Oak Ridge National Laboratory, Oak Ridge, Tenn., January 2010.

[5] MITRE Corporation, *Impacts of Severe Space Weather on the Electric Grid*, JASON report, McLean, Va., November 2011.

[6] North American Electric Reliability Corporation, *2012 Special Reliability Assessment Interim Report: Effects of Geomagnetic Disturbances on the Bulk Power System*, February 2012, available at http://www.nerc.com/files/2012GMD.pdf.

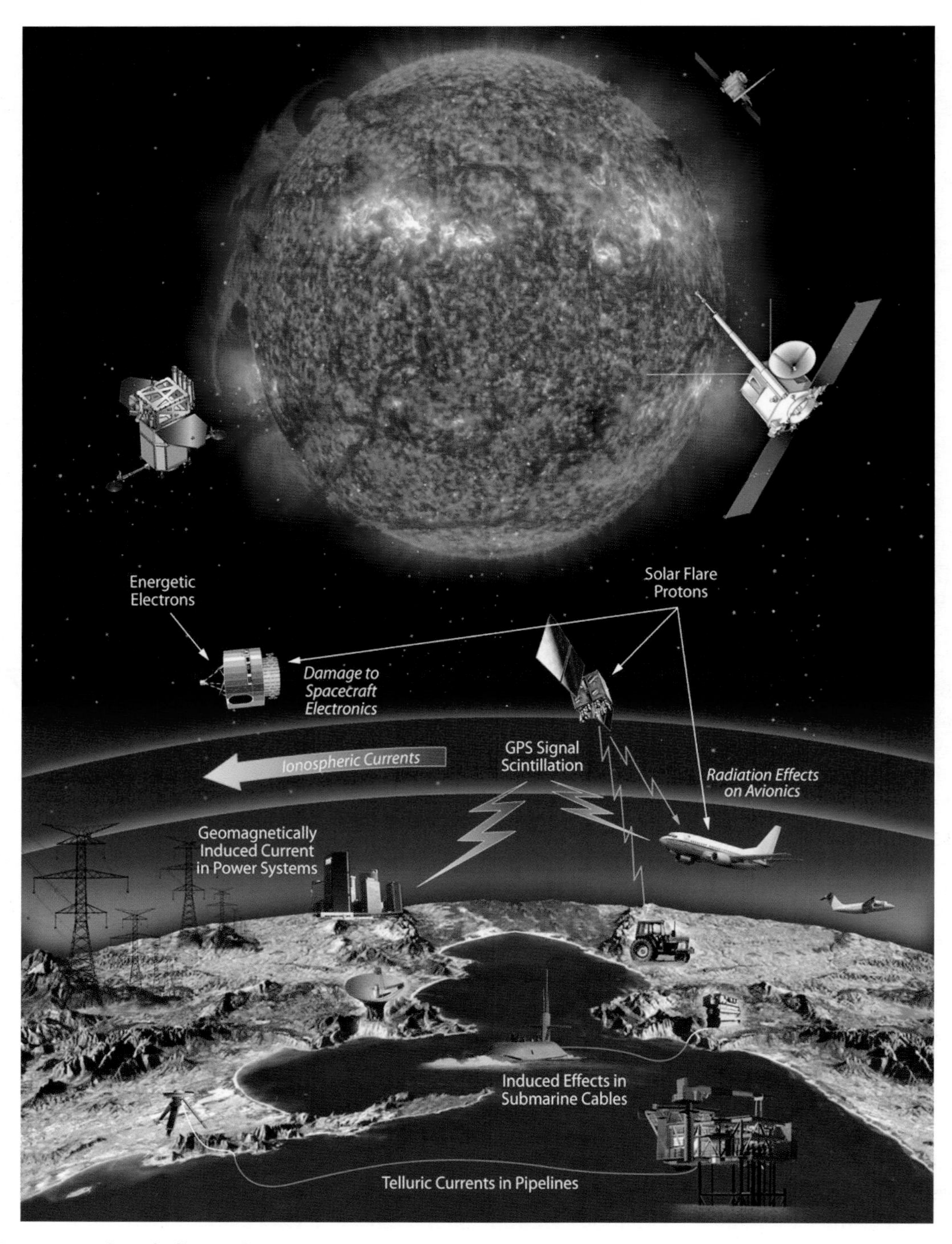

FIGURE 3.4 Examples of effects of space weather on critical infrastructure. SOURCE: NASA, *Heliophysics: The New Science of the Sun-Solar System Connection. Recommended Roadmap for Science and Technology 2005-2035*, NP-2005-11-740-GSFC, NASA, Greenbelt, Md., February 2006, available at http://sec.gsfc.nasa.gov/Roadmap_FINALpri.pdf.

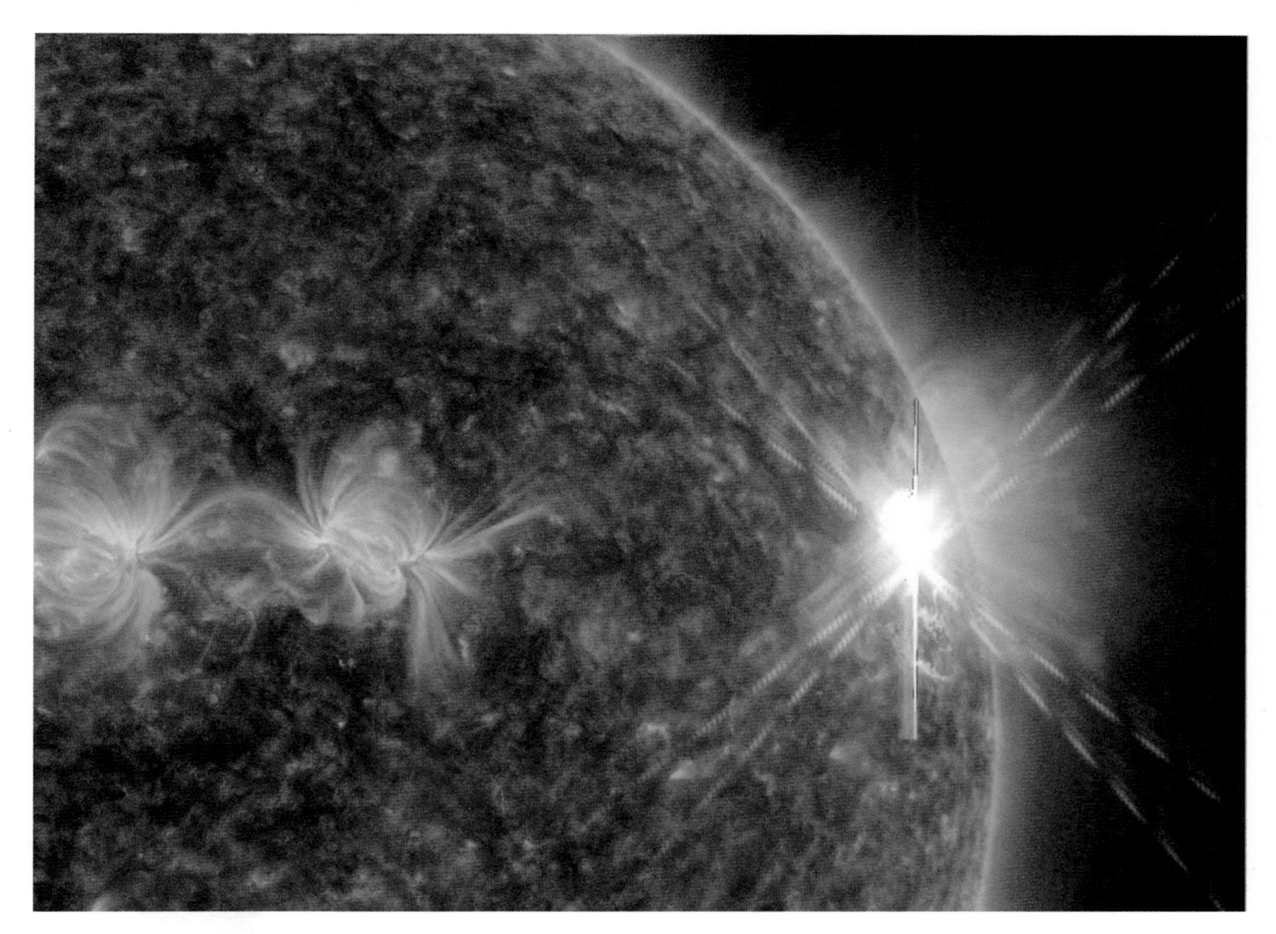

FIGURE 3.5 Solar Dynamics Observatory (SDO) image from August 9, 2011, X7 class flare. An X-class flare began at 3:48 a.m. EDT on August 9, 2011, and peaked at 4:05 a.m. The flare burst from active region AR11263 before it rotated out of view from Earth. The image here was captured by NASA's SDO in extreme ultraviolet light at 131 Å and at the beginning of the event, just before the satellite sensors were overwhelmed by energetic particles. SOURCE: NASA/Solar Dynamics Observatory/ Atmospheric Imaging Assembly; available at http://www.nasa.gov/mission_pages/sunearth/ news/News080911-xclass.html.

electrically charged particles, supersonic flows and shock waves, explosive release of magnetic energy, and solar-driven winds and tides in Earth's atmosphere.

Moreover, the Sun, the heliosphere, Earth, and the planets together constitute a coupled and intertwined system. It is a formidable challenge to understand the detailed individual processes that control the space environment, while also accounting for the global couplings among the various interacting members of the Sun-heliosphere-Earth system and their subelements, such as the neutral atmosphere and ionosphere. Significant progress has accrued during the past few decades from observations made by space missions and ground-based observatories and from theories and models developed to explain the observations. However, owing to the complexity of this variable, coupled system, scientists have not yet achieved a sufficiently reliable predictive capability for when and in what direction major disturbances will be emitted by the Sun; or for how disturbances from the Sun, coupled with inputs from Earth, affect the space environment near Earth; or for what the radiation environment through which astronauts might fly will be; or for exactly how changes on the Sun may affect Earth's climate, atmosphere, and ionosphere.

Despite these challenges, prediction of the space environment is, in principle, a decipherable problem. New ground- and space-based measurements are adding considerable knowledge to enhance understanding of the space environment and its governing processes. In parallel, increasingly sophisticated comprehensive physical models are being developed that run on ever more powerful computers. Given an adequate investment of effort in fundamental scientific research and modeling, the research community should be able to leverage advances in computing capability to develop the predictive models required to specify the extended space environment in order to protect society and advance growing aspirations for the use of space.

4

Recommendations

In previous chapters the survey committee describes challenges and opportunities for the field of solar and space physics and outlines goals for the upcoming decade and beyond. This chapter presents recommendations for accomplishing these goals, addressing both research and applications. Summarized in Tables 4.1 and 4.2, the recommendations are prioritized (and numbered as in the Summary) and, where appropriate, directed to a particular agency to form programs that satisfy fiscal and other constraints. Additional recommendations are offered throughout the chapter. The implementation and budget implications of the research recommendations are described for NSF in Chapter 5 and for NASA in Chapter 6. Chapter 7 expands further on the applications recommendations by presenting the survey committee's vision for a new national program in space weather and space climatology.

TABLE 4.1 Summary of Top-Level Decadal Survey Research Recommendations

Priority	Recommendation	NASA	NSF	Other
0.0	Complete the current program	X	X	
1.0	Implement the DRIVE initiative Small satellites; midscale NSF projects; vigorous ATST and synoptic program support; science centers and grant programs; instrument development	X	X	X
2.0	Accelerate and expand the Heliophysics Explorer program Enable MIDEX line and Missions of Opportunity	X		
3.0	Restructure STP as a moderate-scale, PI-led line	X		
3.1	Implement an IMAP-like mission	X		
3.2	Implement a DYNAMIC-like mission	X		
3.3	Implement a MEDICI-like mission	X		
4.0	Implement a large LWS GDC-like mission	X		

TABLE 4.2 Summary of Top-Level Decadal Survey Applications Recommendations

Priority	Recommendation	NASA	NSF	Other
1.0	Recharter the National Space Weather Program	X	X	X
2.0	Work in a multiagency partnership to achieve continuity of solar and solar wind observations	X	X	X
2.1	Continue solar wind observations from L1 (DSCOVR, IMAP)	X		X
2.2	Continue space-based coronagraph and solar magnetic field measurements	X		X
2.3	Evaluate new observations, platforms, and locations	X	X	X
2.4	Establish a space weather research program at NOAA to effectively transition research to operations			X
2.5	Develop and maintain distinct funding lines for basic space physics research and for space weather specification and forecasting	X	X	X

RESEARCH RECOMMENDATIONS

Baseline Priority for NASA and NSF: Complete the Current Program

The survey committee's recommended program for NSF and NASA assumes continued support in the near term for the key existing program elements that constitute the Heliophysics Systems Observatory (HSO) and successful implementation of programs in advanced stages of development.

NASA's existing heliophysics flight missions and NSF's ground-based facilities form a network of observing platforms that operate simultaneously to investigate the solar system. This array can be thought of as a single observatory—the Heliophysics Systems Observatory (HSO) (see Figure 1.2). The evolving HSO lies at the heart of the field of solar and space physics and provides a rich source of observations that can be used to address increasingly interdisciplinary and long-term scientific questions. Missions now under development will expand the HSO and drive scientific discovery. For NASA, these missions include the following:

- *Radiation Belt Storm Probes (RBSP; Living With a Star (LWS); 2012 launch*[1]*) and the related Balloon Array for RBSP Relativistic Electron Losses (BARREL; first launch 2012).* These missions will determine the mechanisms that control the energy, intensity, spatial distribution, and time variability of the radiation belts.
- *The Interface Region Imaging Spectrograph (IRIS; Explorer program; 2013 launch).* IRIS will deliver pioneering observations of chromospheric dynamics to help reveal their role in the origin of the fluxes of heat and mass into the corona and wind.
- *The Magnetospheric Multiscale Mission (MMS; Solar-Terrestrial Probes (STP) program; 2014 launch).* MMS will address the physics of magnetic reconnection at the previously inaccessible tiny scale where reconnection is triggered.

Missions that are less fully developed but that are part of the assumed baseline program[2] include:

[1] Following its launch on August 30, 2012, RBSP was renamed the Van Allen Probes.

[2] In accordance with its charge, the committee did not reprioritize any NASA mission that was in formulation or advanced development. In June 2010, the committee's charge was modified by NASA to include a request for it to present decision rules to guide the future development of the SPP mission.

- *Solar Orbiter (a European Space Agency-NASA partnership, 2017 launch).* Solar Orbiter will investigate links between the solar surface, corona, and inner heliosphere from as close as 62 solar radii (i.e., closer to the Sun than Mercury's nearest approach).
- *Solar Probe Plus (SPP; LWS program; 2018 launch).* Solar Probe Plus will make mankind's first visit to the solar corona to discover how the corona is heated, how the solar wind is accelerated, and how the Sun accelerates particles to high energy.

The powerful fleet of space missions that explore our local cosmos will be significantly strengthened with the addition of these missions. However, their implementation as well as the rest of the baseline program will consume nearly all of the resources anticipated to be available for new starts within NASA's Heliophysics Division through the midpoint of the overall survey period, 2013-2022.

For NSF, the previous decade has seen the initial deployment of the Advanced Modular Incoherent Scatter Radar (AMISR) in Alaska, a modular, mobile radar facility that is being used for studies of the upper atmosphere and space weather events, and the initial development of the Advanced Technology Solar Telescope (ATST), a 4-meter-aperture optical solar telescope—by far the largest in the world—that will provide the most highly resolved measurements ever obtained of the Sun's plasma and magnetic field. These new NSF facilities join a broad range of existing ground-based assets (see Figure 1.2) that provide an essential global synoptic perspective and complement space-based measurements of the solar and space physics system. With adequate science and operations support, they will enable frontier research, even as they add to the long-term record necessary for analyzing space climate over solar cycles.

The success of these activities at NASA and NSF is fundamentally important to long-term scientific progress in solar and space physics. The survey committee concluded that, with prudent management and careful cost-containment, support for and completion of the ongoing program constitute precisely the right first step for the next decadal interval and as such represent the baseline priority.

First Research Recommendation [R1.0], for NASA, NSF, and Other Agencies—Implement the DRIVE Initiative

The survey committee recommends implementation of a new, integrated, multiagency initiative (DRIVE—Diversify, Realize, Integrate, Venture, Educate) that will develop more fully and employ more effectively the many experimental and theoretical assets at NASA, NSF, and other agencies.

Relatively low-cost activities that maximize the science return of ongoing projects and enable new ones are both essential and cost-effective. However, too often recommendations regarding such activities are relegated to background status or referred to in general terms that are difficult to implement. With this in mind, the survey committee raises as its highest new priority for both NASA and NSF the implementation of an integrated, multiagency initiative (DRIVE; see Figure 4.1) that strengthens existing programs and develops critical new capabilities to address the complex science issues that confront the field. DRIVE is an initiative unified not by a central management structure, but rather through a comprehensive set of multiagency recommendations that will facilitate scientific discovery.

This integrative approach is motivated by a sea-change in the way breakthrough science is done. Innovative science is often about breaking down disciplinary boundaries, and nowhere is this more evident than in solar and space physics where, increasingly, a deep understanding of multiply connected physical systems is required to make significant progress. Such system science requires new types and configurations of observations, as well as a new cadre of researchers who can cross disciplinary boundaries seamlessly

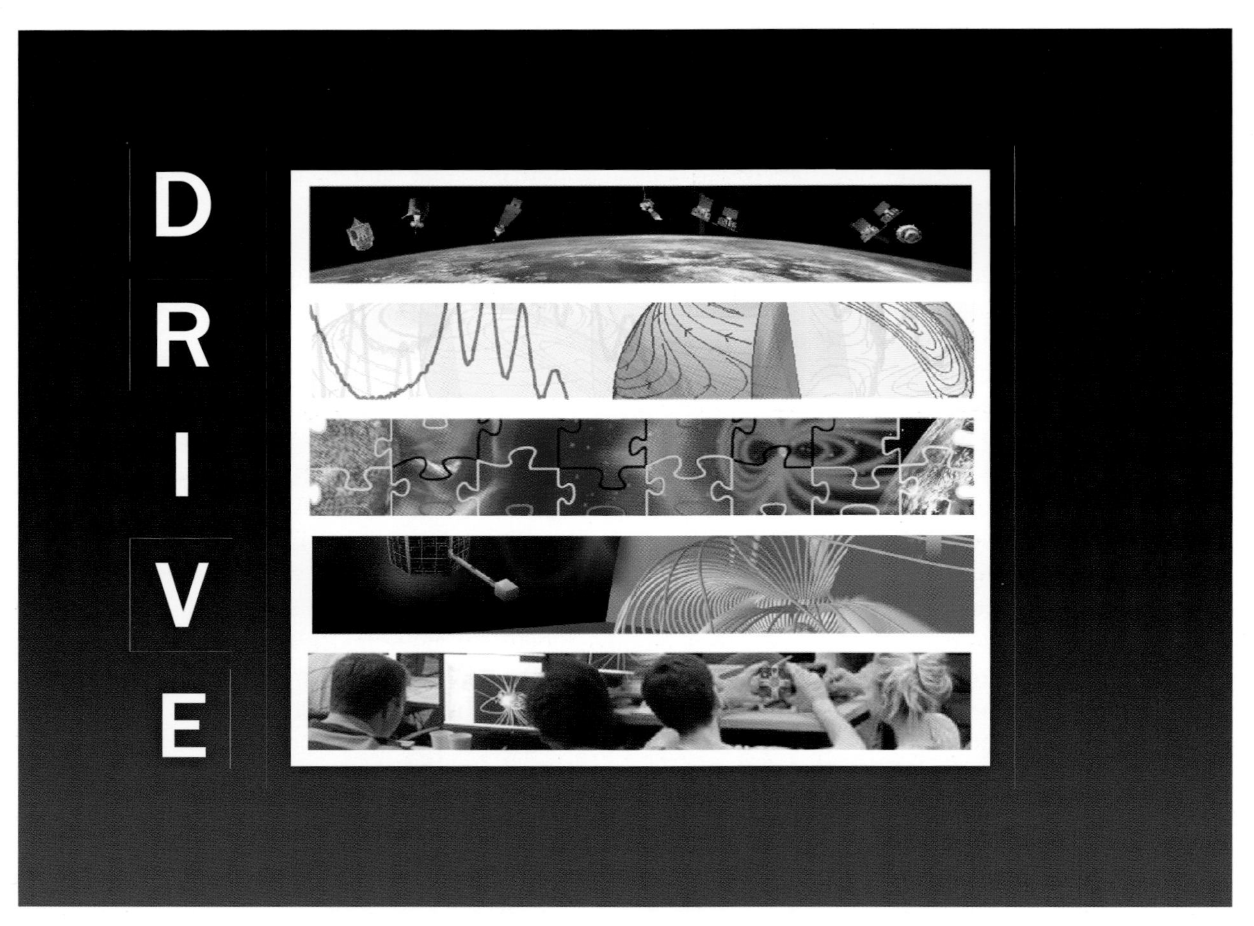

FIGURE 4.1 A relatively small, low-cost initiative, DRIVE provides high leverage to current and future space science research investments with a diverse set of science-enabling capabilities. The five DRIVE components are as follows:

- **D**iversify observing platforms with microsatellites and midscale ground-based assets.
- **R**ealize scientific potential by sufficiently funding operations and data analysis.
- **I**ntegrate observing platforms and strengthen ties between agency disciplines.
- **V**enture forward with science centers and instrument and technology development.
- **E**ducate, empower, and inspire the next generation of space researchers.

and develop theoretical and computational models that extract the essential physics from measurements made across multiple observing platforms.

The survey committee concluded that a successful solar and space physics scientific program over the next decade is one that balances spaceflight missions of various sizes with supporting programs and infrastructure investments. The goal for the next decade is to:

- Aggressively pursue innovative technological and theoretical advances,
- Build tools for the research community that enable new breakthroughs, and
- Implement an exciting program that addresses key science opportunities while being mindful of fiscal and other constraints.

The five DRIVE components are defined in Figure 4.1, and specific, actionable sub-recommendations for each of these components are presented below. In recommending the DRIVE initiative, the survey committee is cognizant that in a constrained budget environment funding for this program, while modest, will come at the expense of NASA missions. In the case of the NASA DRIVE sub-recommendations, the survey committee has therefore provided explicit costing to ensure that the initiative, along with the other program recommendations, fits within the projected NASA budget envelope (Chapter 6). For NSF, the survey committee provides a more general discussion of expected costs, here and in the NSF program implementation discussion in Chapter 5. The survey committee views the implementation of the DRIVE initiative as crucial to accomplishing the proposed program of research in solar and space physics over the next decade.

Diversify: Diversify Observing Platforms with Microsatellites and Midscale Ground-Based Assets

Exploration of the complex heliospheric system in the next decade requires the strategic use of diverse assets that range from large missions and facilities, through Explorers and mid-size projects, to small CubeSats and suborbital flights (Figure 4.2). The field is entering an era of opportunities for multipoint and multiscale[3] measurements made with an increasingly diverse set of platforms and technologies (rockets, balloons, CubeSats, arrays, commercial and international launchers and satellites, and so on). For more information, see Appendix C, "Toward a Diversified, Distributed Sensor Deployment Strategy." As part of the DRIVE initiative, the survey committee particularly urges that NASA and NSF develop ongoing small flight opportunities and midscale ground-based projects. Such platforms enable the direct engagement and training of a new generation of experimentalists who gain end-to-end experiences ranging from concept formation to project execution.

Midscale Projects

The current NSF equipment and facilities program supports investments in both small and very large facilities. NSF maintains a major research instrumentation program for instrument development projects (less than $4 million per year), and the Major Research Equipment and Facilities Construction (MREFC) program for large infrastructure projects (greater than ~$90 million per year for Atmospheric and Geospace Sciences Division-sponsored projects). However, this program does not cover midscale projects, many of which have been identified by the survey committee as cost-effective additions of high priority to the overall program. The addition of a midscale funding capability could enable solar and space physics projects with well-developed science and implementation plans. Examples include the proposed Frequency-Agile Solar Radio (FASR) and Coronal Solar Magnetism Observatory (COSMO) telescopes, as well as next-generation ATST instrumentation and other projects that are not yet well developed but are representative of the kind of creative approaches that will be necessary for filling gaps in observational capabilities and for moving the survey's integrated science plan forward (see "Candidates for a Midscale Line" in Chapter 5). A midscale funding line would also have a major impact on existing ground-based facilities, because it would rejuvenate broadly utilized assets by taking advantage of new innovations and addressing modern science opportunities. Finally, it would include essential support for accompanying research—a key requirement for maximizing scientific benefit. This is consistent with the emphasis in the 2010 astronomy and astrophysics decadal survey, which recommended a midscale line as its second priority in large ground-based projects.[4]

---

[3]Multiscale measurements involve the study of plasma dynamics, which involves the interaction of distinct domains with a large range of spatial scales.

[4]National Research Council, *New Worlds, New Horizons in Astronomy and Astrophysics*, The National Academies Press, Washington, D.C., 2010.

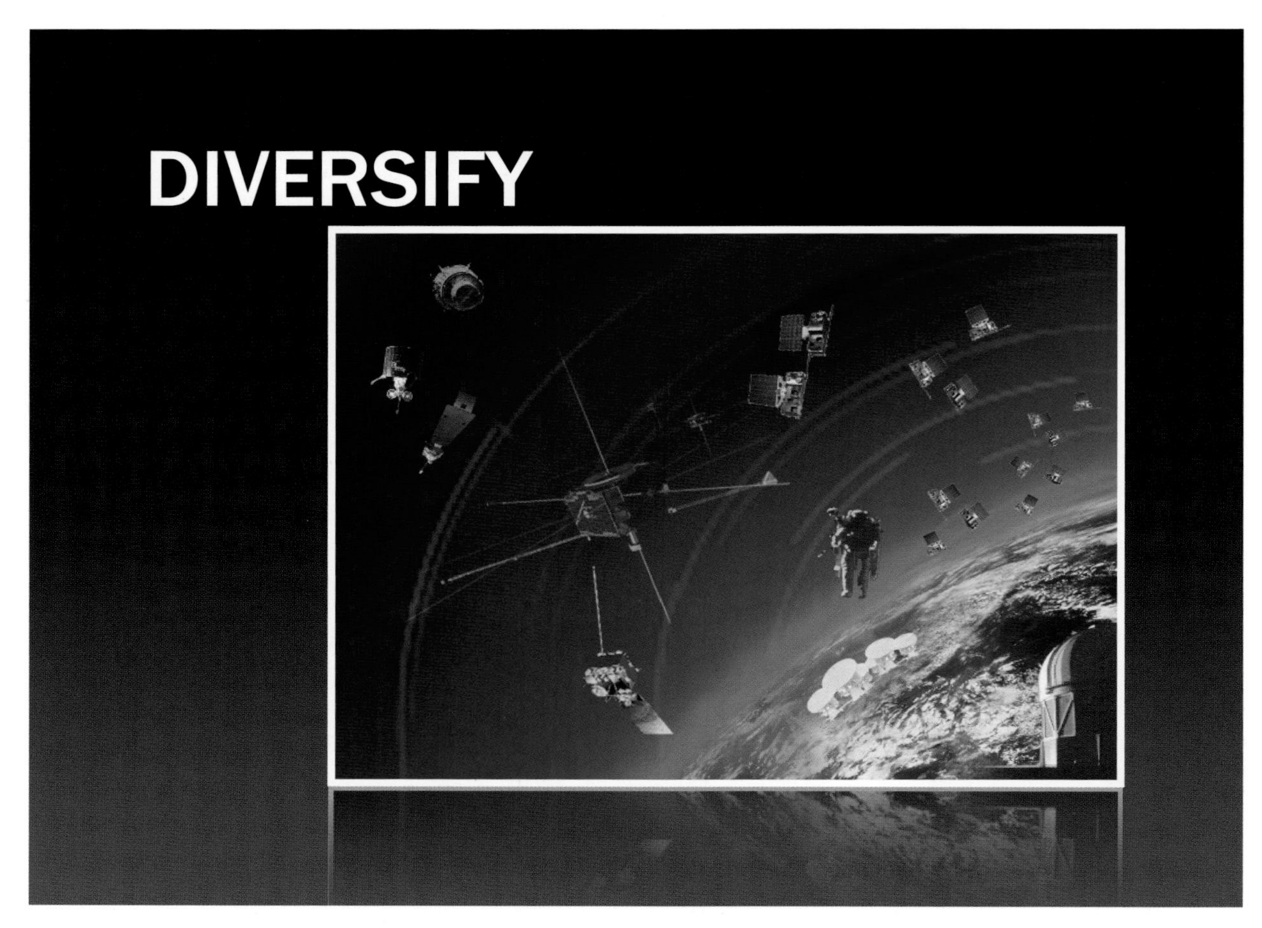

FIGURE 4.2 *D*iversify observing platforms with microsatellites and midscale ground-based assets. A broad range of assets of multiple sizes, both ground- and space-based, are needed to span the heliophysics system.

Recommendation: The National Science Foundation should create a new, competitively selected midscale project funding line in order to enable midscale projects and instrumentation for large projects.

Tiny Satellites

Since the 2003 solar and space physics decadal survey, a new experimental capability has emerged for very small spacecraft, which can act as stand-alone measurement platforms or be integrated into a greater whole. These platforms are enabled by innovations in miniature, low-power, highly integrated electronics and nanoscale manufacturing techniques, and they provide potentially revolutionary approaches to experimental space science. For example, small, low-cost satellites may be deployed into regions where satellite lifetimes are short but where important yet poorly characterized interactions take place. Operation of miniaturized avionics and instrumentation in high radiation environments both spurs technological development and provides valuable knowledge of space weather.

Experiments on very small spacecraft have an important educational impact, as well. A survey of solar and space physics graduate students at NSF's GEM, SHINE, and CEDAR workshops indicated that the opportunity to work on projects that could produce real results within the time frame of a graduate thesis was a great attraction to the field (see Appendix D, "Education and Workforce Issues in Solar and Space Physics").

NSF's CubeSat initiative[5] promotes science done by very small satellites and provides prime educational opportunities for young experimenters and engineers. The education and training value of these programs has been strongly recognized by the university research community, an endorsement that is itself an argument for an increased launch cadence (more than one per year). As CubeSat grows, it is critical to develop best-in-class educational projects and track the impacts of investments in these potentially game-changing assets.

Recommendation: NSF's CubeSat program should be augmented to enable at least two new starts per year. Detailed metrics should be maintained, documenting the accomplishments of the program in terms of training, research, technology development, and contributions to space weather forecasting.

NASA's Low-Cost Access to Space (LCAS) program supports suborbital science missions, and it also provides a unique avenue for graduate-student training and technology development. An increase in the present cadence of sounding rocket investigations and an augmentation by a tiny satellites program, complementary to NSF's CubeSats, will strengthen LCAS and capitalize on new capabilities.

Recommendation: A NASA tiny-satellite grants program should be implemented, augmenting the current Low-Cost Access to Space (LCAS) program, to enable a broadened set of observations, technology development, and student training. Sounding rocket, balloon, and tiny-satellite experiments should be managed and funded at a level to enable a combined new-start rate of at least six per year, requiring the addition of $9 million per year (plus an increase for inflation) to the current LCAS new-start budget of $4 million per year for all of solar and space physics.

Realize: Realize Scientific Potential by Sufficiently Funding Operations and Data Analysis

The value of a mission or ground-based investigation is fully realized, and science goals achieved, only if the right measurements are performed over the mission's lifetime and new data are analyzed fully (Figure 4.3). Realizing the full scientific potential of solar and space physics assets therefore requires investment in their continuing operation and in effective exploitation of data (Box 4.1). Furthermore, a successful investigation should also include a focused data analysis program (Box 4.2) that supports science goals that might span platforms or might change throughout a mission. The following program augmentations expand the potential for new discoveries from data.

Advanced Technology Space Telescope (ATST) Operations

Starting in 2018, NSF's ATST will provide the most highly resolved measurements of the Sun's plasma and magnetic field ever obtained. The ATST is currently under construction, funded as a large project by NSF's MREFC program. To fully realize this investment, funding for its operations and for data analysis has to be identified. In particular, ATST requires adequate, sustained funding from NSF for operation, data processing and analysis, development of advanced instrumentation, and research grant support for ATST users. The National Solar Observatory (NSO) FY 2001-2015 long-range budget estimate for annual ATST operations and data services is approximately $18 million. This amount, in addition to a required $4 mil-

[5]Formally known as the CubeSat-based Science Missions for Space Weather and Atmospheric Research.

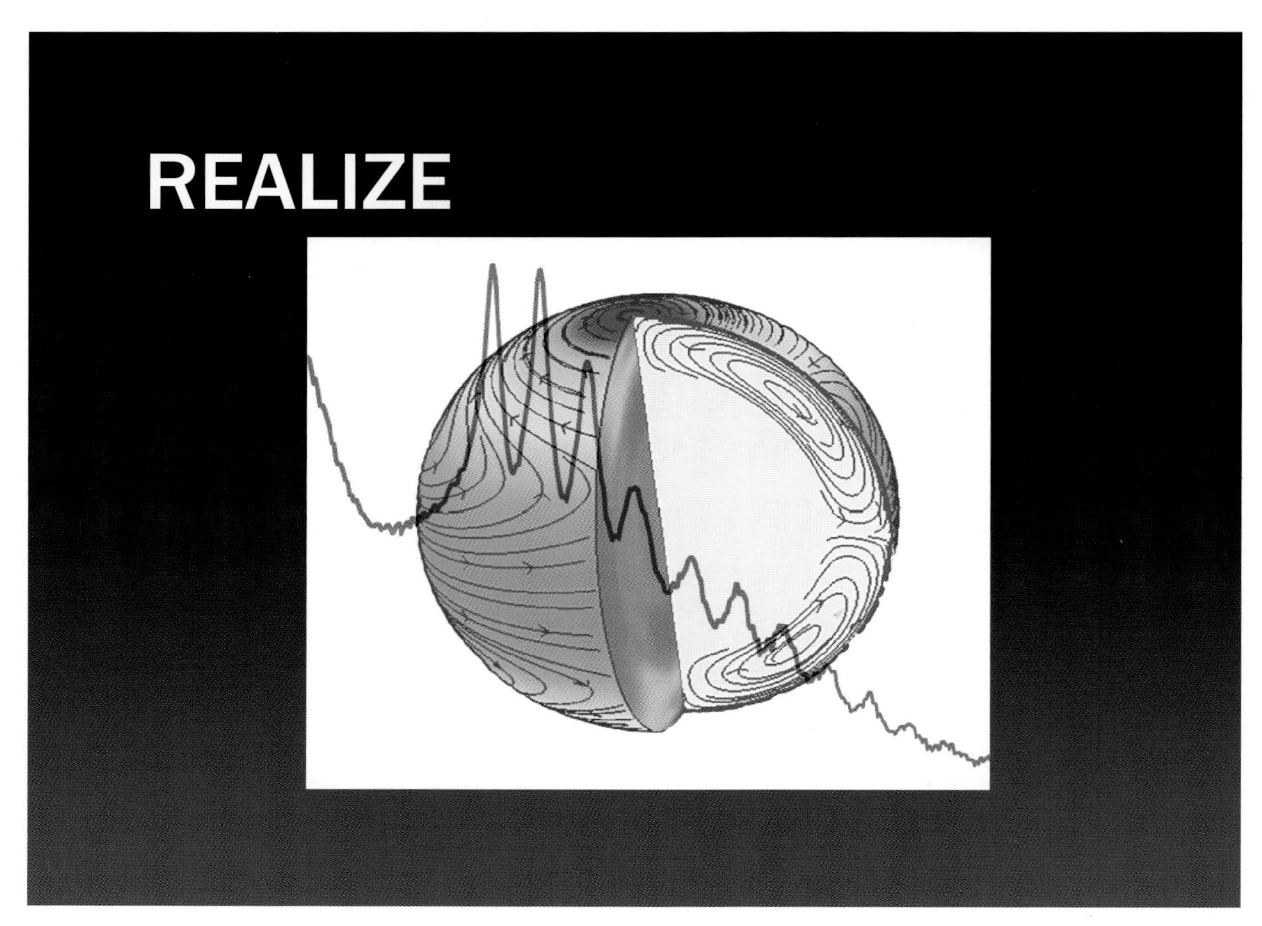

FIGURE 4.3 Realize scientific potential by sufficiently funding operations and data analysis. As an example of how extensive data analysis leads to physical understanding, this figure shows a power spectrum of solar surface oscillations (black line) that can be analyzed via sophisticated helioseismic inversion to determine the nature of plasma flows inside the Sun (two-dimensional color image).

lion per year for NSO synoptic programs, will require a budget augmentation of approximately $12 million above the current NSO operating budget. Research grants and advanced instrumentation development will require additional funds.

Recommendation: NSF should provide funding sufficient for essential synoptic observations and for efficient and scientifically productive operation of the Advanced Technology Solar Telescope (ATST), which provides a revolutionary new window on the solar magnetic atmosphere.

Mission Operations and Data Analysis (MO&DA)

The very successful Heliophysics Systems Observatory is fueled by NASA MO&DA mission extensions and guest investigator (GI) programs, which ensure that the full range of essential data are collected, maintained, and analyzed. The survey committee concluded that a higher level of MO&DA funding is

BOX 4.1 DATA EXPLOITATION

Significant progress has been made over the past decade in establishing the essential components of the solar and space physics data environment. However, to achieve key national research and applications goals requires a data environment that draws together new and archived satellite and ground-based solar and space physics data sets and computational results from the research and operations communities. As discussed in more detail in Appendix B, such an environment would include:

- Coordinated development of a data systems infrastructure that includes data systems software, data analysis tools, and training of personnel;
- Community oversight of emerging, integrated data systems and interagency coordination of data policies;
- Exploitation of emerging information technologies without investment in their initial development;
- Virtual observatories as a specific component of the solar and space physics research-supporting infrastructure, rather than as a direct competitor for research funds;
- Community-based development of software tools, including tools for data mining and assimilation; and
- Semantic technologies to enable cross-discipline data access.

required to exploit the opportunities created by the HSO, especially considering the importance of broad and extended data sets for exploring space weather and space climatology. Moreover, the survey committee concluded that annual competition for support via a stable general GI program is essential. The general GI program is the only funding source for research utilizing data from missions beyond their prime mission phase. Thus, for example, when the data-rich SDO mission finishes its prime phase in 2015, the enormous research potential of this mission should be supported by the general GI program.

Recommendation: NASA should permanently augment MO&DA support by $10 million per year plus annual increases for inflation, in order to take advantage of new opportunities yielded by the increasingly rich Heliophysics Systems Observatory assets and data.

BOX 4.2 NASA FUNDING FOR MISSION DATA ANALYSIS

The hard decisions that have to be made in response to budget limitations have too often resulted in cuts in the mission operations and data analysis (MO&DA) and guest investigator (GI) portions of missions, compromising the scientific potential of the mission. Example 1: The general GI program (supported from MO&DA funds) was completely cut in 2010, with impacts that cascaded through the other Research and Analysis (R&A) programs, causing the Solar and Heliospheric Supporting Research and Technology (SR&T) program to be oversubscribed by a factor of 6.7 (40 percent higher than the previous year). As a result, 36 percent of top-ranked (selectable) proposals were not funded. Example 2: Limited Phase-E funding for data analysis was included in the original mission funding profile of the Solar Dynamics Observatory (SDO). Even less was allocated to the instrument teams after the first 2 years because science funding was planned to come from a mission-specific GI program. The GI program was subsequently scaled back because of budget pressures and little was available in the first year for data analysis for the SDO, a mission that cost around $850 million to build and launch and that is producing unprecedented volumes of data.

Mission Guest Investigators

In addition to MO&DA, a vibrant NASA mission-specific GI program helps ensure mission success by broadening participation and facilitating new discoveries. Future funding should keep pace with new missions.

Recommendation: A directed guest investigator program, set at a percentage (~2 percent) of the total future NASA mission cost, should be established in order to maximize each mission's science return. Further, just as an instrument descoping would require an evaluation of impact on mission science goals, so, too, should the consequences of a reduction in mission-specific guest investigator programs and Phase-E funding merit an equally stringent evaluation.

Integrate: Integrate Observing Platforms and Strengthen Ties Between Agency Disciplines

The frontiers of current solar and space physics research are at the interfaces between traditional disciplines (Figure 4.4). Moreover, increasingly complex observational requirements (multipoint and multiscale) are becoming tractable thanks to platform diversification and other technological advances (see Appendix C). In particular, recent efforts such as the Whole Heliosphere Interval have demonstrated the effectiveness of coordinating multiscale instruments and a multidisciplinary analysis approach to yield cutting-edge science at modest cost (Box 4.3). Similarly, solar and space physics can benefit science investigations in related fields, such as planetary science, Earth science, physics, and astrophysics, and can also benefit from novel experimental (e.g., laboratory) and theoretical tools that provide end-to-end integration.

Dedicated Laboratory Experiments

Laboratory studies probe fundamental plasma physical processes and produce chemical and spectroscopic measurements that support satellite measurements and atmospheric models. They provide benchmarks for integrating theory and modeling with observation in solar and space physics (Box 4.4). Such laboratory experiments should be funded in a multiagency fashion.

Recommendation: NASA should join with NSF and DOE in a multiagency program on laboratory plasma astrophysics and spectroscopy, with an expected NASA contribution ramping from $2 million per year (plus increases for inflation), in order to obtain unique insights into fundamental physical processes.

Multidisciplinary Science

Solar and space physics is a multidisciplinary science, in terms of both the range of topics within its subfields and its interfaces with physics, chemistry, astronomy, and planetary and Earth science. Understanding how solar structure, evolution, and dynamics relate to those of other stars, how the heliosphere relates to astrospheres, and how Earth's magnetosphere compares to those of other planets illustrates why research in each of the subdisciplines of solar, planetary, stellar, galactic, and heliospheric physics benefits the others. Solar-irradiance variations and other indirect forcings of Earth, such as top-down coupling of solar ultraviolet heating and photochemistry in the stratosphere and the possible influence of galactic cosmic rays, affect global climate changes and also regional changes due to subtle shifts in the atmosphere, ocean circulation patterns, and cloud cover. Breakthrough science can arise at the interfaces of these disciplines.

The multidisciplinary nature of solar and space physics is reflected in its placement within multiple divisions and directorates at NSF. However, the survey committee concluded that this organizational structure may be limiting in particular for science lying at the interfaces between solar and space physics and research in Earth science and astrophysics. A key example of the consequences of the current NSF

FIGURE 4.4 *I*ntegrate observing platforms and strengthen ties between agency disciplines. Diverse space- and ground-based assets have to be routinely combined to maximize their multiscale potential for understanding the solar and space physics system as a whole. Likewise, developing connections between field and related scientific disciplines strengthens insight into shared, fundamental physical processes.

arrangement concerns the ongoing exploration of the boundary regions of the heliosphere—currently one of the most fruitful and exciting areas of research, but one that has very little focus at NSF.

Recommendation: NSF should ensure that funding is available for basic research in subjects that fall between sections, divisions, and directorates, such as planetary magnetospheres and ionospheres, the Sun as a star, and the outer heliosphere. In particular, research on the outer heliosphere should be included explicitly in the scope of research supported by the Atmospheric and Geospace Sciences Division at NSF.

Solar and space physics has a clear home in NASA's Heliophysics Division. However, there remain important scientific links between the Heliophysics Division and the Astrophysics, Planetary Sciences, and Earth Sciences divisions. The survey committee concluded that multidisciplinary collaborations between the solar, heliosphere, Earth science, and climate change communities are valuable, and it recognizes in

BOX 4.3 WHOLE HELIOSPHERE INTERVAL

The Whole Heliosphere Interval (WHI) was an international observing and modeling effort to characterize the three-dimensional, interconnected, heliophysical system, utilizing dozens of space- and ground-based solar, heliospheric, geospace, and upper-atmosphere observatories and instruments. WHI was the largest period of focus of the International Heliophysical Year (IHY) (2007-2008), which was inspired by the 50th anniversary of the International Geophysical Year (IGY) (in 1957-1958) and the subsequent 50 years of space exploration. The goal of studying the structure and dynamics originating from one solar rotation (March 20-April 16, 2008) was to describe a complete narrative from the Sun to Earth and beyond at solar minimum. A broad range of data analysis and modeling was achieved, with multidisciplinary and international collaborations aided by a central website (http://ihy2007.org/WHI) and WHI special sessions at international meetings. Many papers relating to the WHI period have been published to date, including 27 that made up a 2011 topical issue of *Solar Physics:* "The Sun–Earth Connection near Solar Minimum."[1]

[1] M.M. Bisi, B.A. Emery, and B.J. Thompson, eds., "The Sun–Earth Connection near Solar Minimum," *Solar Physics*, Volume 274, 2011.

particular the importance of collaborations between the NASA Heliophysics and Earth Science programs. Similarly, the survey committee endorses collaborations across the Heliophysics, Astrophysics, and Planetary Sciences divisions.

Coordinated Observations

Data from diverse space- and ground-based instruments need to be routinely combined in order to maximize their multiscale potential. In fact, such coordinated investigations are likely to be a crucial ele-

BOX 4.4 LABORATORY EXPERIMENTS RELEVANT TO HELIOPHYSICS

Some important problems in solar and space physics will always be difficult to solve from spacecraft observations alone, where remote sensing introduces observational biases and in situ measurements are limited to a small number of trajectories in a complex, time-variable environment. In contrast, dedicated laboratory experiments offer the advantage of a controlled environment where detailed reproducible measurements are possible.

An example of a problem on which laboratory experiments have had a significant science impact is magnetic reconnection. The transition between resistive magnetohydrodynamic and kinetic regimes is one of the most fundamental issues in the study of magnetic reconnection. This transition has important implications for the solar atmosphere and the magnetosphere, but it cannot be tested with direct satellite measurements because plasmas in the magnetosphere and solar wind are nearly collisionless. Laboratory experiments over the past decade have provided confirmation of the collisional-to-kinetic transition, allowing researchers to more confidently predict the reconnection dynamics in various regions of the solar atmosphere and magnetosphere.

Although most laboratory experiments are directed toward understanding basic plasma physics issues, there are also important experiments whose results are used directly to facilitate the interpretation of satellite observations, e.g., spectroscopy measurements of molecules and highly charged ions and modeling of solar wind interactions with airless bodies and dusty plasmas. The measurements of the cross sections associated with ionization, charge exchange, and direct and dielectronic recombination are ongoing and provide key input to the interpretation of satellite spectral measurements and the benchmarking of models.

ment of future breakthrough science and to provide new pathways for translating scientific knowledge into societal value. The idea of coordinating multiscale observations resonates both with the types of system-science questions identified by the survey's disciplinary panels and with the heliophysics science centers described in the next section ("Venture"). Examples might include extending "World Day" coordination of NSF radars to other ground-based and mission data collections, combining data from CubeSat arrays and larger spacecraft, GPS receiver hosting, development of distributed arrays of ground-based instruments (potentially funded by an NSF midscale program), and ground-based and space mission solar observational support for the ATST (see Appendix C).

Recommendation: NASA, NSF, and other agencies should coordinate ground- and space-based solar-terrestrial observational and technology programs and expand efforts to take advantage of the synergy gained by multiscale observations.

Venture: Venture Forward with Science Centers and Instrument and Technology Development

The future of solar and space physics depends on the ability to venture into transformative technologies and capabilities (Figure 4.5). Development of new and innovative means of pursuing grand challenge science is also critical, taking full advantage of the progress that comes from collaborations between theorists, modelers, computer scientists, and observers (Box 4.5). In fact, new mechanisms are required to facilitate development at the frontiers of science and technology.

Grand Challenge Research

The survey committee concluded that a mechanism is needed for bringing together critically sized teams of observers, theorists, modelers, and computer scientists to address the most challenging problems in solar and space physics. The scope of theory and modeling investigations supported by the NSF CEDAR, GEM, and SHINE programs or the NASA Supporting Research and Technology (SR&T), Targeted Research and Technology (TR&T), and GI programs should be expanded so as to enable deep and transformative science. The survey committee's proposed heliophysics science centers would bring scientists together for significant collaborations to address the most pressing scientific issues of heliophysics. Centers should consist of multidisciplinary teams with two to three primary institutions that include theorists, modelers, algorithm developers, and observers. Resources should be focused on the core institutions to avoid spreading the resources too broadly and to achieve a focused investigation of the topic. The centers should be designed to highlight the exciting science problems of the field, to bolster the interest of faculty at universities, and to attract top students into the field. Success would be measured according to the progress the centers make in addressing these science problems.

Recommendation: NASA and NSF together should create heliophysics science centers to tackle the key science problems of solar and space physics that require multidisciplinary teams of theorists, observers, modelers, and computer scientists, with annual funding in the range of $1 million to $3 million for each center for 6 years, requiring NASA funds ramping to $8 million per year (plus increases for inflation).

NASA's Heliophysics Theory Program (HTP) nurtures the formation of small groups of theorists to collaborate intensively on larger targeted projects than are possible using SR&T and TR&T program funding. The survey committee concluded that the medium-size HTP provides an essential bridge between small grants and the heliophysics science center grand challenge investigations, which by their very nature will be limited to a small handful of topics at any given time. The survey committee endorses the continuation

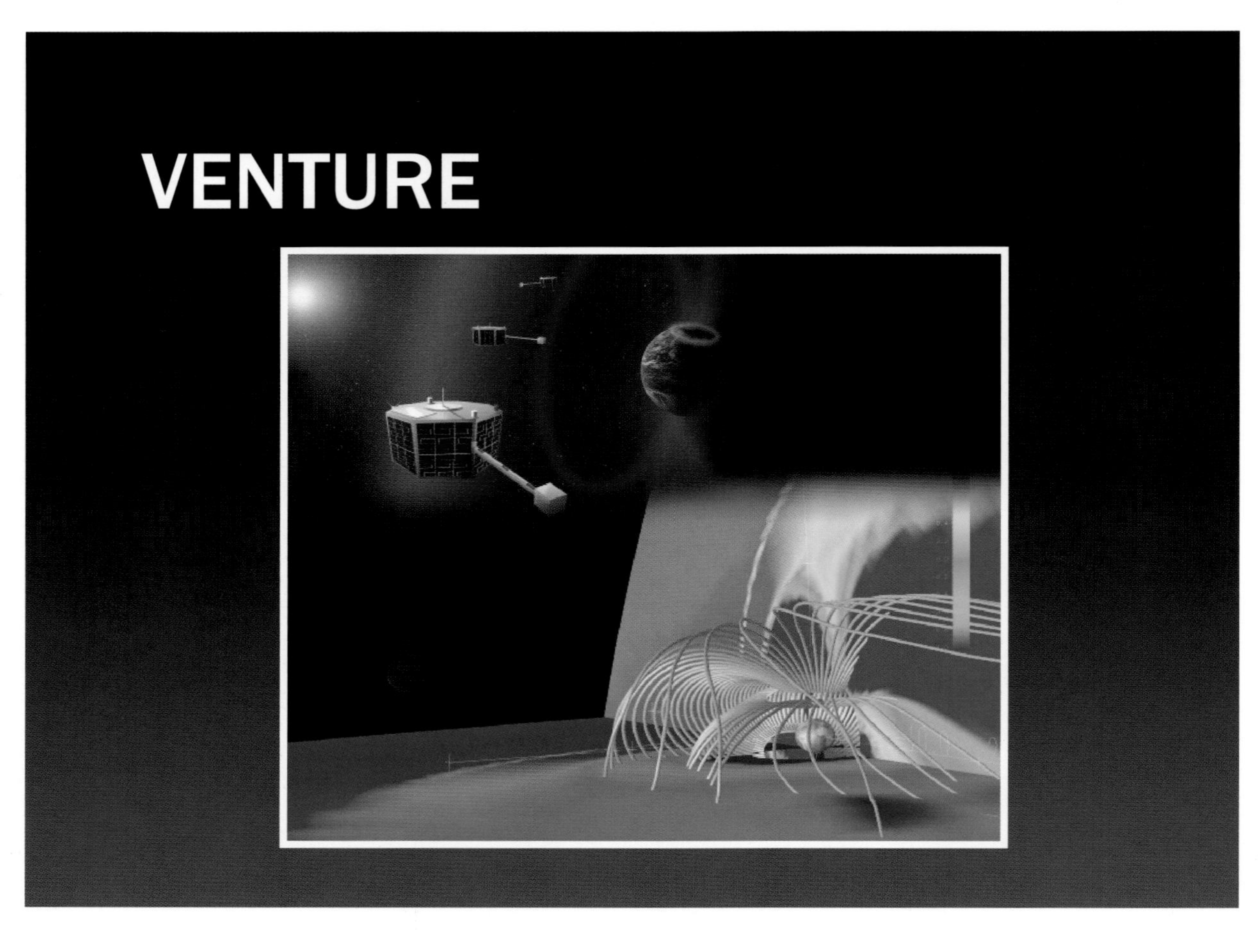

FIGURE 4.5 *V*enture forward with science centers and instrument and technology development. Transformation will come from innovation, both in theory and technology. Shown are examples of state-of-the-art models developed via a coordinated science center (CISM), and an artist's conception of birthday-cake-size microsatellites (NASA).

of the HTP at current funding levels. The committee further concluded that, as the heliophysics science centers are implemented it may be more effective to reduce the total number of HTP awards but increase their average size to the range of ~$400,000 to $600,000 per year.

NASA Instrumentation and Technology

A 2010 National Research Council (NRC) study, *An Enabling Foundation for NASA's Earth and Space Science Missions*,[6] discussed the importance of advanced technology development in all of the science areas of NASA's Science Mission Directorate, and it recommended that instrument and mission technology activities be managed strategically so as to maximize the opportunities to meet each division's strategic

[6]National Research Council, *An Enabling Foundation for NASA's Space and Earth Science Missions*, The National Academies Press, Washington, D.C, 2010.

BOX 4.5 A NEW WAY OF DOING SCIENCE

In Chapter 2, the survey committee discusses the key science challenges for solar and space physics. Embedded in these "grand challenges" are complex questions whose full resolution has remained elusive. Work on the challenges has traditionally been informed by research groups that work mostly independently and employ either observational or theory and modeling-based approaches. Increasingly, major advances in the field are taking place as a result of the close interaction between observers, theorists, and modelers. Thus, a coherent attack on the most challenging problems requires the development of research and analysis (R&A) programs that bring together multidisciplinary teams with a broad range of skills. The heliophysics science centers will facilitate the formation of such diverse teams.

Over the past decade, the ongoing exponential increase in computing power had a significant impact on the process of science discovery—modeling of complex plasma phenomena is now carried out on massively parallel computers and can address physical phenomena on a broad range of spatial and temporal scales. To capitalize on advances in computational architectures and machines, it has become necessary to collaborate in critical-size groups with experts in computer science, algorithm development, and large-scale visualization and analysis tools. At the same time, observations establish ground truth for emerging models. Through these synergies, physical insight can be achieved beyond what is possible with paper-and-pencil models, stimulating new ideas to explore with analytic theory, influencing the interpretation of observations, and motivating the need for new missions.

The level funding of R&A over the past decade, and the subsequent loss of buying power due to inflation, have resulted in increasingly fragmented science, given that individual researchers must rely on multiple proposals to secure adequate funding. This trend toward piecemeal support is happening at a time when advancing the science requires collaboration—but when funding multiple scientists on a single grant at any meaningful level is almost impossible. The formation of several heliospheric science centers will reverse this trend.

goals. Such an approach would more readily enable highly desirable missions that have been deferred to a later decade owing to as-yet immature technology or high cost.

Technologies such as solar sails and constellations of satellites have tremendous potential (see Appendix B, "Instrumentation, Data Systems, and Technology"). Missions reliant on such technologies are not yet feasible, in part because of the constrained budget environment, but also because of a low level of technical readiness. Future progress in solar and space physics hinges on new observational capabilities in state-of-the-art instrumentation, access to unique locations in space, and affordable fabrication and operation of large satellite constellations.

Some of the DRIVE components already discussed for NSF would promote technology development, i.e., CubeSats and a midscale project line. At NASA, current technology development is funded by the SR&T program, Living With a Star (LWS), and LCAS. The survey committee concluded that technologies required for novel mission design and instrumentation need a more coherent and better-funded NASA program than is currently available, one that would emulate the Planetary Instrument and Development Program.

Recommendation: NASA should consolidate the technology funding now in the SR&T, LWS, and LCAS programs into a single heliophysics instrument and technology development program and increase current annual funding levels, ramping to $4 million per year (plus increases for inflation) in order to facilitate urgently needed innovations required for implementation of future heliophysics mission. Further, issues pertaining to implementation of constellation missions (e.g., communications, operations, propulsion, and launch mechanisms) should be explicitly addressed.

Educate: Educate, Empower, and Inspire the Next Generation of Space Researchers

Solar and space physics is a field with global consequences that are both intellectually stimulating and relevant to society. Heliophysics programs empower young scientists and engineers to perform analytical thinking on real science and technology problems, forging a multitalented, creative workforce for the future of the United States. Through education and public outreach, the general public is inspired by the sheer beauty of dynamic solar events and the response seen at Earth in visually stunning auroral forms. However, it is critical to ensure sufficient resources for education and training, and to develop the skills necessary for the next generation of space researchers and for technologically literate workers in many other fields (Figure 4.6).

An analysis conducted under the aegis of the decadal survey (see Appendix D) indicates that improvement can be made in several areas: employment opportunities, education and training, and recruitment and public outreach.

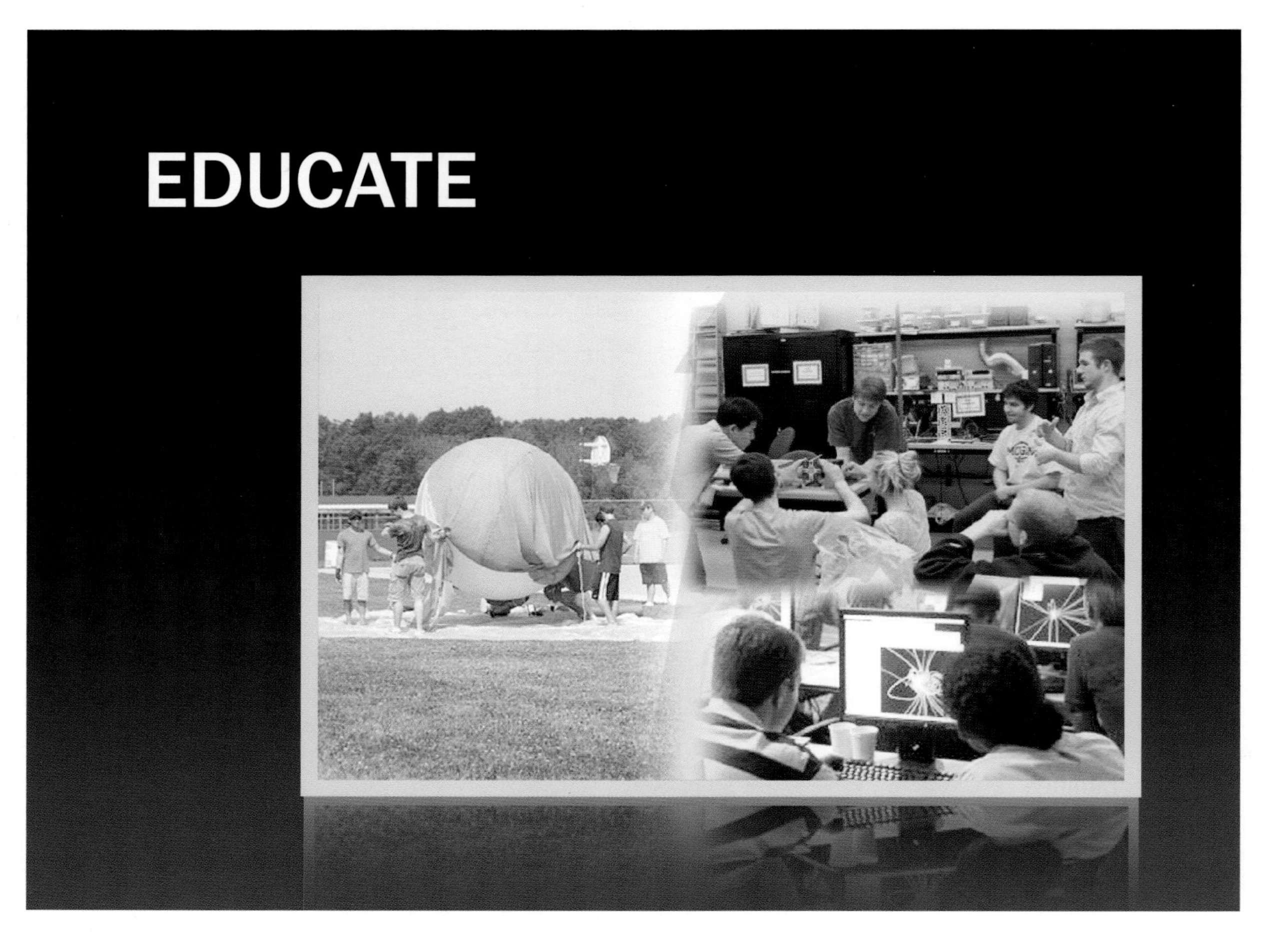

FIGURE 4.6 Educate, empower, and inspire the next generation of space researchers.

Employment Opportunities

Appendix D, "Education and Workforce Issues in Solar and Space Physics," shows that while the Ph.D. production rate for solar and space physics has increased over the past decade, the number of advertised positions in the field, inside and outside academia, has decreased. Indeed, the number of advertised faculty positions reached a decadal low in the last year surveyed, 2010. Although historically many solar and space physics graduates find jobs in areas other than academic research, these oppositely directed trends, of increasing numbers of students being trained versus decreasing hires of faculty to train them, indicate the continued importance of the NSF Faculty Development in the Space Sciences (FDSS) program. This program grew out of a recommendation by the 2003 decadal survey and led to the creation of eight tenure-track faculty positions, most of the holders of which have already become tenured. It is widely viewed, along with NSF CAREER awards, as an exemplary means of sustaining space physics within universities and promoting the science and engineering workforce. To increase the reach of this program and the diversity of students exposed to opportunities in solar and space physics, eligibility for these awards could be expanded to include 4-year institutions, not just Ph.D.-granting research universities. Because many universities have only one faculty member in solar and space physics, curriculum and other educational resources need to be strengthened and shared throughout the United States.

The survey committee concluded that programs supporting solar and space physics faculty and curriculum development are required to maintain a healthy presence in universities and to provide community-wide educational resources. The survey committee endorses ongoing curriculum development efforts and those of NASA's Heliophysics Education Forum. There is a need to further increase the diversity of the solar and space physics community through active encouragement and inclusion of educational institutions that can serve as conduits into underrepresented communities.

Recommendation: The NSF Faculty Development in the Space Sciences (FDSS) program should be continued and be considered open to applications from 4-year as well as Ph.D.-granting institutions as a means to broaden and diversify the field. NSF should also support a curriculum development program to complement the FDSS program and to support its faculty.

Education and Training

Hands-on experience for students is critical to developing a competent workforce (Box 4.6). Most spaceflight programs are far too risk-averse or their duration too long for a graduate student to be directly involved over the life cycle of the program. The LCAS and CubeSat programs, recommended in the "Diversify" section above, provide opportunities for graduate programs to attract and train students in the complete mission life cycle. Complementing hardware training, NASA- and NSF-supported summer schools (summarized in Appendix D) provide training and education in the field. The NSF-supported Center for Integrated Space Weather Modeling summer school, which is coming to an end, was an excellent forum for training young space scientists in modeling and data analysis with a unique, holistic, and integrative emphasis on the entire system from the Sun to Earth.

The committee concluded that the LCAS, CubeSat, and NASA- and NSF-supported summer schools provide important hands-on training for graduate students. The committee found, in addition, that skills needed for becoming a successful scientist go beyond such formal discipline training and include interpersonal and communication skills, awareness of career opportunities, and leadership and laboratory management ability. The community endorses NASA and NSF programs that support postdoctoral and graduate student mentoring. Finally, the survey committee endorses ongoing NASA funding for solar and space physics graduate student research. For 30 years this support was provided under the auspices of the NASA Graduate Student Research Program (GSRP), but, because that program ended in 2012, the survey

BOX 4.6 A TRAINED SOLAR AND SPACE PHYSICS WORKFORCE

A key to the success of the U.S. space program as a whole, and to solar and space physics in particular, is the availability of experimentally oriented scientists and engineers who have been trained with spaceflight hardware. Yet there has been a steady erosion of that workforce, not only at NASA but also across the entire country, and this fact has been decried from many quarters. Several recent National Research Council reports[1] make this case most emphatically. Other technical industries have been able to compensate somewhat by tapping the pool of highly trained immigrants and foreign students, and they have often outsourced work abroad. But spacecraft are ITAR-sensitive items, and so this pool is not available to NASA or to its outside partners, even to universities, because of the constraints of the regulations. All of the space programs at NASA, DOE, NOAA, and DOD feel this shortage acutely. The situation is likely to deteriorate even further if no mitigating actions are implemented.

[1] National Research Council, *Revitalizing NASA's Suborbital Program: Advancing Science, Driving Innovation, and Developing a Workforce*, 2010; National Research Council, *An Enabling Foundation for NASA's Space and Earth Science Missions*, 2010; National Research Council, *NASA Space Technology Roadmaps and Priorities: Restoring NASA's Technological Edge and Paving the Way for a New Era in Space*, 2012 (all published by the National Academies Press, Washington, D.C.).

committee concluded that its replacement, NASA's Earth and Space Science Fellowship (NESSF) program, has an important role to play in maintaining solar and space physics graduate support at historic GSRP levels, and with a strong link between graduate students and NASA mission research.

Recommendation: A suitable replacement for the NSF Center for Integrated Space Weather Modeling summer school should be competitively selected, and NSF should enable opportunities for focused community workshops that directly address professional development skills for graduate students.

Recruitment and Public Outreach

Solar and space physics is for the most part taught at the graduate level, and opportunities to learn about the discipline are limited at the precollege and undergraduate level. One exception is NSF's Research Experiences for Undergraduates (REU) program. As detailed in Appendix D, "Education and Workforce Issues in Solar and Space Physics," approximately 50 percent of the graduate students surveyed at the NSF-supported summer 2011 GEM-CEDAR and SHINE meetings reported an undergraduate research experience in the solar and space physics field akin to the REU. Another successful program has been Los Alamos National Laboratory's (LANL's) post-baccalaureate program to provide recent college graduates the opportunity to explore research experiences in solar and space physics.

Recruitment starts even earlier, however. Appendix D notes that graduate students interviewed at GEM, CEDAR, and SHINE meetings often cited a childhood interest in space and astronomy that first grew through a high school physics course and subsequently into trying out astronomy research as an undergraduate. It is the goal of the community of NASA-supported, discipline-focused education and public outreach (EPO) professionals (e.g., the Heliophysics Education and Public Outreach Forum) to connect the scientists, missions, and results of the solar and space physics research community with active partners in the world of K-12 education, in order to raise the visibility of the field and to develop a diverse solar and space physics workforce.

In summary, a thriving field of solar and space physics requires continued outreach to the general public and in particular to students who will become the next generation of space scientists. The survey committee endorses the NASA requirement that 1 percent of heliophysics mission budgets be devoted to focused EPO efforts that can be evaluated for their effectiveness in enhancing the visibility of the field and in stimulating students to choose solar and space physics as a career. The survey committee found that the Heliophysics Education and Public Outreach Forum plays an important role in ensuring the effectiveness of mission EPO efforts and in informal and formal science education at all levels. The survey committee endorses programs such as NSF's REU and LANL's post-baccalaureate program, which are important recruiting tools for the field. The committee also fully supports the efforts of EPO professionals and physics educators who collaborate with scientists to develop the solar and space physics workforce as well as promote public support and interest. The committee also recognizes the importance of participation in the development of the Common Core standards and the Next Generation Science Standards. The survey committee endorses programs that specifically target enhancing diversity within solar and space physics, akin to NSF's Opportunities for Enhancing Diversity in the Geosciences (OEDG) program. The committee notes, as a final point on this topic, that solar and space physics is not currently listed as a dissertation research area in NSF's annual Survey of Earned Doctorates,[7] a report that influences other rankings, ratings, and demographic surveys such as those done by the National Research Council and the American Institute of Physics (AIP) and is, the survey committee concluded, a significant tool for recruiting students to the field.

Recommendation: To further enhance the visibility of the field, NSF should recognize solar and space physics as a specifically named subdiscipline of physics and astronomy by adding it to the list of dissertation research areas in NSF's annual Survey of Earned Doctorates.

Conclusions Regarding the DRIVE Initiative

The DRIVE initiative capitalizes on the breadth of current programs in solar and space physics and will build capabilities for the future, starting with the current HSO as the foundation. DRIVE as proposed makes the most of these existing assets while enabling advances in science and technology that will fuel progress within realistic cost envelopes. Chapters 5 and 6 discuss implementation of the DRIVE initiative for NSF and NASA, respectively; an explicit, phased budget for the DRIVE recommendations addressed to NASA is included in Chapter 6. By implementing the recommended DRIVE components, NASA and NSF can ensure that the next decade will be rich in new observations made from diverse platforms, new science harvested from missions and projects, new synergies arising between disciplines and platforms, new technologies and theories to enable and inspire future missions and projects, and talented new students to power the future workforce.

NASA Mission Lines

Much progress will be made through the DRIVE initiative, but a properly scoped program with high science return requires new observations from spaceflight missions. Missions explore new frontiers and are rightly the main activity of NASA. Outlined below are the advantages of critical reassessment and creative re-imagination of the solar and space physics mission program, along with ranked recommended science targets for both the STP and the LWS lines.

---

[7] See National Sciencen Foundation, "Survey of Earned Doctorates," updated December 6, 201, available at http://www.nsf.gov/statistics/srvydoctorates/.

Second Research Recommendation [R2.0] for NASA—Accelerate and Expand the Heliophysics Explorer Program

The medium-class (MIDEX) and small-class (SMEX) missions of the Explorer program are ideally suited to advancing heliophysics science and have a superb track record for cost-effectiveness. Since 2001, 15 Heliophysics Explorer mission proposals have received the highest category of ranking in competitive selection reviews, but only 5 have been selected for flight. Thus there is an extensive reservoir of excellent heliophysics science to be accomplished by Explorer missions.

The Explorer Program

As noted in a previous NRC report,[8] the Explorer program's strength lies in its ability to respond rapidly to new concepts and developments in science and to forge a synergistic relationship with ongoing, larger, strategic missions. The Explorer program (Box 4.7) creates a highly competitive environment in which teams led by a principal investigator (PI) rapidly capitalize on advances in technology, enabling cutting-edge science at moderate cost. Over the years, these missions have operated in different management modes, but the common feature is that a PI in partnership with the NASA Explorer Program Office is tasked to ensure the overall success of the mission and is given the authority to make critical decisions to control cost and schedule. New Explorer missions are able to pursue the cutting edge of heliophysics science, because they can make use of emerging technologies that are not available to large facility-class observatories that have longer development times and a more stringent risk posture. Since the 1990s, the decision as to which missions are selected is based on the findings of a competitive process that ensures that the science objectives and implementation approach reflect the frontiers of the field and achieve an appropriate balance between novel technical capabilities and programmatic risk.

Explorer missions provide outstanding science-per-dollar value and often are capable of achieving much more than their baseline science mission. For example, before the ACE mission, first the Interplanetary Monitoring Probes and then the International Sun-Earth Explorers provided crucial measurements of solar wind properties well beyond their design lifetimes. The IMAGE mission provided breakthrough science through 5 years of operations (3 years beyond THEMIS's original design lifetime), including the first simultaneous conjugate observations of the aurora. Two of the five-spacecraft THEMIS magnetospheric Explorer satellites are now orbiting the Moon, enabling an outstanding expansion of mission science beyond its original plan. The TRACE solar Explorer was launched in 1998 and obtained unprecedented high-resolution coronal observations for 12 years. This pattern of Explorers outperforming their as-proposed objectives has tended to be the rule rather than the exception.

In the course of developing Explorer missions, NASA has built an amazing array of capabilities for visiting hostile and exotic environments in space and for making measurements of key properties of the gases and plasma that constitute Earth's environment. These missions have become arguably more successful and visible to the public in the past decade than ever before. They are scientifically productive and often tell a story of space exploration that is pertinent to daily life. Their technical achievements and successful implementation come to fruition at costs not achievable with large flagship missions—and their findings are often true discoveries.

These achievements come from the competitive spirit that the Explorer program encourages, and the tight cost-capped implementations that are forced to carry adequate margins from the earliest phases of

[8]National Research Council, *Solar and Space Physics and Its Role in Space Exploration*, The National Academies Press, Washington, D.C., 2003, p. 36.

BOX 4.7 A BRIEF HISTORY OF THE EXPLORER PROGRAM

The Explorer program is arguably the most storied scientific spaceflight program in NASA's history. Since 1958, when the first U.S. satellite, Explorer I, discovered Earth's radiation belts, the program has produced a wealth of information about the nature of Earth's space environment and properties of the universe. UHURU, IMP, ISEE, DE, SAMPEX, ACE, TRACE, and COBE are some of the well-known missions that have yielded enormous science return for the investment, often operating for years beyond their expected lifetimes. Science from Explorer missions contributed to three of the Nobel Prizes awarded for NASA-directed space science. Many of these missions have provided critical scientific measurements beyond their design lifetime to become cornerstones of the Heliophysics Systems Observatory, indispensable to basic research as well as to space weather operations.

Examples of Explorers launched since 2000 that have made major contributions to scientific understanding pertinent to heliophysics are IMAGE, RHESSI, THEMIS, AIM, and IBEX. Each of these have made fundamental discoveries spanning the full range of the discipline, from the edge of the heliosphere (IBEX) to flare and reconnection physics on the Sun (RHESSI) to the explosive releases of energy taking place in Earth's magnetosphere (THEMIS) to the enigmatic formation of ice clouds in Earth's polar regions (AIM). In addition, the Explorer program is the home for Missions of Opportunity, including SNOE, CINDI, and TWINS, that produce science benefits far beyond their cost. The Explorer program is a cornerstone of heliophysics and has repeatedly proven to be one of the most cost-effective and best cost-controlled avenues for implementing space science missions.

---

concept studies to launch. This approach requires lean management teams, strong collaboration between institutions and NASA centers, and continuous attention of the PI to the balance of risk and scientific reward.

The survey committee recommends that NASA accelerate and expand the Heliophysics Explorer program, the most successful and impactful mission line in the Heliophysics program.

New Worlds, New Horizons in Astronomy and Astrophysics[9] highly recommended an increase in the Astrophysics Explorer budget for many of the same reasons that the present survey committee does.

Augmentation of Explorer Line to Restore MIDEX

The rate of Explorer satellite development has slowed remarkably since the 2003 solar and space physics decadal survey.[10] This decrease in selection rate is due to a major reduction in funding for the Explorer program that occurred in 2004 rather than to any drop in the number of compelling proposals for Explorer missions rated as selectable by NASA. The sharp funding cuts necessitated a reduction of Explorer competitions for the SMEX class in order to preserve even a minimal overall selection cadence. The MIDEX class historically has offered an opportunity to resolve the highest-level science questions (e.g., IMAGE addressed science that was originally identified for a larger solar-terrestrial probe), but this line of Explorer competition has not been possible under the current Explorer budget, particularly given the scarcity of medium-class launch vehicles and the cost of alternatives. A stable of competitively selected

[9]National Research Council, *New Worlds, New Horizons in Astronomy and Astrophysics*, The National Academies Press, Washington, D.C., 2010.

[10]National Research Council, *The Sun to the Earth—and Beyond: A Decadal Research Strategy in Solar and Space Physics*, The National Academies Press, Washington, D.C., 2003; and National Research Council, *The Sun to the Earth—and Beyond: Panel Reports*, The National Academies Press, Washington, D.C., 2003.

PI-led missions will let NASA's Heliophysics Division continue the innovative and cost-effective scientific spaceflight missions that are the hallmark of the Explorer program.

The survey committee recommends that the current Heliophysics Explorer program budget be augmented by $70 million per year, in fiscal year 2012 dollars, restoring the option of Mid-size Explorer (MIDEX) missions and allowing them to be offered alternately with Small Explorer (SMEX) missions every 2 to 3 years.

NASA Missions of Opportunity

Fundamental science can be achieved at a fraction of the cost of stand-alone missions by hosting payloads through partnering with other agencies, nations, or commercial spaceflight providers. The Hinode (Solar-B), TWINS, CINDI, and SNOE missions all demonstrate the benefits of such collaborations. The Solar-C mission now confirmed by Japan is an example of a future opportunity for the United States to provide instrumentation to a major foreign mission and in so doing to obtain high science return for relatively low cost. (Solar-C is discussed in detail by the Panel on Solar and Heliospheric Physics in Chapter 10.)

NASA's primary means of utilizing alternate platforms is via Missions of Opportunity and the current Stand Alone Missions of Opportunities Notices (SALMONs). However, the challenge of multi-organization coordination and the short time line for response to commercial opportunities call for a regular cadence and an expeditious mission proposal, review, and selection process. The survey committee concluded that a SALMON line needs to evolve in response to both community input and short-term opportunities more rapidly than the cadence of decadal surveys or even that of larger Explorers (MIDEX and SMEX). It needs to be flexible enough to allow proposal topics ranging from instruments on hosted payloads to a university-class Explorer satellite.

The survey committee recommends that, as part of the augmented Explorer program, NASA should support regular selections of Missions of Opportunity, which will allow the research community to quickly respond to opportunities and leverage limited resources with interagency, international, and commercial partners.

Third Research Recommendation [R3.0] for NASA—Restructure Solar-Terrestrial Probes as a Moderate-Scale, Principal-Investigator-Led Line

Initially conceived as a program to implement moderate-scale programs, the STP line has evolved into a large-mission program, dominated by NASA centers, with cost growth over the past decade that threatens its future viability. The survey committee concluded that restructuring the STP line is necessary if it is to address heliophysics science goals cost-effectively and offer flight opportunities at an acceptable cadence.

STP as a Moderate-Scale Mission Line

The NASA Planetary Science Division has demonstrated success in implementing larger missions as competed, cost-capped, PI-led investigations via the Discovery and New Frontiers programs. These are managed in a manner similar to Explorers, and they have superior cost-performance histories relative to those of flagship missions (see Appendix E, "Mission Development and Assessment Process"). Many of the most important heliophysics science objectives are too challenging to be addressed even by MIDEX Explorers but would fit within the cost profile of a moderate-scale mission.

The survey committee concluded that a successful moderate mission line has to include the following elements:

- The missions must be executed through a funding line with a fixed budget profile.
- The missions must be cost-capped, with a specified ceiling on full life-cycle costs.
- The missions must be PI-led, with the PI fully empowered and motivated to make scientific and mission trade-offs necessary to remain within the cost cap.
- The NASA center and/or other organization with management capabilities to which project management responsibility is assigned will be selected competitively. The goal is to assist the PI in the successful execution of the mission for minimum cost.
- Management of the STP program will be assigned by NASA Headquarters to a NASA center using a competitive process. The center must demonstrate ability to successfully execute the program for the minimum cost. Program management must also be reviewed periodically by Headquarters and be subject to reassignment in the event of unsatisfactory performance.
- Missions must be confirmed with reserves adequate for remaining within the cost cap and should have descope options in case the cost cap is breached. If there is cost growth beyond the control of the PI, the impacts have to be absorbed within the funding line, with no additional liens on other program elements in the Heliophysics Division.
- The missions will be selected competitively. However, each selection will be restricted to a specific science goal in order to achieve the prioritized strategic objectives described in this survey.

The survey committee recommends that NASA's Solar-Terrestrial Probes program be restructured as a moderate-scale, competed, principal-investigator-led (PI-led) mission line that is cost-capped at $520 million per mission in fiscal year 2012 dollars including full life-cycle costs.

Recommended STP Science Targets [R3.1, R3.2, and R3.3]

Although the new STP program would involve moderate missions being chosen competitively, the survey committee recommends that their science targets be ordered as follows so as to systematically advance understanding of the full coupled solar-terrestrial system.

1. The first new STP science target is to understand the outer heliosphere and its interaction with the interstellar medium, as illustrated by the reference mission[11] Interstellar Mapping and Acceleration Probe (IMAP). Implementing IMAP as the first of the STP investigations will ensure coordination with NASA Voyager missions. The mission implementation also requires measurements of the critical solar wind inputs to the terrestrial system.

2. The second STP science target is to provide a comprehensive understanding of the variability in space weather driven by lower-atmosphere weather on Earth. This target is illustrated by the reference mission Dynamical Neutral Atmosphere-Ionosphere Coupling (DYNAMIC).

3. The third STP science target is to determine how the magnetosphere-ionosphere-thermosphere system is coupled and how it responds to solar and magnetospheric forcing. This target is illustrated by the reference mission Magnetosphere Energetics, Dynamics, and Ionospheric Coupling Investigation (MEDICI).

---

[11] In this report, the committee uses the terms "reference mission" and "science target" interchangeably, given that the mission concepts were developed specifically to assess the cost of addressing particular high-priority science investigations. The concepts presented in this report underwent an independent cost and technical analysis by the Aerospace Corporation, and they have been given names for convenience; however, the actual recommendation from the committee is to address the science priorities enumerated in the reference mission concept.

IMAP

IMAP Overview

The past decade has seen breakthroughs in knowledge of the outer boundaries of the heliosphere and the interaction between the Sun and its local galactic neighborhood. These advances range from the crossing of the termination shock by the Voyager spacecraft to the images captured by IBEX of enhanced energetic neutral atom emission from a localized "ribbon" that encircles the heliosphere (Figure 4.7). The scientific motivation for a more advanced mission to image the heliospheric boundary and measure the key components of the interstellar gas is compelling but also urgent, because the Voyager spacecraft will operate only through this decade. The survey committee therefore recommends as a high priority the Interstellar Mapping and Acceleration Probe reference mission.

IMAP would orbit the inner Lagrangian point (L1) with comprehensive, highly sophisticated instruments to make the key observations that answer the following fundamental questions:

1. What is the spatio-temporal evolution of heliospheric boundary interactions?
2. What is the nature of the heliopause and of the interaction of the solar and interstellar magnetic fields?
3. What are the composition and physical properties of the surrounding interstellar medium?
4. How are particles injected into acceleration regions and what mechanisms energize them throughout the heliosphere and heliosheath?

The unique location of IMAP would also provide a platform from which to pursue the question of what are the time-varying physical inputs at L1 into the Earth system.

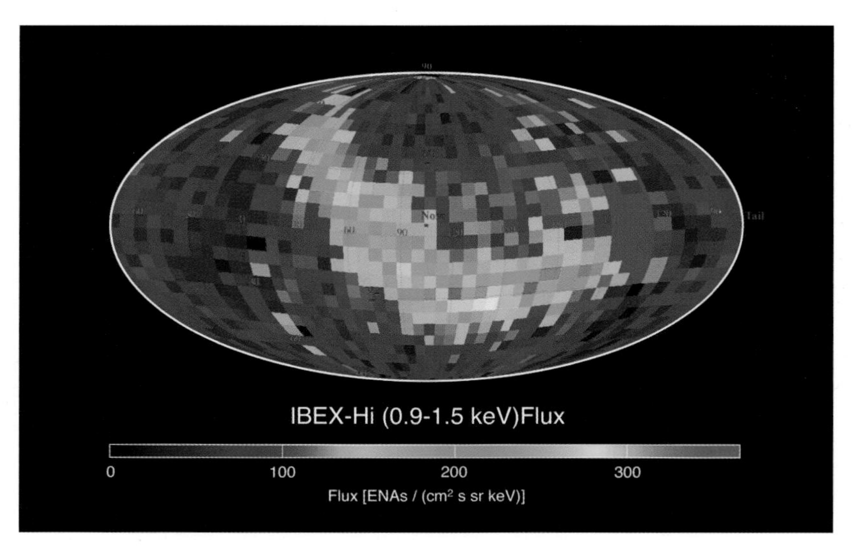

FIGURE 4.7 IBEX-Hi map of energetic neutral atoms in the energy range of 0.9-1.5 keV from the outer heliosphere. SOURCE: Interstellar Boundary Explorer Mission Team.

The mission would focus on making ENA maps and sampling local cosmic-ray particles concurrently with in situ Voyager measurements of the heliospheric boundary region. IMAP enables the understanding of particle acceleration through:

- Measurements of energetic (suprathermal) ions that originate from the solar wind, interstellar medium, and inner heliosphere with unprecedented sensitivity and time resolution;
- Environmental monitoring of pickup ion[12] (PUI) distributions that is critical for effective background evaluation and removal from ENA maps; and
- Comprehensive interplanetary particle and field monitoring in support of geospace interaction studies and space weather observations at the ideal location, L1.

IMAP Mission Concept

A notional spacecraft and instrument implementation for IMAP is based largely on ACE and IBEX. IMAP is a Sun-pointed spinner, with spin axis readjustment every few days to provide all-sky maps every 6 months. The L1 placement avoids magnetospheric ENA backgrounds and enables continuous interplanetary observations. Mission goals are achieved with a 2-year baseline, including transit to L1, with possible extension to longer operation (which would be particularly beneficial for long-term L1 monitoring). IMAP combines the measurement capabilities shown in Table 4.3, all of which are feasible based on extrapolations of current instrument technologies.

IMAP Contribution to the Heliophysics Systems Observatory

Observations from many spacecraft in the HSO contribute dramatically to understanding solar energetic particle events, the importance of suprathermal ions for efficient further energization, the sources and evolution of solar wind, solar-wind and energetic-particle inputs into geospace, and evolution of the solar-heliospheric magnetic field. These observables are controlled by a myriad of complex and poorly understood physical effects acting on distinct particle populations. IMAP combines highly sensitive PUI and suprathermal-ion sensors to provide the critical species, spectral coverage, and temporal resolution to address these physical processes. As an L1 monitor, IMAP also would fill a critical hole in Sun-Earth system observations by measuring the solar wind input, knowledge of which is essential to studying magnetospheric and upper atmospheric processes.

DYNAMIC (Dynamical Neutral Atmosphere-Ionosphere Coupling)

DYNAMIC Overview

DYNAMIC is designed to answer the question: How does lower-atmosphere variability affect geospace? To understand how lower-atmosphere variability drives neutral and plasma variability in the IT system, a mission must address wave coupling with the lower atmosphere. The representative mission developed and studied for this survey is designed to do two things. First, it will reveal the fundamental processes (e.g., wave dissipation, interactions between flow of different species) that underlie the transfer of energy and momentum into the IT system (especially within the critical 100- to 200-km height regime). Second, it will measure the resultant thermospheric and ionospheric variability that these waves incur at higher altitudes. It will do these on a global scale, with high-inclination satellites launched into orbits separated by 6 hours

---

[12] Pickup ions are formed when interstellar neutral atoms interact with the solar wind plasma and become ionized. The now charged particles are carried (thus the origin of the term "pickup") outward by the Sun's magnetic field to the solar wind termination shock.

TABLE 4.3 IMAP Key Parameters to Be Measured from Space

Instrument	Key Parameters	Measurements Requirements
ENA cameras, lower and higher energy ranges	Energetic neutral flux from heliospheric boundary	0.3-20 keV, 3-200 keV, 2-day sampling period
ISM neutral atom camera	ISM flow of H, D, He, O, and Ne	5-1000 eV, pointing knowledge of 0.05°
PUI sensor	Distributions of interstellar H^+, $^3He^+$, $^4He^+$, N^+, O^+, $^{20}Ne^+$, $^{22}Ne^+$, and Ar^+ and inner source C^+, O^+, Mg^+, and Si^+, also providing solar-wind heavy-ion composition	100 eV to 100 keV/e
Suprathermal-ion sensor	Composition and charge state for H through ultraheavy ions	Composition (0.03-5 MeV/nucleon) and charge state (0.03-1 MeV/e) for H through ultraheavy ions, 1-min cadence for H and He
Solar-wind and interplanetary monitoring suite	Background for ENA observations and real-time solar wind and cosmic-ray monitoring	SW ions (0.1-20 keV/e) and electrons (0.005-2 keV) every 15 s; the IMF $\leq$ 1 nT and 16 Hz; and SEP, anomalous cosmic-ray, and galactic cosmic-ray electrons and ions (H-Fe) over 2-200 MeV/nucleon.

NOTE: Parameters listed are those that must be measured to achieve the main objectives and answer the science questions defining the mission. These include two ENA cameras to produce observations of the heliospheric boundary over an extended energy range and with significantly improved sensitivity and spatial, energy, and time resolution compared with prior observations. An ISM neutral atom camera and the first dedicated pickup ion (PUI) sensor will take coordinated high-sensitivity observations of the interstellar gas flow through the inner solar system. Overlapping with the PUI sensor, a suprathermal-ion sensor will provide composition and charge state for H through ultraheavy ions. The solar-wind and interplanetary monitoring suite serves to mitigate backgrounds for high-sensitivity ENA observations and provide societally important real-time solar wind and cosmic-ray monitoring.

of local time, providing the coverage necessary to resolve critical atmospheric tidal components and the effects of wave-wave interaction. The name coined for this mission is DYNAMIC (Figure 4.8).

The DYNAMIC mission addresses the following fundamental science questions:

1. How does lower-atmosphere variability affect geospace?
2. How do neutrals and plasmas interact to produce multiscale structures in the AIM system?
3. How does the IT system respond over global, regional, and local scales to changes in magnetospheric inputs?
4. How is magnetospheric electromagnetic energy converted to heat and momentum drivers for the AIM system?
5. How is our planetary environment changing over multi-decadal scales, and what are the underlying causes?

It is important to note that while DYNAMIC's primary focus is to address the question of meteorological driving of geospace, the orbital sampling and the instruments that are flown also address broader science questions. For instance, the composition, temperature, and wind measurements will enable researchers to understand the relative roles of upwelling, advection, and thermal expansion in determining latitude-time evolution of the O/N_2 ratio during changing geomagnetic conditions, which affects total mass density and plasma density concentrations. Measurement of winds, plasma drifts, and plasma densities at high

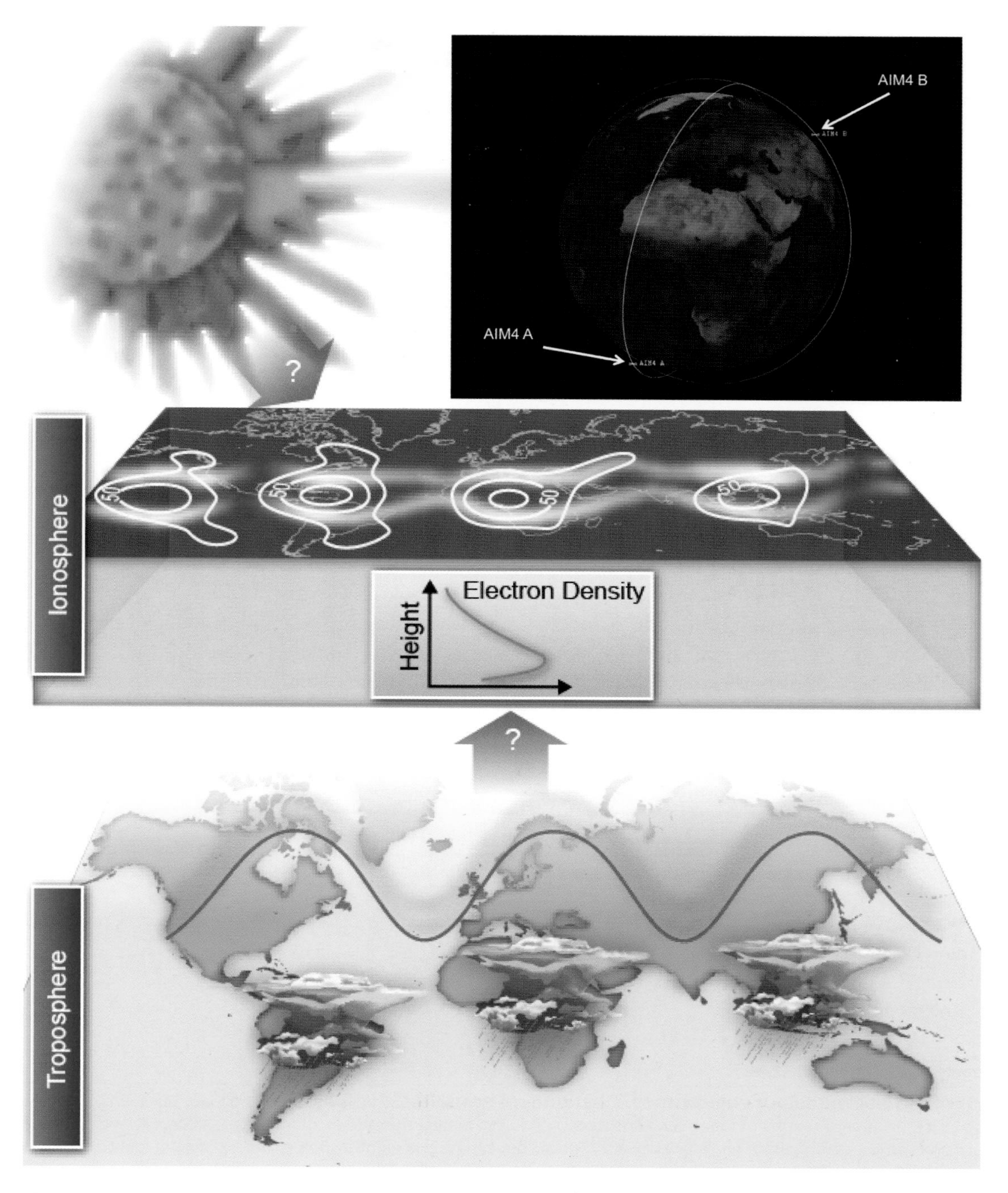

FIGURE 4.8 DYNAMIC targets the effects of lower atmospheric processes on conditions in space, characterizing how the energy and momentum carried into this region by atmospheric waves and tides interact and compete with solar and magnetospheric drivers. Full spatial and temporal resolution of the wave inputs is accomplished by using two identical, high-inclination, space-based platforms in similar orbits, offset by 6 hours of local time. SOURCE: Composite courtesy of Thomas Immel, Space Sciences Laboratory, University of California, Berkeley.

latitudes will lead to estimates of Joule heating, as well as a number of other plasma-neutral interactions at high and low latitudes. In addition, the simultaneous measurement of lower-altitude thermosphere winds and plasma drifts at higher altitudes will enable delineation of the disturbance dynamo in addition to the tidal-driven dynamo.

DYNAMIC Mission Concept

The above science focus translates to a mission involving instruments that remotely sense the lower and middle thermosphere while also collecting in situ data at higher altitudes. A key mission driver is the need to address atmospheric thermal tides, which demands measurements over all local times. Because satellite orbits generally take weeks to months to precess through 24 hours of local time, one must trade off latitude coverage against local time precession rate, or possibly consider multiple satellites. Since the research seeks to include important wave sources at high latitudes (e.g., weather systems and stratospheric warmings) and moreover to distinguish aurorally generated waves from those originating in the lower atmosphere, a high-inclination (75°-90°) orbit is required. However, for these orbital inclinations 24-hour local time precession occurs over a time period that exceeds that of important variability that has to be captured. Taking these factors into account, the recommended strategy is to employ two identical satellites in 80° inclination orbits at 600-km altitude, with their orbital planes spaced about 6 hours apart in local time. Assuming measurements are made at four local times over all longitudes in 1 day, all zonal (longitudinal) components of the diurnal (24-hour) tide would be fully characterized once per day, and semidiurnal (12-hour) tides as well as the diurnal mean would be acquired about once every 20 days. Gravity waves would be measured throughout each orbit, and planetary waves would easily be extracted with 1-day resolution. With the exception of semidiurnal tides, all wave-wave interactions could be explored, and on a 20-day timescale the interactions between the wave field and the mean state could be explored. In situ plasma and neutral responses at 600 km to these wave inputs would be measured over similar timescales (Table 4.4).

TABLE 4.4 DYNAMIC Key Parameters to Be Measured from Space

Instrument[a]	Key Parameters	Altitude Range
Limb Vector Wind and Temperature Measurement WIND (1 unit includes 2 telescopes)	Vn(z) – vector T(z)	80-300 km 80-300 km
Far Ultraviolet Imager (FUV)	Altitude Profiles: O, N_2, O_2, H, O^+ Maps: Q, E_o, O/N_2, O^+, Bubbles	110-300 km 200-600 km
Ion Velocity Meter (IVM)[b]	Vi	In situ
Neutral Wind Meter (NWM)[b]	Vn – vector	In situ
Ion Neutral Mass Spectrograph (INMS)[b]	O^+, H^+, He^+, O, N_2, O_2, H, He	In situ

NOTE: Parameters listed are those that must be measured to achieve the main objectives and answer science questions defining the mission. It includes an instrument to measure horizontal winds and temperatures from about 80 to 250 km, day and night, with horizontal and vertical resolutions of order 100 km and 2-10 km, depending on height. Instruments consisting of flight heritage components approaching this capability are thought to exist at the TRL 5 level, but flight test opportunities are required to establish their true capabilities. A flight-tested FUV imager already exists, and this would provide key measurements of neutral and ionized constituents in the lower and middle thermosphere regime. In situ instruments exist to make the required in situ measurements of neutral and ion composition, winds, and drifts, but further technology developments are underway to enhance performance and reduce size, power, and weight; it is important that these technology developments be supported, because these types of instruments are likely to be flown on almost any terrestrial or planetary ionosphere-thermosphere mission.

[a]All instruments have extensive flight heritage. Technology investments will improve their performance and provide additional capabilities.

[b]The IVM, NWM, and INMS are on the ram and anti-ram sides of the spacecraft. Only one operates at a time.

DYNAMIC Contribution to the Heliophysics Systems Observatory

By resolving the fundamental question of meteorological influences from below, DYNAMIC will firmly connect the ionosphere-thermosphere (IT) system to Earth's lower atmosphere, capturing a critical, missing component of scientific understanding of geospace and providing a critical new capability to the HSO at an important boundary in near-Earth space. In establishing the relative importance of thermal expansion, upwelling, and advection in defining total mass density changes, DYNAMIC will also provide information fundamental to understanding the global IT response to forcing from above. This investigation of the contribution of the lower atmosphere to the mean structure and dynamics of the IT system reflects a scientific appreciation of the importance of these drivers gained since the 2003 solar and space physics decadal survey.

MEDICI (Magnetosphere Energetics, Dynamics, and Ionospheric Coupling Investigation)

MEDICI Overview

MEDICI (Figure 4.9) is a new, strategic, notional mission concept aimed at determining how the complex magnetosphere-ionosphere-thermosphere system is coupled and responds to external solar and internal magnetospheric forcing. Regions of geospace are intrinsically interconnected over diverse scales of space and time. Plasma and fields in the ionosphere and magnetosphere interact, and multiple processes compete simultaneously. Observation of the relationships among components is critical to understanding and characterizing the collective behavior of this complex system across a broad range of spatial scales.

By combining and improving crucial elements from several prior missions, MEDICI takes a major step forward in geospace imaging. The first multispectral, stereo geospace plasma imaging will reveal the three-dimensional structure of cardinal geospace plasmas and will provide conjugate views of the northern

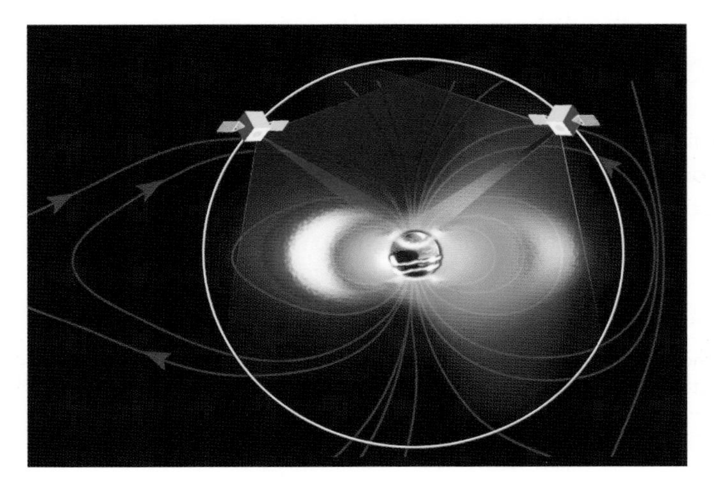

FIGURE 4.9 MEDICI targets complex, coupled, and interconnected multiscale behavior of the magnetosphere-ionosphere-thermosphere system by providing high-resolution, global, continuous three-dimensional images of the ring current (orange), plasmasphere (green), aurora, and ionospheric-thermospheric dynamics and flows, as well as multipoint in situ measurements. SOURCE: Courtesy of Jerry Goldstein, Southwest Research Institute.

and southern aurorae. Simultaneous two-point in situ measurements are closely coordinated with plasma imaging from state-of-the-art instruments to uncover transport and electrodynamic connections at different spatial scales throughout the magnetosphere-ionosphere-thermosphere system, enabling major new insights into cross-scale dynamics and complexity in geospace.

MEDICI will examine how the magnetosphere-ionosphere-thermosphere system is coupled and responds to solar and magnetospheric forcing. In particular, MEDICI would provide definitive, comprehensive answers to two overarching, fundamental science questions that have been outstanding for decades. Each question contains a set of subtopics:

1. How are magnetospheric and ionospheric plasma transported and accelerated by solar wind forcing and magnetosphere-ionosphere (MI) coupling?
    a. How is the cross-scale, dynamic, three-dimensional plasma structure of the ring current, plasmasphere, and aurora reshaped by acceleration and transport?
    b. What controls when and where ionospheric outflow occurs?
    c. What are the cross-scale effects on the system?
2. How do magnetospheric and ionospheric plasma pressure and currents drive cross-scale electric and magnetic fields, and how do these fields in turn govern the plasma dynamics?
    a. What are the cross-scale, interhemispheric structure and timing of currents and fields that mediate MI coupling?
    b. How do these MI coupling electromagnetic fields feed back into the system to affect the plasmas that generated them?

Each of these two main questions and five subquestions focuses on a crucial aspect of the coupled dynamics of geospace. The first question set looks at plasma transport, and the second question set targets the electrodynamics of MI coupling. Previous missions such as IMAGE or TWINS have provided substantial steps toward addressing these problems; however, only MEDICI's comprehensive instrumentation will supply the necessary complete set of measurements to answer these questions.

MEDICI Mission Concept

MEDICI is a cross-scale science mission concept that uses both high-resolution stereo imaging and multipoint in situ measurements. It also incorporates an array of contemporaneously existing ground-based and orbiting observatories. MEDICI employs two spacecraft that share a high circular orbit (see Figure 4.9), each hosting multispectral imagers, magnetometers, and particle instruments. Building on knowledge obtained from TWINS, the MEDICI science payload captures the dynamics of the ring current, plasmasphere, aurora, and ionospheric-thermospheric plasma redistribution through a comprehensive set of measurements. In combination with ground-based and low-Earth-orbit data that yield detailed information on field-aligned currents, ionospheric electron densities, temperatures, and flows, MEDICI's imagers and onboard in situ instruments will provide the means to link the global-scale magnetospheric state with detailed ionospheric conditions, and will yield data for global-model validation. The MEDICI instrumentation is summarized in Table 4.5.

The MEDICI mission uses two nadir-viewing spacecraft, each with an identical spacecraft bus, in a shared 8-R_E circular polar orbit with adjustable orbital phase separation (between 60° and 180° separation along track) to enable global stereo, multispectral imaging, and simultaneous in situ observations. Circular orbits that avoid the most intense radiation environments provide continuous imaging and in situ measurements and enable a long-duration (up to 10 years) lifetime well beyond the required 2-year mission.

TABLE 4.5 MEDICI Key Parameters to Be Measured from Space

Instrument	Key Parameters[a]	Measurement Requirements
ENA imager	Three-dimensional ring current and near-Earth plasma sheet pressure-bearing ion densities	Temporal and spatial resolution: 1 minute, 0.5 R_E
EUV imager	Evolution of plasmasphere density	30.4 nm, temporal/spatial resolution: 1 minute, 0.05 R_E
FUV cameras (1 spacecraft only)	Precipitating auroral particle flux, ionospheric electron density and conductivity, and thermospheric conditions	LBH long and short wavelengths; 5- to 10-km resolution
Ion and electron plasma sensors	In situ electron and ion plasma densities, temperatures, and velocities	Helium, oxygen, protons, electrons from a few electronvolts to 30 keV; ~1-minute resolution
Magnetometer	In situ magnetic fields	Vector B and delta-B (dc and ac); ~ 1-second resolution

NOTE: Parameters listed are those that must be measured to achieve the most important objectives and to address the key science questions that motivate the selection of this reference mission. MEDICI requires no new technology development and all of the instruments have high heritage. Each of the mission's three measurement goals contributes essential information about cross-scale geospace dynamics: the first is to continuously image the three-dimensional distribution of two critical inner magnetospheric plasmas; the second is to image and measure the ionosphere-thermosphere system at multiple wavelengths in the far ultraviolet (FUV); and the third is to measure, in situ, the critical near-Earth plasmas and magnetic field in the cusp and near-Earth plasma sheet plasma. The instrument configuration shown here illustrates one realization of MEDICI; alternative configurations are possible that would not impact cost significantly, for example, the addition, to an otherwise identical spacecraft, of a second set of FUV Lyman-Birge-Hopfield (LBH) long- and short-wavelength cameras (see Appendix E).

Complementing and augmenting the high-altitude observations, MEDICI includes funded participation for significant low-altitude components: measurements from a large range of resources including DMSP or its follow-on Defense Weather Satellite System, IRIDIUM/AMPERE current maps, radar arrays from high to midlatitudes (SuperDARN, Millstone Hill, AMISR), GPS TEC maps, and magnetometer and ground-based auroral all-sky camera arrays. The result will be global specification of the ionospheric electric field and electric current patterns in both hemispheres, essentially completing observational constraints on the electrodynamic system at low altitude.

MEDICI Contributions to the HSO

MEDICI will both benefit from and enhance the science return from almost any geospace mission that flies contemporaneously, such as upstream solar wind monitors, geostationary satellites, and low-Earth-orbit missions. In particular, by providing global context and quantitative estimates for magnetospheric-ionospheric plasma and energy exchange, MEDICI has significant value for missions investigating ionospheric conditions, outflow of ionospheric plasma into the magnetosphere, energy input from the magnetosphere into the ionosphere, and AIM coupling in general. Thus it will add value to a host of possible ionospheric strategic missions, Explorers, and rocket and balloon campaigns. Further, with continuous imaging and in situ observations from two separate platforms, it would provide indispensable validating observations of system-level interactions and processes that feed geospace predictive models. The likely long duration of the notional MEDICI mission will allow it to provide a transformative framework into which additional future science missions can naturally fit.

Fourth Research Recommendation [R4.0] for NASA—Implement a Large Living With a Star Mission

A very important distinction is made by the survey committee between the restructured STP program and the Living with a Star (LWS) mission. Certain scientific problems can be addressed only by missions that are relatively complex and costly. Solar Probe Plus, which will travel closer to the Sun than any previous spacecraft, is an example of this type of mission; future constellation missions that would utilize multiple spacecraft to provide simultaneous measurements from broad regions of space (so as to be able to separate spatial from temporal effects and reveal the couplings between adjacent regions of space) are another. As research evolves naturally from the discovery-based mode to one focused increasingly on quantification and prediction, missions benefit strongly from an integrative approach, whereby the knowledge obtained from prior research can be combined with new, innovative measurements for the development of understanding of the global machinery of the system. This effort may naturally require a larger mission, and it also accords with the societal-relevance theme of the LWS program. In the survey committee's plan, major missions are thus appropriately undertaken via NASA's LWS program and would continue to be executed by NASA centers, whereas the STP program should be considered a community program, like the Explorer program.

Besides the flight program, there are other integral thematic elements essential to the LWS science and technology program, such as the TR&T program that conducts relevant focused research, and strategic capabilities that establish essential underlying technical capabilities. Also included in LWS are heliophysics summer schools. The survey committee concluded that the unique LWS research, technology, and education programs remain of great value.

The survey committee's recommended science target for the next major LWS mission, as demonstrated by the reference mission Geospace Dynamics Constellation, would provide crucial scientific measurements of the extremely variable conditions in near-Earth space.

Recommended LWS Science Target

The survey committee recommends that, following the launch of RBSP and SPP, the next LWS science target focus on how Earth's atmosphere absorbs solar wind energy. The recommended reference mission is Geospace Dynamics Constellation (GDC).

Geospace Dynamics Constellation

GDC Overview

During geomagnetic storms, solar wind energy is deposited in Earth's atmosphere, but only after being transformed and directed by a number of processes in geospace. The primary focus of the Geospace Dynamics Constellation reference mission is to reveal how the atmosphere, ionosphere, and magnetosphere are coupled together as a system and to understand how this system regulates the response of all geospace to external energy input. Using current and foreseeable technologies, GDC implements a systematic and robust observational approach to measure all the critical parameters of the system in optimally spaced orbital planes, thus providing unprecedented coverage in both local time and latitude. Moreover, spacecraft in the GDC constellation will orbit at relatively low altitudes where both neutral and ionized gases are strongly coupled through dynamical and chemical processes. The GDC reference mission brings a new focus to critical scientific questions:

1. How does solar wind/magnetospheric energy energize the ionosphere and thermosphere?
2. How does the IT system respond and ultimately modify how the magnetosphere transmits solar wind energy to Earth?

3. How is solar wind energy partitioned into dynamical and chemical effects in the IT system, and what temporal and spatial scales of interaction determine this partitioning?
4. How are these effects modified by the dynamical and energetic variability of the ionosphere-upper atmosphere introduced by atmospheric wave forcing from below?

The observational problem is such that global dynamics cannot be captured by any number of probes on a single satellite. When averaged over a sufficiently long period of time, data from a single satellite provide a useful climatology as a function of latitude and longitude. However, such data are static and do not show the physical coupling inherent in the continuously evolving density and velocity patterns (dynamics) as they respond at all local times to the many drivers of the AIM system.

For example, electromagnetic flux and energetic particle precipitation are highly structured and variable in latitude and local time. The dynamical response includes hydrodynamic atmospheric waves propagating from high to low latitudes, but differently during day and night due to the large difference in neutral-ion drag. During major storms, the large-scale upper atmospheric wind patterns are greatly disturbed and constantly altered by the penetrating and dynamo electric fields that exhibit strong local time variations. Further, the chemical mixing of the upper atmosphere by auroral heating expands to low latitudes and depletes the ionospheric plasma, reducing an important source of fuel for a geomagnetic storm, but again in a highly local-time-dependent manner. These phenomena exemplify why a new approach must be taken to advance understanding of the AIM system and how Earth's upper atmosphere and ionosphere regulate the response of geospace to significant solar wind energy inputs.

GDC would be a constellation of identical satellites in low Earth orbit providing simultaneous, global observations of the AIM system over roughly the range of local times over which magnetospheric drivers (and thus AIM responses) are organized. The satellites would have high-inclination circular orbits in the 300- to 450-km altitude range. Table 8.1 in Chapter 8 summarizes the science objectives, science merit, and space weather relevance of GDC and how it relates to the overall decadal survey strategy.

GDC Mission Concept

The nominal plan for GDC is to have six identical satellites that will be spread individually into equally spaced orbital planes separated by 30° longitude, thus providing measurements at 12 local times, with a resolution of 2 hours local time, as shown in Figure 4.10. The satellites will nominally have an inclination of 80°, in order to use precession to help separate the local time planes, while maintaining adequate coverage of the high-latitude region.

GDC Contributions to the HSO

GDC will make measurements critical to understanding how the IT system regulates the response of geospace to external forcing (Table 4.6). The constellation of satellites will provide a complete picture of the dynamic exchange of energy and momentum that occurs between ionized and neutral gases at high latitudes, providing the HSO a critical capability for measuring the response and electrodynamic feedback of Earth's IT system to drivers originating in the solar wind and magnetosphere. GDC will also determine the global response of the IT system to magnetic activity and storms and expose how changes in the system at different locations are related. Finally, it will determine the influence of forcing from below on the IT system, by measuring the global variability of thermospheric waves and tides on a day-to-day basis with the spatial resolution that only a constellation of satellites can provide.

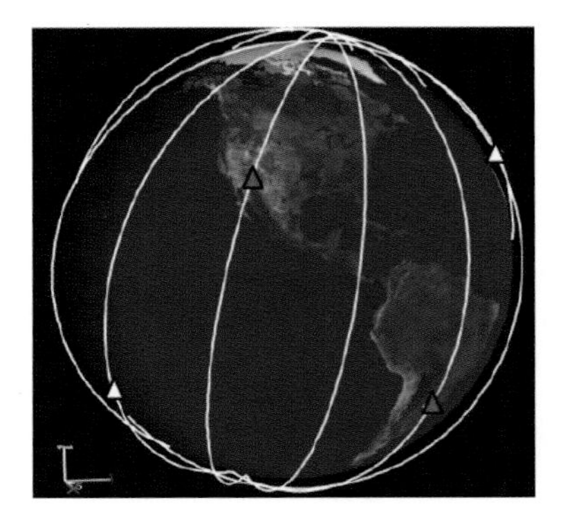

(a) Full global coverage

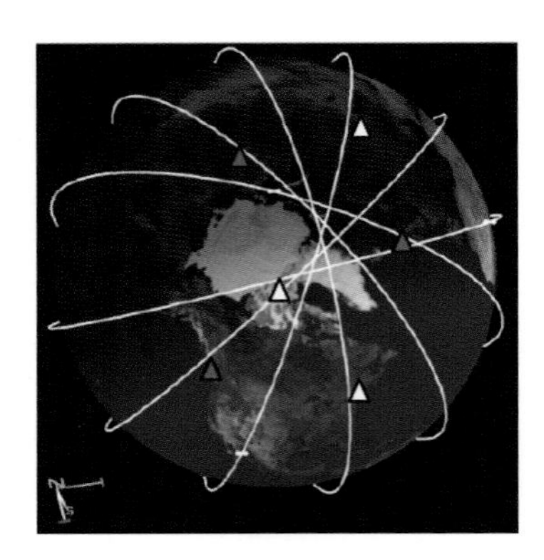

(b) 6-spacecraft high-latitude "armada"

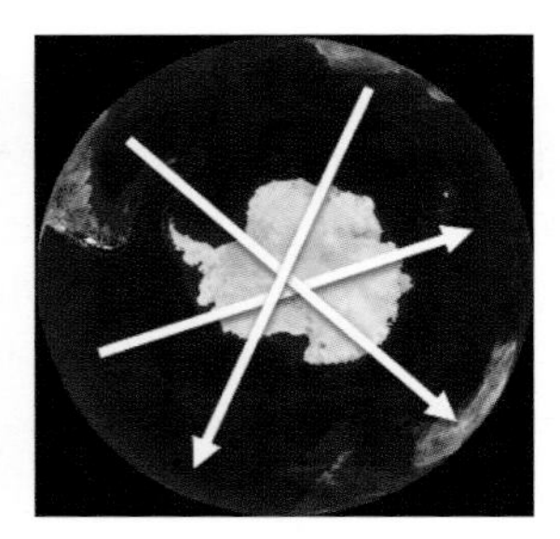

(c) 3 spacecraft simultaneously sampling each pole

+60°
0°
-60°

(d) 6 s/c cross the equator every 45 min.

FIGURE 4.10 Potential orbital configurations for Geospace Dynamics Constellation. Three main orbital configurations have been considered thus far: (a) spacecraft fully spread out in latitude to provide continuous, global coverage; (b) spacecraft configured as a group with simultaneous, dense coverage at high latitudes alternating between polar cap regions every 45 minutes; and (c) satellites configured to orbit in two, separated three-satellite armadas, such that three are in the Northern Hemisphere while three are in the Southern Hemisphere, providing simultaneous coverage of both polar regions every 45 minutes. Configurations (b) and (c) both gather consolidated measurements at the mid- and low latitudes, with simultaneous crossings of the equator by all six satellites every 45 minutes (d), while the entire globe is sampled every 90 minutes at 12 local times (as is the case for configuration (a). Minimal amounts of propellant relative to the baseline capacity are needed to alternate between these configurations, and the station-keeping time to change and maintain these configurations is on the order of hours.

The most cost-effective launch approach (given available launchers) would be to launch all six satellites with one vehicle. In this case, the satellites are first placed in a highly elliptical orbit plane (e.g., with perigee of 450 and apogee of 2,000 km) with an 80° inclination. As they then precess, these satellites will spread out in equally separated local time planes (requiring ~12 months). Note that propulsion can be used to decrease this deployment time. In this scenario, the apogee of one satellite is immediately lowered to provide an initial 450-km circular orbit, while the remaining five satellites precess in local time. After about 2.5 months, in which the satellites have precessed 30 degrees in local time, a second satellite orbit is changed to 450-km circular orbit. The process continues until all six are spread out equally and converted to 450-km circular orbits. During the time required to establish the final distribution in local time of the six satellites, this initial observing phase permits "pearls-on-a-string" observations by the satellites along highly elliptical orbits. SOURCE: (a-c) Orbital configurations courtesy of the Aerospace Corporation using Earth images provided by Living Earth, Inc.; (d) NASA Goddard Space Flight Center.

TABLE 4.6 GDC Key Parameters to Be Measured from Space

Notional Instrument	Key Parameters	Nominal Altitude
Ion Velocity Meter (includes RPA)	Vi, Ti, Ni, broad ion composition	300-400 km
Neutral Wind Meter (NWM)	Un, Tn, Nn, broad neutral composition	300-400 km
Ionization Gauge	Neutral density	300-400 km
Magnetometer	Vector B, Delta B, currents	300-400 km
Electron Spectromenter	Electron distributions, pitch angle (0.05 eV to 20 keV)	300-400 km

NOTE: The measurements needed to achieve the main objectives of the mission and answer the science questions linked to them are itemized. Each satellite includes an identical suite of "notional" instruments, as listed here. Instruments to measure both neutral and ionized state parameters, including dynamics, are included. Also included are a magnetometer and an energetic particle detector for measuring energy and momentum forcing from the magnetosphere. All instruments have extensive flight heritage.

APPLICATIONS RECOMMENDATIONS: SPACE WEATHER AND SPACE CLIMATOLOGY

Space weather is receiving increased attention as the importance of its effects on society is more broadly recognized. Previous NRC reports[13] and the National Space Weather Program (NSWP) Strategic Plan 2010[14] document the nation's need for increased capability to specify and predict the weather and climate of the space environment. The past decade has seen the growth of new services and technologies, including GPS location and timing services, aircraft flights over polar regions, electric power transmission systems, and a growing space tourism industry, all of which increase the need to consider vulnerabilities that can result from space weather conditions. Space weather affects our lives directly and indirectly. In the extreme case of an event of historic proportion, space weather may even lead to catastrophic disruptions of society.

From economic and societal perspectives, reliable knowledge about and forecasting of conditions in the space environment are important on a range of timescales for multiple applications. Prominent among them are radio signal utilization (which enables increasingly precise navigation and communication) and mitigation of the drag on Earth-orbiting objects that alters the location of spacecraft, threatens their functionality as a result of collisions with debris, and impedes reliable determination of reentry. Energetic particles can damage assets and humans in space. Currents induced in ground systems can disrupt and damage power grids and pipelines.

All national space weather forecasting entities (Box 4.8) currently rely on potentially threatened operational space assets and critical data from limited-term research missions, and they require a better-supported and cost-effective research-to-operations pathway for models. As such, the future of even the status quo is threatened—at a time when national space weather requirements are continuously growing.

The U.S. and international space physics communities are poised to make significant advances in space weather and space climate science. There is already a vibrant, cooperative enterprise of study along with a strong culture of student development across the three major skill areas: instrument development, data analysis, and theory and modeling. The future is therefore highly promising. However, the key is long-term

[13] See National Research Council, *Severe Space Weather: Understanding Societal and Economic Impacts* (2009), and National Research Council, *Limiting Future Collision Risk to Spacecraft: NASA's Meteoroid and Orbital Debris Programs* (2011), both published by the National Academies Press, Washington, D.C.

[14] Committee for Space Weather, Office of the Federal Coordinator for Meteorological Services and Supporting Research, "National Space Weather Program Strategic Plan," FCM-P30-2010, August 17, 2010, available at http://www.ofcm.gov/nswp-sp/fcm-p30.htm.

BOX 4.8 SPACE WEATHER TODAY

As the nation's official source of space weather alerts and warnings, NOAA's Space Weather Prediction Center (SWPC) delivers space weather products and services that meet evolving national needs. Within NOAA, SWPC works especially closely with the National Environmental Satellite, Data, and Information Service (NESDIS), which provides satellite services and includes the National Geophysical Data Center (NGDC), which archives and disseminates NOAA's data. As part of the National Weather Service, SWPC provides actionable alerts, forecasts, and data products to space weather customers and works closely with partner agencies, including NASA, the Department of Defense (DOD), the Department of Energy, the U.S. Geological Survey, the Department of Homeland Security, and the Federal Aviation Administration. In addition to these agencies, SWPC also works with commercial service providers and the international community to acquire and share data and information needed to carry out its role in serving the nation with space weather products and services.

The Air Force Weather Agency (AFWA) has the lead for providing space weather data, products, and services to DOD users worldwide. Missions supported range from ground- and sea-based users of HF and satellite communications and GPS navigation systems, to space surveillance and tracking radars, to on-orbit satellite operations. AFWA's Space Weather Operations Center collects observations in real time, operates specification and forecast models, and disseminates mission-tailored information to users via Web services, Web pages, and dedicated communications. AFWA, in close collaboration with Air Force Space Command and the Air Force and Naval Research Laboratories, has been a strong proponent of transitioning research to operations, and it routinely leverages sensor data and models provided by NOAA, NASA, and DOD-funded research efforts. AFWA collaborates closely with the Air Force Research Laboratory Space Weather Center of Excellence and with NASA's Space Weather Laboratory.

NASA is addressing its own space weather needs through a combination of two collaborating centers. The Space Radiation Analysis Group (SRAG) has as its focus the protection of humans in space, from low Earth orbit to destinations beyond. Combining data from NASA, NOAA, and partner sensors, SRAG provides state-of-the-art assessments tailored to the needs of NASA's human spaceflight program. The Space Weather Laboratory (SWL) provides services to NASA's robotic missions, based on execution of the largest set of space weather forecasting models in existence to date. Data from a large variety of sources and model output are gathered and disseminated by an innovative space weather analysis system that is accessible not only by NASA but also by interests worldwide. Both SRAG and SWL collaborate closely with each other, as well as with entities inside and outside the United States.

Recent years have seen the emergence of a vibrant commercial sector that is engaging in space weather operations and providing new services and products for customers ranging from agencies and commercial aerospace to consumers. University organizations have established operational centers that produce space weather data for commercial users (e.g., Utah State University's Space Weather Center). An important focus of these commercial activities is the provision of derivative products to improve communications and navigation. The American Commercial Space Weather Association was formed in 2011 to represent private-sector commercial interests nationally and internationally; its formation represents a milestone at the end of the first decade of a maturing commercial space weather enterprise. Member companies also supply advisory services related to space weather to government agencies and commercial organizations.

space environment measurements and modeling, with many commonalities between space weather and space climate needs.

First Applications Recommendation [A1.0]—Recharter the National Space Weather Program

The survey committee concluded that, in addition to agency-appropriate activities by NASA, NSF, NOAA, and DOD to support space weather model research and development, validation, and transition to operations, a comprehensive plan for space weather and climatology is also needed to fulfill the requirements presented in the June 2010 U.S. National Space Policy[15] and envisioned in the 2010 National Space Weather Program Strategic Plan.[16] However, implementation of such a program would require funding well above what the survey committee assumes to be currently available; the committee advises that an initiative in space weather and climatology proceed only if its execution does not impinge on the development and timely completion of the other recommended activities that are described in this chapter and shown in Figure 6.1.

In Chapter 7, the survey committee presents a vision for a renewed national commitment to a comprehensive program in space weather and climatology that would provide long-term observations of the space weather environment and support the development and application of coupled space weather models to protect critical societal infrastructure, including communication, navigation, and terrestrial meteorological spacecraft. Realization of the committee's plan, or a similar plan, will require action across a number of agencies and coordination at an appropriately higher level in government.

The charter for the NSWP dates to its inception in 1995. Rechartering the NSWP would provide an opportunity to review the program and to consider the issues raised here, especially those pertaining to program oversight and agency roles and responsibilities. This need informs the following recommendation:[17]

As part of a plan to develop and coordinate a comprehensive program in space weather and climatology, the survey committee recommends that the National Space Weather Program be rechartered under the auspices of the National Science and Technology Council. With the active participation of the Office of Science and Technology Policy and the Office of Management and Budget, the program should build on current agency efforts, leverage the new capabilities and knowledge that will arise from implementation of the programs recommended in this report, and develop additional capabilities, on the ground and in space, that are specifically tailored to space weather monitoring and prediction.

---

[15] *National Space Policy of the United States of America*, June 28, 2010, available at http://www.whitehouse.gov/sites/default/files/national_space_policy_6-28-10.pdf.

[16] Committee for Space Weather, Office of the Federal Coordinator for Meteorological Services and Supporting Research (OFCM), *National Space Weather Program Strategic Plan*, FCM-P30-2010, August 17, 2010, available at http://www.ofcm.gov/nswp-sp/fcm-p30.htm.

[17] A similar recommendation was made in a review of the NSWP program that was published in 2006. In that review, it is recommended that,

> The NSWPC should review and update its now 10-year-old charter to describe clearly its oversight responsibilities. These should include, but not be limited to, the authority to: (1) address and resolve interagency issues, concerns, and questions; (2) reprioritize and leverage existing resources to meet changing needs and requirements; (3) approve priorities and new requirements as appropriate and take coordinated action to obtain the needed resources through each agency's budgetary process; (4) identify resources needed to achieve established objectives; and (5) coordinate and leverage individual organizational efforts and resources and ensure the effective exchange of information.

See, Office of the Federal Coordinator for Meteorological Services and Supporting Research, *Report of the Assessment Committee for the National Space Weather Program*, FCM-R24-2006, June 2006, p. xiii, available at http://nswp.gov/nswp_acreport0706.pdf.

Benefits of Research-to-Operations and Operations-to-Research Interplay

An interplay between research and operations benefits both, as reflected by the operational community's use of real-time data from research spacecraft and by use of operational ground and space assets for science. There are many examples of the value of operational assets to the research community and of research assets to the operational community. Examples include research data from GOES, POES, DMSP, and COSMIC, and operational use of data from ACE, SOHO, SDO, and STEREO. The routine provision of space weather data from science missions is invaluable. Nevertheless, although NOAA's operational mission data are generally available, other data sets, e.g., from DOD's LANL and GPS spacecraft, are not receiving the necessary support or being made available to the scientific community. The operational use of models, and their validation, help to identify model limitations and contribute to future model improvements. Research models and theory are becoming more accessible to operators and are contributing to improved forecasts (e.g., the Community Coordinated Modeling Center). By making output from research and operational models widely available to the research community, broad-based model development and supporting validation and verification are facilitated. Conducting space weather operations in a closely coupled fashion with related research efforts benefits from an infusion of the latest knowledge and technology. Including research personnel in the review of requirements for operational sensors and data maximizes the data's value to both operations and science. Such efforts advance the pace of model development and transition to operations by all participants, ranging from the academic researcher to the SWPC, AFWA, and NASA operators. The SWL at NASA GSFC had many successes in transitioning models to NASA operations. The first major transition of a large-scale physics-based numerical model to SWPC operations, the WSA-Enlil model (developed as part of the NSF Center for Integrated Space Weather Modeling), was achieved in 2011, building on 20 years of scientific research funded by multiple agencies.

Second Applications Recommendation [A2.0]—Work in a Multiagency Partnership to Achieve Continuity of Solar and Solar Wind Observations

Solar, interplanetary, and near-Earth observations of the space environment are the mainstays of the space weather enterprise. Space observations are used for (1) space environment situational awareness; (2) inputs to models that provide spatial and temporal predictions; (3) assimilation into models to improve model accuracy; (4) validation of model performance, both in operations and during model development; (5) building of a historical database for climatology and empirical models; (6) specific tailored products such as, for example, satellite anomaly resolution; and (7) research to improve understanding of the space environment and the physical processes involved in solar-terrestrial interactions.

Most of these observations are needed in real time to be useful for operations. They come from NOAA or DOD operational missions, from NASA's science missions, or from science activities at other agencies. Examples include NOAA's Geostationary Operational Environmental Satellites (GOES), the Air Force's Solar Observing Optical Network (SOON), and H-alpha monitoring via NSO's Global Oscillation Network Group (GONG), or NASA missions such as SDO, SOHO, ACE, and STEREO. They also come from other agency programs, such as ground-based magnetometers operated by the U.S. Geological Survey, or from international collaborations, such as the proposed collaboration between the United States and Taiwan to jointly fund and operate the COSMIC-2 spacecraft constellation. For some uses—such as resolving satellite anomalies, validating operational model performance, and providing data for long-term climate purposes—data are not needed in real time.

The space environment from the Sun to Earth and other planets is vast, in terms not only of the volume that has to be monitored but also the parameters that have to be measured and the spectral ranges that need to be covered. In the past decade, space weather needs have been served by improved observations from both operational and science missions, such as GOES, STEREO, and ACE. At the same time, there are deficiencies, for example, the likely absence of energetic-particle sensors on future NOAA low-altitude polar-orbiting satellites and uncertainty regarding their presence on future Air Force defense weather satellites.

Recently, there has been admirable progress as well as interagency cooperation among NASA, the U.S. Air Force, and NOAA to develop a long-sought and much-needed replacement for the aging ACE solar wind satellite, which was launched into a halo orbit around the L1 libration point in 1997. However, it now appears that the replacement spacecraft, DSCOVR, will not carry the coronagraph that is needed to replace observations from the SOHO spacecraft, which was launched in 1995. Moreover, instruments on DSCOVR, like those currently on ACE, have significant limitations in monitoring the highest-velocity events. IMAP, described above, provides an opportunity to apply to a next generation of solar wind instruments what has been learned over the past two solar cycles about solar wind variability.

In summary, the survey committee found that NOAA, DOD, and other agencies play an important role in maintaining and expanding the observational foundation of accurate and timely data used for space weather operations and space climatology. The survey committee also concluded that it is important that NASA continue to make science mission data available in cases in which those data contribute to timely space weather operations.

The survey committee recommends that NASA, NOAA, and the Department of Defense should work in partnership to plan for continuity of solar and solar wind observations beyond the lifetimes of ACE, SOHO, and STEREO. In particular,

- **[A2.1] Solar wind measurements from L1 should be continued, because they are essential for space weather operations and research. The DSCOVR L1 monitor and IMAP STP mission are recommended for the near term, but plans should be made to ensure that measurements from L1 continue uninterrupted into the future.**
- **[A2.2] Space-based coronagraph and solar magnetic field measurements should likewise be continued.**
- **[A2.3] The space weather community should evaluate new observations, platforms, and locations that have the potential to provide improved space weather services. In addition, the utility of employing newly emerging information dissemination systems for space weather alerts should be assessed.**

Models and the Transition of Research to Operations

Research, the foundation for future improvements in space weather services, is necessary for progress and improvement in models that are just beginning to reach a level of maturity such that they can benefit space weather customers. But just as the first Sun-to-Earth model (WSA-Enlil) is being implemented at the National Centers for Environmental Prediction (Figure 4.11), space weather specialists are witnessing a major decline in support for model improvement and new model development. Continued support is critical to developing models that are useful for operational forecasts, focused primarily on addressing operational needs and prepared with the goal of making the transition from the research environment to operational service. Further, model development aimed at improving operational forecasts requires a dedicated effort focused on validation and verification. Finally, the pursuit of models that are meant to be more than just science tools is most effective when the end purpose is usefulness in operations. This in turn requires a close working relationship between the end users of space weather forecasts and the research community.

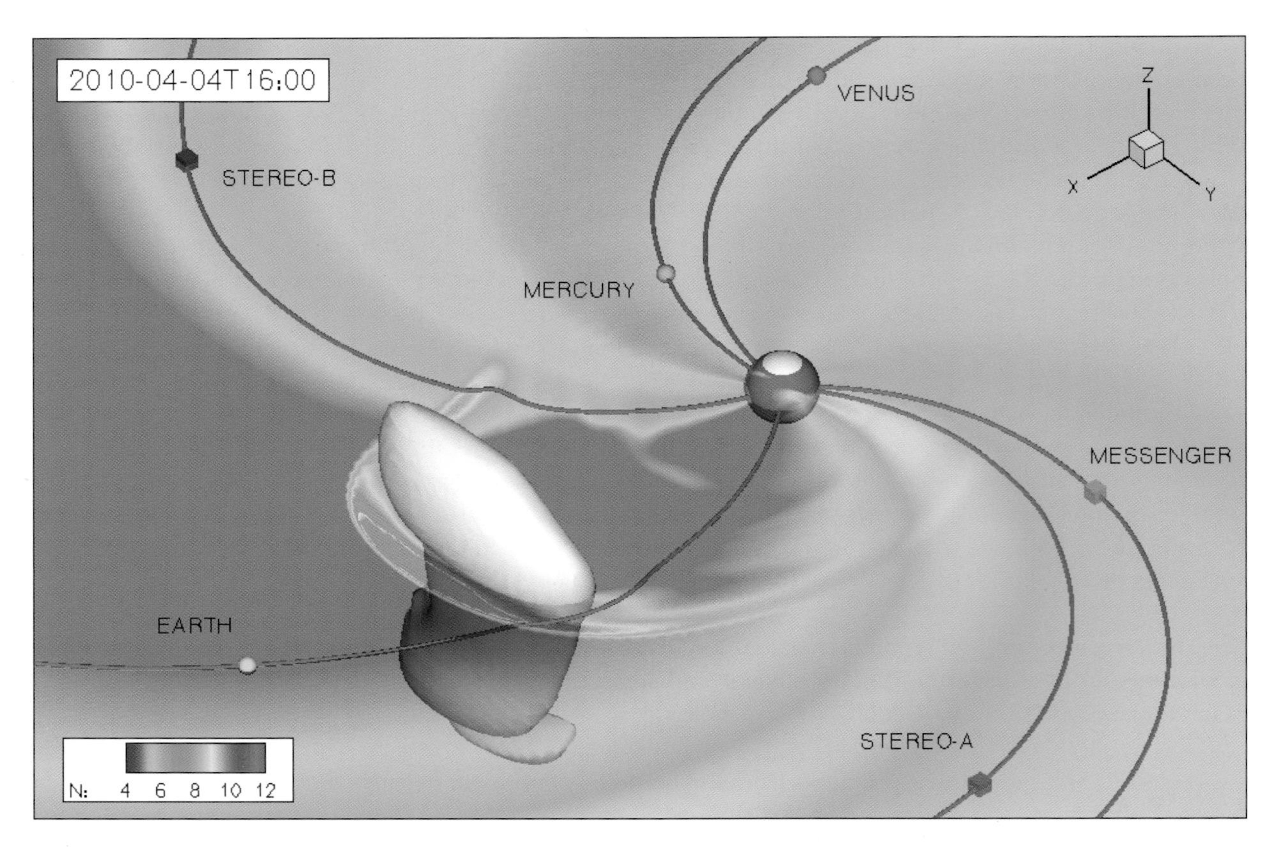

FIGURE 4.11 WSA-Enlil simulation of solar wind transient associated with Galaxy 15 failure in April 2010. Solar wind density is indicated by color in the heliospheric equatorial plane, and a coronal mass ejection approaching Earth is shown by the three-dimensional white structure. This ejecta drives a moderate interplanetary shock with a speed of >900 km/s. Interplanetary magnetic field lines are shown by red lines. SOURCE: Courtesy of Dusan Odstrcil, George Mason University.

From an operational perspective, it is clear that models are essential to predict the state of the system, to specify the current conditions, and to provide information at locations not served by sensors (see Figure 4.11). Recent results of "metric challenges"[18] and validation efforts demonstrate that models must be improved in order to meet the demands of both now-casting and forecasting. Although substantial progress has been made over the past decade in understanding the fundamental physics of space weather, leading to better physics-based, integrated models of the dynamic space environment, users can benefit from this improved understanding only if it is incorporated in operationally useful forecast tools. Transitioning to

[18]Acquiring quantitative metrics-based knowledge about the performance of various space physics modeling approaches is central for the space weather community. Quantification of the performance helps the users of the modeling products to better understand the capabilities of the models and to choose the approach that best suits their specific needs. See A. Pulkkinen, M. Kuznetsova, A. Ridley, J. Raeder, A. Vapirev, D. Weimer, R.S. Weigel, M. Wiltberger, G. Millward, L. Rastätter, M. Hesse, H.J. Singer, and A. Chulaki, Geospace Environment Modeling 2008-2009 Challenge: Ground magnetic field perturbations, *Space Weather* 9:S02004, doi:10.1029/2010SW000600, 2011.

operations requires time, resources, a dedicated effort, and a mind-set different from that brought to the problem by the science community.

The survey committee concluded that a national, multifaceted program is needed to transition research to operations more effectively by fully leveraging labor from different agencies, universities, and industry and by avoiding duplication of effort. Such a program could coordinate the development of models across agencies, including advance planning for developing and coupling large-scale models. The survey committee further concluded that each operations agency should develop the appropriate processes and funding opportunities to facilitate model transition to operations, which must include validation and establishment of metrics and skill scores that reflect operational needs. To a large extent, the value of research and operational codes can be gauged by whether end-user needs are being met. The survey committee concluded that further efforts are needed to better identify what these needs currently are and to anticipate what they will be in the future.

It is also important for NOAA to maintain a level of research expertise needed to work together with its partners, to provide professional forecasts and products, to define requirements, to understand possibilities for supporting customer needs, and to make wise and cost-effective choices about new models and data to support space weather customers.

On a broader level, the survey committee concluded that distinct funding lines for basic space physics research and for space weather specification and forecasting need to be identified and/or developed. This will require maintaining and growing the research programs at NSF, NASA, AFOSR, and ONR, and it will provide a more effective transition from basic research to space weather forecasting applications.

Accordingly, the survey committee recommends:

- **[A2.4] NOAA should establish a space weather research program to effectively transition research to operations.**
- **[A2.5] Distinct funding lines for basic space physics research and for space weather specification and forecasting should be developed and maintained.**

5

NSF Program Implementation

In this chapter the survey committee discusses its recommendations to the National Science Foundation (NSF) in the context of the committee's recommended program in solar and space physics. Where appropriate, the chapter also addresses connections to the 2010 astronomy and astrophysics decadal survey, *New Worlds, New Horizons in Astronomy and Astrophysics*,[1] which also made recommendations concerning NSF ground-based solar physics facilities and programs. Cost implications are considered, but because the recommendations to NSF are not fit to a specific budget, the committee does not prioritize its recommendations.

GROUND-BASED OBSERVATIONS

The committee's baseline priority for NSF is to support existing ground-based facilities and to complete programs in advanced stages of implementation. These programs are described in Chapter 1 and illustrated in Figure 1.2 (the Heliophysics Systems Observatory). The global nature of solar and space physics is such that it requires a synergistic complement of space- and ground-based observational approaches, which both support and are supported by theory and modeling. Ground-based observations are also increasingly used in near-real-time data-driven models of the heliosphere and space weather. Synoptic and long-term measurements from ground-based instruments are essential for capturing the complex dynamics of geospace and observing long-term trends. Ongoing ground-based observations of the Sun likewise facilitate studies of long-term variations, as well as revealing solar features with the finest spatial resolution and presenting unique views of solar eruptions.

Maintaining ground-based observatories also requires that NSF maintain and develop, as necessary, systems for accessing, archiving, and mining synoptic and long-term data sets (see Appendix B and Box 4.1). Furthermore, in DRIVE "Integrate," the committee describes the importance of expanding and formalizing the ground-based program's contribution to the success of NASA Explorer- and strategic-class science so

[1] National Research Council, *New Worlds, New Horizons in Astronomy and Astrophysics*, The National Academies Press, Washington, D.C., 2010.

that increasingly important synergies between ground- and space-based observations can be fully realized (see below, e.g., the subsection "A Heterogeneous Ionospheric Facility Network").

Advanced Technology Solar Telescope

When it begins operation in 2018, the 4-meter ATST will be, by far, the largest optical solar telescope in the world. Its ability to reach down to the fundamental photospheric density scale-length as a magnetometer, and to remotely sense coronal magnetic fields where they have never been measured, is revolutionary. However, despite facility closures by the National Solar Observatory (NSO), a significant increase in NSO base funding will be required to fully exploit the capabilities of the ATST. The NSO long-range plan estimates that ATST operations and data services will require at least $18 million per year, plus $4 million per year for NSO synoptic programs. Research grants and advanced instrumentation development would require additional funds.

The committee's DRIVE initiative "Realize" recommendations emphasize the importance of NSF providing the ATST with base funding sufficient for operation, data analysis and distribution, and development of advanced instrumentation for the ATST in order to realize the scientific benefits of this major national investment. This emphasis agrees with the 2010 astronomy and astrophysics decadal survey recommendation regarding the need to develop a funding model for ATST operation, instrumentation, and scientific research.

Midscale Funding Line

Important research is often accomplished through midscale research projects that are larger in scope than typical single principal investigator (PI)-led projects (MRIs) and smaller than facilities (MREFCs). The Advanced Modular Incoherent Scatter Radar (AMISR) is an example of a midscale project widely seen as having transformed research in the ground-based AIM community. Although different NSF directorates have programs to support unsolicited midscale projects at different levels, these programs may be overly prescriptive and uneven in their availability, and practical gaps in proposal opportunities and funding levels may be limiting the effectiveness of midscale research across NSF. It is unclear, for instance, how projects like the highly successful AMISR would be initiated and accomplished in the future. Mechanisms for the continued funding of management and operations at existing midscale facilities are also not entirely clear.

The NSF Committee on Programs and Plans formed a task force to study how effectively it supports midscale projects, how flexible the funding is, how uniformly it is administered across NSF, and how well such projects serve the interests of education and public outreach. The resulting report affirmed the importance of strongly supporting midscale instrumentation but did not recommend any new or expanded NSF-wide programs. Nevertheless, as described in Chapter 4 in the "Diversify" recommendations of DRIVE, the committee strongly endorses the creation of such a competitively selected midscale project line for solar and space physics. This approach is also consistent with the 2010 astronomy and astrophysics decadal survey, which recommended a midscale line as its second priority in large ground-based projects.[2]

---

[2] In that report, the recommendation for NSF to establish a "Mid-Scale Innovations Program" was accompanied by the following:

> New discoveries and technical advances enable small- to medium-scale experiments and facilities that advance forefront science. A large number of compelling proposed research activities submitted to this survey were highly recommended by the Program Prioritization Panels, with costs ranging between the limits of NSF's Major Research Instrumentation and MREFC programs, $4 million to $135 million. The committee recommends a new competed program to significantly augment the current levels of NSF support for midscale programs. An annual funding level of $40 million per year is recommended—just over double the amount currently spent on projects in this size category through a less formal programmatic structure. The principal rationale

Candidates for a Midscale Line

This survey committee's white-paper process and the subsequent disciplinary panel studies brought forward a number of important heliophysics projects that would require a new midscale funding line. The examples below illustrate the kind of science that the line could enable. The survey committee chose not to explicitly rank these projects but notes that the first two have well-developed science and implementation plans and have already been vetted by NSF. These projects are seen as being central to the integrated science program outlined in this report and as highly synergistic with the ATST as well as NASA flight programs.

The Frequency-Agile Solar Radiotelescope (FASR)

Designed specifically for observing the Sun, FASR will produce high-quality images of radio emissions in the 50-MHz to 21-GHz band with fine spatial, spectral, and time resolution. The radio emissions of interest convey unique, otherwise inaccessible information about the solar atmosphere and the acceleration of energetic particles. Discoveries in the areas of quiet sun physics, the evolution of coronal magnetic fields, solar flares, and space weather drivers are anticipated with the undertaking of this project. FASR was ranked highly by both the 2003 solar and space physics decadal survey[3] and by the 2010 astronomy and astrophysics decadal survey.[4]

The Coronal Solar Magnetism Observatory (COSMO)

COSMO will make continuous synoptic measurements of the corona and chromosphere, investigating solar eruptive events that are central to space weather and other solar-cycle-timescale and long-term coronal phenomena. Observations will show how the coronal magnetic field behaves across the sunspot cycle and how the polarity reversal of the global field affects the heliosphere. COSMO data will provide information about interactions between magnetically closed and open regions that determine the changing structure of the heliospheric magnetic field. The large field of view and continuous observations of COSMO will complement high-resolution, but small field-of-view, coronal magnetic field observations that may be made by the ATST.

Additional Projects

In addition, the committee identified four other projects that would be suitable for the midscale line. These projects are not yet well developed but represent the kind of creative approaches that will be necessary for filling the gaps in observational capabilities and for moving the survey's integrated science plan forward. They are the following.

---

for the committee's ranking of the Mid-Scale Innovations Program is the many highly promising projects for achieving diverse and timely science.

See National Research Council, *New Worlds, New Horizons in Astronomy and Astrophysics*, 2010, p. 23.

[3]National Research Council, *The Sun to the Earth—and Beyond: A Decadal Research Strategy in Solar and Space Physics*, The National Academies Press, Washington, D.C., 2003; and National Research Council, *The Sun to the Earth—and Beyond: Panel Reports*, The National Academies Press, Washington, D.C., 2003.

[4]National Research Council, *New Worlds, New Horizons*, 2010.

An All-Atmosphere Lidar Observatory

The most significant discoveries in the AIM discipline over the past decade involve increased appreciation of the influence of neutral atmospheric waves and instabilities on ionospheric structure and dynamics. An impediment to further research is the lack of direct, ground-based observations of the dynamics and thermodynamics of the mesosphere and, crucially, the thermosphere. Recent technical developments in the areas of high-power Rayleigh lidar and new resonance lidars now offer the possibility of wind and temperature measurements from the ground well into the thermosphere for the first time. A lidar observatory capable of observing gravity waves and tides and associated phenomenology in the mesosphere and lower thermosphere would accelerate discovery across the AIM discipline.

A Heterogeneous Ionospheric Facility Network

Processes central to AIM science are multiscale in nature, with global features that extend from the equator to the poles together with local features such as embedded small-scale irregularities that intermittently affect communications. Examples include traveling ionospheric disturbances, regions of storm-enhanced density, and the ionospheric response to sudden stratospheric warming events. Capturing these phenomena will require the deployment of an autonomous network of heterogeneous instruments, using optical and radio remote sensing techniques to measure neutral winds and temperatures, plasma densities, and plasma irregularities. Such a network would become a valuable facility in its own right, comparable to an EarthScope USArray[5] for heliophysics, and would also be the ground-based counterpart to space-based investigations, complementing everything from CubeSat projects to NASA strategic missions.

A Southern-Hemisphere Incoherent Scatter Radar

The AMISR phased-array incoherent scatter radar has proven to be a most incisive instrument for measuring the state properties of the ionosphere with panoramic coverage and high precision. Unknown, however, is the degree of inter-hemispheric conjugacy that can be assumed. The next logical step is to deploy an AMISR face in the Southern Hemisphere, expanding the latitudinal coverage of the heterogeneous network further. A deployment in the Antarctic region in particular would allow for the first ground-based assessment of conjugacy of geomagnetic storms.

Next-Generation Ground-Based Instrumentation

There is a need to support continuing instrumentation and technology development for ground-based solar physics in both the national facilities and the universities. Support for advanced instrumentation and seeing-compensation techniques for the ATST and other solar telescopes is necessary to keep ground-based solar physics at the cutting edge. At the same time it is necessary to ensure that adequate support is available to nurture young scientists and engineers in the field of solar instrumentation. That implies a need for adequate funding and good career opportunities, including the opportunity to work on exciting new instrumentation projects.

[5]See EarthScope, "USArray Instrumentation Network," available at http://www.earthscope.org/observatories/usarray.

CUBESATS

The efforts by NSF's Atmospheric and Geospace Sciences Division (AGS) to support student CubeSats have engendered an enormous amount of interest from universities and partner institutions. As of October 2011, eight CubeSat projects were underway. Launches have been scheduled (between 2011-2013) for all but two of the projects. NASA's ELaNA program is instrumental in obtaining launch opportunities and serves as a model for other small-satellite projects discussed by the survey.

All of the projects have been deemed by peer review to have well-defined, important science objectives and to provide unique data sets. All involve entirely new flight hardware and carry the promise of precedent-setting measurements. The limitations imposed by the small platforms demand a high degree of technical innovation in terms of power, control, storage, and downlink. CubeSats provide a unique platform for technological innovation whereby technical readiness can be developed to levels appropriate for application on larger spacecraft. Furthermore, most of the CubeSat hardware is designed, built, and tested by student teams under faculty and professional engineering supervision. Students in fact participate in every aspect of a CubeSat project.

Each CubeSat project requires approximately $0.4 million of funding annually. NSF is targeting a continuous queue of six CubeSat projects, with two new starts and two launches each year. This plan will require approximately $2.5 million of sustained annual funding. Current AGS budgets allow for approximately $1.5 million annually. There is therefore a shortfall of about $1 million per year.

The CubeSat program has clearly moved beyond its initial trial phase and has demonstrated great success, particularly in areas of education. As described in Chapter 4 in the "Diversify" recommendations of DRIVE, the survey committee believes that the program deserves its own line of funding at the level necessary to sustain two starts per year. The committee also recommends specific metrics to be employed for assessing the adequacy of the size of the program going forward. The committee is enthusiastic about the prospects of the CubeSat program for contributing to technology development as well as basic research.

EDUCATION

Faculty and Curriculum Development

As recommended in Chapter 4 in the "Educate" component of DRIVE, the committee endorses the continuation of the successful NSF Faculty Development in Space Sciences (FDSS) program, as well as the development of a complementary curriculum development program. The committee also recommends that 4-year institutions of higher education should be considered eligible for FDSS awards as a means to further broaden and diversify the field, subject to the burden of proof that program objectives pertaining to research education are achievable by the proposing institution. As existing FDSS awards come to term, the program is expected to change, with new awards being staggered to avoid boom-bust faculty hiring cycles. The number of junior faculty in the FDSS queue will likely remain the same, and so the burden on other AGS programs will also remain constant.

Undergraduate and Graduate Training

The NSF Research Experiences for Undergraduates program is an excellent means to attract talented undergraduates to the field, and the committee has endorsed it in the "Educate" component of DRIVE, along with the various summer school offerings supported by NSF. Currently, these include the annual Polar Aeronomy and Radio Science (PARS) summer school, the AMISR school on incoherent scatter, and CISM. The total allocation for these schools is $200,000 per year. Additional schools take place at the

National Solar Observatory and as part of the annual CEDAR, GEM, and SHINE meetings. The committee notes the particular need for a replacement for the CISM school and also the desirability of providing opportunities for professional development of graduate students via community workshops. In addition, the skills needed to become a successful scientist go beyond such formal discipline training and include interpersonal and communication skills, awareness of career opportunities, and leadership and laboratory management ability. The committee endorses NSF programs that support postdoctoral and graduate student mentoring and recommends that NSF enable opportunities for focused community workshops that directly address professional development skills for graduate students. Finally, the committee endorses programs that specifically target enhancing diversity within solar and space physics, such as the NSF Opportunities for Enhancing Diversity in the Geosciences program.

MULTIDISCIPLINARY RESEARCH

Solar and space physics is intrinsically multidisciplinary and appears in more than one NSF division or directorate. The National Solar Observatory and the ATST are currently within the Astronomy Division of the Mathematical and Physical Sciences Directorate. The AGS Division within the Geosciences Directorate manages ionospheric and magnetospheric science, but also solar-heliospheric and space weather science. AGS is also the home of the National Center for Atmospheric Research (NCAR) and its High Altitude Observatory (HAO), which supports a broad range of research topics ranging from the Sun to Earth.

Funding Cross-Cutting Science

The placement of solar and space physics in multiple divisions and directorates arises from the cross-cutting relevance of the science. However, funding for basic research on subjects that are not clearly aligned with one division, and thus have no clear home at NSF, can be difficult to obtain. For example, sun-as-a-star and planetary magnetospheric research falls between the AGS and Astronomy divisions. Another timely example is the science of the outer heliosphere. Recent observations of the outer heliosphere by NASA satellites raise fundamental science questions pertaining to the structure of shocks, where and how magnetic reconnection takes place, and how particles are accelerated, all of which are subjects integral to the Sun-Earth-heliosphere system science program. The survey committee recommends in the "Integrate" element of the DRIVE initiative that NSF ensure that funding is available for basic research in subjects that fall between sections, divisions, and directorates, and that in particular the outer heliosphere be considered within the scope of the AGS Division. The committee further calls attention to the importance of maintaining a laboratory program to probe fundamental plasma physics.

Heliophysics Science Centers

Another way to promote cross-disciplinary research is via critical-mass groupings of observers, theorists, modelers, and computer scientists who together target grand-challenge questions in the field of heliophysics, as recommended in "Venture" of the DRIVE initiative. The periodic competition for heliophysics science centers (HSCs) with substantial funding (at the level of $1 million to $3 million per year) will focus attention on the field in a way that is not possible with the present programmatic mix. The NSF Physics Frontier Centers are successful examples that have become highly competitive in the university community and might serve as models for the HSCs. They also have great potential for attracting faculty and students via their focus on exciting and challenging science.

Solar and Space Physics at NSF

The assets across NSF for solar and space physics are significant. The Astronomy Division of the Directorate for Mathematical and Physical Sciences is the home of the National Solar Observatory, with ongoing synoptic observations and the ATST under construction, and of the National Radio Astronomy Observatory, which includes some solar researchers. The AGS Division of the Geosciences Directorate has championed the CubeSats program, arguably the most innovative development in spaceflight over the past decade. AGS is also the home of solar and space physics research at NSF, both in the Geospace Section and in the NCAR/Facilities Section at HAO, which also runs the Mauna Loa Solar Observatory. AGS has increased its responsibility for the Arecibo Radio Observatory, which remains the largest-aperture telescope in the world for astrophysical, planetary, and atmospheric studies. It has pioneered the utilization of hosted payloads through its involvement with Iridium and Iridium NEXT, serving as a model for NASA in that respect.

The 2010 astronomy and astrophysics decadal survey[6] considered the future of NSF-supported solar research in view of its likely expansion in the ATST era. The current funding split, with the majority of grant funding coming from AGS and with the facilities funding divided between AGS and AST, was noted for being unusual and differing from the space-based solar research model. The 2010 report concluded that large facilities like the ATST would benefit from a more unified approach to how the two NSF divisions develop and support ground-based solar physics. It further encouraged NSF to work with the solar, heliospheric, stellar, planetary, and geospace communities to find a way to ensure a coordinated, balanced ground-based solar astronomy program able to maintain multidisciplinary ties. The relevance and importance of these recommendations have not diminished in the intervening time.

A more unified approach to solar and space physics at NSF would help establish the field as a professional discipline. The FDSS program also works in this direction, and the committee has emphasized in DRIVE "Educate" the need for NSF to make solar and space physics an officially recognized subdiscipline of physics and astronomy. Currently it is not listed as a dissertation research area within NSF's Annual Survey of Earned Doctorates, an omission that influences other rankings, ratings, and the demographic surveys done by the National Research Council and the American Institute of Physics. Ultimately, recognition of solar and space physics as an official subdiscipline will enhance its visibility and the ability to recruit future space scientists.

INTERNATIONAL COLLABORATIONS

A comprehensive investigation in solar and space physics cannot take place in isolation but should be part of an international effort, with different countries able to bring to bear unique geographic advantages, observing platforms, and expertise. The research community in the United States is poised to participate in and take advantage of a number of emerging international initiatives that could contribute to the fulfillment of the overall strategy recommended by the survey committee. For example, the international incoherent scatter radar consortium EISCAT is embarking on the EISCAT3D project, a very large, distributed, multistatic, transceiving array that will be able to measure ionospheric state variables in three dimensions through incoherent scatter. The technological, analytical, and logistical challenges that must be addressed to realize EISCAT3D are daunting but could be overcome more easily with the participation of U.S. researchers, who would benefit enormously from access to this prototypical instrument. Another example is the International Space Weather Meridian Circle Program headquartered in China. This ambitious program seeks to fully instrument the 120E and 60W meridian in order to provide a global picture of unfolding space weather

[6]National Research Council, *New Worlds, New Horizons,* 2010.

events. Because they already maintain extensive arrays of space weather monitoring instruments in North and South America, researchers in the United States are natural partners for the program. The addition of the Asian half of the meridian will be useful for distinguishing local-time from storm-time space weather phenomenology.

While participation in international solar and space research projects could be accomplished through numerous individual, bilateral initiatives and agreements, the overall impact would be increased by coordinated agency involvement. The NSF in particular is well situated to help organize U.S. participation in these and other international projects.

6

NASA Program Implementation

THE NASA HELIOPHYSICS CORE PROGRAM

The systems approach for the study of the coupled Sun-heliosphere-Earth system that is advocated in this report requires flight opportunities that can provide observations of phenomena throughout the domain as well as adequate support of research and analysis activities to ensure that the full potential of the data is realized. With current projections showing at best level funding for solar and space physics over the coming decade, the survey committee was challenged to prescribe a program that would effectively "do more with less." Among the key considerations in this effort are the appropriate mix and repeat-cycle (cadence) of small-, medium-, and large-class missions, and matching these efforts with a robust program of research and analysis and theory and modeling activities that will exploit, complement, add value, and, in effect, extend experimental observations. Further, in order to take advantage of future opportunities, there is an evident need to develop a well-trained workforce

The survey committee's recommended core program for heliophysics includes the following elements, in descending order of priority:

1. Completion of the implementation of the NASA missions that are currently selected, with commitment to maintaining agreed-to costs and schedule commitments. These missions include RBSP (renamed the Van Allen Probes), MMS, Solar Probe Plus, and Solar Orbiter, along with IRIS and other already-selected Explorers.

2. Implementation of the DRIVE initiative as an augmentation to the existing enabling research program. The DRIVE components provide for operation and exploitation of the Heliophysics Systems Observatory for effective research programs. The community must be equipped to take advantage of new innovative platforms.

3. Execution of a robust Explorer program with an adequate launch rate, including Missions of Opportunity. The cadence should be accelerated to accomplish the important science goals that do not require larger missions and to provide access to space for all parts of the discipline.

4. Launch of strategic missions in the reinvigorated Solar-Terrestrial Probes (STP) line and in the Living With a Star (LWS) line to accomplish the survey committee's highest-priority science objectives. This element includes first the notional IMAP investigation and then DYNAMIC and MEDICI in the STP program and GDC as the next larger-class LWS mission.

Figure 6.1 shows a proposed implementation of the core program, in which each of the assets required to achieve the goals of the survey committee's recommended solar and space physics program have their

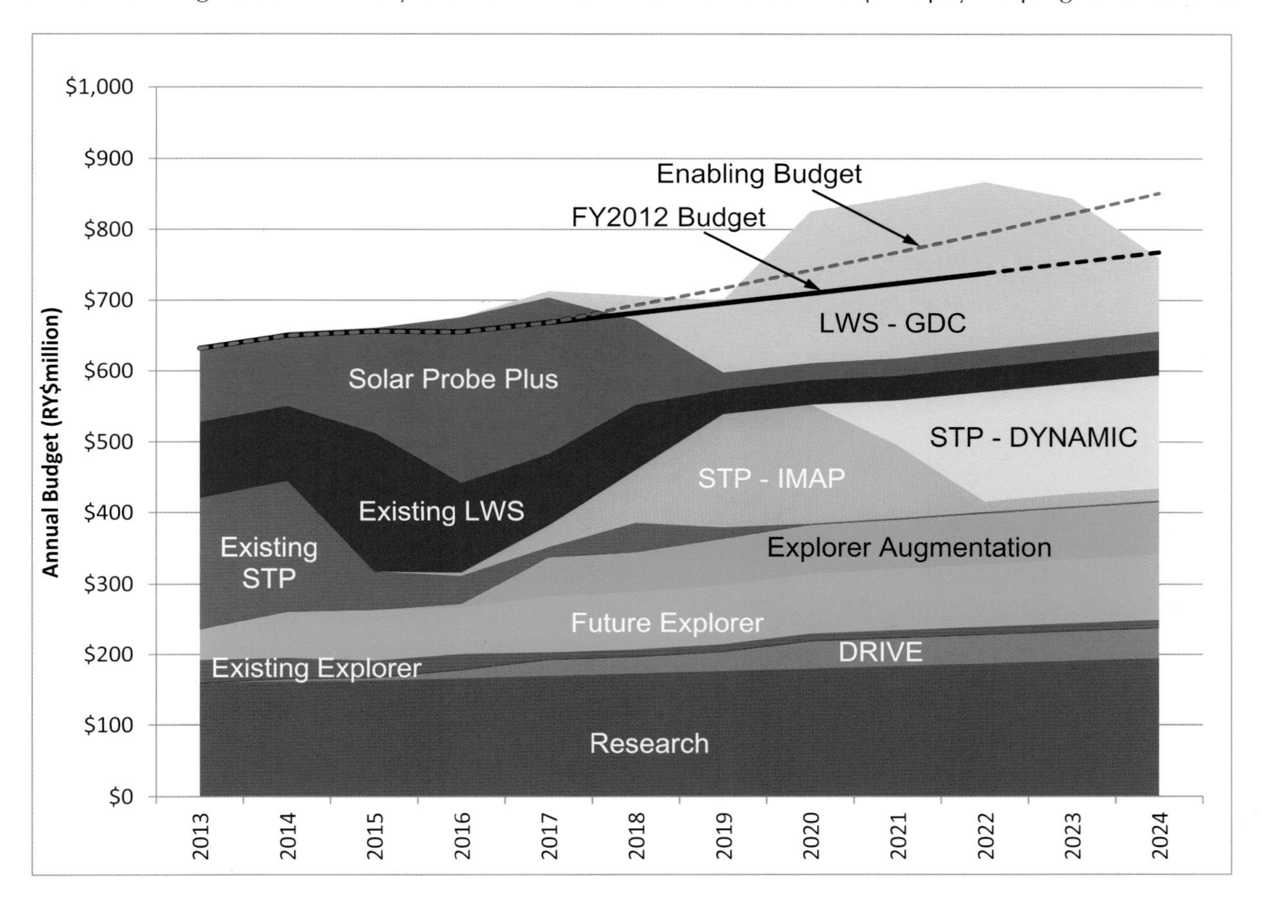

FIGURE 6.1 Heliophysics budget and program plan by year and category from 2013 to 2024. The solid black line indicates the funding level from 2013 to 2022 provided to the survey committee by NASA as the baseline for budget planning, and the black dashed line extrapolates the budget forward to 2024. After 2017 the amount increases with a nominal 2 percent inflationary factor. Through 2016 the program content is tightly constrained by budgetary limits and fully committed for executing existing program elements. The red dashed "Enabling Budget" line includes a modest increase from the baseline budget starting in 2017, allowing implementation of the survey-recommended program at a more efficient cadence that better meets scientific and societal needs and improves optimization of the mix of small and large missions. From 2017 to 2024 the Enabling Budget grows at 1.5 percent above inflation. (Note that the 2024 Enabling Budget is equivalent to growth at a rate just 0.50 percent above inflation from 2009.) Geospace Dynamics Constellation, the next large mission of the LWS program after Solar Probe Plus, rises above the baseline curve in order to achieve a more efficient spending profile, as well as to achieve deployment in time for the next solar maximum in 2024. NOTE: LWS refers to missions in the Living With a Star line, and STP refers to missions in the Solar-Terrestrial Probes line.

proper cadence, within a budget profile that should be attainable. The recommended program addresses in a cost-effective manner many of the most important and interesting science objectives, but the anticipated budget significantly constrains what can be accomplished. Built on top of the existing research foundation, the core program recommended here ensures that a proper distribution of resources is achieved. In particular, it restores a balance between small, medium, and large missions.

Figure 6.2 illustrates how, in the past decade, the number of large missions has increased at the expense of medium and small missions. With implementation of the committee's recommended program, the bal-

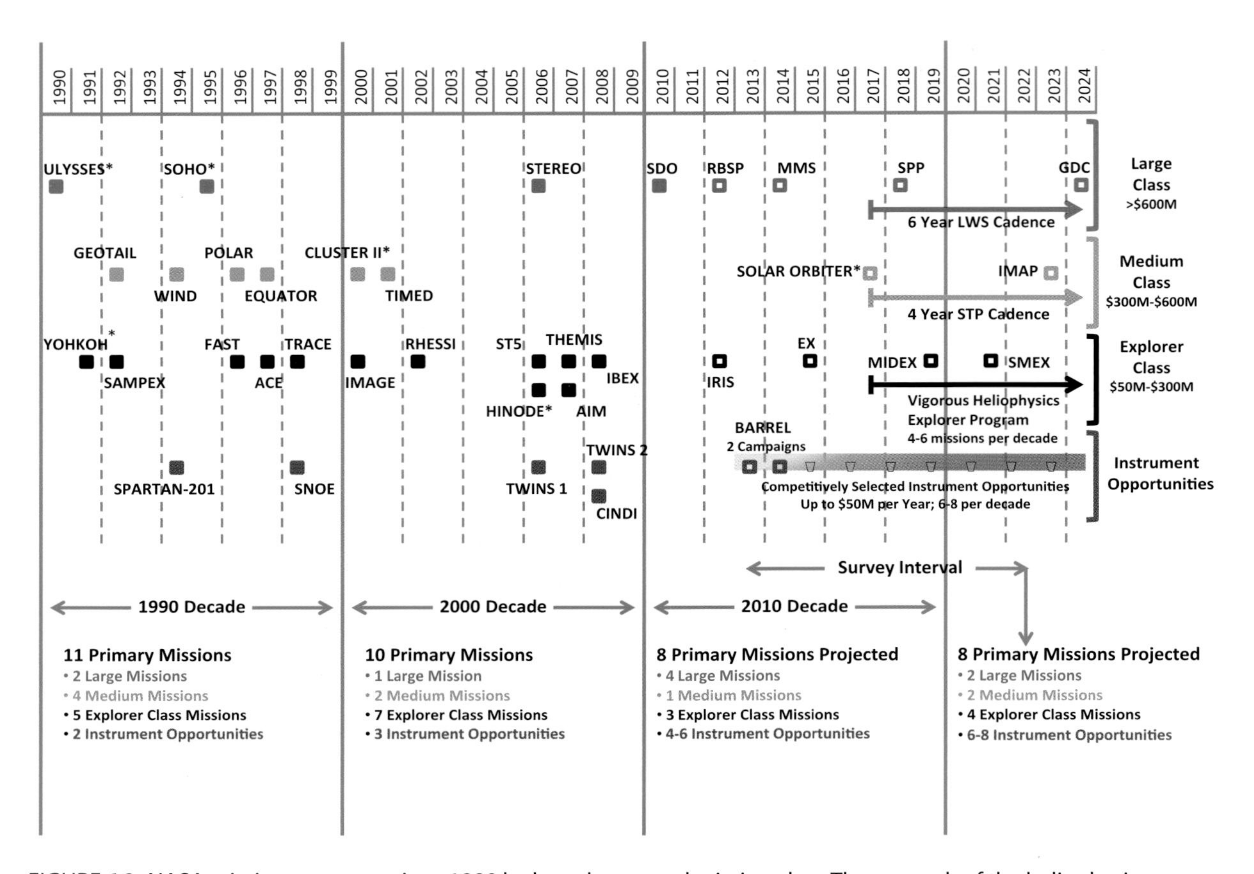

FIGURE 6.2 NASA mission sequence since 1990 by launch year and mission class. The strength of the heliophysics program over the past two decades has been the regular cadence of missions in a variety of sizes. It can be seen that the 1990 and 2000 decades each had 13 missions, with the Medium- and Explorer-class categories having 9 missions in each decade. A trend toward a loss of balance can be seen in the 2010 decade, in which the mission complement tipped toward fewer missions with a bias toward the Large category.

A key objective for the next survey interval is to restore the number of Medium- and Explorer-class missions such that, in combination with competitively selected instrument opportunities on hosted payloads (Missions of Opportunity), a higher cadence can be achieved that is capable of maintaining the vitality of the science disciplines. As discussed in the text and described in Figure 6.1, funding constraints affect the restoration and rebalance of the programs such that realization of the survey committee's recommended strategy cannot begin until after 2017.

Missions denoted by an asterisk (*) demonstrate the importance of international collaborations in which the heliophysics community has a long, active, and very fruitful mission history. Open boxes indicate missions that are in development or are planned but have not yet flown.

ance between mission size and enabling research is restored. As shown in Figure 6.3, at the beginning of the 2013-2024 decade large missions from both the LWS and the STP lines dominate the budget. By the end of the decade, a balance between large, mid-size, Explorer, and enabling assets (baseline research plus DRIVE) has been achieved. The phasing that leads to this rebalance is implemented based on the priorities accorded to the elements of the core program. During the early part of the decade, when there is very little

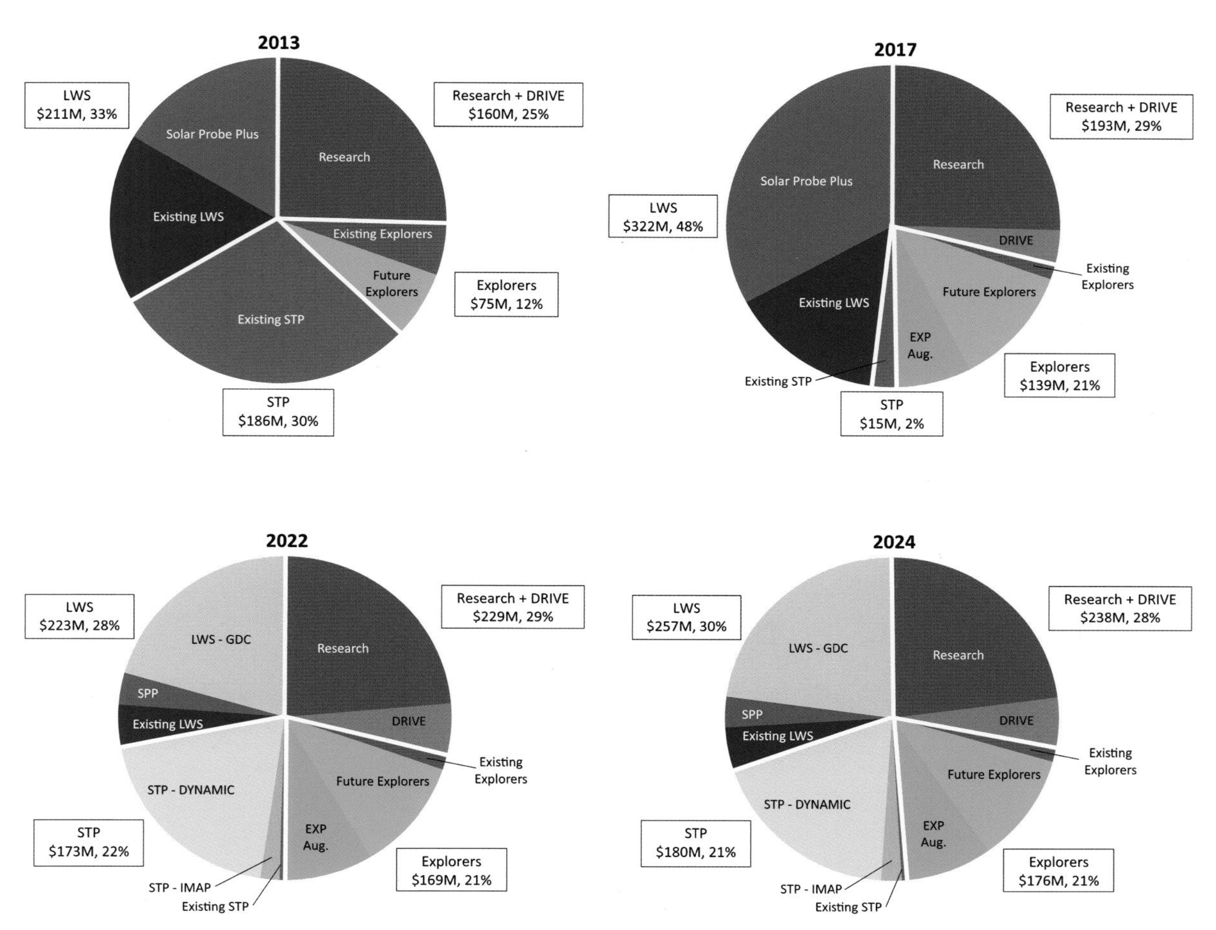

FIGURE 6.3 Effects of strategic rebalancing. These pie charts illustrate the evolution of program balance among four core program elements as the survey-recommended plan is implemented over the survey interval, 2013-2022, plus 2 years. The charts reflect that much of the NASA heliophysics budget from 2013 to 2017 is already locked in. The year 2017 is effectively the start for an enabling trajectory. An important result apparent in 2017 is the effect of implementing the DRIVE initiative and the restructured Explorer program. It can be seen that the sum of Research + DRIVE and Explorers (R + D + E) increases from approximately 37 percent to the intended value of 50 percent of the total program budget.

The year 2022 represents the endpoint of the survey interval and occurrence of the overall rebalancing of the program with the maintenance of the 50 percent R+D+E funding but also a sustainable division between the STP and LWS programs with the remaining 50 percent of the budget. The year 2024 represents the planned endpoint of the rebalancing initiative as well as the legacy profile for the next survey. An important legacy result is how the program stably maintains the balance moving forward from 2022. Note that the 2022 and 2024 budgets are represented by the enabling-trajectory budget and do not include the Geospace Dynamics Constellation bump, which is considered a plus-up from the base program.

flexibility in the NASA heliophysics program, the focus is on completing the implementation of existing missions, as well as on maintaining the baseline level of enabling research programs and Explorers. The first new recommendations of this survey to be acted on pertain to the DRIVE initiative. Early implementation of DRIVE ensures the fastest possible return on new investments for the decade (Figure 6.4).

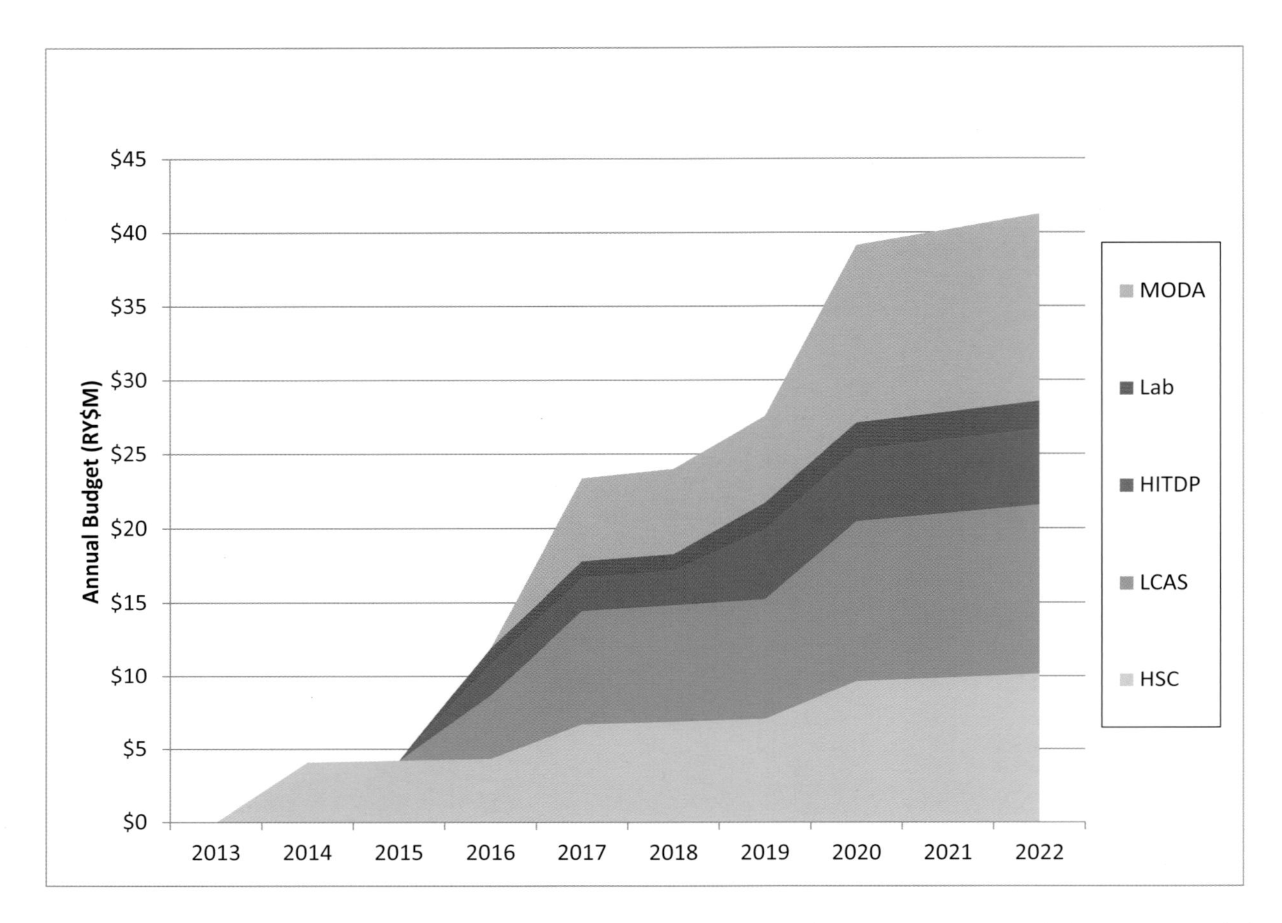

FIGURE 6.4 NASA DRIVE implementation. For the cost of a small mission, the DRIVE initiative includes recommended augmentations to NASA mission-enabling programs that have been carefully chosen to maximize the effectiveness of the program overall. Six of the DRIVE sub-recommendations made in Chapter 4 have a cost impact for NASA. Of these, a NASA mission guest investigator program would require a cost allocation within the Solar-Terrestrial Probes and Living With a Star missions of ~2 percent of total mission cost for a directed guest investigator program. Shown above are expected costs for the other five: a NASA tiny-satellites augmentation to the Low-Cost Access to Space program (LCAS), Mission Operations and Data Analysis augmentation (MODA), heliophysics science centers (HSCs), a heliophysics instrument and technology development program (HITDP), and a multiagency laboratory experiments program (Lab). They have been phased with a slow start because of budget constraints, and in sequence that allows for time to develop and ramp up new programs. Note that the MO&DA augmentation begins in 2016, at a time when the Solar Dynamics Observatory will have moved out of prime mission, adding greatly to data covered by the general guest investigator program. Implementation of the NASA portion of DRIVE ramps up by 2022 to an augmentation to existing program lines that is equivalent to approximately $33 million in current (2013) dollars. Note that in developing this figure, the survey committee assumed a 2.7 percent rate of inflation, which is what NASA currently assumes as the inflation factor to be used for its new starts.

The next priority element is the Explorer augmentation, and this is also to be acted on as early in the decade as possible. Explorer missions are cost-effective means of strategically pursuing new and exciting science. The recommended augmentation of the Explorer line allows for a restored MIDEX line to be deployed alternately with SMEX missions at a 2- to 3-year cadence and for regular selection of Missions of Opportunity.

The committee's highest-ranked STP and LWS missions, IMAP and GDC, are implemented next, exploring the outer heliosphere and near-Earth space, respectively. The minimum recommended cadence for the STP missions is 4 years, and so IMAP, as a moderate-size mission, could launch in 2021 and be followed by DYNAMIC and MEDICI, the next-highest-ranked STP science targets (Figure 6.5). The recommended cadence for LWS is 6 years (Figure 6.6), and so only GDC could begin in this decade. Moreover, launching GDC even as early as 2024, as shown in Figure 6.1, would require an increase in the heliophysics budget,

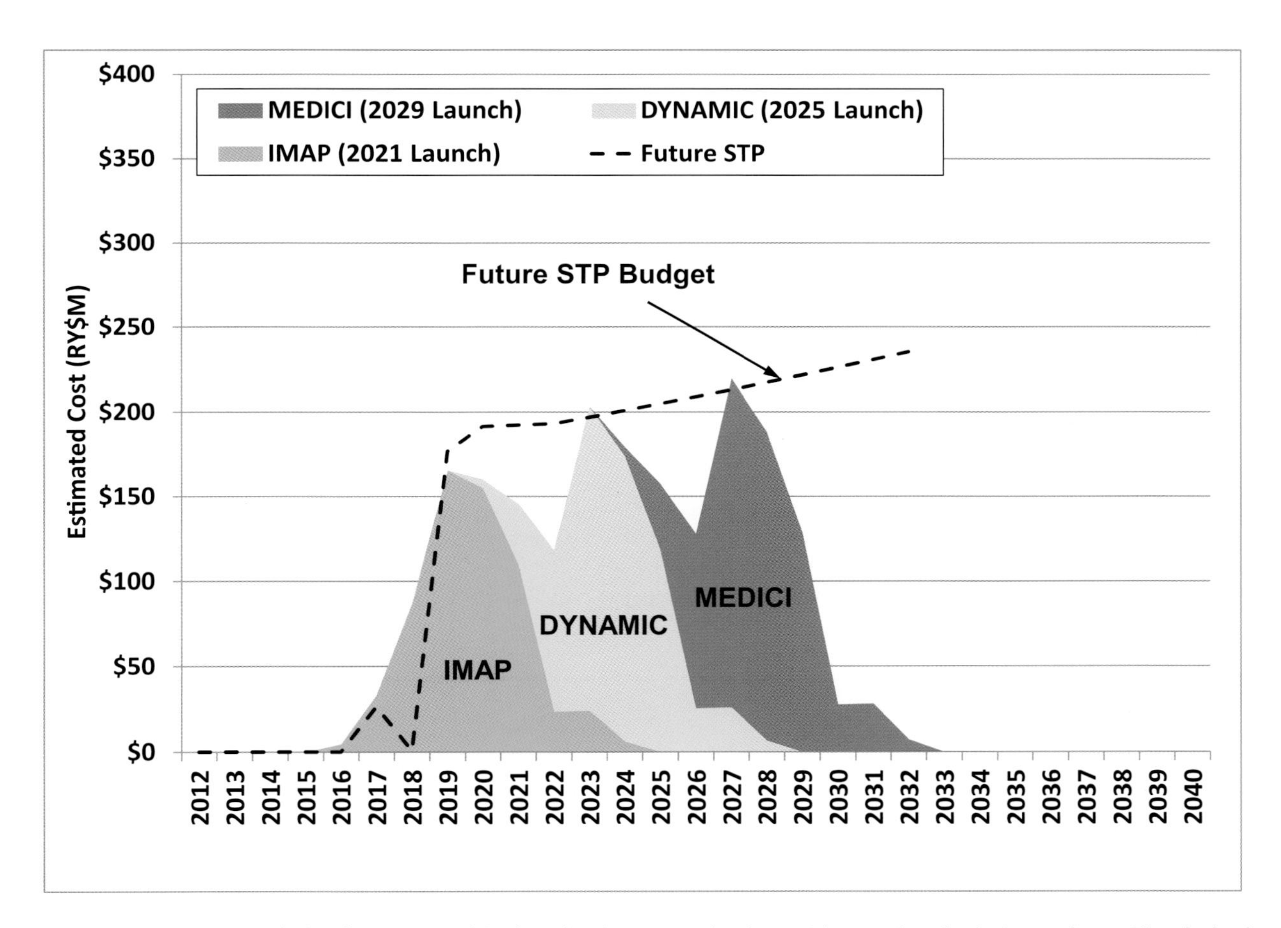

FIGURE 6.5 Recommended Solar-Terrestrial Probes (STP) program budget with associated mission cadence. The dashed line represents the recommended funding level for the STP program including a 2 percent inflation slope. Funding at the full level occurs in approximately 2019, with a 4-year mission cadence beginning with IMAP and then continuing with DYNAMIC and MEDICI. The dips in funding between missions are intentional, with the objective of allowing for both technology development and mission extensions.

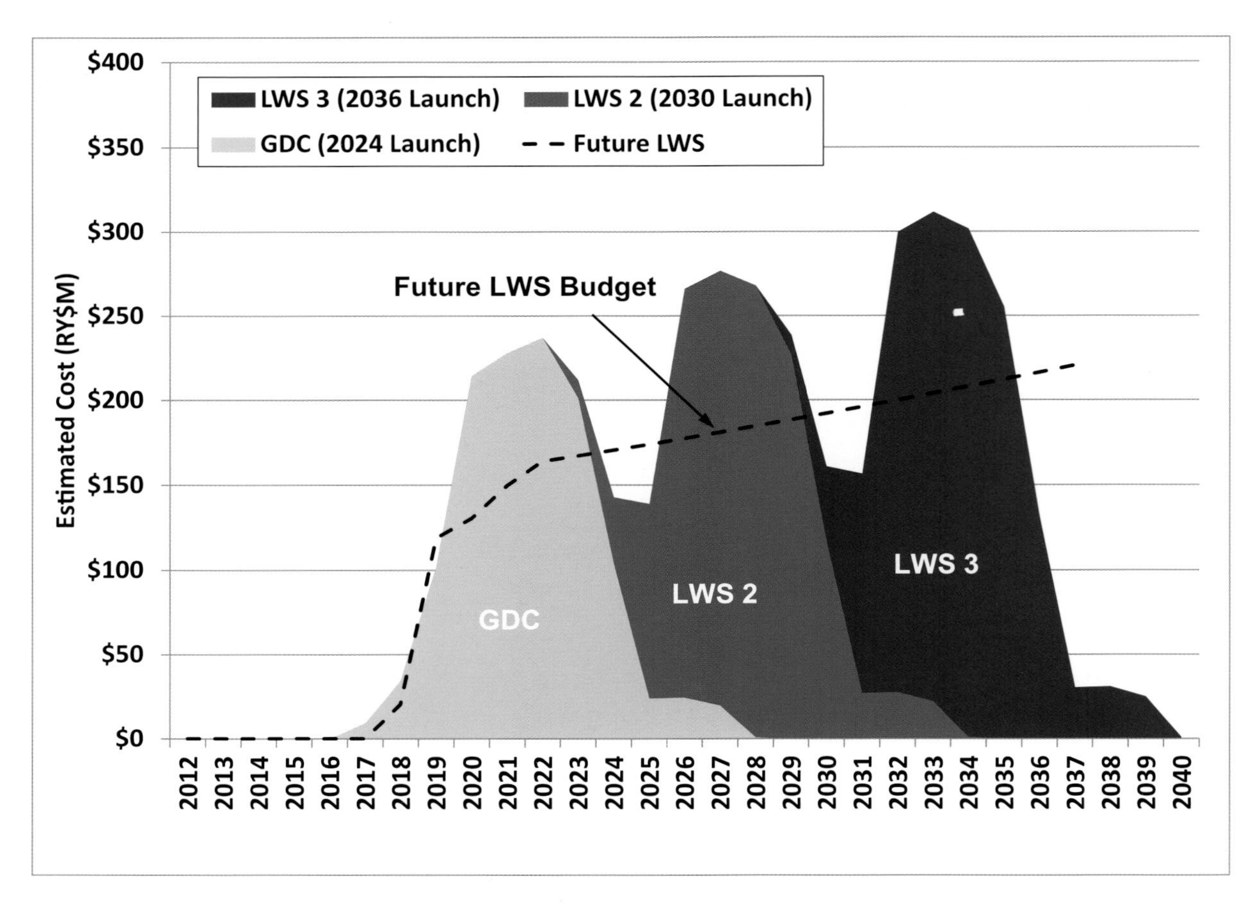

FIGURE 6.6 Recommended Living With a Star (LWS) program budget with associated mission cadence. The dashed line represents the recommended base funding level for the LWS program, including a 2 percent inflation slope. Funding at the full level occurs near the end of the decade, with a 6-year mission cadence beginning with GDC. The bumps above the future LWS budget line in mission funding are intentional in order to achieve the most cost-effective implementation at the preferred cadence. The actual bump will be based on mission cost and the associated development plan.

beyond a simple extension of the current budget. The increase would permit the timely execution of a major mission in coordination with related STP missions. However, if the required increase is not possible, it is appropriate to delay both the start and the launch date of GDC.

DECISION RULES AND AUGMENTATION PRIORITIES

The recommended program for NASA addresses important and highly compelling science objectives in a cost-effective manner. However, the survey committee recognizes that an already significantly constrained program could face further budgetary challenges. Decision rules are required to preserve an orderly and effective program in the event that less funding than anticipated is available, or some other disruptive event occurs. The decision rules need to ensure that, under any funding profile, balanced progress can be made across the subdisciplines of solar and space physics and that the only adjustment that is possible, i.e., the cadence of NASA assets, is properly applied. To further guide the allocation of resources, the committee recommends the following decision rules.

Recommended Decision Rules for Maintaining Balance Under More Constrained Budgets

The decision rules should be applied in the order shown to minimize disruption of the higher-priority program elements.

Decision Rule 1. Missions in the STP and LWS lines should be reduced in scope or delayed to accomplish higher priorities (see below for explicit triggers for review of the Solar Probe Plus mission[1]).

Decision Rule 2. If further reductions are needed, the recommended increase in the cadence of Explorer missions should be reduced, with the current cadence maintained as the minimum.

Decision Rule 3. If further reductions are still needed, the DRIVE augmentation profile should be delayed, with the current level of support for elements in the NASA research line maintained as the minimum.

Rule 1 calls for the reduction in scope or delay of missions in the LWS or STP lines as the first line of defense against budget stress. However, the committee is aware that in the early years of the decade there is very little flexibility in the NASA heliophysics program. Only Solar Probe Plus (SPP) is not already in design and development, Phase C/D. SPP science remains important and timely (Box 6.1), but the mission is costly and technically challenging. Significant cost growth beyond the current cap threatens to disrupt the balance of the total program and should not be accepted without careful consideration. While a low-cost delay imposed early in the decade may have minimal impact, the committee otherwise recommends the following specific triggers for NASA to initiate a review of the SPP mission in order to maintain program balance during the first 5 years:

Trigger 1. A decrease in the heliophysics budget expected to cause an interruption in the current cadence of Heliophysics Explorers lasting more than 1 year, or that would impact the remainder of the core research program.

Trigger 2. A decrease in the heliophysics budget that prevents the Heliophysics Division from launching SPP and at least one new STP mission before 2022.

Trigger 3. An increase in the total SPP life-cycle cost above NASA's projected $1.23 billion level,[2] irrespective of where the cost growth occurs.

---

[1] In accordance with its charge, the survey committee did not reprioritize any NASA mission that was in formulation or advanced development. In June 2011, the survey committee's charge was modified by NASA to include a request for it to present "decision rules" to guide the future development of the Solar Probe Plus mission.

[2] The committee understood the life-cycle costs of SPP to be approximately $1.23 billion as of September 2011, which is consistent with the President's FY 2012 budget request announced in February 2011. NASA informed the survey committee that funding already appropriated for the advanced technologies required for SPP have retired substantial technical risk. Note: Life-cycle costs do not include potential mission extensions.

BOX 6.1 SOLAR PROBE PLUS—A MISSION TO THE CORE OF THE HELIOSPHERE

Solar Probe Plus (SPP) will travel closer to the Sun than any other spacecraft and explore the innermost region of our solar system—the corona. The 2003 decadal survey[1] recommended a solar probe intended to "determine the mechanisms by which the solar corona is heated and the solar wind is accelerated and to understand how the solar wind evolves in the innermost heliosphere." The present survey found that the scientific rationale for a solar probe remains compelling and concluded that SPP meets that challenge. SPP will study the streams of charged particles the Sun hurls into space from a vantage point where the processes that heat the corona and generate the solar wind actually occur. SPP will repeatedly sample the near-Sun environment and, using in situ measurements, reveal the mechanisms that produce the fast and slow solar winds, coronal heating, and the transport of energetic particles.

The region inward of 0.3 AU is one of the last unexplored frontiers in our solar system. Discovering how the solar wind originates and evolves in the inner heliosphere requires in situ sampling of the plasma, energetic particles, magnetic field, and waves as close to the solar surface as possible. SPP measurements will determine how energy flows upward in the solar atmosphere, heating the corona and accelerating the solar wind. SPP will also reveal how the solar wind evolves with distance in the inner heliosphere. During the past decade remote observations have revealed much about particle acceleration, heating, plasma turbulence, waves, and the flows of mass and energy in the corona. In the survey committee's view, these observations only increase the need for measurements from this critical region.

The current solar probe differs in several respects from the mission envisioned in the 2003 survey. The closest approach for SPP is 9.5 solar radii (R_S) instead of 4 R_S (Figure 6.1.1). The loss in proximity is significant but is more than compensated for by the opportunity to spend far more time close to the Sun and gather observations spanning half a solar cycle. This latter feature in particular makes the timing of the mission with respect to the solar activity cycle a less significant issue. SPP orbits in the ecliptic plane instead of a polar plane and thus leaves the polar regions unexplored. Nevertheless, fast solar wind streams typical of polar coronal holes will still be adequately sampled, albeit at larger radial distances from the Sun. Only one remote sensing instrument remains on SPP, but the most significant impairment comes from the lack of in situ solar wind composition measurements; the survey committee urges NASA to consider restoring this capability. However, in most other aspects SPP offers the prospect of a substantially enhanced science return compared to the earlier solar probe concept.

The four science investigations provide a combination of in situ and remote sensing capability. Three in situ instruments will make comprehensive measurements of the solar wind ion and electron thermal plasma, of suprathermal and energetic particles, and of magnetic and electric fields from DC to high frequencies. A side-looking imager will provide both global context and quasi-in situ measurements of density and dust in the corona from 2.2 to 20 R_S.

[1] National Research Council, *The Sun to the Earth—and Beyond: A Decadal Research Strategy in Solar and Space Physics*, The National Academies Press, Washington, D.C., 2003; and National Research Council, *The Sun to the Earth—and Beyond: Panel Reports*, The National Academies Press, Washington, D.C., 2003.

The committee emphasizes that what is called for is a review by NASA, in which the expected outcome will be actions by NASA with regard to SPP that preserve a program of balanced progress in heliophysics throughout the decade.

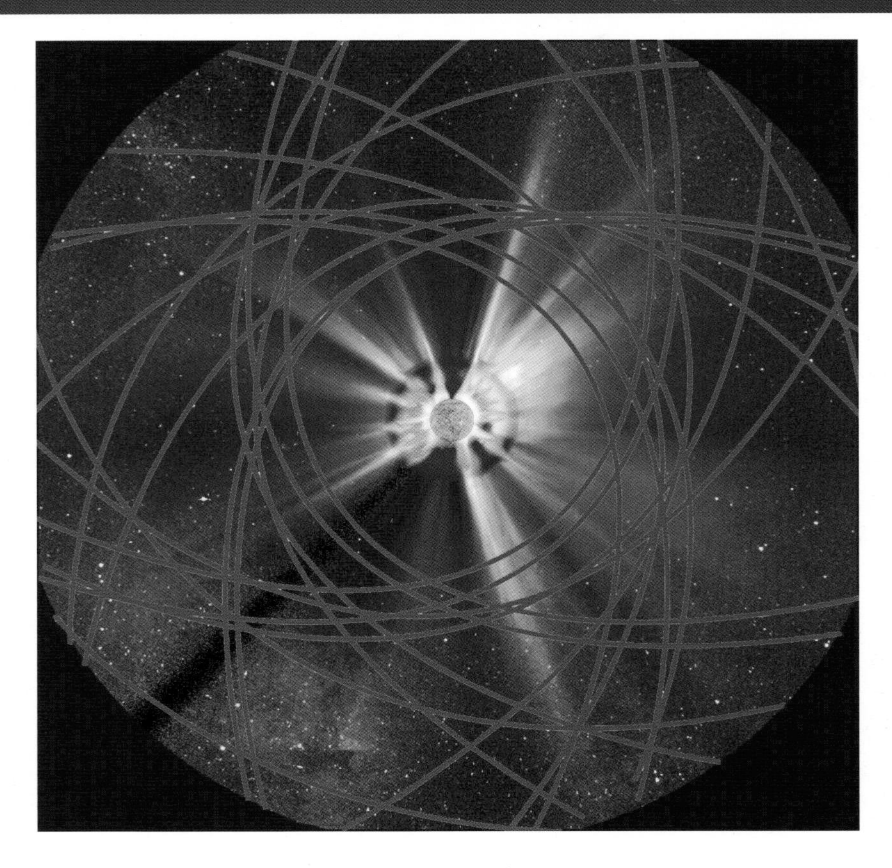

FIGURE 6.1.1 A simulated view of the Sun and inner heliosphere seen from above the pole illustrates the revamped trajectory of Solar Probe Plus during its 19 near-Sun passes inside 30 R_S. The spacecraft will spend nearly 1,000 hours within 20 R_S of the Sun and 30 hours within 10 R_S. The 3-month near-ecliptic orbit allows repeated measurements of the slow wind from the streamer belt as well as of the fast wind from equatorial coronal holes. Terrestrial observatories, from both ground and space, will provide global context measurements. SOURCE: NASA, *Solar Probe Plus: Report of the Science and Technology Definition Team*, NASA/TM-2008-214161, NASA Goddard Space Flight Center, Greenbelt, Md., July 2008.

The survey committee notes that the resources assumed in crafting this decadal survey's recommended programs are barely adequate to make the required progress; with reduced resources, progress will be inadequate. It is also evident that with increased resources, the cadence of the assets by which the nation pursues its program could be increased with a concomitant increase in the pace of scientific discovery and societal value.

Recommended Augmentation Priorities for Maintaining Balance Under More Favorable Budgets

The survey committee recommends the following augmentation priorities to aid in implementing a program under a more favorable budgetary environment:

Augmentation Priority 1. If even modest additional financial resources can be made available early in the decade, the implementation of the DRIVE initiative should be accelerated.

Augmentation Priority 2. Given sufficient funds throughout the decade, the Explorer line should be further augmented so as to increase the cadence and amount of funding available for missions including Missions of Opportunity.

Augmentation Priority 3. Given further budget augmentation, the cadence of STP missions should increase to allow the recommended third-priority mid-size science target (MEDICI) to be initiated in this decade.

Augmentation Priority 4. The major-mission-line recommendation (GDC) should be implemented in the most cost-effective manner, if possible with a funding bump as shown in Figure 6.1.

INTERNATIONAL COLLABORATIONS

Collaborations between NASA and foreign space agencies have historically achieved major science at a relatively low cost to the United States. Missions labeled with an asterisk in Figure 6.2 are examples of such international collaborations. These include SOHO, Ulysses, Yohkoh, CLUSTER, Hinode, and Solar Orbiter. Collaborations with the Japanese space agency have been particularly fruitful, with investments of Explorer-size funds having resulted in full U.S. participation in major missions (Yohkoh, Hinode). Such missions leverage U.S. investments while simultaneously sustaining a U.S. leadership role in science.[3]

Opportunities continue to arise for NASA to collaborate with other nations, and in so doing to obtain a high science return at a relatively low cost. The Solar-C mission, a follow-on to Yohkoh and Hinode that will study the magnetic coupling of the lower solar atmosphere and the corona, has been confirmed by Japan and would greatly benefit from contributions of instrumentation from the United States. A mission concept has also been developed in Japan, Canada, and Europe that involves a fleet of spacecraft performing simultaneous in situ measurements of fields and plasmas at key locations in the magnetosphere. Finally, NASA support for the U.S.-Taiwanese FORMOSAT-3/COSMIC microsatellite science mission for weather, climate, space weather, and geodetic research illustrates the range of possible collaborative opportunities that might arise. The augmented Explorer line is the recommended source of funding for participation in most such international collaborations.

[3] At the same time, collaborations with international partners add complexity and risk that must be actively managed. See, for example, National Research Council, *U.S.-European Collaboration in Space Science*, National Academy Press, Washington, D.C., 1998.

7

Space Weather and Space Climatology: A Vision for Future Capabilities

MOTIVATION—ECONOMIC AND SOCIETAL VALUE

The availability of timely and reliable space-based information about our environment underpins elements of the infrastructure that are critical to a modern society. From economic and societal perspectives, reliable knowledge on a range of timescales of conditions in the geospace environment (including the mesosphere, thermosphere, ionosphere, exosphere, geocorona, plasmasphere, and magnetosphere) is important for multiple applications, prominent among them utilization of radio signals (which enables increasingly precise navigation and communication), as well as mitigation of deleterious effects such as drag on Earth-orbiting objects (which alters the location of spacecraft, threatens their functionality by collisions with debris, and impedes reliable determination of reentry). In addition, energetic particles can damage assets and humans in space, and currents induced in ground systems can disrupt and damage power grids and pipelines, a topic that has been a focus of recent research (Box 7.1).

Moreover, society expects and relies on instant coverage of events, the ability to communicate to remote corners of the world, and the availability of geospatial information needed for national security and other purposes. However, the space-based technologies that provide this information are vulnerable to the conditions in the dynamic and complex space environment through which radio waves propagate and where satellites orbit. In this chapter, the survey committee presents its vision for a comprehensive program consisting of observations, models, and forecasting, enhanced beyond current capabilities, to help protect these critical technologies.

Understanding space weather and climate is a prerequisite for fulfilling at least two directives of U.S. national space policy:[1]

1. "Take necessary measures to sustain the radiofrequency environment in which critical U.S. space systems operate." Societal use of the radio wave spectrum is growing dramatically, but its reliability and

[1] "National Space Policy of the United States of America," June 28, 2010, available at http://www.nasa.gov/pdf/649374main_062810_national_space_policy.pdf.

BOX 7.1 PREDICTING GEOMAGNETICALLY INDUCED CURRENTS ON THE POWER GRID: AN EXAMPLE OF A CRITICAL NATIONAL NEED

A geomagnetic storm is caused by energetic streams of particles and magnetic flux that originate from the Sun and impact and distort Earth's magnetic field. The transient changes in Earth's magnetic field interact with the long wires of the power grid, causing electrical currents to flow in the grid. The grid is designed to handle AC currents effectively, but not the DC currents induced by a geomagnetic storm. These currents, called geomagnetically induced currents (GICs; also known as ground-induced currents), cause imbalances in electrical equipment, reducing its performance and leading to dangerous overheating.[1]

Solar and space physicists, working with bulk power grid engineers, have helped to create the capability to model the effects of GICs on electricity transmission and distribution systems. This crucially important work relies on a body of knowledge built up over years of study. Today, sophisticated modeling software is used to assess the response of the electrical power system to geomagnetic storms, to assess the system's vulnerabilities, and to develop mitigation strategies, an important example of which is work to develop sensors that can detect transformer saturation (via harmonic detection) and overheating. With this information, operators can take steps to protect costly (on the order of $10 million) and difficult-to-replace transformers. In addition, in response to the prediction of intense geomagnetic disturbances, utilities will be able to pre-position replacement equipment at key locations of high vulnerability. Such measures are critical to restoration of bulk power capabilities after disruption from a possibly crippling space weather event.

The electric power industry continues to rely on the latest developments in space weather forecasting and thus would benefit directly from implementation of the research- and applications-related programs that are recommended in this report.

[1] Adapted from National Research Council, *Severe Space Weather Events—Understanding Societal and Economic Impacts: A Workshop Report*, The National Academies Press, Washington, D.C., 2008.

---

precision depend fundamentally on conditions in the ionosphere that alter the paths and properties of radio waves of all frequencies, including Global Positioning System (GPS) signals.

2. Preserve the space environment, in part by pursuing "research and development of technologies and techniques . . . to mitigate and remove on-orbit debris, reduce hazards, and increase understanding of the current and future debris environment" and by leading "the continued development and adoption of international and industry standards to minimize debris." Satellite drag is relevant to orbit and reentry prediction and to long-term mitigation of orbital debris. The recent inability, for example, to forecast the demise of the Upper Atmosphere Research Satellite (UARS) spacecraft underscores limitations in current capabilities for modeling and understanding the interaction of Earth-orbiting objects with the upper atmosphere. Space junk now exceeds 22,000 objects larger than a softball (Figure 7.1); collisions are expected to become more frequent (and may have propelled the UARS satellite into a less stable orbit).

Previous National Research Council reports[2] and the interagency *National Space Weather Program Strategic Plan*[3] document the need for increased U.S. capability to specify and predict the weather and

[2] National Research Council reports *Severe Space Weather: Understanding Societal and Economic Impacts: A Workshop Report* (2008) and *Limiting Future Collision Risk to Spacecraft: NASA's Meteoroid and Orbital Debris Programs: An Assessment of NASA's Meteoroid and Orbital Debris Programs* (2011), both published by The National Academies Press, Washington, D.C.

[3] Committee for Space Weather, Office of the Federal Coordinator for Meteorological Services and Supporting Research, *National Space Weather Program Strategic Plan*, FCM-P30-2010, August 17, 2010, available at http://www.ofcm.gov/nswp-sp/fcm-p30.htm.

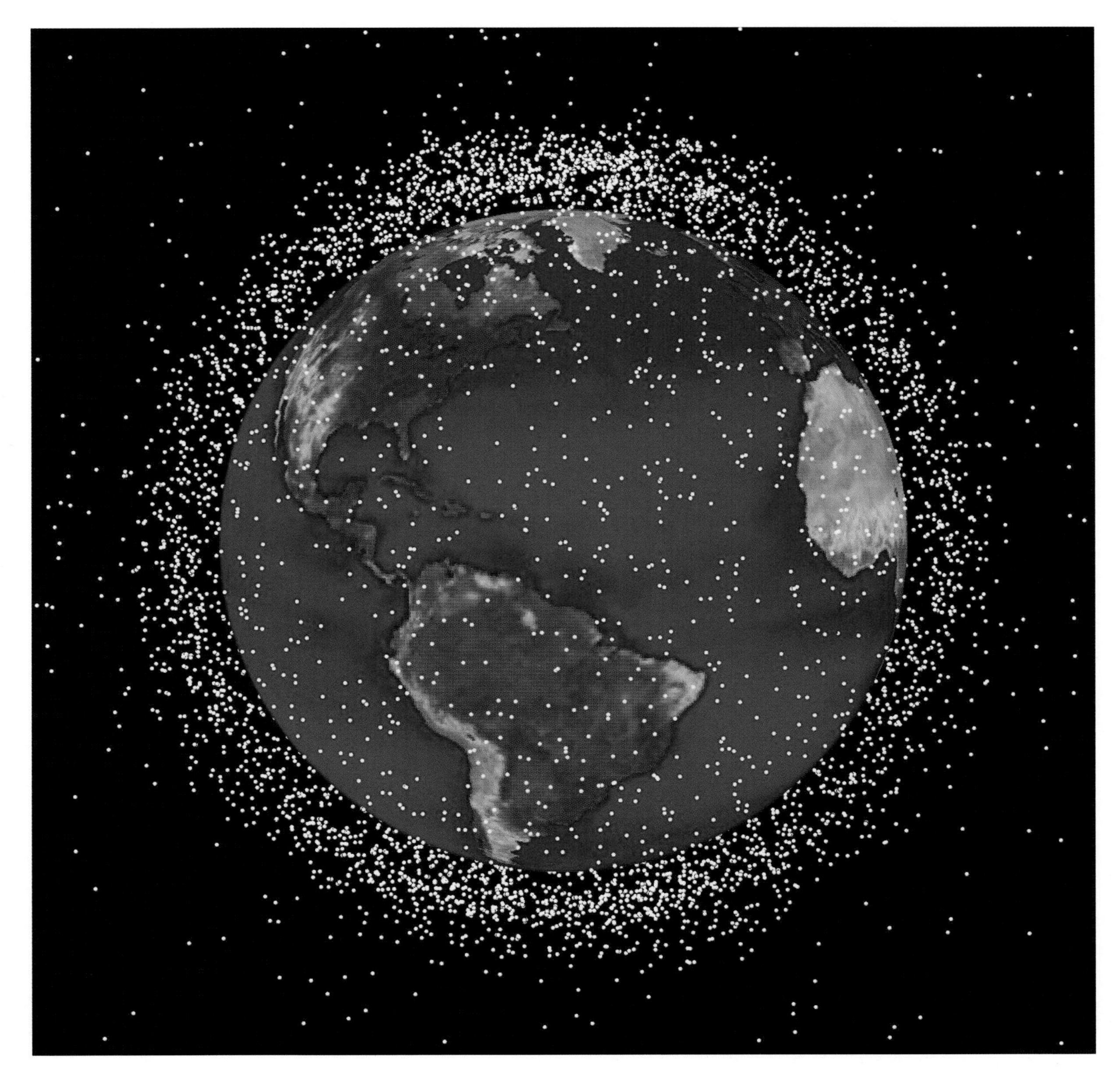

FIGURE 7.1 Snapshot of debris larger than 10 cm in low Earth orbit on May 1, 2001. SOURCE: Hugh Lewis, University of Southampton.

climate of the space environment. Growth in the number of space weather customers since the National Oceanic and Atmospheric Administration (NOAA) initiated a customer subscription service in 2004 (Figure 7.2) is another important indicator of the increasing U.S. need for monitoring and characterization of space weather. Moreover, the number of customers has been growing rapidly despite the deep and long-lasting solar minimum for much of the time period shown.

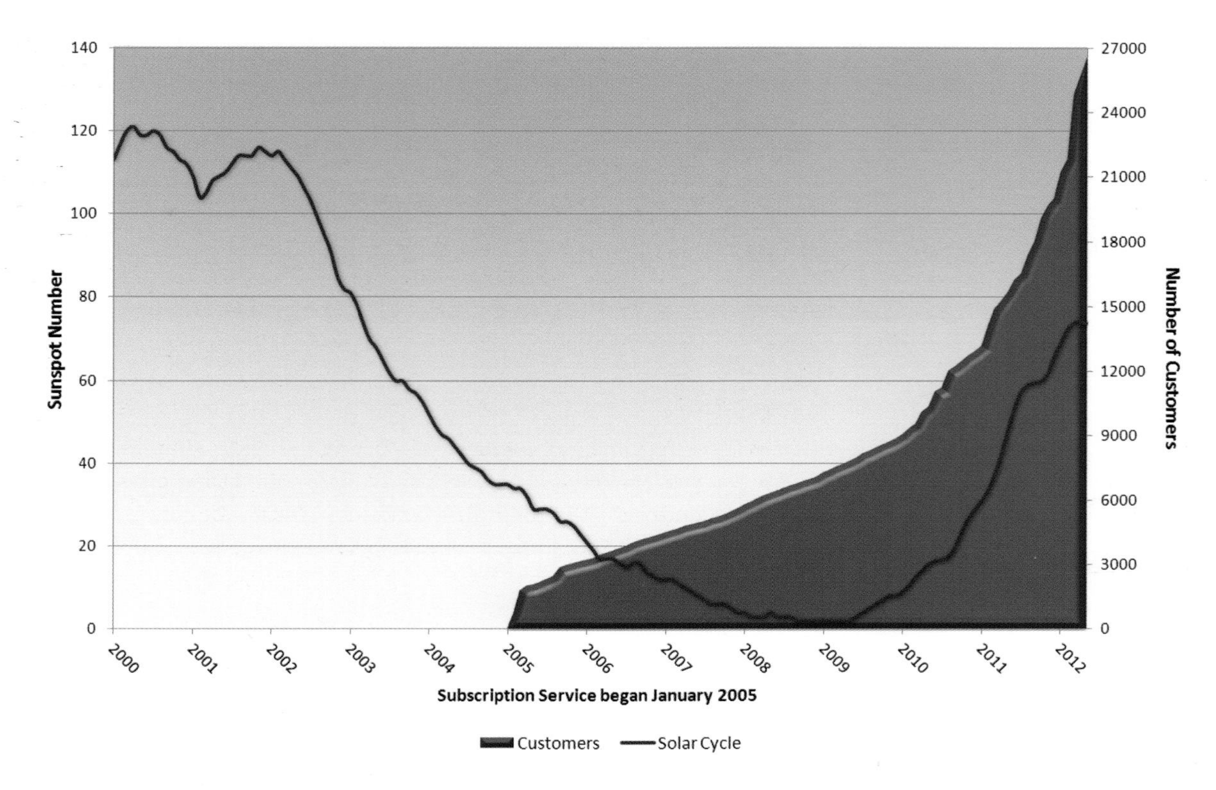

FIGURE 7.2 Number of unique customer subscribers routinely receiving the Space Weather Prediction Center's space weather services electronically. SOURCE: Updated from E. Hildner, H. Singer, and T. Onsager, Space weather workshop: A catalyst for partnerships, *Space Weather* 9:S03006, doi:10.1029/2011SW000660, 2011.

STRENGTHENING THE NATIONAL CAPABILITY FOR OBTAINING SPACE WEATHER AND CLIMATE INFORMATION

Current National Space Weather Program

U.S. space-based operational environment monitoring, currently based on data collected by NOAA Geostationary Operational Environmental Satellite, NOAA Polar Operational Environmental Satellite (POES), and Department of Defense (DOD) Defense Meteorological Satellites Program (DMSP) satellites, is of recognized fundamental importance to both the space weather operational and the space weather research communities. However, despite the well-documented vulnerability of essential societal, economic, and security services, space environment monitoring remains resource challenged.[4] For example, key energetic particle measurements now made by the POES and DMSP spacecraft are not currently slated to continue with the next generation of low-Earth-orbiting weather satellites. To address this and related problems will require, in the survey committee's view, a National Space Weather Program (NSWP) that is

[4]See, for example, *National Space Weather Program Strategic Plan*, 2010, available at http://www.ofcm.gov/nswp-sp/fcm-p30.htm.

strengthened through organizational changes and an infusion of resources. The committee also envisions a greater role for NASA in the multiagency NSWP.[5]

Research Sources of Space Weather Information

NASA research satellites such as the Advanced Composition Explorer (ACE), Solar and Heliospheric Observatory (SOHO; with the European Space Agency [ESA]), Solar Terrestrial Relations Observatory (STEREO), and Solar Dynamics Observatory (SDO), which are designed for scientific studies, have over the past decade or more provided critical measurements essential for specifying and forecasting the space environment system, including the outward propagation of eruptive solar events and solar wind conditions upstream from Earth. Although these observational capabilities have become essential for space environment operations, climatological monitoring, and research, NASA currently has neither the mandate nor the budget to sustain these measurements into the future. A growing literature has documented the need to provide a long-term strategy for monitoring in space and has elucidated the large number of space weather effects, the forecasting of which depends critically on the availability of suitable data streams.[6] An example is the provision of measurements of particles and fields at the L1 Lagrange point[7] (or, using technologies such as solar sails, closer to the Sun on the Sun-Earth line), which is critical for short-term forecasting of such harmful effects of space weather as damage to Earth-orbiting satellites, reduction of GPS accuracy, and potentially deleterious geomagnetically induced currents on the power grid.

The survey committee found that the existing ad hoc approach to providing space weather-related capabilities is inadequate. To help ensure a stronger approach, the committee articulates a vision for an enhanced national commitment by partnering agencies to continuous measurements of critical space environment parameters, analogous to the monitoring of the terrestrial environment being conducted by NASA in collaboration with a number of other agencies such as NOAA and the U.S. Geological Survey (USGS). Anticipating that the criticality of such a program will grow relative to other societal demands, the survey committee envisions NASA as utilizing its unique space-based capabilities as the basis for a new program that could provide sustained monitoring of key space environment observables to meet pressing national needs (see Box 7.1).

In addition to ensuring the continuity of critical measurements, robust space environment models capable of operational deployment are also necessary for the prediction and specification of conditions where observations are lacking. The survey committee anticipates that it will take decades to achieve an infrastructure for monitoring and characterizing space environment weather and climatology that is equivalent to current capabilities in the modeling and forecasting of terrestrial weather and climate; thus, it is necessary to start immediately. Achievement of critical continuity of key space environment parameters, their utilization in advanced models, and application to operations constitute a major endeavor that will require unprecedented cooperation among agencies in the areas in which each has specific expertise and unique capabilities.

---

[5]The committee's objectives are also consistent with the recommendations contained in the 2010 NSWP strategic plan. The plan's recommendations for the next decade included the following: (1) establish a NSWP focal point in the Executive Office of the President; (2) ensure continuity of critical data sources; (3) strengthen the science-to-user chain; and (4) emphasize public awareness of space weather critical needs (Committee for Space Weather, Office of the Federal Coordinator for Meteorological Services and Supporting Research, *National Space Weather Program Strategic Plan*, FCM-P30-2010, August 17, 2010, available at http://www.ofcm.gov/nswp-sp/fcm-p30.htm).

[6]For example, see National Research Council, *Severe Space Weather Events—Understanding Societal and Economic Impacts: A Workshop Report*, The National Academies Press, Washington, D.C., 2008, and D.N. Baker and L.J. Lanzerotti, A continuous L1 presence required for space weather, *Space Weather* 6:S11001, doi:10.1029/2008SW000445, 2008.

[7]A description of the Lagrange points is available at http://map.gsfc.nasa.gov/mission/observatory_l2.html.

A ROBUST SPACE WEATHER AND CLIMATOLOGY PROGRAM

Core Elements

Like Earth's near-surface environment, where climate and weather occur, the extended operational environment that encompasses space weather and space climate varies continuously on multiple timescales in response to forcing from the Sun, the heliosphere, and the underlying atmosphere. To advance space weather and space climatology capabilities, it is essential to improve, and design appropriately, the temporal and spatial coverage of space- and ground-based measurements. A mix of assets is needed: (1) space-based measurements that provide the coverage necessary for detecting space weather hazards, some of which cannot be discovered from the ground, and (2) ground-based measurements that provide more extensive spatial coverage and a link to historical measurements. In Box 7.2 the survey committee lists the highest-priority additional data needed, an initial step toward describing a notional new program for NASA, building on the unique strengths of that agency.

New Elements

Essential components of a robust space environment operational program that will complement what exists today or, in some cases, provide much needed continuity of critical capabilities, include the following:

- Monitor the variable solar-heliospheric photon, particle, and magnetic field inputs with satellites at L1 and L5.
- Monitor the geospace global and regional responses to the varying solar-heliospheric inputs with Earth-orbiting satellites, one in a high-altitude orbit (geostationary Earth orbit [GEO], for ionospheric imaging) and one in a low-altitude orbit (low Earth orbit [LEO], for detailed regional sensing and radiation belt monitoring).
- Develop, validate, test, and transition to operations physical and assimilative models of coupled solar, heliospheric, and geospace properties for specification and forecasting of the extended operational environment.
- Integrate relevant research efforts with operational activities to achieve seamless research to operations/operations to research and identify emerging needs and advances.
- Leverage the strength of NASA's community by taking advantage of principal-investigator-led missions, hosted payloads, and other innovative approaches such as the use of microsatellites.
- Coordinate with other complementary agency missions such as those of the National Science Foundation (NSF; supporting model development and ground-based observations), DOD and NOAA (providing operational forecasts and space weather monitoring), Department of Energy (DOE; supporting modeling and monitoring), and USGS (supporting ground-based magnetic observations).

From their quasi-stable orbits at the Earth-Sun L1 libration point, which is approximately 1.5 million kilometers from Earth, instruments on NASA's ACE and the NASA/ESA SOHO spacecraft continuously monitor the solar wind and provide solar coronagraph imaging, respectively. Information from ACE is used to provide approximately 1 hour of warning of a geomagnetic storm. To sample solar wind structures 5 days before they reach Earth and to provide global coverage of disturbances moving Earth-ward through the inner heliosphere, a spacecraft could be located at L5, the gravitationally stable location approximately 60 degrees behind Earth in its orbit as seen from the Sun. From L5, solar activity behind the limb rotating Earth-ward could be observed; in addition, in situ sampling of solar wind structure at a longitude distinct from

BOX 7.2 CONTINUOUS MEASUREMENTS FOR SPACE WEATHER AND CLIMATOLOGY: COMPLEMENTING AND PRESERVING OBSERVATIONS OF THE SPACE ENVIRONMENT

Solar-Heliosphere System Forcing

- Solar X-ray, extreme ultraviolet, and ultraviolet spectral irradiance and images, magnetograms
- Solar coronal-heliospheric images
- Solar wind (speed, density, temperature, ion composition)
- Interplanetary magnetic fields
- Energetic particles (solar and galactic)

Atmosphere-Ionosphere-Magnetosphere System Response and Variability

- Neutral temperature
- Electron density profile and total electron content
- Total mass density
- Neutral winds
- Electric and magnetic fields
- Ionospheric scintillation (amplitude, phase, morphology)
- Composition of major species (especially O, and O^+)
- Concentration of minor species (especially radiatively active gases)
- Wave activity on all scales (gravity and planetary waves, tides)
- Energetic particles (protons, electrons, ions, neutrals)
- Auroral morphology

that accessible at L1 and rotating Earth-ward is possible. An L5 mission would build on experience using STEREO B coronagraph measurements for space weather forecasting. Finally, in addition to new space-based elements, a comprehensive and sustained program of measurements in geospace would include ground-based measurements, supported by NSF and the Air Force, as well as space-based measurements from NOAA and DOD, with NASA taking on a new monitoring role with new resources, coordinating with operational agencies. New models are also needed to satisfy the demands posed by increased user diversity.

An Illustrative Scenario

The survey committee envisions a national commitment to a new program in solar and space physics that would provide long-term observations of the space weather environment and support the development and application of geospace models to protect critical societal infrastructure, including communication, navigation, and terrestrial weather spacecraft, through accurate forecasting of the space environment. Because NASA has a long history of conducting collaborative forefront space weather research in concert with researchers in academia, the commercial sector, and other government laboratories, the survey committee envisions an expanded role for NASA in a future space weather and climatology program. The strengths inherent in the NASA community, combined with the benefit of synergy between forefront research and space weather operations, should be brought to bear to meet U.S. needs for accurate, reliable monitoring and forecasting of space weather and climatology.

A space weather and climatology program would encompass long-term planning for critical measurements, such as the L1 solar and solar wind measurements currently acquired from ACE and SOHO. The survey committee endorses DSCOVR as a temporary interagency solution to the current lack of continuity beyond ACE, which was launched in August 1997, of L1 plasma and field measurements essential to current space weather models, while advocating the need to have a plan beyond DSCOVR for continuous and comprehensive L1 coverage. NASA is uniquely qualified to develop, build, launch, and operate spacecraft in Earth orbit and beyond, including the recommended measurements at L1 and L5 (as described in a number of white papers[8]). Operating such spacecraft requires use of the Deep Space Network, for example.

Below, the survey committee describes new (notional) agency-specific activities that are needed to develop the required capabilities. The program builds on critical existing capabilities, which are described in Chapter 4 for all participating agencies. Organization of the new activities is not discussed here; however, a key recommendation is to recharter the NSWP at a level of the federal government appropriate for strategic support and coordination (see Chapter 4). Availability of new resources is assumed for the notional space weather and climatology program implementation elements for each agency.

NSF

NSF would be enabled to provide real-time monitoring by means of its set of ground-based facilities.[9] Ground-based facilities include radars, lidars, magnetometers, and solar observatories such as the Global Oscillation Network Group, Synoptic Optical Long-term Investigations of the Sun, and the Advanced Technology Solar Telescope. NSF would also be enabled to provide key data streams from platforms such as Iridium and to support space weather model development.

NOAA and DOD

Both NOAA and DOD would be enabled to transition models, developed as part of this vision, into operations, and they would be enabled to utilize any new data streams provided by an implementation of this vision to enhance operational services.

DOE

DOE, in cooperation with DOD, would be enabled to provide both continuity of, as well as access to, real-time data streams from space weather sensors on geosynchronous and GPS satellite platforms.

Commercial Sector

The United States would have a healthy commercial sector enabled to develop tailored space weather products for specific applications.

---

[8]The survey committee and panels reviewed 288 mission concept white papers, which were submitted in response to an invitation to the research community. The survey's request for information is reproduced in Appendix H, and the submissions that were received in response are listed in Appendix I and supplied on the CD that contains this report.

[9]Important ground-based assets are operated by the Department of the Interior U.S. Geological Survey's Geomagnetism Program, which provides high-quality, ground-based magnetometer data continuously from 13 observatories distributed across the United States and its territories. The program collects, transports, and can disseminate these data in near-real time, and it also has significant data-processing and data-management capacities.

An Expanded Role for NASA

Today, NASA missions such as ACE, SOHO (a collaboration with ESA), STEREO, and SDO provide space weather information without which the forecasting of solar eruptions and their heliospheric propagation would not be possible. NASA and partner agencies NSF, the Air Force Office of Scientific Research, and the Office of Naval Research support research to develop the most advanced space environment models. Many of these models can be applied to space weather forecasting, with a potentially dramatic increase in forecasting capabilities. NASA already conducts operational activities in a number of areas, including space weather forecasting for its own human and robotic missions, human spaceflight activities, and communications such as those based on the Tracking and Data Relay Satellite System. NASA also routinely provides societally relevant information obtained from models and space-based measurements, such as from the Moderate Resolution Imaging Spectroradiometer Earth-monitoring system for the atmosphere, land, and oceans.

NASA's Heliophysics Division has developed exceptional capabilities for continuous measurement of critical space environment parameters but does not have a program that sustains the observational capabilities that are required for meeting societal needs. Consequently, it currently does not support long-term heliophysics monitoring (with continuously evolving skill) in a manner analogous to the monitoring of the terrestrial environment and surface climate that the Earth Science Mission Directorate has implemented over the past two decades.

Implementation Concept

Below is a concept to develop the space weather and climatology (SWaC) capability outlined above. In addition, Table 7.1 lists space missions enhanced to provide needed SWaC data, and Table 7.2 presents an illustrative funding scenario for a NASA SWaC program. This new program could be started as soon as fiscal year 2014, in response to a demonstrated national need. Assuming the availability of the necessary new funding, the following steps could be taken during the first 5 years:

- Year 1

  —Initiate development of an operational solar wind and solar monitoring (L1) mission.

  —Initiate NASA center activities to provide real-time data streams from missions and models, to evaluate and test models, and to continue operating space-weather-relevant research missions.

  —Initiate a grants program to develop and advance to operational readiness space environment specification and forecasting models.

  —Initiate continued coordination with space weather forecasting organizations at DOD, NOAA, NASA, and the commercial sector for transition of relevant observations and models to operations.
- Year 2

  —Conduct the development and build phase of an operational L1 mission.

  —Expand NASA center activities to coordinate acquisition of space weather data with other agencies (NOAA, DOD, DOE) and incorporation of data into a proposed new space weather clearinghouse, providing access to previously unavailable space weather data.

  —Expand the grants program to develop and advance to operational readiness space environment specification and forecasting models.
- Year 3

  —Conduct the build phase of an operational L1 mission.

  —Begin development of a solar and solar wind monitoring mission for L5.

TABLE 7.1 Enhanced Space Missions for Space Weather and Climatology

Spacecraft and Key Regions	Key Instruments/Observations	Utility	Heritage/Status
L5 Mission • Solar • Heliosphere • Solar wind	• Solar coronagraph • Solar X-ray, extreme ultraviolet, and magnetic field imagers • Solar irradiance • Heliospheric imager • Solar wind parameters (e.g., B, T, v, n composition)	Advanced warning of solar activity, background solar wind, and solar disturbances (e.g., coronal mass ejections) aimed at Earth and other locations in the solar system	• STEREO satellite demonstrated importance • No plans for future L5 observations
L1 Mission • Solar • Solar wind	• Solar coronagraph • Solar X-ray, extreme ultraviolet, and magnetic field imagers • Solar irradiance • Solar wind parameters (e.g., B, T, v, n composition)	Provides high-accuracy, ~ 30- to 45-min advanced warning of impending geomagnetic storms, validates models, and detects coronal mass ejections	Solar • NASA/ESA SOHO demonstrated value • NASA concepts exist, and NOAA Compact Coronagraph under evaluation, but no funding identified Solar Wind • ACE and WIND demonstrated value • ACE data currently in use • NOAA/interagency DSCOVR satellite under development • IMAP recommended
GEO Mission • Geosynchronous orbit, geospace remote imaging of ionosphere/ thermosphere	• Ultraviolet, extreme ultraviolet, and energetic neutral atom Earth imaging	Provides instantaneous global distribution of geospace neutral and electron densities (important, e.g., for GPS and satellite drag)	• No existing measurements
LEO Mission • Low-Earth-orbit geospace in situ and remote sensing of ionosphere/ thermosphere	• Electron and neutral density • Temperature • Electric and magnetic fields • Winds	Provides high spatial resolution and regional detail of conditions important, e.g., for GPS and satellite drag	• TIMED GUVI demonstrated value of remotely sensing temperature and densities • Current observations from Air Force/DMSP • No plans for future measurements

NOTE: ACE, Advanced Composition Explorer; DMSP, Defense Meteorological Satellites Program; DSCOVR, Deep Space Climate Observatory; ESA, European Space Agency; GUVI, Global Ultraviolet Imager; IMAP, Interstellar Mapping and Acceleration Probe; NOAA, National Oceanic and Atmospheric Administration; SOHO, Solar and Heliospheric Observatory; STEREO, Solar Terrestrial Relations Observatory; and TIMED, Thermosphere-Ionosphere-Mesosphere Energetic and Dynamics.

—Evaluate the effectiveness of the proposed new space weather clearinghouse in meeting multi-agency operational forecast needs.

—Continue the grants program to test the operational readiness of space environment specification and forecasting models and coordinate with DOD and NOAA partners.

- Year 4

—Build, integrate, and launch an operational L1 mission.

—Continue development of a solar monitoring mission for L5.

—At a NASA center, begin to integrate and distribute operational L1 measurements.

TABLE 7.2 Illustrative Funding Scenario for a NASA Space Weather and Climatology Program (in $millions)[a]

	Year 1	Year 2	Year 3	Year 4	Year 5	Year 6	Year 7	Year 8	Year 9	Year 10
L1	50	100	100	100	25	25	25	25	25	25
L5	0	0	50	50	100	100	100	100	50	25
Earth orbiting	0	0	0	0	25	25	25	25	75	100
NASA centers	25	25	25	25	25	25	25	25	25	25
Grants program[b]	25	25	25	25	25	25	25	25	25	25
Total	100	150	200	200	200	200	200	200	200	200

[a] Assumes L1 launch in year 4, $500 million over 10 years, start over year 10; assumes L5 launch year 8, $575 million over 9 years, start over year 12; assumes 5-year multi-satellite Earth orbit technology development in years 5-9, launch year 10.
[b] Model development, data assimilation

—Continue the grants program to transition to operations the environment specification and forecasting models for use by NASA and NOAA and DOD operations.

- Year 5

—Operate the operational L1 mission, and initiate a concept study of a follow-on mission.

—Continue development of a solar monitoring mission for L5.

—Initiate a geospace monitoring mission concept study.

—At a NASA center, plan for integration of operational L5 and geospace monitoring measurements into space environment specification and forecasting models.

—Continue the grants program to facilitate a transition to operational readiness of space environment specification and forecasting models that will include new data sets as they become available.

A new plan is also needed that synthesizes and capitalizes on the strengths of the participating agencies listed above as well as opportunities in the commercial sector, such as Iridium/AMPERE. The committee sees NASA as assuming a leading role in creating a clearinghouse for coordinating the acquisition, processing, and archiving of underutilized real-time and near-real-time ground- and space-based data needed for space weather applications. For example, highly valued energetic particle measurements made by GPS and Los Alamos National Laboratory GEO satellites for specification of the radiation belts are not now routinely provided. Likewise, model development has been supported by individual agencies rather than being coordinated across relevant stakeholders.

The survey committee also foresees NASA assuming a leading role in coordinating model development through a center such as the Community Coordinated Modeling Center, which both serves as a repository for models and coordinates model development and transition to operations at NOAA, NASA, and DOD. Additional funding will be required by NOAA and DOD to support the integration of data acquired and models developed by the envisioned new NASA program in order to address the specific needs of the user communities. The illustrative scenario shown in Table 7.2 for a NASA SWaC program incorporates support for the proposed data clearinghouse and modeling effort. Combined with existing and recommended activities, a program such as the proposed SWaC effort can put the nation's space weather forecasting and space weather and space climate monitoring program on a solid footing.

The program described above would help meet the growing space weather needs of the United States. However, given scarce resources, the survey committee recommends implementation only under circumstances that would not delay the development or the timely execution of the recommended programs for NASA that are shown in Figure 6.1.

SUMMARY COMMENTS

NASA is the appropriate agency to ensure that critical data sources for space weather forecasts and operations are sustained. A suitable vehicle may be a new heliophysics space weather and climatology program whose primary focus would be on obtaining societally relevant data and observations. This program could be implemented in concert with relevant research programs underway in academia and in government laboratories, and it would leverage opportunities in the commercial sector and capitalize on the strength of NASA's science community. Recognizing the importance of modeling to the forecasting of terrestrial weather, the program would also support model development and model applications to space weather forecasting, applying the latest advances in modeling capabilities and the most advanced data sources to drive models. The new activity could make its space weather information available to interests in the United States and beyond, and coordinate its relevant activities with DOD and NOAA operations, as well as the commercial and international space weather communities.

Part II

Reports to the Survey Committee from the Discipline Panels

The decadal survey committee's assessment and recommendations for the field of solar and space physics, Part I of this report, were informed to a great degree by the extensive scientific discussion and technical input of the survey's three science discipline panels. Themes for these panels were chosen to emphasize interactions between physical domains, with the goal to further the integration of the overall research across traditional discipline boundaries. The Panel on Solar and Heliospheric Physics (SHP), Panel on Solar Wind-Magnetosphere Interactions (SWMI), and Panel on Atmosphere-Ionosphere-Magnetosphere Interactions (AIMI) were charged to summarize scientific progress and to identify the most compelling science questions emerging as targets for research within the next 10 years. The panels also were chartered to develop a prioritized approach to addressing those questions in the most productive manner, and they were encouraged to investigate and report on the broader context of their proposed research, for example, how it pertains to societal needs, and to identify technological needs and means to address the most compelling science questions.

Panel deliberations drew on information gathered at town hall meetings, three face-to-face 2.5-day panel meetings, and weekly teleconferences. The panels also made extensive use of community input received through the white papers that were submitted as part of the survey committee's request for information[1] and from briefings from other decadal survey activities, such as those involving the five cross-disciplinary working groups.[2]

Panel interactions with the survey committee were numerous. Each panel was assigned a liaison member who was, at the same time, also a member of the survey committee. Survey committee members also attended panel meetings to stay informed of emerging developments. Panel leads (chairs and vice chairs)

[1] The survey's website, http://sites.nationalacademies.org/SSB/CurrentProjects/SSB_056864, includes links to the request for information (RFI) and to the nearly 300 submissions received in response. The RFI is also reprinted in Appendix H of this report, and a list of responses is given in Appendix I.

[2] The topics for the five working groups were Theory, Modeling, and Data Exploitation; Explorers, Suborbital, and Other Platforms; Innovations: Technology, Instruments, and Data Systems; Research to Operations/Operations to Research; and Education and Workforce.

also participated in most survey committee teleconferences and face-to-face meetings.[3] Notably, panel leads were full participants in the survey committee meetings that developed the overarching scientific motivations and key science goals for the decade, the latter of which are described in detail in Chapter 1 and highlighted below:

Motivation 1. To understand our home in the solar system.
Motivation 2. To predict the changing space environment and its societal impact.
Motivation 3. To explore space to reveal universal physical processes.

Key Science Goal 1. Determine the origins of the Sun's activity and predict the variations in the space environment.
Key Science Goal 2. Determine the dynamics and coupling of Earth's magnetosphere, ionosphere, and atmosphere and their response to solar and terrestrial inputs.
Key Science Goal 3. Determine the interaction of the Sun with the solar system and the interstellar medium.
Key Science Goal 4. Discover and characterize fundamental processes that occur both within the heliosphere and throughout the universe.

The panels cast their scientific prioritization in the form of discipline goals and priorities, from which they derived more detailed scientific imperatives—actions that are needed to make progress—and, finally, implementation scenarios or reference mission concepts. *It is important to recognize that panel-specific imperatives are not equivalent to survey report recommendations, which can be offered only by the decadal survey committee.*[4]

The work of the three discipline panels was fundamental to the decadal survey, and it forms the foundation of Part I of this report. In particular, each of the panels' emphases for research in the field was brought forward to the survey committee for consideration and possible action in the form of a report recommendation. In the further course of this work, a set of spacecraft mission concepts that would achieve particular science goals of each individual panel was developed and evaluated for cost and technical readiness. That evaluation process is described in Part I (Chapter 1) and in Appendix E.

[3]However, panel leadership was excluded at meetings of the survey committee during the final phase of the study when the committee's recommendations were established. These are shown in the Summary and Part I of this report.

[4]The report of the decadal survey committee and its recommendations are found in Part I of this report. Key recommendations of the survey committee are aggregated in the report Summary.

8

Report of the Panel on Atmosphere-Ionosphere-Magnetosphere Interactions

8.1 SUMMARY OF AIMI SCIENCE PRIORITIES AND IMPERATIVES FOR THE 2013-2022 DECADE

Informed by observations and modeling efforts that have occurred during the past decade, there is increasing recognition among scientists of the atmosphere-ionosphere-magnetosphere (AIM) system as a complex and active element of space weather, and as a region where important science questions with broad applicability across our solar system can be answered. Earth's space environment, or geospace, is unique in many ways: the interconnected behavior of the plasma and neutral gas in the AIM system, the strong signature of lower atmospheric conditions in space, and the development of massive plasma structures with embedded variability at multiple scales are just a few examples. The pursuit toward understanding energy transfer and physical manifestations in near-Earth space has yielded and will continue to offer insights into fundamental processes that occur at other planets and bodies in our solar system and indeed throughout the universe. There are also practical reasons to study the AIM system. The space-based assets for observation and communication of human activities all operate in geospace and therefore must be designed or otherwise protected from the hazards and unpredictability that this energetic, nonlinear system produces. Advances in scientific understanding of the AIM system enable the development of a capability for the prediction of geospace conditions.

In this chapter, the decadal survey's Panel on Atmosphere-Ionosphere-Magnetosphere Interactions (AIMI) articulates its science goals and aspirations for the decade ahead and suggests an implementation strategy to achieve that vision. Building on the significant accomplishments of the previous decade, the panel presents an interlinked and achievable research program to address the most compelling science questions in the field. Summarized below are the AIMI panel's science priorities, imperatives, and recommendations to the survey committee for the 2013-2022 decade.

The three major AIMI science priorities for the 2013-2022 decade are as follows:

AIMI Science Priority 1. Determine how the ionosphere-thermosphere system regulates the flow of solar energy throughout geospace.

AIMI Science Priority 2. Understand how tropospheric weather influences space weather.

AIMI Science Priority 3. Understand the plasma-neutral coupling processes that give rise to local, regional, and global-scale structures and dynamics in the AIM system.

These priorities emerge from the five AIMI science goals described in Section 8.4, "Science Goals and Priorities for the 2013-2022 Decade," and from the panel's assessment of the resources required to address them, and lead to the following imperatives:

AIMI Imperative 1. Close a critical gap in the NASA Heliophysics Systems Observatory with a mission that determines how solar energy drives ionospheric-thermospheric variability and that lays the foundation for a space weather prediction capability.

AIMI Imperative 2. Provide a broad and robust range of space-based, suborbital, and ground-based capabilities that enable frequent measurements of the AIM system from a variety of platforms, categories of cost, and levels of risk.

AIMI Imperative 3. Integrate data from a diverse set of observations across a range of scales, coordinated with theory and modeling efforts, to develop a comprehensive understanding of plasma-neutral coupling processes and the theoretical underpinning for space weather prediction.

AIMI Imperative 4. Conduct a theory and modeling program that incorporates accumulated understanding and extends the legacy of observations into physics-based models that are utilized for new scientific insight and operational specification and forecast capabilities.

These imperatives represent a balanced strategy for addressing the panel's priority science; they are not listed in any particular order. The requirements underlying the imperatives span five categories: spaceflight missions; Explorers, suborbital, and other platforms; ground-based facilities; theory and modeling; and enabling capabilities. Priorities within each of these categories are summarized in turn below.

8.1.1 Spaceflight Missions

No new NASA missions are under development or are currently planned for the future that address any of the AIMI science priorities articulated above. This deficiency represents an acute imbalance in the study of the Sun-Earth system that impedes researchers' ability to resolve complex AIM system behavior that impacts geospace dynamics and the operation of ground- and space-based assets on which society depends. A critical AIMI imperative therefore is that a mission addressing the response of the ionosphere-thermosphere (IT) system to variable forcing be put forth as the highest priority of the solar and space physics decadal survey.

The most compelling AIM science questions of the coming decade are best addressed with a Geospace Dynamics Constellation (GDC) mission nominally consisting of six identical satellites in high-latitude equally spaced circular orbits, with the goal of understanding how winds, temperature, composition, chemistry, charged particles, and electric fields interact to regulate the observed global response of the IT. This mission will also provide new insights into the IT response to dynamical coupling with the lower atmosphere. If this mission must be delayed at all due to budgetary constraints, then a revitalized heliophysics instrument and technology development program must support GDC's implementation later in this decade. In that case, the AIMI panel suggests that the DYNAMIC (Dynamical Neutral Atmosphere-Ionosphere Coupling) mission be put forth as the decadal survey's number-one priority for the 2013-2022 decade. DYNAMIC is a pair of satellites in low-Earth orbits separated by 6 hours of local time, carrying the instruments to measure the critical energy inputs to the AIM system from the spectrum of waves entering from below. Although the primary focus is to understand how lower-atmosphere variability drives IT variability, DYNAMIC will also measure important properties of the IT response to variable magnetospheric forcing.

Additional NASA missions that address another high-priority science challenge of the next decade—understanding the two-way interaction between the ionosphere-thermosphere and the magnetosphere—are also described in this chapter. These missions and the associated science are also potential candidates for the Explorer program.

8.1.2 Explorers, Suborbital, and Other Platforms

The relative proximity of the AIM system makes it amenable to observational strategies involving a wide variety of platforms. This attribute is a significant strength in crafting a program that is responsive to budgetary realities and to the changing climate of programmatic risk factors. The following AIMI panel priorities reflect this crucial flexibility:

- *Explorer program enhancement (highest priority).* Enhance the Heliophysics Explorer line to execute a broad range of science missions that can address important AIMI science challenges. Mission classes should range from a tiny Explorer that takes advantage of miniaturized sensors and alternative platforms and hosting opportunities, up to a medium Explorer that could address multiple science challenges for the decade.
- *Constellations of satellites.* Develop the means to effectively and efficiently implement constellation missions, including proactive development of small-satellite capabilities and miniaturized sensors and pursuit of cost-effective alternatives such as commercial constellations.
- *Suborbital research.* Maintain a strong suborbital research program. Continue development of observatory-class capabilities, such as a high-altitude sounding rocket and long-duration balloons, and expand funding for science payload development for these platforms.
- *Strategic hosted payloads.* Develop a strategic capability to make global-scale AIMI imaging measurements from host spacecraft, notably those in high Earth orbit and geostationary Earth orbit, as is currently done in support of solar (GOES SXT) and magnetospheric (TWINS, GOES, LANL) research.

8.1.3 Ground-Based Facilities

New ground-based instrumentation and associated research programs can also address an array of AIMI science questions in this decade. These facilities will play a major role in an overall strategy to understand the origins of plasma-neutral structures over local (tens to hundreds of kilometers), regional (hundreds to thousands of kilometers), and global scales (thousands to tens of thousands of kilometers), as well as the interactions between structures over these different scales. In particular, several prospective facilities are particularly compelling for advancing AIMI panel science priorities:

- *Autonomous American sector network.* Develop, deploy, and operate a network of 40 or more autonomous observing stations extending from pole to pole through the (North and South) American longitudinal sector. The network nodes should be populated with heterogeneous instrumentation capable of measurements, including winds, temperatures, emissions, scintillations, and plasma parameters, for study of a variety of local and regional ionosphere-thermosphere phenomena over extended latitudinal ranges.

- *Whole-atmosphere lidar observatory.* Create and operate a lidar observatory capable of measuring gravity waves, tides, wave-wave and wave-mean flow interactions, and wave dissipation and vertical coupling processes from the stratosphere to 200 km. Collocation with a research facility such as an incoherent scatter radar (ISR) installation would enable study of a number of local-scale plasma-neutral interactions relevant to space weather.
- *NSF medium-scale research facility program.* The above two facilities are candidates for support by the NSF Geospace Program and would require that a medium-scale (~$40 million to $50 million) research facility funding program be instituted at NSF to fill the gap between the Major Research Instrumentation (MRI; <$4 million) and Major Research Equipment and Facilities Construction (MREFC; >$100 million) programs.
- *Southern-Hemisphere expansion of incoherent scatter radar (ISR) network.* In addition to the two facilities listed above, expansion of the now proven Advanced Modular Incoherent Scatter Radar (AMISR) technology to southern polar latitudes (i.e., Antarctica) would provide for the first-ever view of detailed ionosphere processes in the southern polar hemisphere, thus contributing a critical missing component to the Heliophysics Systems Observatory.
- *Ionospheric modification facilities.* Fully realize the potential of ionospheric modification techniques through colocation of modern heating facilities with a full complement of diagnostic instruments including incoherent scatter radars. This effort requires coordination between NSF and DOD agencies in the planning and operation of existing and future ionospheric modification facilities.

8.1.4 Theory and Modeling

Cross-scale coupling processes are intrinsic to AIM system behavior. Phenomena and processes that are highly structured in space and time (e.g., wave dissipation, turbulence, electric field fluctuations) can produce effects (e.g., wind circulations, chemical transport, Joule heating, respectively) over much larger scales. At the same time, larger-scale phenomena create local conditions that can either promote or suppress development of rapidly changing structures at small spatial scales (e.g., instabilities and turbulence). The observational strategies presented in this report place high priority on understanding how local, regional, and global-scale phenomena couple to produce observed responses across scales. These strategies call for complementary development of theory and numerical modeling capabilities that enable comprehensive treatment of cross-scale coupling processes, together with new data synthesis technologies that combine multiple, hetero-scale data sources into a common framework for understanding critical aspects of the AIM system.

Therefore, to support the synergistic program of space-based investigations and ground-based facilities, the AIMI panel has the following priorities regarding theory and modeling:

- *Model development.* Comprehensive models of the AIM system would benefit from the development of embedded grid and/or nested model capabilities, which could be used to understand the interactions between local- and regional-scale phenomena within the context of global AIM system evolution.
- *Theory.* Complementary theoretical work would enhance understanding of the physics of various-scale structures and the self-consistent interactions between them.
- *Assimilative capabilities.* Comprehensive models of the AIM system would benefit from developing assimilative capabilities and would serve as the first genre of space weather prediction models.

Further priorities concerning theory and modeling are provided in Section 8.5.4, "Theory and Modeling."

8.1.5 Enabling Capabilities

The missions and initiatives described above require additional capabilities and infrastructure that enable cheaper and more frequent measurements of the AIM system, that transform measurements into scientific results, that maintain the health of the scientific community, and that serve the needs of 21st-century society. These enabling capabilities (i.e., working group priorities) fall into the following categories:

- Innovations: technology, instruments, and data systems;
- Theory, modeling, and data exploitation;
- Research to operations, and operations to research; and
- Education and workforce.

The panel's priorities in these areas are detailed in Section 8.5.5, "Enabling Capabilities."

8.2 MOTIVATIONS FOR STUDY OF ATMOSPHERE-IONOSPHERE-MAGNETOSPHERE INTERACTIONS

Electromagnetic radiation from the Sun is the source of energy for photosynthesis and life. However, the Sun's other energetic outputs produce conditions and events that can be disruptive and even catastrophic to society. Hurricanes and tornadoes are examples of extreme and dangerous terrestrial weather events that occur on the surface of Earth. But our planet is also embedded in the streaming plasma and magnetic field of the Sun's outer corona (Figure 8.1), which can lead to hazardous weather in space with similarly catastrophic consequences. Although Earth's magnetic field serves as a protective cocoon that is difficult for the Sun's plasma and magnetic field to penetrate, transmission of a few percent of this energy into near-Earth space can produce large effects.

Reconnection between the magnetic fields of the Sun and Earth causes electric fields, currents, and energetic particles to be created. The source of magnificent auroral displays, energetic particles, can penetrate satellite electronics and solar cells and disrupt or sometimes even terminate their operation. Electric currents flowing through the auroral ionosphere heat the atmosphere and produce global changes in upper-atmosphere density that make it difficult to predict the future locations of satellites and potential collisions between them. Electrical connections between the near-Earth space environment and the ionosphere can also disrupt the operation of communications and navigation systems, and cell phones, and even induce dangerous levels of currents in the U.S. power distribution system. Energetic particle precipitation into the upper atmosphere can also initiate a chain of events that lead to massive depletions of stratospheric ozone in the polar regions. These are only a few of the consequences that emerge from a complex web of interactions occurring within this active region called geospace and that motivate us to understand our home in the solar system (M1) and to predict the changing space environment and its societal impact (M2).[1]

The focus of the AIMI panel and the subject of this chapter is the region of geospace where atmosphere-ionosphere-magnetosphere interactions occur. That region extends from roughly the top of the stratosphere (at about 50 km) to several thousand kilometers, where the presence of the neutral atmosphere ceases to exert any significant control over the system. As will be discussed in more detail in this chapter, this

[1] The motivations referred to in this section are those outlined in the introduction to Part II of this report: M1. Understand our home in the solar system; M2. Predict the changing space environment and its societal impact; M3. Explore space to reveal universal physical processes.

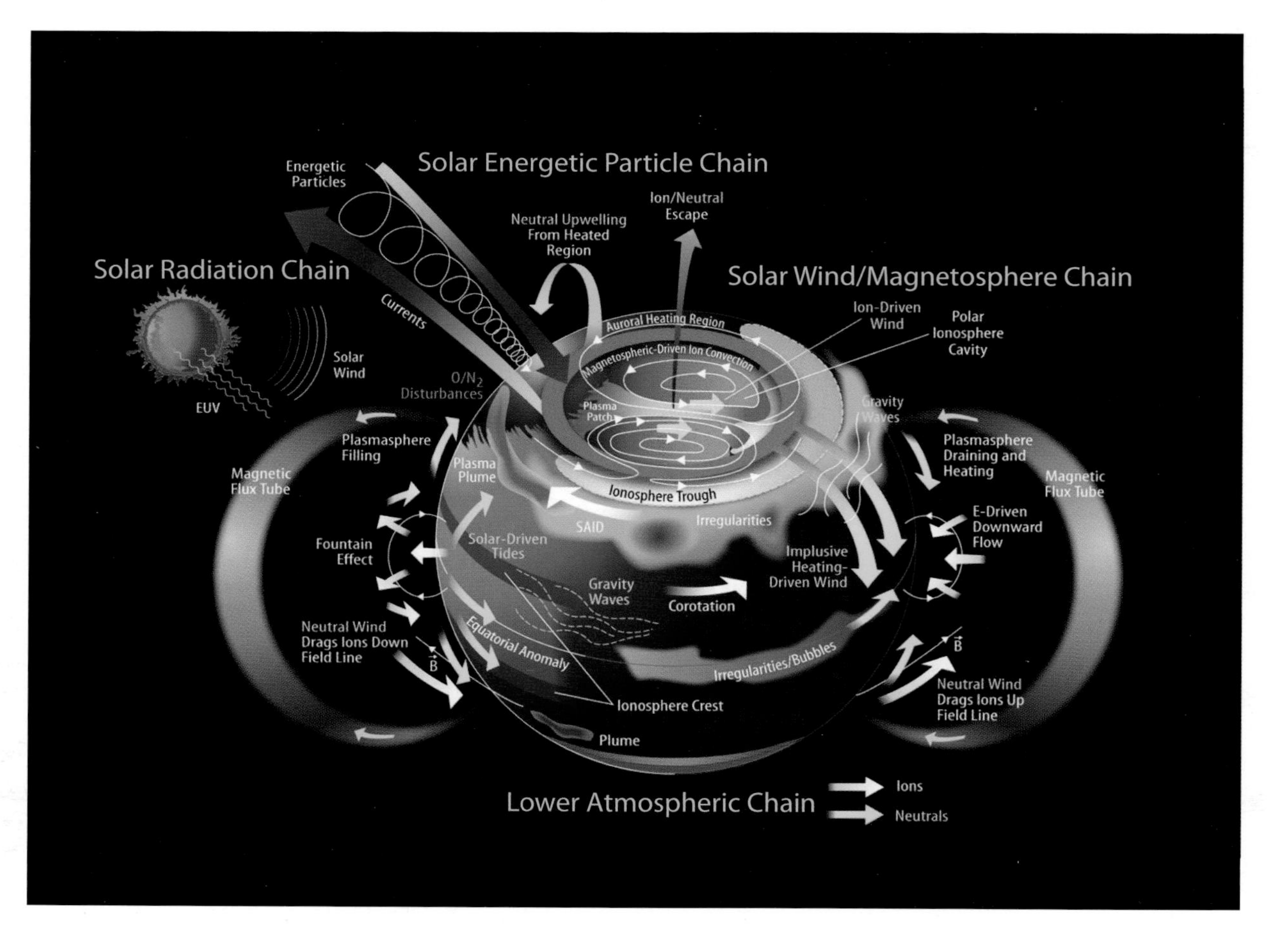

FIGURE 8.1 A depiction of the atmosphere-ionosphere-magnetosphere (AIM) system and the major processes that occur within that system. Absorption of short-wavelength solar radiation accounts for a large fraction of the heat input. Energetic particles, mostly from the magnetosphere, enhance the ionospheric conductance at high latitudes and modify the electrical currents that flow between the ionosphere and magnetosphere. Magnetospheric convection imposes electric fields that drive currents in the lower part of the ionosphere and set the ionospheric plasma into motion at higher altitudes, with a portion escaping into geospace and beyond. These injections of energy drive a global thermospheric circulation that redistributes heat and molecular species upwelling from the heated regions and also excites a spectrum of waves that redistribute energy both locally and globally. Planetary waves, tides, and gravity waves propagate upward from the lower atmosphere, deposit momentum into the mean circulation, and generate electric fields via the dynamo mechanism in the lower ionosphere. Dynamo electric fields are also created by disturbance winds. Neutral winds and electric fields from these combined sources redistribute plasma over local, regional, and global scales and sometimes create conditions for instability and production of smaller-scale structures in neutral and plasma components of the system. SOURCE: Courtesy of Joe Grebowsky, NASA Goddard Space Flight Center.

region of geospace possesses several distinguishing characteristics that define it as a domain for compelling scientific inquiry and warrant the attention of a decadal survey with an explicit focus on solar and space physics' connections to a technological society. Notably, this region serves as a "final link" in the transfer of energy within the solar-terrestrial chain. The primary drivers for variability in the region consist of direct solar energy in the form of extreme ultraviolet (EUV) and ultraviolet (UV) radiation, solar energy transformed into the charged particles and fields that permeate the magnetosphere, and solar-driven waves propagating upward from the lower atmosphere (see Figure 8.1).

Responses to these drivers are determined by interacting dynamical, chemical, and electrodynamic processes that occur over a wide range of spatial and temporal scales, and moreover are strongly influenced by the presence of a strong magnetic field. Often these processes involve nonlinearity and feedback, and it is thus evident that this complex system can often exhibit *emergent behavior*.[2] In fact, scientific investigations of this geospace region resolve and interpret the system's response to variable forcing, and ultimately unravel the complex chains of events leading to the observed, emergent behavior. (Several examples of emergent behavior are provided in this chapter.) Given this complexity, one can appreciate the difficulties of predicting the variability of neutral and plasma densities to the accuracies required to support orbital, reentry, communications, and navigation systems in operational settings. Thus, as this chapter unfolds, it will become evident that the study of atmosphere-ionosphere-magnetosphere interactions presents challenging scientific problems that are fundamental to understanding planetary atmospheres and exospheres and that underlie the ability to predict environmental conditions that serve operational needs. In addition, the processes studied in this context can often be translated to other planetary bodies, and in this way geospace serves as a local laboratory to reveal and study universal physical processes (M3).

The following section of this chapter summarizes the main scientific achievements of the past decade, reflecting back on the recommendations of the previous decadal survey. This lays the foundation for the subsequent section, which sets forth the science agenda for 2013-2022. The section after that addresses the various assets, resources, and strategies needed to advance AIM science most productively and presents a prioritized program for doing so.

8.3 SIGNIFICANT ACCOMPLISHMENTS OF THE PREVIOUS DECADE

Understanding of atmosphere-ionosphere-magnetosphere (AIM) interactions has advanced through a number of vigorous programs, ranging from national, international, and multiagency programs to smaller-scale programs. Examples of programs that have helped shape the research landscape over the past decade are NASA's Living with a Star (LWS) and Heliophysics Geospace Science programs, the NASA TIMED Solar-Terrestrial Probe, the NASA IMAGE Mid-size Explorer (MIDEX) mission, the NASA FAST Small Explorer (SMEX) mission, the NASA THEMIS MIDEX mission, the NASA AIM SMEX mission, and the U.S. Air Force (USAF) C/NOFS mission; the NASA Sounding Rocket Program; the National Space Weather Program; the NSF-sponsored SHINE, GEM, CEDAR, and small-satellite programs and their international counterparts; the NSF major research initiative (MRI) and science and technology centers (STCs); numerous DOD activities; international satellite programs, such as CHAMP, GRACE, and COSMIC; and international science programs, such as CAWSES. These various programs have supported satellite and ground-based instruments and the related data analysis, theory, and modeling efforts. Research models and data assimilation schemes have advanced operational space weather prediction and created new models of the Sun-Earth system using a systemic and holistic perspective: Center for Integrated Space Weather Modeling (CISM), Community Coordinated Modeling Center (CCMC), NCAR Whole Atmosphere Community Climate Model (WACCM) development, and solar wind/magnetosphere models coupled with ionosphere/thermosphere global circulation models. Through these targeted programs and the critically important base programs funded by NSF, NASA, NOAA, and DOD, important scientific progress has been made, helping us to clarify needs and identify priorities that form the basis of this panel report.

[2]Emergent behavior results from the interaction of a large number of system components that could not have been anticipated on the basis of the properties of components acting individually.

New supporting technologies, not specifically targeted for AIM research, have also significantly contributed to the field's advancement in the past decade. These include cyberinfrastructure, advanced communications, improved sensors, networking technology, increases in computing power, precision navigation systems, and small satellites. Complementing this technology growth were planned developments in space-borne and ground-based missions, major research instrumentation and facilities development, data assimilation schemes, and whole-atmosphere model development. These technological advancements help accelerate scientific endeavors and enable new science areas to be investigated and understood. The emergence of relocatable incoherent scatter radars (ISRs) based on electronically steerable antenna arrays is an excellent example. These NSF-supported advanced modular ISRs (AMISRs) can be steered on a pulse-to-pulse basis, allowing the simultaneous acquisition of information from multiple directions. The rapid steering capabilities of AMISR-class ISRs provide a unique capability for supporting AIM science objectives. For instance, these instruments can be used to construct three-dimensional views of the evolving plasma state within a volume traversed by a satellite or rocket.

Model development has been facilitated by major advances in instrumentation and measurement techniques, experimental facilities, and observing networks, which are starting to provide unprecedented volumes of data on processes operating across AIM. Together with concurrent progress in computational techniques, these advances have enabled the development of ever more sophisticated, multidimensional models of geospace. These models, along with data assimilation schemes, offer the promise of greater insights into the physical processes at work and improved ability to forecast disruptive events and their potential impacts.

Development of numerical models that extend from Earth's surface to the thermosphere/ionosphere has made significant breakthroughs during the past decade. These whole-atmosphere models are able to generate atmospheric disturbances, such as sudden stratospheric warming and quasi-biennial oscillation internally without having to impose artificial forcing, and to investigate their dynamical and electrodynamical coupling to the upper atmosphere in a self-consistent manner. Models that couple the magnetosphere and the ionosphere/thermosphere have reached the maturity to include feedback interaction between thermospheric neutrals and magnetospheric plasmas, as well as mass and momentum exchanges within geospace. In addition, physics-based data assimilation models of the global ionosphere have been developed that are capable of assimilating multiple data types, for example, to reconstruct the electron density configuration during storms. These models are now running routinely in a test-operational mode for space weather specification.

The adoption and implementation of a systems approach are more realizable today with the rapid expansion of multidimensional databases, increasing computational capabilities and sophistication of numerical tools, and emergence of new sensor technologies. Complementing these technological advancements have been new scientific discoveries that are rooted in a systems perspective of AIM science. What has emerged from this past research is the recognition that many of the natural coupling processes within AIM are linked through system complexity processes of feedback, nonlinearity, instability, preconditioning, and emergent behavior. The following examples of significant accomplishments of the previous decade reflect this overarching recognition.

8.3.1 Magnetosphere-Ionosphere Coupling

A recent discovery in AIM science comes from a fortuitous combination of new measurement capabilities. The explosive increase in the global distribution of GPS receivers both on the ground and in space and the flight of the NASA IMAGE mission to image Earth's magnetosphere-ionosphere (MI) showed a completely new view of ionospheric/magnetospheric coupling during storms. Global GPS maps of ionospheric density showed, for the first time, large-scale dense plumes of plasma extending from middle latitudes to the auroral zone at the onset of magnetic storms (Figure 8.2). During such events, plasmaspheric imaging of He^{+} ions by IMAGE showed corresponding structures in the inner magnetosphere, where plasma was sheared away from the plasmasphere and advected toward the magnetopause. The plasmaspheric structure was never expected to appear in the ionosphere, and the discovery points to a process critical to enhancing auroral ion outflow during storms. Further research results from NASA's FAST and IMAGE satellites revealed that storm-enhanced ionosphere plasma feeds outflows of ionospheric ions into a tornado-like cusp funnel, powered by Alfvén waves generated by the solar wind-magnetosphere interaction. Evidence is accumulating that energy in these small-scale Alfvénic current filaments is deposited over a range of spatial and temporal scales and is converted to heat and momentum through ion-neutral interactions.

These findings highlight the importance of feedback through the AIM system where magnetospherically imposed electric fields redistribute ionospheric plasma, fueling the flux of outflowing ions to the magnetosphere. The outflows, especially heavy-ion outflows, can overwhelm reconnection mass loss in the plasma sheet and, effectively, reduce the cross-polar-cap potential and the concomitant ionospheric electric fields. In some instances, emergent behavior results whereby ~3-hour planetary-scale (sawtooth)

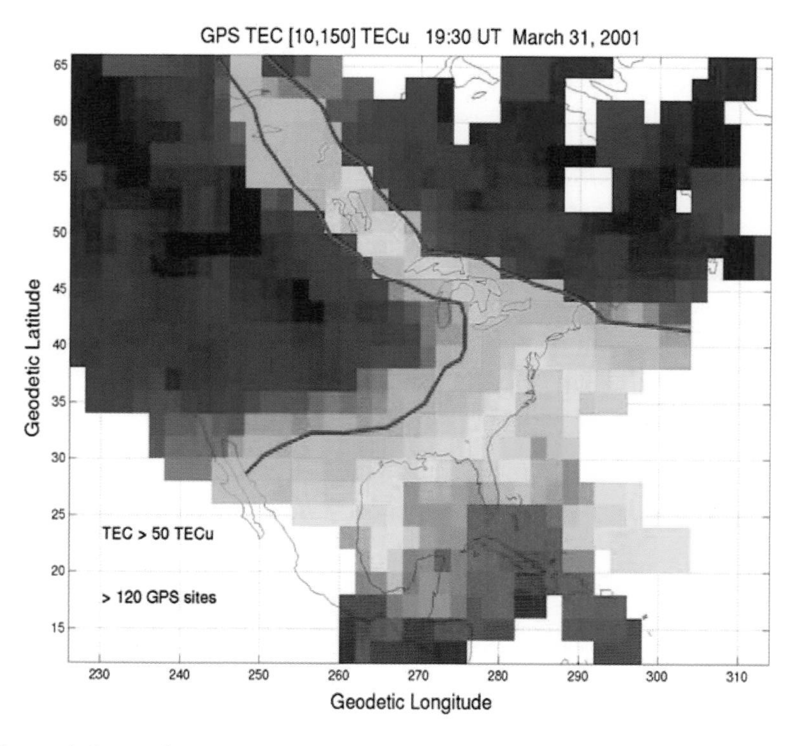

FIGURE 8.2 Storm-enhanced plasma density signatures in total electron content (TEC) observed on March 31, 2001. These are believed to be connected to plasmashere erosion and driven by subauroral electric fields from the inner magnetosphere. SOURCE: J.C. Foster, P.J. Erickson, A.J. Coster, J. Goldstein, and F.J. Rich, Ionospheric signatures of plasmaspheric tails, *Geophysical Research Letters* 29(13):1-1, doi:10.1029/2002GL015067, 2002. Copyright 2002 American Geophysical Union, reproduced by permission of American Geophysical Union.

oscillations in the magnetosphere are observed. These occur when the MI system is strongly driven by a steady solar wind and seem to rely on superfluent nightside outflows of ionospheric O^+.

8.3.2 Solar-AIM Coupling

The past decade marked the 23rd solar cycle on modern record. Notable events included a number of powerful geomagnetic storms, two separate sunspot maxima, and a very deep solar minimum. With observations from an array of space- and ground-based instruments unprecedented in their capabilities, solar cycle 23 is the first cycle since the initial detection of coronal mass ejections (CMEs) in the early 1970s in which a complete record of CMEs, coronal hole distributions, and solar wind data are all available over the whole cycle. The availability of simultaneous space- and ground-based data covering the Sun-Earth space has made solar cycle 23 solar storms and geomagnetic activity one of the best sets of events to analyze. It has been possible to assemble atmospheric, ionospheric, magnetospheric, interplanetary, and solar data on 88 CME storms during solar cycle 23. Many more events of enhanced geomagnetic activity were observed during this cycle associated with corotating interaction regions (CIRs)/high-speed solar wind streams (HSS) related to low-heliolatitude distributions of persistent coronal holes.

A few of the CME storms were considered "great" storms that led to unexpected or emergent behavior in the AIM system. Ionosphere observations indicated the emergence of a daytime super-fountain effect lifting the ionosphere to new heights and increasing its total electron content by as much as 250 percent. Also observed were very large amplitude traveling ionospheric disturbances (Figure 8.3), new ionosphere layers, and very different behavior in equatorial plasma irregularities.

The atmosphere responded with dramatic changes in neutral composition, winds, temperature, and mass density. Thermosphere mass density at 400 km increased by over 400 percent during these great storms while experiencing exceptionally fast recovery times, indicating a unique overcooling effect. The CIR/HSS storms were predominant during the declining and minimum phase of the solar cycle, producing an entirely different response in the AIM system. Where CME storms lasted a few days and were episodic, CIR/HSS storms lasted for more than a week and recurred for many solar rotations—in some instances sustaining common periodicities for an entire year. This has led to the discovery in atmosphere and ionosphere data sets of pervasive periodicities at subharmonics of the ~27-day solar rotation period during solar cycle 23 (Figure 8.4). Unfortunately, although CHAMP, COSMIC, and ground-based platforms provided new discoveries in terms of total neutral and plasma density responses of the AIM system to the various solar disturbances noted above, only sparse measurements were made of the key parameters (e.g., winds, plasma drifts, neutral and ion composition) needed to understand these responses. It is a high-priority goal of the next decade to gain this understanding.

8.3.3 Meteorology-AIM Coupling

One of the most exciting developments in recent years has been a new realization of the direct and strong impact of tropospheric weather and climate on the upper atmosphere and ionosphere. The connection has been elicited, first, from measurements of the ionospheric density near the equator by NASA IMAGE and TIMED satellites, showing large changes in the structure of the ionosphere on seasonal timescales. This signature has subsequently been observed in upper-thermospheric composition and temperature. The clear correspondence demonstrated in this confluence of efforts has energized the study of atmospheric wave coupling to space plasma. Other observations and model studies have unequivocally revealed that Earth's IT system owes a considerable amount of its longitudinal, local time, seasonal-latitudinal, and day-to-day variability to atmospheric waves that begin near Earth's surface and propagate into the upper atmosphere.

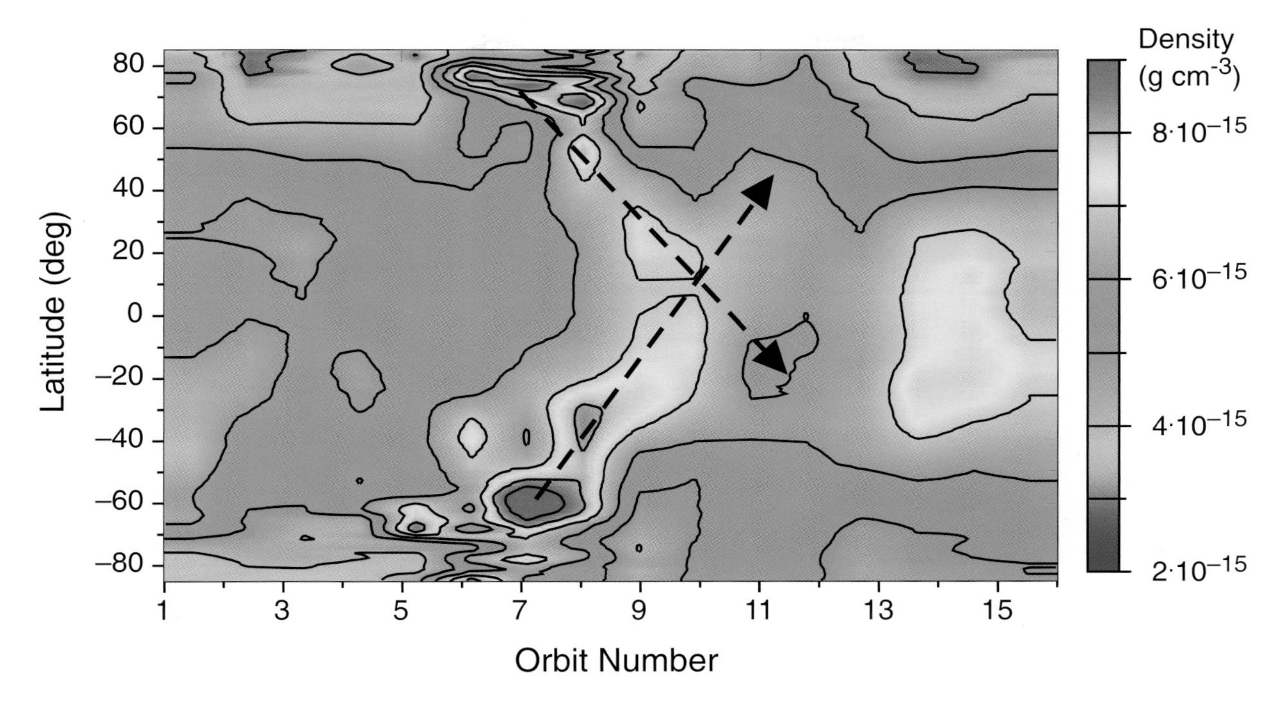

FIGURE 8.3 Illustration of traveling atmospheric disturbances seen in densities near 400 km near local noon measured by the accelerometer on the Challenging Mini-Satellite Payload (CHAMP) satellite, in connection with a geomagnetic disturbance on day 308 of 2003. The data are obtained at nearly constant longitudes about every 1.5 hours, i.e., the time between consecutive "passes" or "orbits" of the satellite. The disturbance was initiated between pass 6 and 7 and took roughly 4.5 hours to reach the equator from both polar/auroral regions. The southward-propagating disturbance appears to pass into the Southern Hemisphere. Researchers do not know the behaviors of these disturbances in other local time sectors, or how dissipation of this disturbance has modified the mean state of the ionosphere-thermosphere system. SOURCE: Courtesy of Sean Bruinsma, Centre National d'Études Spatiales.

Waves propagating upward from the lower atmosphere contribute about equally to the energy transfer in the IT system as direct solar energy in the form of EUV and UV radiation and reprocessed solar energy in the form of particles and fields from the magnetosphere. This unexpected and new realization is important for the space weather of the IT system. It is becoming increasingly clear that understanding wave driving from below is critical for predicting large- and small-scale structures in the IT system, such as ionospheric scintillations important to communication and navigation, and for testing and improving models for orbit propagation and collision warnings.

8.3.4 AIM Coupling and Global Change

Earth is changing, and there are compelling and urgent needs for society to expand and develop basic science research to assess and answer society's concerns in the area of climate change. There is sufficient evidence to believe that any climate change connected to changes in solar activity may involve chemical and dynamical pathways through the upper and middle atmosphere. Furthermore, long-term evolution-

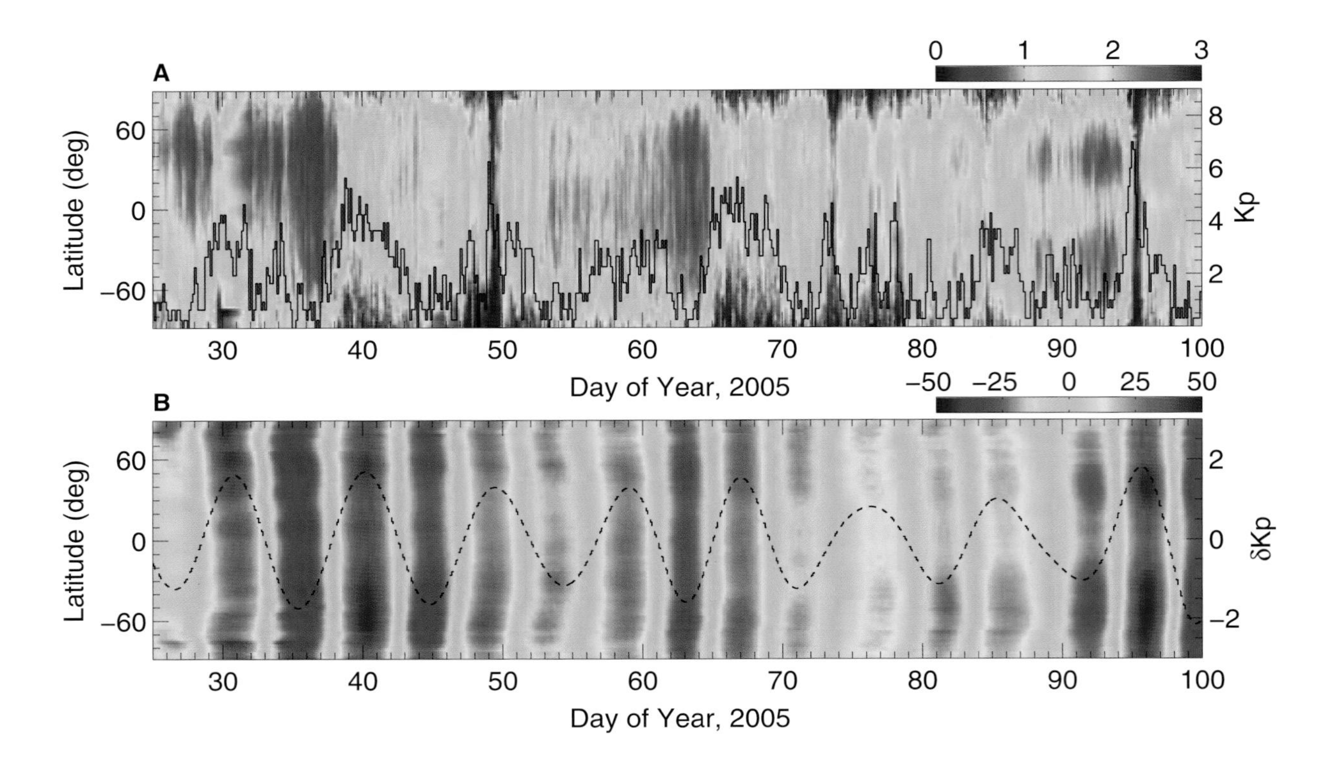

FIGURE 8.4 Quasi-9-day periodicity in the thermosphere densities as a result of recurring high-speed solar wind streams (and associated recurrent geomagnetic activity) originating from longitudinally distributed solar coronal holes. (A) Latitude versus time variations of CHAMP neutral densities (in units of 10^{12} kg/m^3) during days 25-100, 2005. The solid black line denotes the Kp values, corresponding to the right-hand scale. (B) Percent of the band-pass filter density residuals to 11-day running mean during days 25-100, 2005. The band-pass filter was centered at the period of 9 days, with half-power points at 6 and 12 days. The perturbations in Kp obtained from the same band-pass filter are superimposed in the lower panel (dashed line, right-hand scale). SOURCE: J. Lei, J.P. Thayer, J.M. Forbes, E.K. Sutton, and R.S. Nerem, Rotating solar coronal holes and periodic modulation of the upper atmosphere, *Geophysical Research Letters* 35:L10109, doi:10.1029/2008GL033875, 2008. Copyright 2008 American Geophysical Union, reproduced by permission of American Geophysical Union.

ary change in Earth's atmosphere may alter short-term system variability or multiscale temporal response in the AIM system. Studies of these effects have been newly undertaken in the recent decade, with some remarkable findings.

The AIM system can serve as a conduit or amplifier of externally induced climate drivers, coupling different regions via radiative, dynamical, and chemical feedbacks. These pathways are shown schematically in Figure 8.5. Through these feedback processes remote regions highly driven by solar influences are linked to the troposphere.

Yet how and to what extent solar and magnetospheric variability affects atmospheric conditions and climate in such an interconnected system remains an open question. A clear link between space weather and ozone destruction is contained in the large fluxes of aurorally produced nitric oxide (NO) that are observed moving downward into the stratosphere within the winter polar vortex. The circulation of air within the vortex confines the NO to dark high latitudes and rapidly transports it to lower altitudes before

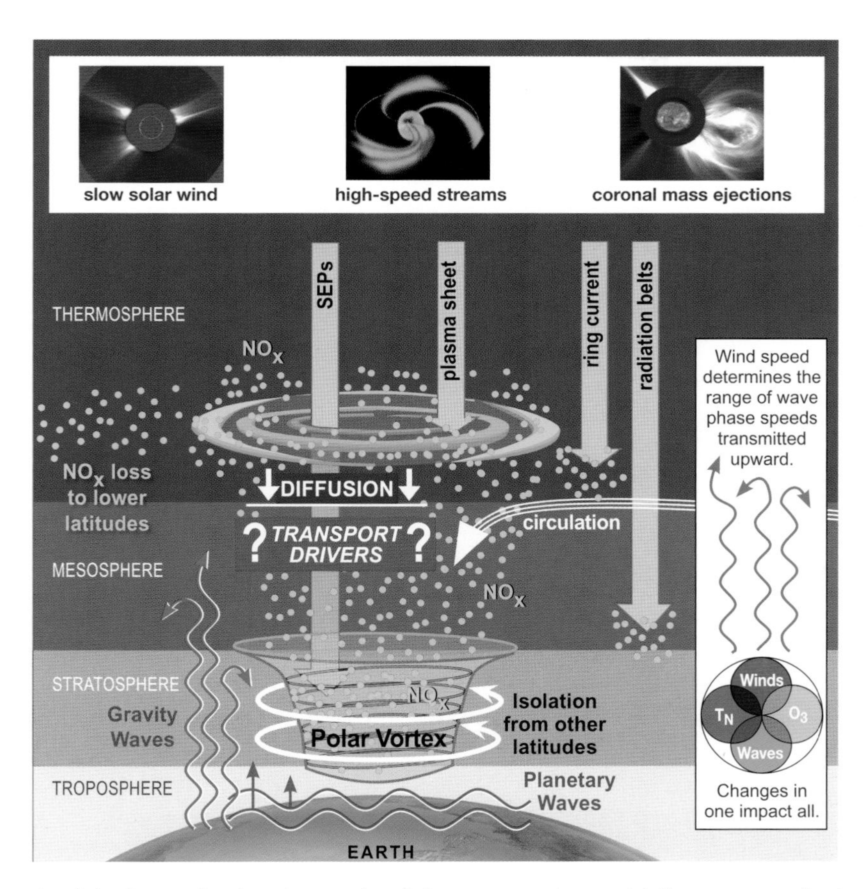

FIGURE 8.5 Schematic of the intertwined pathways that link space weather variability to atmospheric coupling. SOURCE: Courtesy of Cora E. Randall, University of Colorado, Boulder; Janet U. Kozyra, University of Michigan, Ann Arbor; and Scott M. Bailey, Virginia Institute of Technology.

---

it can be destroyed by sunlight. Once in the stratosphere, NO catalytically destroys ozone as the vortex breaks up, influencing temperature and circulation over a broad region in the middle atmosphere. But whether or not the effects of these fluxes are strong enough to be transmitted into the troposphere through resulting large-scale changes in atmospheric circulation remains unknown.

Long-term change in the lower and middle atmosphere can also drive change in the AIM system. For instance, a systematic decrease by several percent per decade in thermosphere mass density is now evident in the record of satellite orbit decay measured since the beginning of the space age. An effect predicted in the 1980s, this change is thought to arise largely in response to the increase in atmospheric CO_2 that acts as a radiative cooler in the upper atmosphere, diametric to its role in the lower atmosphere. This is not itself an effect of the change in climate in the lower atmosphere, but rather a human-influenced change in the upper. An important practical consequence for society is that a less dense upper atmosphere lengthens the residence time of orbital debris.

The AIM community has thus begun to undertake investigations of "space climate." Studies of space climate deal with determining and understanding the average behavior of the coupled geospace system and the elements of that system. The typical behavior of geospace, on a variety of spatial and temporal scales, is of interest to the research and applications community. Gradual changes in solar activity, solar wind, EUV radiation, and Earth's magnetic field each play a significant role in defining the longer-term variation in the geospace environment. For instance, long-term changes in Earth's magnetic field are occurring and producing measurable changes in the ionosphere. From the solar irradiance perspective, the latest solar minimum, from late 2007 to mid-2009, marks the lowest solar EUV fluxes (and heating rates) of the longest duration in the past four solar cycles. This "super-minimum" produced unprecedented low temperatures in the ionosphere and a contracted thermosphere that none of the current numerical models were able to predict or reproduce. This low solar minimum was also accompanied by a weaker than normal interplanetary magnetic field, cosmic rays at record high levels, high tilt angle of the solar dipole magnetic field, and low solar wind pressure. All of these solar surface, solar wind, and interplanetary parameters constitute a change in the space climate and played an integral role in how the AIM system evolved and responded. These differences also had important practical consequences for satellite operations, space debris/hazard prediction, ionospheric forecasts, and airline operations.

8.3.5 International Programs

AIM science is global in nature. International cooperation is thus a key component of the AIMI panel's decadal plan. One of the more important entities for coordinating these international efforts is the Scientific Committee on Solar-Terrestrial Physics (SCOSTEP). SCOSTEP currently sponsors the Climate and Weather of the Sun-Earth System (CAWSES) program. CAWSES was initiated in 2004 as a 5-year program and was extended into a second phase dubbed CAWSES-II covering the period 2008-2012. The CAWSES-II program focuses on four science questions: (1) What are the solar influences on Earth's climate? (2) How will geospace respond to an altered climate? (3) How does short-term solar variability affect the geospace environment? and (4) What is the geospace response to variable waves from the lower atmosphere?

A successful example of international collaboration in space benefiting AIM science is the COSMIC (Constellation Observing System for Meteorology Ionosphere and Climate) project. While the project is jointly sponsored by Taiwan and the United States, its scientific benefits extend to all nations. Since its launch in 2006, the COSMIC mission has observed the tidal influence on total electron content (TEC) and F-region ionosphere, wave-4 signatures in the topside ionosphere/plasmasphere, and a geographically fixed (with the Weddell Sea) ionospheric anomaly, and it is demonstrating the complex structure in ionosphere F-region density and peak altitude.

Contributions also come from serendipitous opportunities. One such example is that from international geodesy programs. The German Challenging Minisatellite Payload (CHAMP) and NASA/German Gravity Recovery and Climate Experiment (GRACE) missions led to improved estimates of the thermosphere mass density and winds from the need to better model and understand Earth's gravity field and its spatial and temporal variability. These thermosphere measurements have led to several discoveries and new insights into the AIM system, such as a neutral density cusp enhancement, an equatorial thermosphere anomaly, tides in the upper thermosphere, and periodic expansion of the thermosphere gas associated with coronal hole distributions on the Sun.

8.3.6 Current and Future Programs

Solving the compelling, remaining mysteries of AIM science requires a continued commitment to a space, ground, and modeling effort. Current and future programs and interagency activities provide context and the boundary conditions for AIM efforts. These include NASA's ACE, SDO, and STEREO missions for studying the Sun and solar wind, and NASA's RBSP (renamed the Van Allen Probes) for studying Earth's radiation belts. The NASA TIMED mission continues to provide valuable IT science, as does the USAF C/NOFS satellite, which was placed in a unique, low-inclination orbit where its instruments gather data related to the unstable equatorial ionosphere. NASA's Sounding Rocket Program supports a variety of cutting-edge investigations in the ionosphere, thermosphere, and mesosphere focused on solving outstanding science questions at high, middle, and low latitudes. International programs such as ESA's SWARM and GOCE missions and Canada Space Agency's e-POP payload on its CASSIOPE satellite will further contribute to AIM science and enrich international collaboration. The NSF AMISR-class incoherent scatter radars are currently deployed near Poker Flat, Alaska (PFISR), and Resolute Bay, Canada (RISR), the latter consisting of two full radar faces, one funded through an international collaboration with Canada. Plans are being developed to deploy an additional radar in Antarctica and to relocate the PFISR facility to La Plata, Argentina, a location magnetically conjugate to Arecibo. NSF also supports the AMPERE project, which utilizes the engineering magnetometers aboard the Iridium communications network to resolve field-aligned magnetospheric currents in the auroral zone.

8.4 SCIENCE GOALS AND PRIORITIES FOR THE 2013-2022 DECADE

The range of intellectually stimulating science questions that arise within the purview of atmosphere-ionosphere-magnetosphere interactions is enormous. However, there are a subset whose connections to the needs of a 21st-century society make them compelling; it is on this basis that the panel defined the science challenges articulated in this section. Moreover, they also form an integral part of the key science goals (see Chapter 1) that this solar and space physics decadal survey has set forth as its prime agenda.

Before proceeding, it is useful to consider the ionosphere-thermosphere from a systems perspective and to describe what aspects of the system behavior need to be understood in order to advance toward a predictive capability. Figure 8.1 provides a useful reference. First, it must be understood how energy and momentum inputs from the magnetosphere are spatially and temporally distributed in the polar and auroral regions, and how the global IT system responds to these inputs. Equally important, it must be understood how energy and momentum are transferred from the lower atmosphere into the IT system, and what this means in terms of IT spatial and temporal variability. Key to the above, it must further be understood how internal processes transform and transfer energy and momentum within the system, regulate responses to external forcing, and control the formation of regional and local structures in both neutral and ionized constituents. The consequences of two-way interactions between the IT system and the magnetosphere must also be considered. In this context, it must be understood how the high-latitude IT system moderates the transfer of energy from the solar wind and magnetosphere and how the inner magnetosphere and plasmasphere interact with the mid-latitude ionosphere and drive its variability.

Achieving the above level of understanding is a multidecade task. The AIMI panel has, however, narrowed the scope of aspirations to five AIMI science goals that have the potential to be comprehensively addressed with current technologies or those under development, and within the 2013-2022 decade. These are enumerated in Figure 8.6 and are expanded on throughout the remainder of this section. In addition, Figure 8.6 maps AIMI science goals into the decadal survey's key science goals 1-4, and the text below explains these mappings further. Finally, the following section sets forth a prioritized set of strategies to address these challenges.

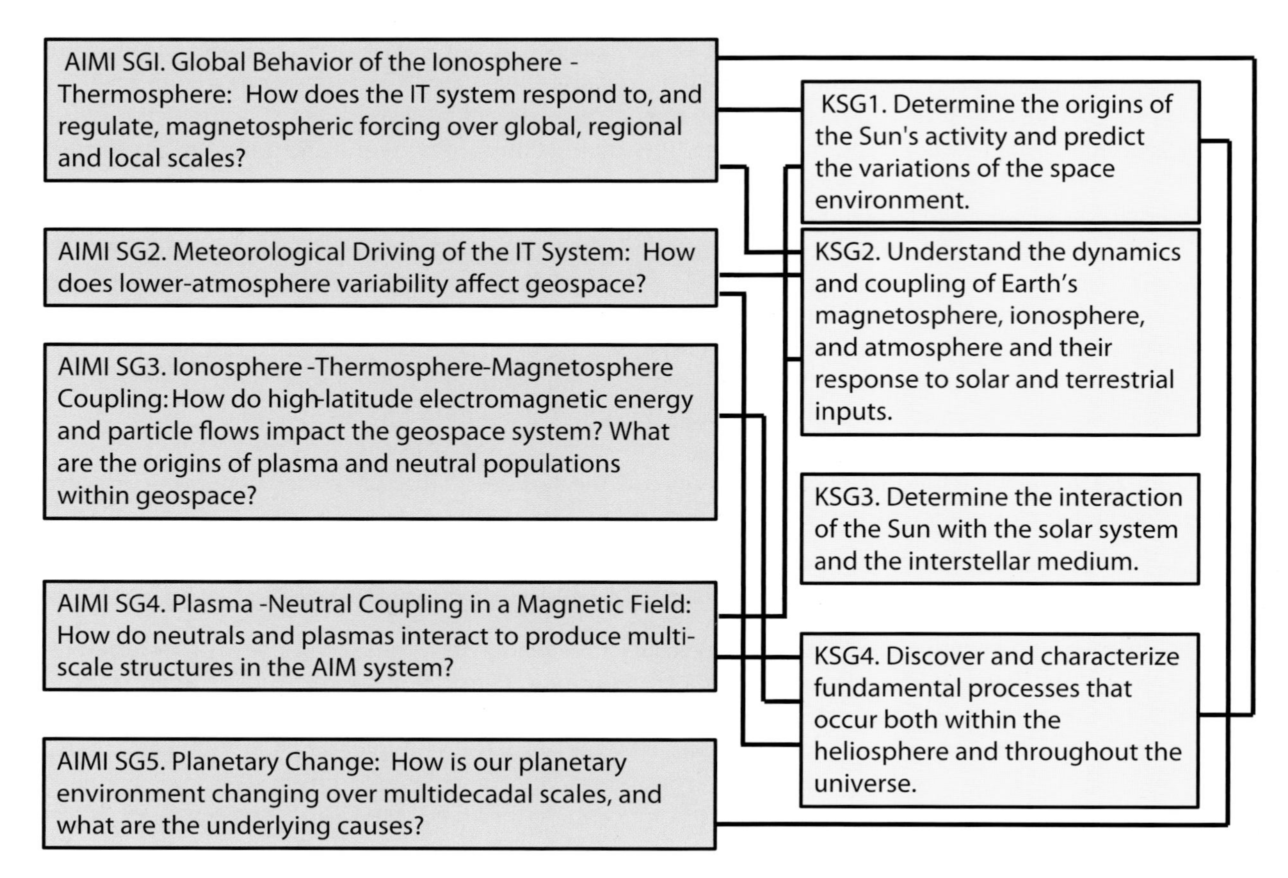

FIGURE 8.6 Five AIMI science goals for the 2013-2022 decade and how they map into the decadal survey's key science goals 1-4.

8.4.1 AIMI Science Goal 1. Global Behavior of the Ionosphere-Thermosphere

How does the IT system respond to, and regulate, magnetospheric forcing over global, regional, and local scales?

At high latitudes, the AIM system directly impacts magnetospheric dynamics through conductivity changes, current closure, and ion outflow. Spatially confined energy input from the magnetosphere can quickly be redistributed to the ionized and neutral gases over much larger scales. A response of the system can occur at locations well removed from the input. Determination of changes in the system that are imposed externally and changes resulting from the internal system response is required to understand the geo-effectiveness of the interaction of the planet with the solar wind. In other words, the ionosphere-thermosphere system both adjusts to the varying input from above, and also feeds back and regulates this exchange.

Inter-hemispheric differences address the asymmetric closure of currents, as well as local ionospheric structuring and electric fields. Indeed, the dissimilarities present in the simultaneously measured magnetic perturbations and derived currents shown in Figure 8.7 may be due to the differences in the high-latitude neutral density and plasma environments, which affect both the return currents in the aurora and the field-aligned potentials that accelerate the particles that create the visible aurora. The compelling science questions that researchers must answer are these: How do field-aligned currents, precipitation, conductivity, neutral winds and density, and electric fields organize a self-consistent, electrodynamic/hydrodynamic

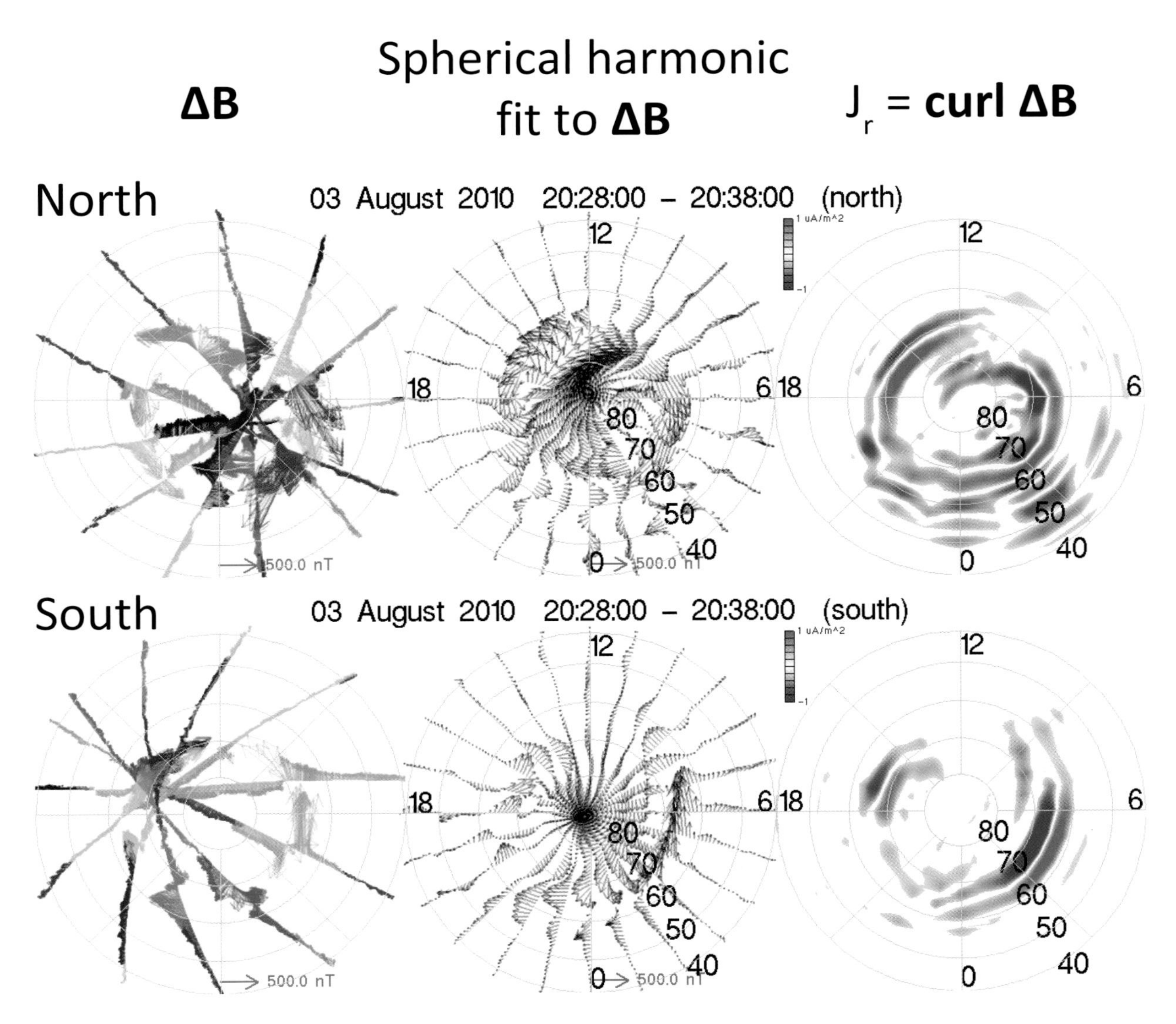

FIGURE 8.7 Simultaneous measurements of currents gathered in the northern and southern polar regions showing a striking difference in the current patterns and hence of energy input. Differences in ionospheric conductivity play an important role in the closure of magnetospheric currents and may have a profound influence on magnetospheric current closure, as the ionosphere and magnetosphere interact to regulate the response of geospace to solar wind input. SOURCE: Active Magnetosphere and Planetary Electrodynamics Response Experiment (AMPERE), Johns Hopkins University, Applied Physics Laboratory.

system at high latitudes? How does such a varying, spatially structured environment feed back on and modify field-aligned current and electric potential patterns imposed from the magnetosphere?

High-latitude heating (mainly below 200 km) causes N_2-rich air to upwell, and strong winds driven by this heating transport N_2 equatorward, which then mixes with ambient O in unknown ways (Figure 8.8).

IT constituents are controlled by gravity, diffusion, chemical reactions, and bulk transport. It is essential to understand how these processes determine global responses in O and N_2 after heating occurs at high latitudes. Since these disturbances are superimposed on a solar EUV-driven circulation system that is mainly ordered in a geographic coordinate frame that varies with local time and season, the interactions can be complex, and IT responses are very different depending on prevailing conditions. The relative abundances of O and N_2 are fundamental to understanding local plasma densities and total mass densities, both of which are key parameters underlying space weather forecast needs. The question then remains, How do winds, temperature, and chemical constituents interact to produce the observed global neutral and plasma density responses of the IT system?

Since the B field plays a major role in controlling the distribution of ionospheric plasma, and since ion-neutral collisions can serve to decelerate or accelerate the neutral gas, the ionospheric plasma can in many ways regulate the IT response to magnetospheric forcing. This occurs mainly through the redistribu-

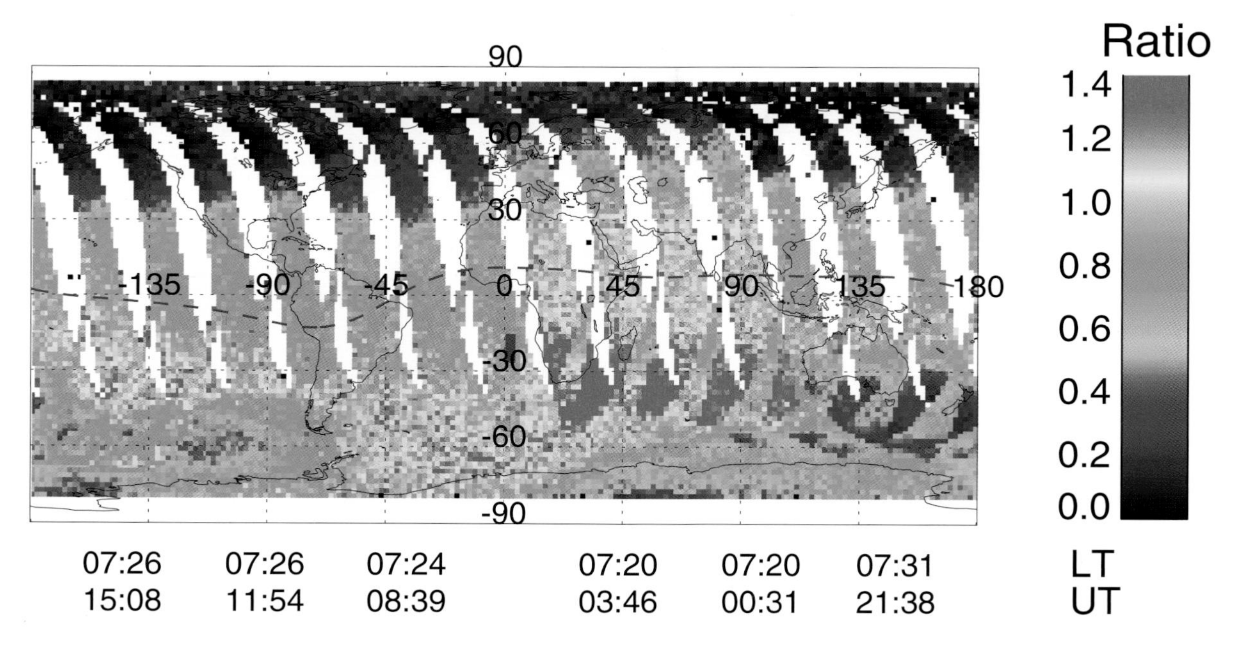

FIGURE 8.8 This image from the Thermosphere-Ionosphere-Mesosphere Energetic and Dynamics (TIMED)/Global Ultraviolet Imager (GUVI) instrument provides the height-integrated O/N_2 density ratio for a single moderately disturbed day in April 2002. This picture varies considerably from day to day, but is available only at a single local time on any given day. Without coincident global measurements of neutral winds, temperature, and total mass density and some measure of localized heating, the causes and consequences of this composition variability cannot be ascertained. The Geospace Dynamics Constellation mission, described below in this chapter, will enable researchers to understand the relationships between these variables and, moreover, will provide this information simultaneously as a function of local time in a single day. SOURCE: Courtesy of Johns Hopkins University, Applied Physics Laboratory.

tion of plasma by electric fields. For instance, in connection with a sudden storm commencement, eastward penetrating electric fields can lift the equatorial ionosphere and accelerate the neutral gas through removal of the drag effect of the ions. A similar effect can occur at middle latitudes when equatorward winds push the plasma up magnetic field lines, lessening the drag on the zonal winds. Large redistributions of plasma occur as the result of subauroral electric fields that couple the inner magnetosphere and plasmasphere to the mid-latitude ionosphere (Figure 8.9). Disturbance winds below 200 km generate electric fields through the dynamo mechanism, which then redistribute plasma that affects the wind system at higher altitudes. As discussed below, there are also tidal-driven electric fields that redistribute plasma as a function of local time, longitude, and season and that modify the interaction between the plasma and neutral components of the IT system. The key question is, How do plasma and neutrals interact to produce the observed response of the IT system, including hemispheric and longitudinal asymmetries?

At high latitudes the IT system and the magnetosphere are engaged in a two-way interaction with each other. Energetic particles from the magnetosphere ionize the upper atmosphere, creating complex conductive pathways that regulate the flow of current from the magnetosphere. Electric fields guide the flow of

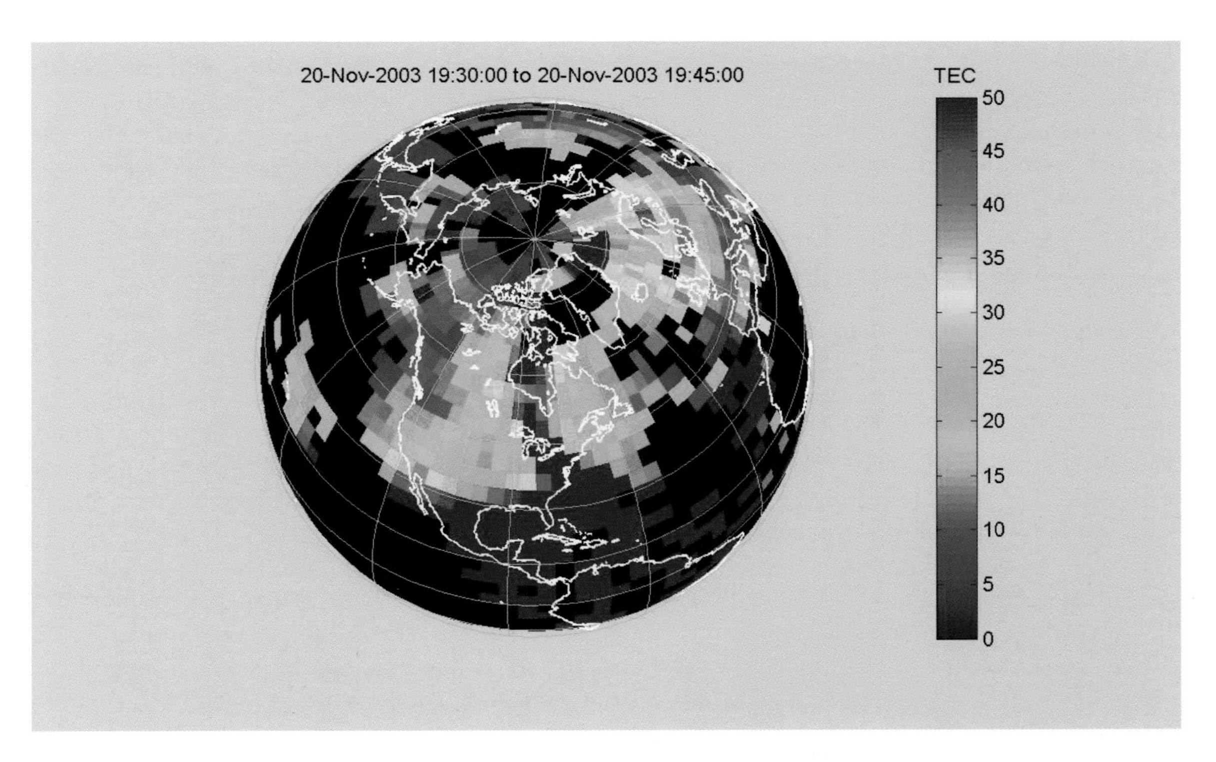

FIGURE 8.9 Storm-enhanced plasma density signatures in total electron content (TEC) observed on November 20, 2003. These signatures are believed to be connected to plasmasphere erosion and driven by subauroral electric fields from the inner magnetosphere. Strong plasma density gradients are observed over North America, the details of which could be observed by a network of ground-based observatories. Spatial and temporal evolution of the global structure would be well observed by a constellation of satellites making in situ measurements. SOURCE: A. Coster and J. Foster, Space weather impacts of the subauroral polarization stream, *Radio Science Bulletin* 321:28-36, 2007. Copyright 2007 Radio Science Press, Belgium, for the International Union of Radio Science (URSI), used with permission.

currents within the ionosphere, leading to Joule heating that depends on the spatial and temporal variability of the E fields as well as their absolute magnitudes. The peak altitude of Joule heating in turn determines the response time of the global thermosphere to this energy input. Energetic particles also initiate a chemical pathway to create nitric oxide, which regulates the response and recovery of the neutral atmosphere through radiative cooling. Local heating of the IT system and ionospheric flows from lower latitudes (see Figure 8.9) serve as sources of O^+ to the magnetosphere, which then regulates how the magnetosphere transfers solar wind energy to the IT system (see further details on magnetosphere-IT interactions under "AIMI Science Goal 3").

The interactions and feedbacks that occur between energy deposition, dynamics, radiative cooling, energetic particles, electric fields, and plasma and neutral constituents and temperatures are how the global IT system *regulates* its response to magnetospheric forcing, and how it also regulates the response of the magnetosphere to solar wind forcing. The complexity of the AIM system is such that emergent behaviors occur, sometimes involving coupling across spatial and temporal scales (see further details under "AIMI Science Goal 4").

The AIMI panel concluded that a major goal of the coming decade, therefore, is to understand how regulation of the IT system occurs, and how connectivity between multiple scales arises within this regulation process.

Making the required coincident multi-parameter measurements of the system over local, regional, and global scales poses major challenges in terms of observational strategies. Strategies that employ an optimal combination of ground-based, suborbital and space-based platforms involving innovative in situ and remote-sensing instrumentation will be required. The panel's implementation strategies are presented in the section "Implementation Strategies and Enabling Capabilities" below.

8.4.2 AIMI Science Goal 2. Meteorological Driving of the IT System

How does lower-atmosphere variability affect geospace?

Numerous observational and modeling studies conducted since the 2003 decadal survey have unequivocally revealed that the IT system owes much of its longitudinal, local-time, seasonal-latitudinal, and day-to-day variability to meteorological processes in the troposphere and stratosphere. The primary mechanism through which energy and momentum are transferred from the lower atmosphere to the upper atmosphere and ionosphere is through the generation and propagation of waves (Figure 8.10).

Owing to rotation of the planet, periodic absorption of solar radiation in local time (LT) and longitude (e.g., by troposphere H_2O and stratosphere O_3) excites a spectrum of thermal tides having periods and zonal (east-west) wavenumbers (or harmonics) defined by the planetary rotation period and longitudinal variability, respectively. Surface topography and unstable shear flows arising due to solar forcing excite planetary waves (PWs) and gravity waves (GWs) extending from planetary to very small (approximately tens to hundreds of kilometers) spatial scales and periods ranging from 2 to 20 days down to minutes. The absorption of solar radiation at the surface and the subsequent release of latent heat of evaporation in convective clouds radiate additional thermal tides, GWs, and other classes of waves. Those waves that propagate vertically grow exponentially with height into the more rarified atmosphere, ultimately achieving large amplitudes. Some parts of the wave spectrum achieve convective instability, spawning additional waves or turbulence. Other parts of the wave spectrum are ultimately dissipated by molecular diffusion in the 100- to 150-km-height region, and some fraction of those waves penetrate all the way to the base of the exosphere (ca. 500-600 km). Along the way, nonlinear interactions between different wave types occur, modifying the interacting waves and giving rise to secondary waves. Finally, the IT wind perturbations

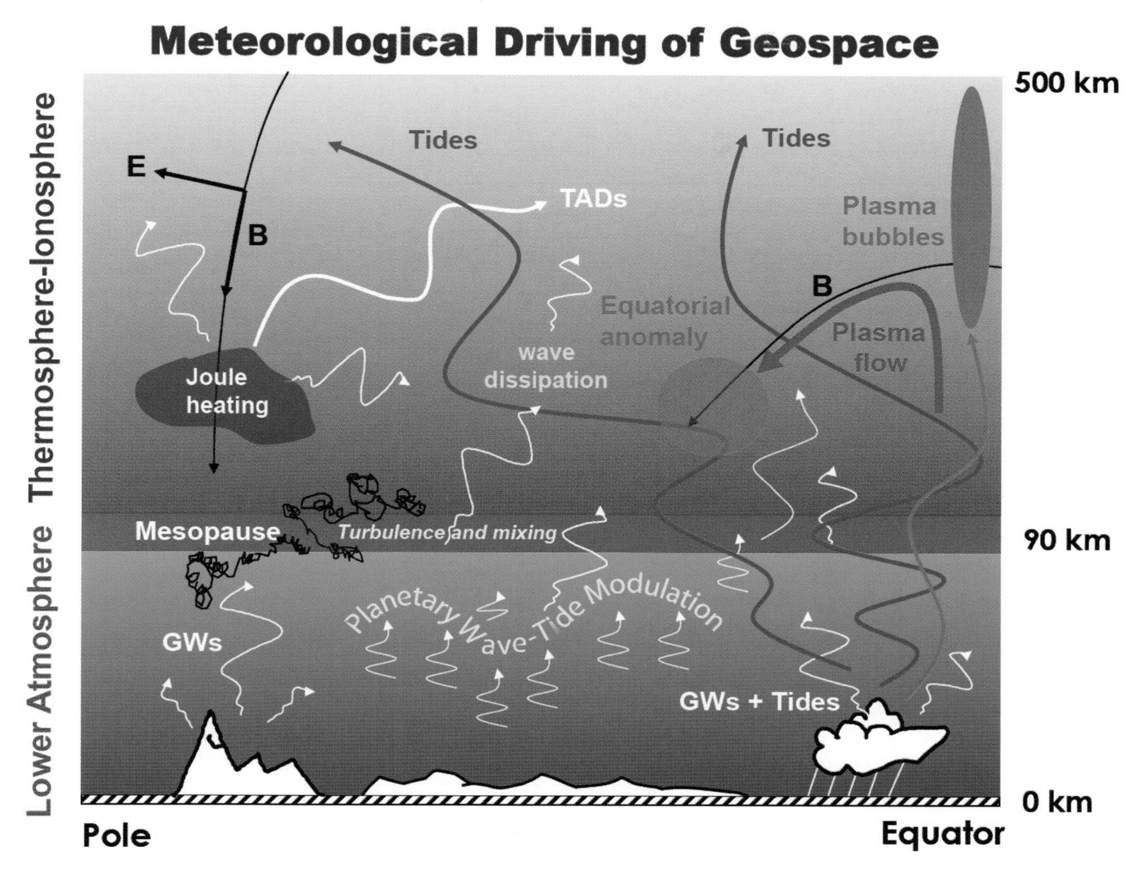

FIGURE 8.10 Schematic of the various mechanisms through which lower-atmosphere processes influence the ionosphere and thermosphere. See text for details. SOURCE: Courtesy of Jeffrey M. Forbes, University of Colorado, Boulder, and David Fritts, Colorado Research Associates.

carried by the waves can redistribute ionospheric plasma, either through the electric fields generated via the dynamo mechanism, or directly by moving plasma along magnetic field lines (Figure 8.11).

Although the presence and importance of waves are without dispute, the relevant coupling processes operating within the neutral atmosphere, and between the neutral atmosphere and ionosphere, involve a host of multiscale dynamics that are not understood at present. The connection between tropical convection and modification of the ionosphere described above is just one example of emergent behavior that typifies the coupling between the lower atmosphere and the IT system. Below, the panel presents its analysis of what are the most pressing science questions that must be addressed on this topic in the coming decade, particularly with respect to developing a capability to predict the space weather of the IT system.

A first and fundamental question is, How does the global wave spectrum evolve temporally and spatially in the thermosphere? The TIMED, CHAMP, and GRACE missions provide approximately 2-month average tidal climatologies below 110 km and above 400 km, respectively (Figure 8.12), but with little information on the intervening region where the tidal and gravity wave spectra evolve with height, dissipate,

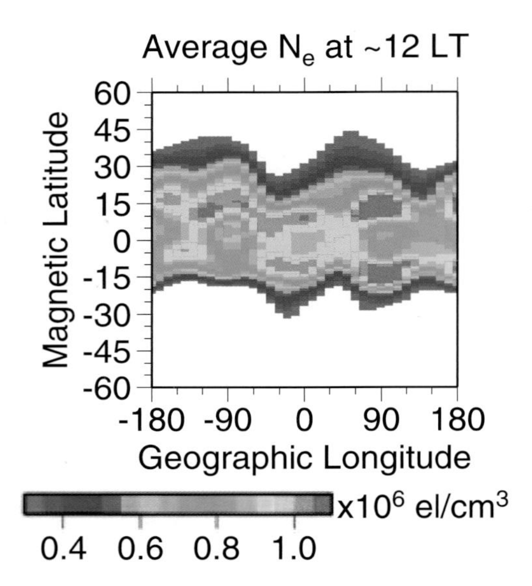

FIGURE 8.11 The 10-day-mean structure in electron density (m^{-3}) at 400 km measured by the CHAMP satellite. The 3-4 maxima in longitude are believed to result from electric fields generated by longitude-dependent atmospheric tides in the dynamo region, with possible contributions from associated composition variations and possibly in situ north-south winds. However, no electric field, wind, or composition measurements were available to understand the interplay between these quantities that results in the displayed structure. Satellite-based measurements are urgently needed to resolve this and many other similar issues in IT science. SOURCE: N.M. Pedatella, J.M. Forbes, and J. Oberheide, Intra-annual variability of the low-latitude ionosphere due to nonmigrating tides, *Geophysical Research Letters* 35:L18104, doi:10.1029/2008GL035332, 2008. Copyright 2008 American Geophysical Union. Reproduced by permission of American Geophysical Union.

and give up momentum to the mean circulation. What are needed are observations between about 100 and 200 km that include the critical dynamo region where electric fields are generated, and that would, moreover, make it possible to answer the question, How does the mean thermosphere state respond to wave forcing? Observations of both the mean state and of the waves are required to elucidate how the waves dissipate, how they relate to the background flow and thermal structure, and how their effects can be parametrized in general circulation models.

It is important to measure the tidal PWs, and GWs together, to be able to understand the interactions between them. For instance, PWs do not penetrate much above 100 km, but instead are thought to impose their periodicities on the IT system by modulating the tidal and GW parts of the spectrum that do penetrate to higher altitudes. This raises the following questions: How are GWs modulated by PWs and tides, and do they effectively map these structures to higher altitudes? and By what mechanisms are electric fields and plasma drifts generated in the dynamo region at PW periods? As one example, recent measurements reveal the fascinating result that stratospheric warmings significantly alter the state of the IT system: a prevailing theory is that enhanced quasi-stationary PWs common to these dynamical events interact nonlinearly with existing tides to produce secondary tides that propagate globally and generate dynamo electric fields in the ionosphere. The electric field subsequently redistributes ionospheric plasma, dramatically changing TEC gradients that are known to degrade communications and navigation systems. This emergent behavior in the system, once completely understood, has the potential to dramatically improve ionospheric predictions

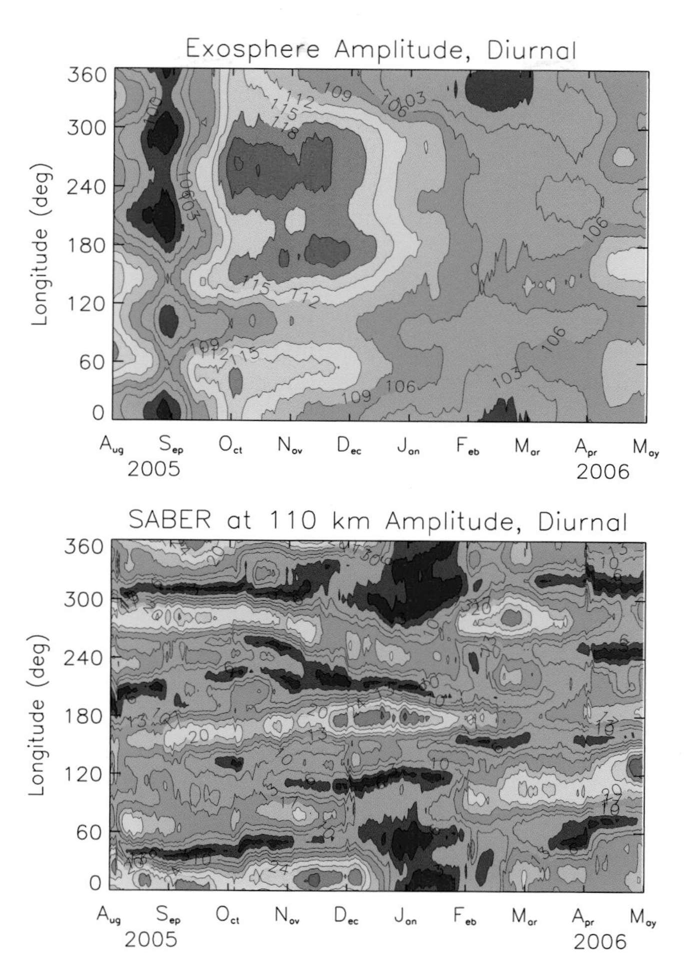

FIGURE 8.12 Equatorial diurnal tidal temperature amplitudes as a function of longitude and month from August 2005 to May 2006. (*Top*) Exosphere temperatures, ranging from 97 K (maroon) to 121 K (red). (*Bottom*) SABER temperatures at 110 km, ranging from 3 K (maroon) to 27 K (orange). The diurnal tidal spectrum evolves with height, with the larger-scale waves penetrating to 400 km, while the shorter-scale waves are absorbed at intervening altitudes, giving up their energy and momentum to the mean atmosphere. Researchers know very little about how the tidal, planetary wave, and gravity wave spectra evolve with height and modify the mean thermal and dynamical structure of the thermosphere. SOURCE: J.M. Forbes, S.L. Bruinsma, X. Zhang, and J. Oberheide, Surface-exosphere coupling due to thermal tides, *Geophysical Research Letters* 36:L15812, doi:10.1029/2009GL038748, 2009, Copyright 2009 American Geophysical Union, reproduced by permission of American Geophysical Union.

since the peak of the ionospheric response occurs several days after the stratospheric warming. In addition, first-principles modeling predicts a thermospheric warming in response to the stratospheric warmings, and resulting changes in thermospheric winds and density that impact satellite drag.

The above wave-plasma interactions focus on electric fields generated by the dynamo mechanism, but one must ask: What other processes compete with dynamo electric fields to modify and redistribute plasma in the F region (~200-600 km)? Recent studies, in fact, show that winds associated with tides that penetrate to high altitudes can significantly modify ionospheric peak heights at low latitudes. Variations in composition also accompany tidal dynamics, thereby introducing chemical influences on ionospheric production and loss with large effects in scale and magnitude. Finally, breaking gravity waves are thought to provide the turbulent mixing at the base of the thermosphere (ca. 90-100 km) that determines the geographical and temporal variation of the turbopause altitude, and hence that of the O/N_2 ratio at higher altitudes. How does the turbopause vary in space and time, and what are the causes and consequences? remains one of the outstanding fundamental questions in aeronomy, and one that can conceivably be addressed in the next decade.

Gravity waves have often been cited as the source for small-scale plasma variability, but the absence of coordinated observations of neutral waves and ionospheric perturbations in the right altitude regions has greatly impeded progress. In particular, a long-standing question that must be answered in the next decade if significant progress is to be made in understanding and predicting how small-scale plasma structures interfere with radio propagation is, What is the role of gravity waves in "seeding" equatorial Rayleigh-Taylor instabilities that lead to plasma bubbles (depletions)?

One hypothesis suggests that the interaction between in situ gravity waves and the steep bottom-side plasma gradient of the post-sunset equatorial ionosphere generates alternating east and west electric fields that can excite this instability. Another theory requires gravity-wave winds only in the E region, which generate electric fields that couple to the F layer. In addition, the tidal and mean wind fields modulate the accessibility of gravity waves to these ionosphere regions, and moreover contribute to instability onset and suppression criteria, and to instability growth rates. Thus, the interactions between small, local, and regional-scale plasma-neutral coupling phenomena are all involved in this complex but highly relevant emergent behavior in the system. Resolving this problem requires high-resolution measurements of neutral and plasma parameters with high spatial and temporal resolution over the 100- to 300-km height region, and further development of the relevant theories and models.

Finally, lightning is known to generate low-frequency electromagnetic waves called whistlers, which can induce precipitation of radiation belt particles into the opposite hemisphere and enhance lower ionosphere densities there. Lightning events also accelerate electrons to very high energies and create strong electric fields in the mesosphere. Gamma-ray flashes observed from space (e.g., from RHESSI) may indeed result from the deceleration of very energetic electrons due to collisions with atmospheric molecules. Luminous optical manifestations of these events are referred to variously as sprites, elves, or blue jets (Figure 8.13). All of these processes raise questions about chemical modification of the mesosphere and electrodynamic coupling between the troposphere, the ionosphere, and all of geospace through these energetic lightning events.

The AIMI panel concluded that a major goal of the coming decade is to understand how tropospheric weather drives space weather.

8.4.3 AIMI Science Goal 3. Ionosphere-Thermosphere-Magnetosphere Coupling

How do high-latitude electromagnetic energy and particle flows impact the geospace system?

What are the origins of plasma and neutral populations within geospace?

The IT-magnetosphere interaction at high latitudes is catalyzed by convective flows, which transport and mix plasma and neutral gases across subauroral, auroral, and polar regions, and by magnetic field-aligned flows of plasma and electromagnetic energy, which couple the collisionless magnetosphere to the collisional ionosphere-thermosphere boundary layer. Researchers now recognize that the active response

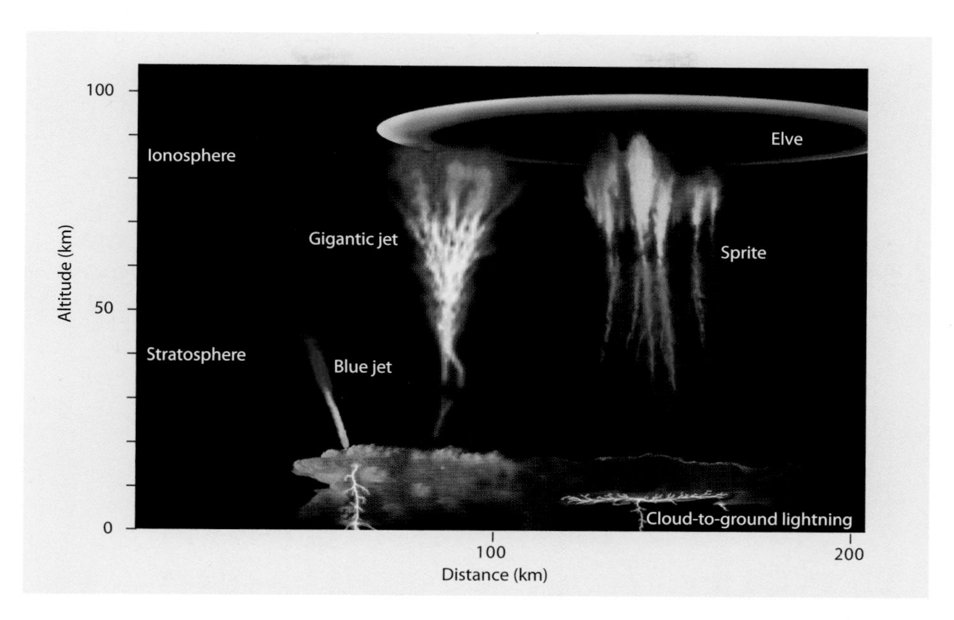

FIGURE 8.13 Illustration of transient luminous events (including elves, sprites, and jets) that occur at stratospheric and mesospheric/lower-ionospheric altitudes and are directly related to electrical activity in underlying thunderstorms. Effects on the upper atmosphere and ionosphere of transient electric fields, electromagnetic waves, and high-energy electrons produced by these events remain unknown. SOURCE: Reprinted by permission from Macmillan Publishers Ltd: *Nature,* V.P. Pasko, Atmospheric physics: Electric jets, *Nature* 423:927-929, 2003, doi:10.1038/423927a. Copyright 2003.

---

of the IT to solar wind-magnetosphere forcing, together with the response of the collisionless high-latitude region spanning the topside ionosphere and the low-altitude magnetosphere up to altitudes of ~10^4 km, introduces feedback and coupling between IT and magnetosphere system elements. Determining the processes that control this coupling is critical in understanding geospace dynamics and for development of accurate predictive capabilities. Knowledge of auroral acceleration processes and of auroral electrodynamics derived from satellite missions such as FAST, POLAR, and IMAGE is now fairly mature, but placing these processes in the context of IT-magnetosphere system dynamics is forcing the need to confront larger-scope questions: How is electromagnetic energy converted to particle energy? What controls the conversion rates and the spatial-temporal distributions of Joule heating, particle precipitation, and ionospheric outflows at high latitudes? How do these distributions and their spatial gradients, combined with neutral-wind feedback, regulate ionosphere-thermosphere-magnetosphere dynamics?

Answering these questions over the next decade will require combining model results with new multipoint in situ and remote-sensing measurements. The relationships are shown schematically in Figure 8.14.

Measurements at two or more points along magnetic flux tubes in the collisionless region above the topside ionosphere will be required to determine the mechanisms through which electromagnetic energy is converted to particle energy, and their rates; conjugate measurements at lower altitudes are essential for determining the impacts of precipitating and outflowing particles on the ionosphere and thermosphere and, in turn, the influence of the resulting IT activity on the source populations of outflowing ions and on the development of gradients (for example, in conductivity) that moderate electrical current flow and electro-

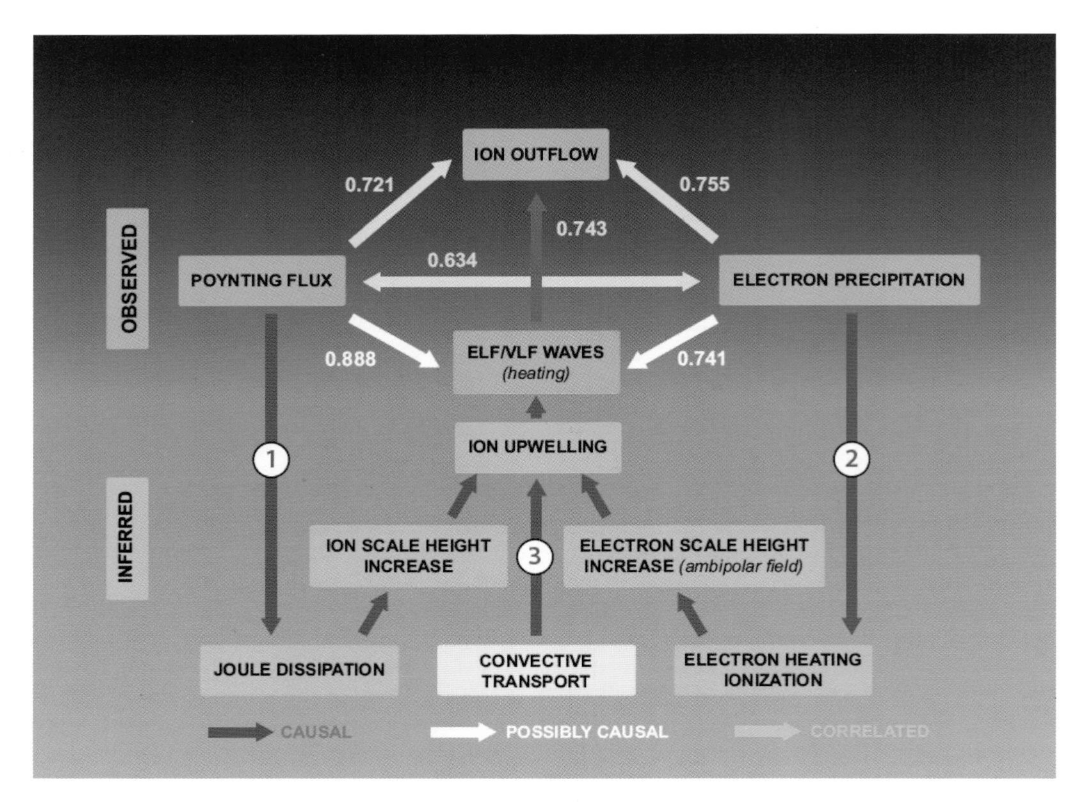

FIGURE 8.14 Observed correlations from FAST satellite data and inferred causal relationships among ionosphere-thermosphere-magnetosphere processes leading to the outflow of ionospheric ions. Researchers lack firm empirical knowledge of the relative importance of the inferred processes, all of which occur in the ionosphere. The observed correlations from FAST satellite data suggest causal relationships, but determining causality among these processes requires, at a minimum, two-point measurements along magnetic flux tubes. SOURCE: Adapted from R.J. Strangeway, R.E. Ergun, Y.-J. Su, C.W. Carlson, and R.C. Elphic, Factors controlling ionospheric outflows as observed at intermediate altitudes, *Journal of Geophysical Research* 110:A03221, doi:10.1029/2004JA01082, 2005. Copyright 2005 American Geophysical Union. Modified by permission of American Geophysical Union.

magnetic energy conversion and absorption. Combining optical measurements with in situ measurements is needed to provide contextual information, in particular, how locally inferred acceleration processes influence, and are influenced by, larger-scale structure and dynamics. The synthesis required to connect these measurements with solar wind and magnetospheric drivers will require development and application of increasingly realistic models for global and regional dynamics.

Without the photo-ionization present in the dayside ionosphere, the nightside ionosphere is susceptible to structuring and modulation by variable fluxes of charged particles precipitating from the magnetosphere. Gradients in the resulting ionization cause ionospheric currents to be diverted into field-aligned currents. Recent studies have revealed the unexpected possibility that the accompanying ionospheric flow structures are mirrored in the plasma sheet by the formation of fast flow channels and by steep plasma pressure gradients in the outer ring current. These ionospheric flow structures form at steep gradients in ionospheric conductivity, which remains one of the most poorly diagnosed and vitally important ionospheric variables.

While empirical models of electron precipitation have become increasingly sophisticated, knowledge of the associated conductivity dynamics on spatial scales down to 1 km is still lacking. Even less is known about conductivity enhancements due to ionospheric turbulence—effects that have been theoretically predicted to be capable of doubling the total height-integrated conductivity during disturbed geomagnetic conditions. A compelling question is thus, What are the spatial and temporal scales of ionospheric structure and associated conductivity that determine energy deposition, plasma and neutral flows, and electrical current flow in the ionosphere-thermosphere interaction?

Plasma of ionospheric origin mixes with solar wind plasma to populate the plasma sheet, ring current, and plasmasphere. During episodic events such as storms and substorms, the presence of ionospheric plasma in these regions can be a controlling factor in geospace dynamics. For example, dense, convecting plasmaspheric plumes are thought to modulate dayside magnetic reconnection upon contacting the magnetopause. What are the processes that cause the plume structure to appear as storm-enhanced densities in the ionosphere? Ionospheric outflows emerging from the dayside cleft ion fountain and nightside Alfvénic acceleration regions can dominate both the density and the pressure of the plasma sheet during superstorms and the energy density of the ring current. Plasma in the inner magnetosphere is composed of protons and He^+ and O^+ ions of ionospheric origin. The relative abundance of these ions influences the plasma wave intensities that are responsible for the scattering and loss of radiation belt electrons. Recognition that ionospheric plasma is a critical agent in regulating the geospace system is accompanied by the humbling reality that researchers do not know what controls the abundance or distribution of ionospheric plasma in the magnetosphere. How does the flow of ionospheric plasma into the magnetosphere during storms change as a result of IT plasma and neutral redistributions?

The AIMI panel concluded that an additional major goal of the coming decade is to understand how the IT and magnetosphere interact to regulate their coupled response to solar wind forcing.

8.4.4 AIMI Science Goal 4. Plasma-Neutral Coupling in a Magnetic Field

How do neutrals and plasmas interact to produce multiscale structures in the AIM system?

An intriguing aspect of the IT system is the transfer of energy and momentum that occurs between the plasma and neutral components of the system, and how electric and magnetic fields serve to accentuate and sometimes moderate this interchange. The pathways through which ions and neutrals interact are of course fundamental to space physics, as they occur all over our solar system. Addressing the compelling science questions described within previous sections also presents many opportunities to employ the IT system as a local laboratory to expand understanding of plasma-neutral coupling processes that have broad applicability across the solar system. In particular, these interactions occur over local, regional, and global scales, and in many cases cross-scale coupling exists. Some insight into the range of topics that can be addressed is provided in the following section, which begins with the low latitudes and then moves toward the polar regions.

The equatorial IT system represents a rich laboratory for investigation of plasma-neutral coupling in the presence of a magnetic field. The unique features are the quasi-horizontal orientation of the B field, the plasma instabilities that arise from this configuration, the ability of winds to generate electric fields through the E- and F-region dynamo mechanisms, the change in plasma-neutral collision frequency with height, the unimpeded ability of neutral winds to move plasma along field lines, and the relatively rapid change in magnetic inclination with latitude. Combined with a variety of chemical processes, interactions between the plasma and neutral gases in the above environment produce emergent behaviors in the neutral and plasma densities, their bulk motions, and their temperatures. One example of emergent structures in neutral density is provided in Figure 8.15.

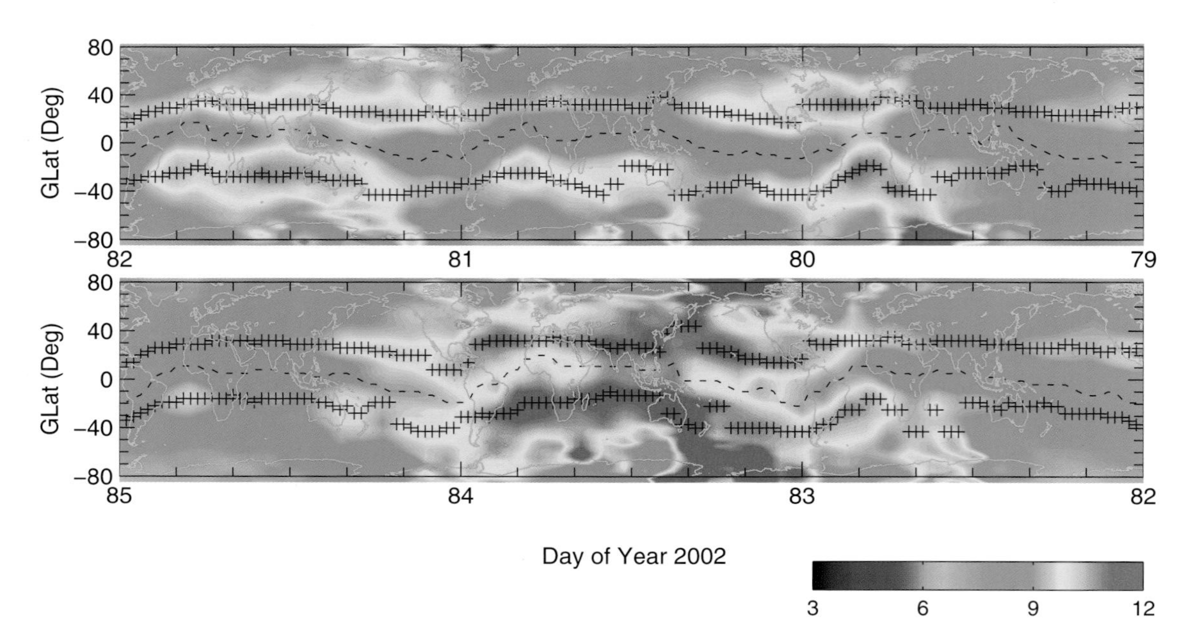

FIGURE 8.15 Neutral mass density (kg/m^3) structures measured by the accelerometer on the CHAMP satellite near 400 km and 1800 local time during days 79-85, 2002. The universal time day runs from right to left, so as to display the data on top of geographic maps. Crosses (dashed lines) mark the locations of the equatorial temperature anomaly (ETA) crests (troughs). Magnetic activity was mostly quiet, except on day 83 when Kp reached values of 4-6. The longitudinal alignment of the ETA trough and crests suggests a connection with a magnetic coordinate system and hence with plasma densities. Researchers do not know how quiet-time and disturbance wind, electric field, and composition variations interacted to produce the observed changes. With just single-satellite sampling, temporal variations cannot be separated from longitude variations; moreover, measurements are made only at two local times. A constellation of satellites would remove the longitude-universal time ambiguity and would reveal how these structures varied in local time. SOURCE: J. Lei, J.P. Thayer, and J.M. Forbes, Longitudinal and geomagnetic activity modulation of the equatorial thermosphere anomaly, *Journal of Geophysical Research* 115:A08311, doi:10.1029/2009JA015177, 2010. Copyright 2010 American Geophysical Union. Reproduced by permission of American Geophysical Union.

The longitudinal alignment is reminiscent of the plasma feature referred to as the equatorial ionization anomaly (EIA), but the EIA maxima are less widely spaced in latitude and do not respond to changes in geomagnetic activity to the same degrees as do the neutral structures. Although theories exist that involve plasma and neutral transport and temperature and density responses due to adiabatic heating and cooling terms in the thermodynamic equation, the absence of coincident wind, temperature, electric field, and composition measurements over a range of spatial and temporal scales precludes a definitive interpretation.

What is needed to dispel speculation are simultaneous measurements of neutral and ion densities, temperatures, winds, and plasma drifts (E fields) so that physical connections can be explored and model simulations can be constrained. The measurement and modeling strategies proposed in the section "Implementation Strategies and Enabling Capabilities" will do exactly this. In fact, the proposed measurements will enable investigation of several low-latitude phenomena whose origins can be elucidated only through simultaneous multi-parameter measurements. These phenomena include, for instance, the equatorial temperature and wind anomaly, the midnight temperature maximum, the post-sunset atmospheric jet, terminator waves, and atmospheric superrotation. In summary, the overarching question that captures most low-latitude IT phenomena is thus, How are gas temperatures and densities at low latitudes modified by momentum transfer between neutrals and ions in the presence of a magnetic field?

Middle latitudes serve as a laboratory for different kinds of plasma-neutral interactions that also exhibit emergent behavior. For instance, consider the TEC and 6,300 Å airglow emissions depicted in Figure 8.16. The TEC data are obtained from a network of ground-based GPS receivers, and airglow measurements are obtained from ground-based all-sky imagers. The spontaneous emission of 6,300 Å airglow at nighttime originates from excited atomic oxygen atoms that result from the dissociative recombination of molecular ions with electrons. Similar TEC measurements have been made using a GPS receiver network over the continental United States. The depicted waves, which are predominantly south-westward propagating, are thought to originate as neutral density waves at high latitudes, which then interact with the mid-latitude ionosphere to create the observed structures; however, the occurrence of these waves is curiously unrelated to level of magnetic activity.

It is hypothesized that the southwestward directionality of the waves, at least at nighttime, is aligned in the direction of weakest Joule damping as predicted by the Perkins instability. This hypothesis is supported by measurements that indicate the existence of electric fields within the wave structures, which are

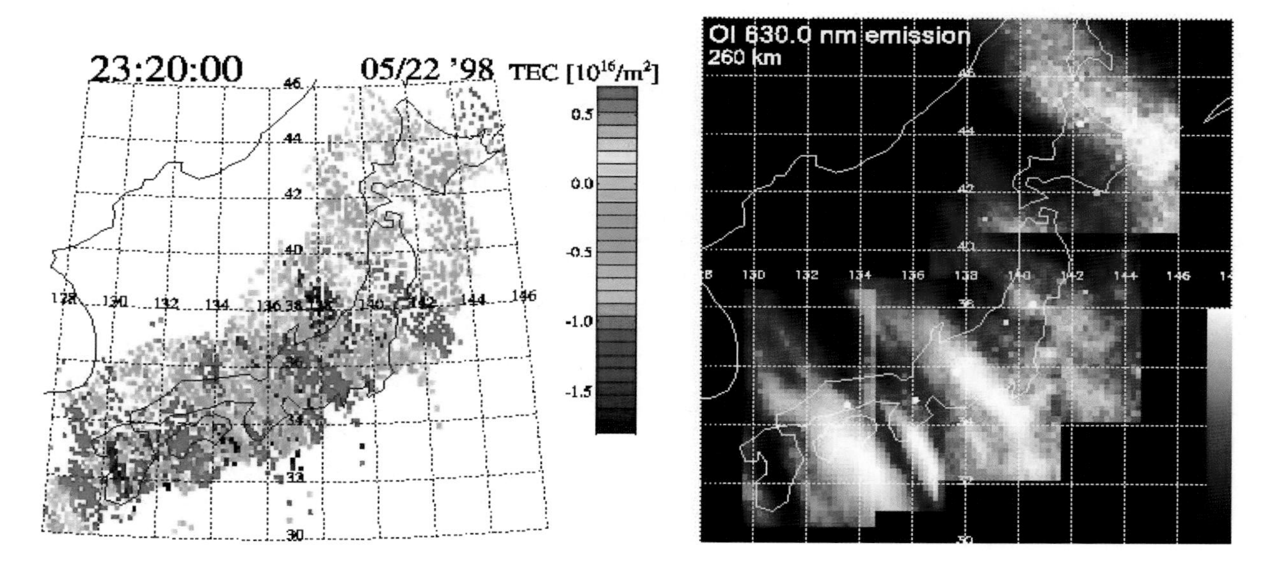

FIGURE 8.16 Two-dimensional distribution of total electron content (TEC) (*left*) and 6,300 Å airglow emission (*right*) near midnight over Japan on May 22, 1998, during passage of a traveling ionospheric disturbance. The TEC measurements were obtained using an array of GPS receivers, and the airglow measurements were made with five all-sky charge-coupled device cameras. The airglow emission is from excited neutral atomic oxygen atoms produced as a result of dissociative recombination of molecular ions and electrons. Note that the maxima in TEC mostly correspond with the airglow peaks, consistent with this interpretation. A North American ground-based observing network would enable investigation of these and many other regional-scale space weather phenomena involving fundamental plasma-neutral coupling processes, while a complementary satellite mission would provide insights into coupling on a global scale. SOURCE: A. Saito et al., Traveling ionospheric disturbances detected in the FRONT Campaign, *Geophysical Research Letters* 28(4):689-692, 2001. Copyright 2001 American Geophysical Union. Reproduced by permission of American Geophysical Union.

thought to arise from winds blowing across Pedersen conductivity gradients associated with the waves. It also appears that the physics behind the ionospheric manifestation of the waves may be different during nighttime and daytime, and that their directionality varies with season. This may reflect different generation mechanisms and/or the influence of the background large-scale wind circulation on the wave propagation. Note that explanation of this phenomenon involves coupling between instability, local, regional, and global-scale processes in ways that scientists do not understand, leading to the question, How do plasmas and neutrals interact across local, regional, and global scales to produce the operationally important density variations referred to as space weather?

Plasma-neutral interactions at high latitudes are strongly coupled to solar wind and magnetospheric dynamics. This coupling is regulated to a large extent by the ionospheric conductance, which is dependent on the ion and neutral gas densities. Despite their importance in the AIM interaction, the spatial distributions of these densities and their time variability are among the most poorly measured parameters of the IT system. Consequently, the interplay between neutral and ionized gas constituents and electromagnetic activity are not well understood. At high latitudes, ion motions driven by the interaction of Earth with the solar wind provide the strongest forcing to the neutral atmosphere. The temporal and spatial scales of the neutral atmosphere response are quite different at different altitudes and quite different from those imposed by the driver. The driven neutral gas motions persist long after the driving fields change, and the neutral gas convects and diffuses well beyond the region of ion forcing. This complex interaction changes the energy deposited in the atmosphere, which causes changes in the global temperature, composition, and density that cannot yet be predicted. For example, in both the cusp region and its counterpart in the night-side ionospheric convection throat, the average mass density of the neutral atmosphere near the F-region peak is observed to be significantly higher than that predicted by the empirical reference thermosphere (MSIS90) (Figure 8.17).

This discrepancy has stimulated a search for the causative mechanisms. The neutral density enhancements are statistically collocated with observed regions of soft electron precipitation, Joule heating, ion upflows and outflows, dispersive Alfvén waves, and small-scale field-aligned currents. Determining cause and effect among these variables and advancing an operational capability to predict regional enhancements in neutral density will require simultaneous, multivariable measurements in the topside and bottomside ionosphere. Model results indicate that thermospheric upwelling strongly influences the scale height of O^+ ions in the topside ionosphere and their escape flux into the magnetosphere. Thus the population of the magnetosphere by ionospheric outflows is also dependent on plasma-neutral interactions. Breakthroughs in the next decade in understanding the dynamic interaction between the magnetosphere and the IT system must therefore confront the question, How do activities in neutral and ionized gases and electromagnetic fields interact to produce observed magnetic field-aligned structure and motions of the thermosphere and ionosphere?

The AIMI panel concluded that a major goal of the 2013-2022 decade is to understand the plasma-neutral coupling processes that give rise to local, regional, and global-scale structures in the AIM system, particularly those relevant to society.

8.4.5 AIMI Science Goal 5. Planetary Change

How is our planetary environment changing over multidecadal scales, and what are the underlying causes?

The preceding discussions indicate why achieving an understanding of how the whole atmosphere system is coupled to the geospace environment remains a singular challenge for AIM research. Addressing

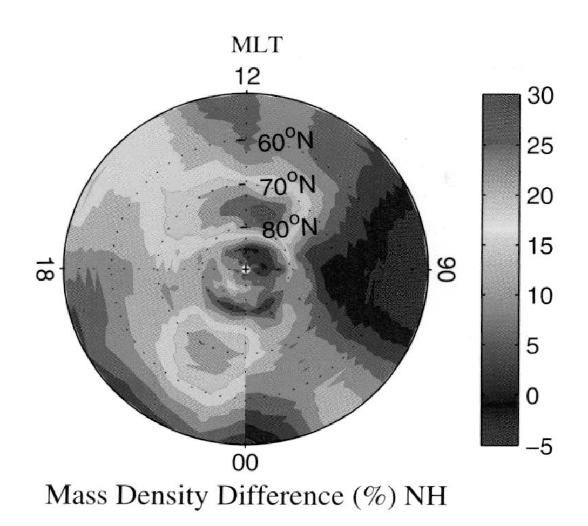

FIGURE 8.17 Magnetic-latitude and local-time distribution of a 1-year average of the percent difference between the thermospheric mass density derived from CHAMP satellite measurements and MSIS90 at 400-km altitude in the northern polar region during quiet conditions (Kp = 0-2). Model ionospheric convection streamlines are superposed. Large differences occur in the cusp and midnight sectors. SOURCE: Adapted from H. Liu, H. Lühr, V. Henize, and W. Köhler, Global distribution of the thermospheric total mass density derived from CHAMP, *Journal of Geophysical Research* 110:A04301, doi:10.1029/2004JA010741, 2005. Copyright 2005 American Geophysical Union. Reproduced by permission of American Geophysical Union.

---

this challenge will provide vital information not only on the variability of Earth's near-space environment, but also on terrestrial climate variability and change.

Human activities, particularly the introduction of greenhouse gases (e.g., CO_2 and CH_4) into the atmosphere, are changing the global climate. One of the many manifold demonstrations that the change in Earth's climate is due to the rise in CO_2 concentrations is that the lower atmosphere is warming while the upper atmosphere is cooling. This demonstrable fact is fully consonant with the well-understood role of CO_2 as an effective radiator of energy in the upper atmosphere. A systematic decrease by several percent per decade near 400-km altitude in thermosphere mass density has recently been identified, evident in the record of satellite orbit decay measured since the beginning of the space age (Figure 8.18).

One result of an increase in the average temperature in the lower atmosphere is that the amount of water vapor, and consequently the available latent heat, may increase. Possible consequences include changes in the strength of the lower-atmosphere tides that, as noted above, modify the longitudinal structure of the ionosphere. Gravity wave, Kelvin wave, and planetary wave fluxes are also likely to be affected. Given recent findings of an El Niño-Southern Oscillation-related signature in low-latitude ionospheric structure, scientists should expect its structure to change in response to other climate changes in the lower atmosphere. Another aspect of lower-atmosphere changes is an expected increase in the number of severe storms. If ionospheric instabilities are seeded by tropospheric gravity waves propagating into the upper atmosphere, this may have an impact on the frequency of these events. Another source of gravity waves is the flow of tropospheric winds over topographic features. If lower atmospheric circulation patterns are altered, this too may change the spectrum and frequency of occurrence of gravity waves, and ionospheric instabilities.

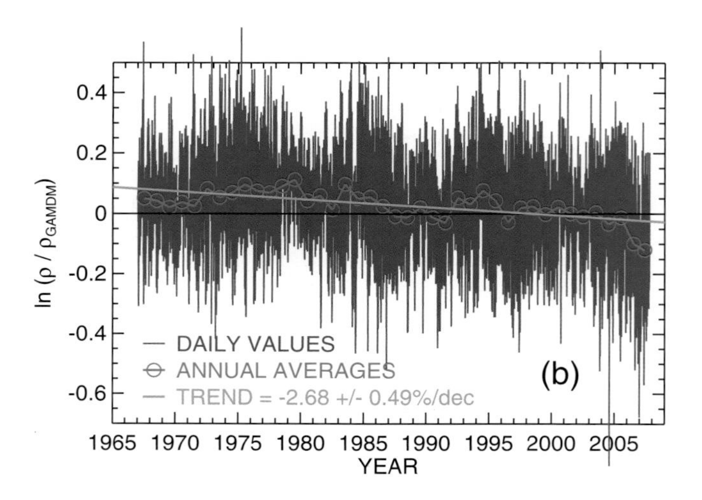

FIGURE 8.18 Analyses of decades of satellite drag data indicate a long-term trend of decreasing thermosphere densities. SOURCE: J.T. Emmert, J.M. Picone, and R.R. Meier, Thermospheric global average density trends, 1967-2007, derived from orbits of 5000 near-Earth objects, *Geophysical Research Letters* 35:L05101, doi:10.1029/2007GL032809, 2008. Copyright 2008 American Geophysical Union. Reproduced by permission of American Geophysical Union.

Model simulations and decades of observations indicate that changes in ozone and greenhouse gases have produced long-term changes in the wind fields of the stratosphere and mesosphere that serve as the environment through which tropospherically excited waves propagate. Waves propagating through the middle atmosphere and into the base of the thermosphere (ca. 100 km) must therefore also undergo change, and this is confirmed by trends seen in long-term wind and magnetic perturbations between 80 and 120 km observed from the ground. Since dissipation of waves in this region helps drive the mean circulation of the mesosphere and lower thermosphere, these effects must feed back to modify the circulation system that resulted in the changed wave spectrum in the first place. In addition, there is the question of downward control and whether the adiabatic heating and cooling effects of the wave-driven vertical motion field extend far enough down in altitude to have practical consequences.

One of the most spectacular signatures of the coupling of the lower and upper atmosphere is the existence of polar mesospheric clouds (PMCs). These water ice clouds at the edge of space (~84 km) are seen in the summer hemisphere. Stratospheric and mesospheric water vapor is created largely through the oxidation of methane (CH_4). Water vapor is trapped at the tropopause while methane is mixed into the upper atmosphere, thus changing the hydrogen chemistry of the upper atmosphere. Local temperature is also key to the existence of PMCs, and at these altitudes is determined largely by the adiabatic heating and cooling accompanying the wave-driven circulation discussed above. In the coming decade, the AIM community will build on the work of previous missions like the NASA Aeronomy of Ice Mission and the Thermosphere Ionosphere Mesosphere Energetics and Dynamics (TIMED) mission to understand how geospace is influenced by the lower atmosphere.

One of the urgent, unresolved heliophysics questions is how feedback processes in the Earth system amplify the effects of small changes in solar energy output, leading to disproportionately large changes in atmospheric parameters. The atmospheric response to energetic particle precipitation (EPP) is a key component of this unexplored frontier and provides a natural means of probing the coupling mechanisms that redistribute solar and magnetospheric energy at Earth. Determining how redistribution of precipitating particle energy influences atmospheric composition and structure, and how nonlinear coupling processes amplify these impacts beyond those expected from the absolute energy input, would represent a fundamental advance in heliophysics.

Coupling of different regions via radiative, dynamical, and chemical feedbacks controls weather and climate throughout the atmosphere. Through these feedback processes the remote regions highly driven by solar influences are linked to the troposphere where human activities are concentrated. Yet how solar and magnetospheric variability affects atmospheric conditions and climate in such an interconnected system remains an open question. Only with a deeper understanding of the natural variability, and of the way external influences perturb the coupling processes, will researchers be able to model this complex system and thereby develop a predictive capability for future generations.

EPP refers to energetic electrons and protons impinging on Earth's atmosphere after they have been accelerated by solar and magnetospheric activity. Figure 8.5 in the section above titled "Significant Accomplishments of the Previous Decade" depicts current, incomplete understanding of the atmospheric response to EPP. Solar magnetic variability produces changes both in irradiance and in the plasmas and magnetic fields that permeate interplanetary space, driving space weather at Earth. The resulting EPP occurs at all times during the solar cycle but has different characteristics depending on particle sources, which are broadly associated with different geomagnetic activity levels. Through ionization and dissociation EPP produces NO and NO_2, collectively referred to as NO_x, the primary catalytic destroyer of ozone (O_3) in most of the stratosphere. NO_x produced by EPP (EPP-NO_x) can be created directly in the stratosphere by high-energy particles, or in the mesosphere and lower thermosphere (MLT) by lower-energy particles. EPP-NO_x descends from the MLT into the stratosphere during the polar winter (see Figure 8.19, for example). The downward transport is controlled by an unknown combination of diffusion, large-scale circulation, and confinement in the polar vortex, with winds and waves modifying these elements.

NO_x is an atmospheric coupling agent because of its impact on O_3, either through the NO_x catalytic loss cycle, or by interfering with other catalytic O_3 loss cycles. Changes in O_3 can alter temperature gradients, and thereby influence circulation. Perturbations to these three entities are intertwined and can trigger nonlocal changes, e.g., by perturbing propagation of waves to the upper atmosphere or weather and climate in the lower atmosphere. Thus EPP-NO_x-induced perturbations in O_3 are communicated via effects on temperature and circulation upward to the MLT and potentially downward to the troposphere, thereby triggering the redistribution of particle energy throughout the atmosphere. These results suggest a mechanism for EPP indirectly affecting even tropospheric climate, evidence for which is provocative but tenuous. Tropospheric perturbations themselves—either natural or human-induced—can be communicated to the middle and upper atmosphere, thereby altering the atmospheric response to EPP. The coupling pathways described here form a critical yet poorly understood link between heliophysical forcing and Earth's climate system. A complete description of the Sun-Earth system requires understanding of the pathways by which EPP transmits signals of solar magnetic variability through Earth's atmosphere, specifically the amplifying coupling processes triggered by EPP-NOx. It is then important to ask: How do EPP-initiated chemical changes translate to thermal and dynamical changes throughout the atmosphere?

The AIMI panel concluded that a major goal for the upcoming decade is to determine how our planetary environment is changing over multi-decadal scales, and to understand how the changes are embodied in or transmitted through the AIM system.

This goal triggers two major implications. First, long-term observations that provide the best information about the degree of change, and that best serve to constrain models, must be protected if scientists are to understand and predict long-term change in the AIM system. Second, fundamental processes and

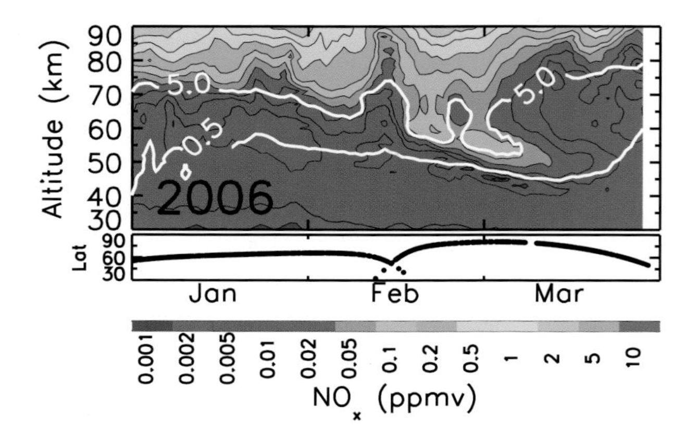

FIGURE 8.19 ACE measurements of NO_X (color contours) during the Northern Hemisphere winter of 2006. White lines are CO mixing ratios, a tracer that indicates descent. The origin of the NO_X is energetic particle precipitation. In 2006, meteorological conditions were favorable for NO_X descent. The upper-stratosphere vortex was particularly strong, trapping the NO_X in the polar region as it descended farther into the stratosphere where it has a longer chemical lifetime. SOURCE: Adapted from C.E. Randall et al., Enhanced NO_X in 2006 linked to strong upper stratospheric Arctic vortex, *Geophysical Research Letters* 33:L18811, 2006. Copyright 2006 American Geophysical Union. Modified by permission of American Geophysical Union.

mechanisms that underlie long-term change, such as wave-mean flow interactions, must be identified and understood. The latter are also important in terms of understanding and predicting the behavior of the AIM system over the short term, and toward this end wave coupling between the lower and upper atmosphere has emerged in this panel report as one of a few core high-priority research areas for the future decade.

8.4.6 Science Priorities

The AIMI panel's science priorities for the 2013-2022 decade are presented here. These overarching priorities reflect, and to an extent cut across, the science goals discussed above. The panel's strategy for addressing these science priorities reflects the need to allocate scarce resources optimally and the desire for a program that will have high societal benefit. The science priorities for 2013-2022 are as follows:

1. Determine how the ionosphere-thermosphere system regulates the flow of solar energy throughout geospace.
2. Understand how tropospheric weather influences space weather.
3. Understand the plasma-neutral coupling processes that give rise to local, regional, and global-scale structures and dynamics in the AIM system.

8.5 IMPLEMENTATION STRATEGIES AND ENABLING CAPABILITIES

The following section focuses on strategies and enabling capabilities to address the AIMI science goals outlined in the previous section, with particular emphasis on the three science priorities just enumerated. The AIMI panel's four imperatives as summarized in Section 8.1 fall under the categories of spaceflight missions; Explorers, suborbital, and other platforms; ground-based facilities; theory and modeling; and enabling capabilities. Priorities within each are indicated below.

8.5.1 Spaceflight Missions

As stated previously, the nature of AIM science is that it requires a synergistic complement of space-based and ground-based observational approaches with complementary theory and modeling activities. However, there are no NASA missions in development or approved for development that address these priorities, and therefore there is a void in the current capacity to understand the solar-terrestrial system and to contribute to the space weather needs of 21st-century society. It is the AIMI panel's view that the Heliophysics Systems Observatory should be continued with a mission that determines how solar energy drives ionosphere-thermosphere variability, and that lays the foundation for a space weather prediction capability. There are two complementary notional missions that are put forth as the panel's highest-priority imperatives to satisfy this need: GDC and DYNAMIC. Out of the many different means of measuring the AIM system, they are representative of the types of missions that will provide the needed global view unattainable in any other way (i.e., Explorer-class missions, which are less costly). GDC (a large-class mission) and DYNAMIC (a moderate-class mission) are described immediately below. Some level of detail is provided to enable the reader to understand the scope of mission required to achieve its science goals and to relate this to the broad cost categories described in Chapter 1 of this report. Thus, although some details are provided, these missions are not prescriptive; the AIM community will ultimately decide on the optimal implementation to achieve the science goal. Two additional notional missions, ESCAPE (Energetics, Sources and Couplings of Atmosphere-Plasma Escape) and MAC (Magnetospheric-Atmosphere Coupling), are also described below; each addresses very high priority science topics (i.e., regulation of the IT-magnetosphere interaction and fundamental plasma processes).

8.5.1.1 GDC (Geospace Dynamics Constellation) Mission (Intermediate Class)

Overview

Assuming resources enable a new start early in the decadal survey interval, the AIMI panel sees the Geospace Dynamics Constellation (GDC) as the optimal way to significantly advance both solar-terrestrial and AIM science. GDC would be a fundamental contributor to the Heliophysics Systems Observatory while also enabling measurements that are highly relevant for research aimed at understanding and developing a predictive capability for space weather. The primary focus of GDC is to gather the necessary data to reveal how the IT operates as a system and how it regulates its response to external forcing. With current and foreseeable technologies, this requires a robust, systematic observation approach using in situ probes to gather data at all local times and latitudes simultaneously. Furthermore, these satellites must gather data at sufficiently low altitudes where both the neutral and ionized gases are sufficiently dense and inherently coupled.

The observational problem is that such global dynamics cannot be captured by a single satellite regardless of the number of its instrument probes. When averaged over a sufficiently long period of time, data from a single satellite provides a useful climatology as a function of latitude and longitude. However, such data are static and do not show the physical coupling inherent in the continuously adjusted density contours and velocity patterns (dynamics) that, by their very nature, respond at all local times to the interconnected processes that define the AIM system. On the other hand, a constellation of identical, multiple satellites in low Earth orbit, such as proposed here as the Geospace Dynamics Constellation, would provide the necessary global, simultaneous observations covering all latitudes and local times while orbiting Earth in

TABLE 8.1 Geospace Dynamics Constellation (GDC) Science and Relevance to Space Weather

Solar and Space Physics Motivations	1 Understand Our Home in the Solar System 2 Predict the Changing Space Environment and Its Societal Impact 3 Explore Space to Reveal Universal Physical Processes	
GDC Primary Objective	Characterize and understand how the ionosphere-thermosphere behaves as a system, responding to, and regulating, solar wind/magnetosphere energy input.	
GDC Measurements and Description	Gather simultaneous, global measurements of plasma and neutral gases and their dynamics, and magnetosphere energy/mass input, using 4-6 multiple platforms in 80° inclination, circular orbits (320 to 450 km) equally spaced in local time.	
GDC Science Objectives	**GDC Scientific Merit**	**GDC Space Weather Relevance**
Understand the dynamic, energy/momentum exchange between ionized/neutral gases at high latitudes and their coupling and feedback to the magnetosphere and solar wind.	Will determine how the global IT system participates as an active element in the evolution of storms.	Enable prediction of how high-latitude structures in the ionosphere and thermosphere are driven by magnetosphere input and then propagate to mid and low latitudes.
Determine the global response to the AIM system to magnetic activity and storms.	GDC will determine how winds, temperature, and chemical constituents interact to produce the observed global neutral and plasma density responses of the global IT system. Simultaneous measurements of global neutral and plasma parameters will for the first time provide all of the information required to expose how changes in the system at different locations are related.	Enable prediction of how neutral and plasma densities and motions respond to magnetic activity and storms.
Determine the influence of forcing from below on the ionosphere/ thermosphere system.	GDC will measure the global variability of thermosphere tides on a day-to-day basis for the first time. GDC will show how waves/tides of tropospheric origin contribute to the mean structure, dynamics, and electrodynamics of the ionosphere and upper thermosphere.	Provide a means to predict how the ionosphere and thermosphere will react to strong tidal and gravity wave forcing from below.

circular orbits within the altitude region (300-450 km) at high inclination. Table 8.1 summarizes the science objectives, scientific merit, and space weather relevance of GDC, and how it relates to motivations and related questions of the current decadal survey.

Mission Configuration

The basic approach of GDC is straightforward: a suite of six satellites will gather simultaneous, global measurements of key ionospheric and thermospheric parameters using identical instruments on high-inclination platforms, executing circular orbits along planes evenly distributed in local time. Each satellite includes an identical suite of notional instruments, as listed in Table 8.2. The instruments include those that measure the neutral and ionized gases and their motions and hence are used to fully describe the global dynamics of the co-existing ionized and neutral fluids that define the IT system. Also included are a magnetometer and energetic particle detector in order to measure energy and momentum drivers from the magnetosphere. The satellites, in their final configuration, will have circular orbits that are initially at 450 km. The satellites will then slowly decay in altitude due to atmospheric drag. When the altitude decays

TABLE 8.2 Geospace Dynamics Constellation (GDC) Key Parameters to Be Measured from Space

Notional Instrument	Key Parameters	Nominal Altitude (km)
Ion Velocity Meter (includes RPA)	Vi, Ti, Ni, broad ion composition	300-400
Neutral Wind Meter (NWM)	Un, Tn, Nn, broad neutral composition	300-400
Ionization Gauge	Neutral density	300-400
Magnetometer	Vector B, Delta B, currents	300-400
Electron Spectrometer	Electron distributions, pitch angle (0.05 eV to 20 keV)	300-400

NOTE: All instruments have extensive flight heritage.

to about 320 km, each satellite will then use on-board propulsion to re-boost to 450 km, a maneuver that will be completed in a single orbit. Such a re-boost is a routine maneuver expected to be required approximately once every 6 months, depending on solar activity.

The nominal plan for GDC is to have six identical satellites that will spread into six equally spaced orbital planes separated by 30° longitude, thus providing measurements at 12 local times (LTs), with a resolution of 2 hours of LT, as shown in Figure 8.20a. The satellites will nominally have an inclination of 80°, in order to use precession to help separate the local time planes, while maintaining adequate coverage of the high-latitude region.

Three main orbital configurations were considered by the AIMI panel:

1. Spacecraft fully spread out in latitude to provide continuous, global coverage (see Figure 8.20a);
2. Spacecraft configured as an "armada" with simultaneous, dense coverage at high latitudes alternating between polar cap regions every 45 minutes (Figure 8.20b); and
3. Satellites configured to orbit in two separated three-satellite armadas, such that three are in the Northern Hemisphere while three are in the Southern Hemisphere, providing simultaneous coverage of both polar regions every 45 minutes (Figure 8.20c).

Both configuration 2 and configuration 3 gather consolidated measurements at the mid- and low latitudes, with simultaneous crossings of the equator by all six satellites every 45 minutes (Figure 8.20d), while the entire globe is sampled every 90 minutes at 12 local times (as is the case for configuration 1). The panel notes that only minimal amounts of propulsion are needed to alternate between these configurations and that the "station keeping" time to change and maintain these configurations is very short.

Initial Deployment Phase and Operational Strategy

The most cost-effective launch approach (given available launchers) would be to launch all six satellites with one launcher. In this case, the satellites are first placed in a highly elliptical orbit plane (e.g., with perigee of 450 and apogee of 2,000 km) with an 80° inclination. As they then precess, these satellites will spread out in equally separated local time planes (requiring ~12 months). Note that propulsion can be used to decrease this deployment time. In this scenario, the apogee of one satellite is immediately lowered to provide an initial 450-km circular orbit, while the remaining five satellites precess in local time. After about 2.5 months, in which the satellites have precessed 30 degrees in local time, a second satellite orbit is changed to 450 km circular. The process continues until all six are spread out equally and converted to 450-km circular orbits. During the time required to establish the final distribution in local time of the six satellites, this initial observing phase permits "pearls-on-a-string" observations by the satellites along

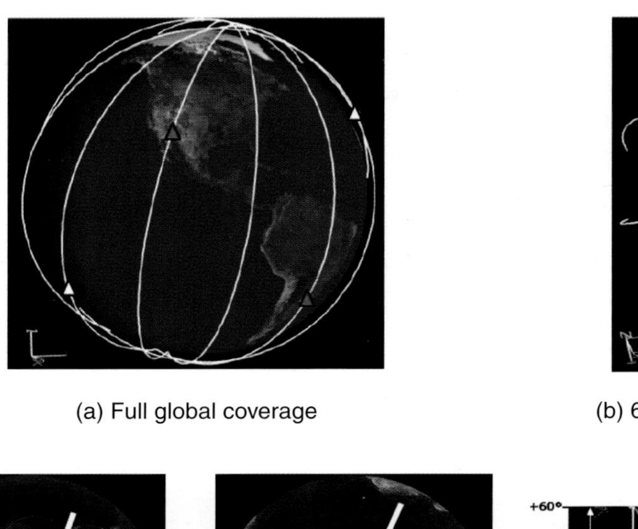

(a) Full global coverage

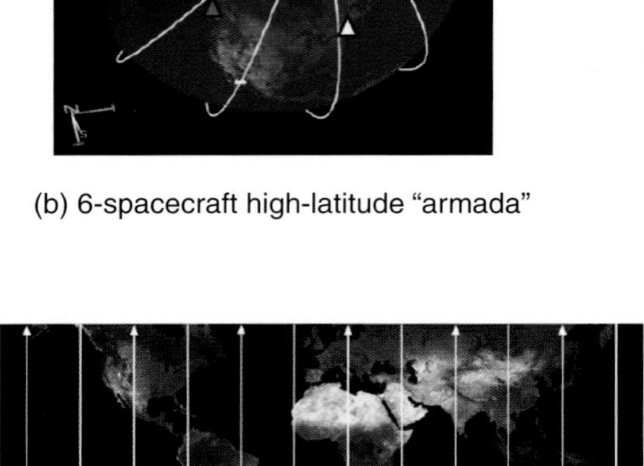

(b) 6-spacecraft high-latitude "armada"

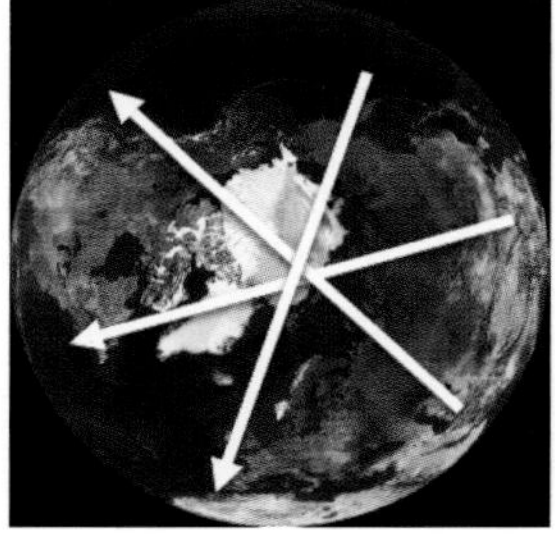

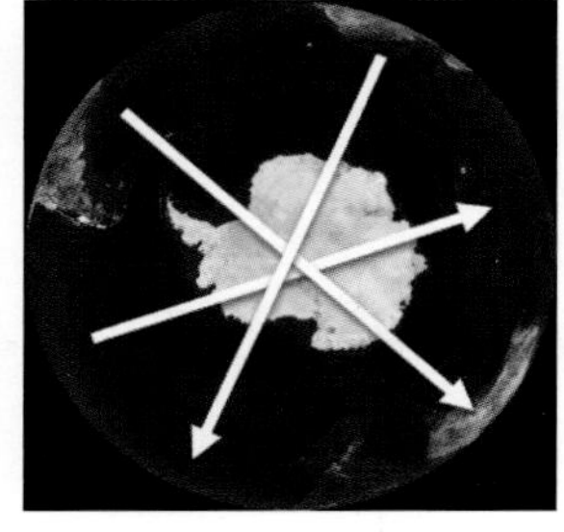

(c) 3 spacecraft simultaneously sampling each pole

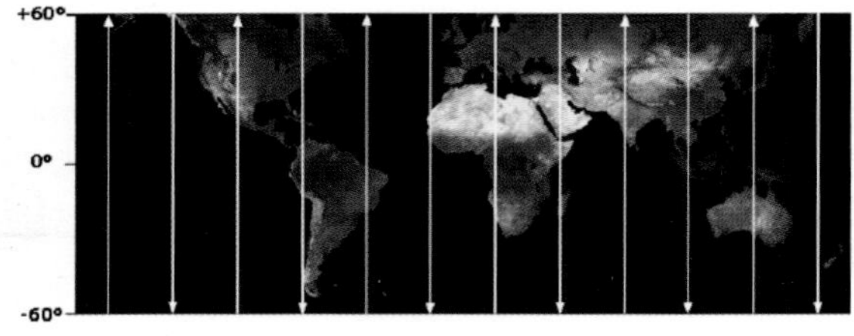

(d) 6 s/c cross the equator every 45 min.

FIGURE 8.20 Various orbital configurations and sampling options offered by a six-satellite constellation. SOURCE: (a-c) Orbital configurations courtesy of the Aerospace Corporation using Earth images provided by Living Earth, Inc. (d) NASA Goddard Space Flight Center.

highly elliptical orbits that will gather data that will address important science questions under AIMI science goals 2, 3, and 4, in addition to science goal 1, thus cutting across all three AIMI science priorities for the 2013-2022 decade.

Baseline Mission and Possible Descope

The GDC baseline mission can fully meet its science objectives with a complement of six satellites as discussed above. These satellites will gather data along six orbit planes, sampling 12 local times with 2-hour spacing, providing continuous global measurements as they orbit Earth every 90 minutes. This temporal coverage is sufficient for studies of large-scale processes at high latitudes and is more than adequate to resolve major changes to the IT system during magnetic storms, whose main phases typically last 6-8 hours. The longitude/local time coverage of six satellites is sufficient to resolve tidal and planetary wave effects, including the resolution of terdiurnal tides. Another advantage of six satellites is that three can be placed in the Northern Hemisphere high-latitude region, while three are simultaneously observing in the Southern Hemisphere high-latitude region, providing data that address inter-hemispherical asymmetries and energy input along six local time planes within each polar region. In addition, six satellites can be launched by one launcher (in a manner similar to the launch and separation of the six COSMIC satellites on one launcher), and operations and telemetry transmissions are very straightforward. If the estimated cost of the nominal GDC configuration with six satellites falls outside of available budget limits, the panel also considered a GDC constellation with only four satellites. In this case, the prime GDC science objectives can still be met, yet in a less robust sense. Coverage becomes more regional and less local, because only eight local time planes can be supported. Nevertheless, such an observing scenario would provide important strides forward in understanding of the AIM system.

Expected Outcomes

The IT constellation proposed here will provide a major advance in the field of heliophysics, addressing fundamental physical processes and providing a new level of understanding of geospace. Specifically, through its global and simultaneous measurements of interconnected state variables, GDC will provide (1) breakthroughs in understanding of feedbacks between field-aligned currents, ion drifts (electric fields), conductivities, neutral densities, and winds that result from the interaction between the atmosphere-ionosphere and the magnetosphere; (2) fundamental discoveries of global ion-neutral coupling and feedback processes active in the geospace-atmosphere system; (3) unprecedented knowledge about how the ionosphere-thermosphere system at global, regional, and local scales responds to variations in solar EUV irradiance, tropospheric forcing, and solar wind (magnetospheric) driving; and (4) the data required to advance space weather models of the AIM system to the next level of sophistication.

8.5.1.2 DYNAMIC (Dynamical Neutral Atmosphere-Ionosphere Coupling) Mission (Moderate Class)

Overview

A highly compelling, complementary, and somewhat less expensive way to advance knowledge of the AIM system would be to devote a mission to answering the question, How does lower-atmosphere variability affect geospace? The primary goal of DYNAMIC is to address wave coupling with the lower atmosphere, and to come to near-closure on understanding how lower-atmosphere variability drives neutral and plasma variability in the IT system. It concentrates on revealing and understanding the processes (i.e., wave dissipation, mean-flow interactions) that underlie the transfer of energy momentum into the IT system (especially within the critical 100- to 200-km-height regime) and the thermosphere and ionosphere variability that these waves incur at higher altitudes.

Mission Configuration

The above science focus translates to a mission involving instruments that remotely sense the lower and middle thermosphere while also collecting in situ data at higher altitudes. A key mission driver is the need to address atmospheric thermal tides, which demand measurements over all local times. Since satellites generally take weeks to months to precess through 24 hours of local time, one therefore must trade latitude coverage against local time precession rate, or possibly consider multiple satellites. To include important wave sources at high latitudes such as weather systems and stratospheric warmings, and moreover to separate aurorally generated waves from those originating in the lower atmosphere, a high-inclination (75°-90°) satellite is required. However, for these orbital inclinations, 24-hour local time precession occurs over a time period that exceeds that of important variability that needs to be captured. Taking these factors into account, the preferable strategy is two identical satellites in 80° inclination orbits at 600-km altitude, with their orbital planes spaced about 6 hours apart in local time. Assuming measurements are made at four local times over all longitudes in one day, all zonal (longitudinal) components of the diurnal (24-hour) tide would be fully characterized once per day, and semidiurnal (24-hour) tides as well as the diurnal mean

TABLE 8.3 DYNAMIC Key Parameters to Be Measured from Space

Instrument	Key Parameters	Altitude Range
Limb Vector Wind and Temperature Measurement WIND (1 unit includes 2 telescopes)	Vn(z) - vector T (z)	80-300 km 80-300 km
Far Ultraviolet Imager (FUV)	Altitude profiles: O, N_2, O_2, H, O^+	110-300 km
	Maps: Q, Eo, O/N_2, O^+, bubbles	200-600 km
Ion Velocity Meter (IVM)	Vi	In situ
Neutral Wind Meter (NWM)	Vn - vector	In situ
Ion Neutral Mass Spectrograph (INMS)	O^+, H^+, He^+ O, N_2, O_2, H, He	In situ

NOTE: The IVM, NWM, and INMS are on the ram and anti-ram sides of the spacecraft. Only one operates at a time.
All instruments have extensive flight heritage. Technology investments will improve performance and provide additional capabilities.

would be acquired about once every 20 days. Gravity waves would be measured throughout each orbit, and planetary waves would easily be extracted with 1-day resolution. With the exception of semidiurnal tides, all wave-wave interactions could be explored, and on a 20-day timescale the interactions between the wave field and the mean state could be explored. In situ plasma and neutral responses at 600 km to these wave inputs would be measured over similar timescales.

Table 8.3 lists the key parameters that have to be measured to achieve the main objectives and science questions defining the mission, which are tabulated in Table 8.4. Table 8.3 assumes an instrument to measure horizontal winds and temperatures from about 80 to 250 km, day and night, with horizontal and vertical resolutions on the order of 100 km and 2-10 km, depending on height. Instruments consisting of flight heritage components approaching this capability are thought to exist at TRL 5,[3] but flight test opportunities are required to establish their true capabilities. A flight-tested FUV imager already exists, and this would provide key measurements of neutral and ionized constituents in the lower and middle thermosphere regime. In situ instruments exist to make the required in situ measurements of neutral and ion composition, winds, and drifts, but further technology developments are underway to enhance performance and reduce size, power, and weight; it is important that these technology developments be supported, given that these types of instruments are likely to be flown on almost any terrestrial or planetary ionosphere-thermosphere mission.

It is important to note that while DYNAMIC's primary focus is to address the question of meteorological driving of geospace, the orbital sampling and instruments that are flown also address several of the science questions of GDC. For instance, the composition, temperature, and wind measurements will enable understanding of the relative roles of upwelling, advection, and thermal expansion in determining latitude-time evolution of the O/N_2 ratio during changing geomagnetic conditions, which affects total mass density and plasma density concentrations. Measurement of winds, plasma drifts, and plasma densities at high latitudes will lead to estimates of Joule heating, as well as to estimates of a number of other plasma-neutral interactions at high and low latitudes. In addition, the simultaneous measurement of lower-thermosphere winds and plasma drifts at higher altitudes (or, equivalently, electric fields) will enable delineation of the disturbance dynamo in addition to the tidal-driven dynamo.

[3]For an explanation of technology readiness levels, see J.C. Mankins, NASA Advanced Concepts Office, Office of Space Access and Technology, "Technology Readiness Levels: A White Paper," April 6, 1995; available at http://www.hq.nasa.gov/office/codeq/trl/trl.pdf.

TABLE 8.4 DYNAMIC Science and Relevance to Space Weather

Solar and Space Physics Motivations	1 Understand Our Home in the Solar System 2 Predict the Changing Space Environment and Its Societal Impact 3 Explore Space to Reveal Universal Physical Processes	
DYNAMIC Primary Objective	Characterize and understand how the lower atmosphere drives IT variability.	
DYNAMIC Secondary Objectives	Characterize and understand the IT response to change magnetosphere forcing Neutral-plasma interactions in the presence of a magnetic field	
DYNAMIC Measurements and Description	Key neutral and ion state variables; high-resolution remote sensing and in situ Cost category A, 2 spacecraft mission, orthogonal circular 80° LEO orbits	
DYNAMIC Science Questions	DYNAMIC Scientific Merit	DYNAMIC Space Weather Relevance
How and to what extent do waves from the lower atmosphere determine the variability and mean state of the IT system?	"Meteorological influences from below" is a new discovery and a fundamental problem; come to closure on this question.	Enable prediction of large- and small-scale structures in the ionosphere and thermosphere driven by waves. Enable prediction of regions that would seed ionospheric instabilities.
How does the global wave spectrum evolve in the thermosphere, and how does the mean thermosphere state respond to this wave forcing?	Wave coupling, dissipation, and forcing are fundamental to all planetary atmospheres. Defining and understanding the mean state of the IT system is a fundamental question.	How the IT system responds to variable forcing depends on the mean state of the system.
How do neutral-plasma interactions produce neutral and ionospheric density changes over local, regional, and global scales?	Provide the first comprehensive view of the dynamo process over multiple scales. Understand how chemical processes, winds and electric fields combine to drive ionosphere variability.	Provide knowledge of plasma gradients and other spatial and temporal variability key to radio-based operational needs.
What is the role of gravity waves in "seeding" equatorial Rayleigh-Taylor instabilities that lead to plasma bubbles (depletions)?	Cross-scale plasma-neutral processes in the equatorial ionosphere are fundamental. Dispel existing controversies on the origins of plasma bubbles that lead to plasma irregularities and radio scintillations.	Develop the basis for forecasting ionospheric scintillations.
What is the relative importance of thermal expansion, upwelling, and advection in defining total mass density changes?	This question is fundamental to understanding the global IT response to magnetospheric forcing.	Develop a better physical basis for empirical drag prediction models.

8.5.1.3 ESCAPE (Energetics, Sources and Couplings of Atmosphere-Plasma Escape) Mission (Medium Class)

Overview

With the recognition that outflows of ionospheric ions can have profound effects on the AIM system, it has become abundantly evident that understanding of the outflow process is severely lacking, especially during episodic space weather events when O^+ ion outflows become superfluent. An important step forward in remedying this deficiency can be accomplished with a dual spacecraft mission identified here as ESCAPE. Its configuration and its instrumentation resemble those of the SWMI panel mission MISTE, but, with its closer vertical separations and lower-altitude apogee, ESCAPE achieves relatively high-accuracy magnetic alignments at altitudes below the ion baropause (~2,000-km altitude), where upflowing ionospheric ions become sufficiently energized to escape coulomb collisions and gravity. The primary goal of the ESCAPE mission is to answer the question, How are ionospheric outflows energized?

Despite the many successes of satellite missions devoted to the physics of auroral and polar-region particle acceleration, the physical processes of conversion of electromagnetic energy into particle energy, the evolution of energy conversion and particle acceleration along magnetic field lines, and the control of ionospheric outflows by plasma-neutral interactions are yet to be discovered because two spacecraft have never been accurately positioned along magnetic field lines while measuring ionized and neutral gas properties and simultaneously imaging the aurora at their ionospheric footpoints. A key aspect of such measurements, relevant to space weather prediction, is to determine how the outflow flux and other properties such as composition, density, and energy vary with electromagnetic and precipitating particle energy inputs into the outflow source region; e.g., the efficiency of energy conversion may be quantitatively expressed as an intensive transport relation between electromagnetic energy flux and particle energy flux. Such relationships are crucial elements of simulation models of AIM dynamics, yet little reliable information is available on their form.

By combining two-point ESCAPE measurements with solar wind and interplanetary magnetic field measurements; ground-based radar, lidar, imaging, and TEC measurements; and global geospace simulations, the ESCAPE mission can also address two related, global questions directly aligned with AIMI panel science priorities 1 and 3: How do interplanetary and AIM conditions control outflows, their distributions, and fluxes? How does the AIM system respond to ionospheric outflows?

ESCAPE science objectives and their connections to the heliophysics decadal survey science themes are summarized in Table 8.5, together with assessments of the mission's scientific merit and relevance to space weather applications.

TABLE 8.5 ESCAPE Science and Relevance to Space Weather

Solar and Space Physics Motivations	1 Understand Our Home in the Solar System 2 Predict the Changing Space Environment and Its Societal Impact 3 Explore Space to Reveal Universal Physical Processes	
ESCAPE Primary Objective	Determine how ionospheric outflows are energized, the processes controlling their fluxes and distributions, and how they affect the AIM system.	
ESCAPE Orbital Configuration and Key Measurements	Acquire FUV auroral images and in situ measurements of charged particles, neutral gases, and electromagnetic fields at two points along magnetic flux tubes using 2 spacecraft in 84° inclination, coplanar elliptic orbits with collinear lines of apsides, 180° phase difference in apogees. Two orbital phases: (1) topside perigee, 500 km × 2,500 km orbits; (2) bottomside perigee, 200 km × 2500 km orbits.	
ESACPE Science Objectives	**ESCAPE Scientific Merit**	**ESCAPE Space Weather Relevance**
Understand how ionospheric outflows are energized.	Promises breakthroughs in physics of charged-particle acceleration, plasma-neutral interactions, and planetary atmospheric escape.	Proved transport relations needed in geospace forecast models: out flow responses as functions of drivers.
Understand how interplanetary and AIM conditions control outflows, their distributions and fluxes.	Determines causal mechanisms of upper atmospheric, ionospheric variability, and planetary outflows.	Develop outflow climatology: empirical basis for predicting IT disturbances, magnetosphere mass compsition.
Determine how the AIM system responds to ionospheric outflows.	Resolves wave and particle couplings between collisional and collisionless media and their regulation of AIM system dynamics.	Enable prediction of IT cavitation, upwelling in outflow processes: affects drag prediction and radio propagation.

TABLE 8.6 Key Parameters to Be Measured by ESCAPE

Package	Instrument	Key Parameters	
Fields	Double probe	Vector E, δE (dc to 8 MHz)	3D
	Magnetometer	Vector B, δB (dc to 8 MHz)	3D
	Langmuir probe	Plasma density, temperature	
Plasma	Thermal particle spectrometers	Ion, electron core distribution (0.1-20 eV)	~3D
	Superthermal particle	e^-, H^+, He^+, He^{++}, O^+ distributions (5 eV-30 keV)	2D
Gas	Ionization gauge	Neutral density	
	Neutral wind sensor	Vector winds (within ±1000 m/s)	
	Mass spectrometer	Ion, neutral composition: O^+, H^+, He^+, O, N_2, O_2, H, He	
Remote	FUV imager—1356, LBH-S	Auroral Q, E_0; O/N_2 (30° FOV, 0.5° res., <1 min)	
	Ionospheric sounder	Electron density profile	

NOTE: Versions of all instruments have extensive flight heritage.

Mission Configuration

ESCAPE achieves the necessary measurements with two identically instrumented, three-axis stabilized spacecraft (MISTE employs despun platform segments), nominally in 84°-inclination, coplanar elliptical orbits with collinear lines of apsides and apogees of the two spacecraft 180° out of phase. The perigee of the nominal initial orbits (500 km × 2,500 km) is in the topside ionosphere. When one spacecraft is at higher altitude, it measures electromagnetic and precipitating particle energy inputs and properties of outflowing ions, while the magnetically aligned low-altitude spacecraft measures the properties of the ion and neutral gas source region and physical attributes of the energy conversion process. A wide range of vertical separations is achieved, including over the equatorial ionosphere, as the lines of apsides complete a full rotation in about 4.5 months. With additional propellant, the perigees of the spacecraft are lowered midway through the mission from 500 km to 200 km, thereby providing source region measurements in the bottomside ionosphere during the second orbital phase (200 km × 2,500 km) of the mission.

Table 8.6 lists the key parameters that are required to achieve closure of primary mission science. The cost envelope of the complete ESCAPE mission is estimated to be like that of a mid-range Solar Terrestrial Probe. It achieves many science targets of the 2009 Heliophysics Roadmap STP#5 called ONEP (Origins of Near-Earth Plasmas). Descoped versions of ESCAPE would carry fewer instruments and could be designed for a single orbital phase. As an example, if the mission were configured for only the higher-altitude perigee phase (500 km × 2,500 km), one or more of the neutral-gas instruments might be eliminated. With compromises in the time resolution of FUV images, both spacecraft could also be deployed as spinning platforms, which would lessen the instrument complexity and mass overhead required for dc electric, dc magnetic, and thermal particle measurements. Further descopes might include eliminating one or both imagers and/or the ionospheric sounder. A suitably descoped but still scientifically vital ESCAPE mission would likely fit in the Medium-class Explorer program envelope.

8.5.1.4 MAC (Magnetosphere-Atmosphere Coupling) Mission (Medium Class)

Overview

Recent advances resulting from studies of Earth's upper and lower atmosphere and magnetosphere have revealed the importance of the dynamic connection between these regions. At high latitudes the ionosphere and thermosphere interact with energy inputs from the magnetosphere that can be characterized by the field-aligned Poynting flux and the flux of the precipitating charged particles. Variations in these inputs are expected to be among the largest sources of variability in the ionosphere and thermosphere. However, they are poorly understood and poorly represented in global models due largely to the inability so far to identify the most important spatial and temporal scales that characterize the interaction. Magnetospheric energy inputs, which can vary with timescales of a few minutes and be input over spatial scales of 100 km, are redistributed as heat and momentum in the ionosphere and thermosphere, producing changes in the dynamics and effective conductance over spatial scales of thousands of kilometers and temporal scales of many hours. Changes in the neutral atmosphere dynamics and conductance also change the internally generated electric fields and the coupling processes to the magnetosphere. Thus the ability to specify the energy input and view the dynamic state of a large volume at middle and high latitudes over time periods ranging from less than 1 minute to many tens of minutes is necessary to determine how magnetosphere-atmosphere coupling processes affect the behavior of both regions.

This challenge can be efficiently met with a Magnetosphere-Atmosphere Coupling (MAC) mission. With two spacecraft spaced in the same orbit in the ionosphere and a single satellite imaging the sampled volume from high altitude, it is possible to identify coherent spatial features in the input drivers from the magnetosphere and the temporal and spatial scales over which the ionosphere and thermosphere respond. The MAC mission objective, science questions, and their connections to the heliophysics decadal survey science motivations are summarized in Table 8.7, together with assessments of the mission's scientific merit and relevance to space weather applications.

Mission Configuration

MAC provides the necessary measurements with two identically instrumented, three-axis stabilized spacecraft, nominally in the same circular orbit with altitude near 400 km and inclination 78°. Each space-

TABLE 8.7 MAC Science and Relevance to Space Weather

Solar and Space Physics Motivations	1 Understand Our Home in the Solar System 2 Predict the Changing Space Environment and Its Societal Impact 3 Explore Space to Reveal Universal Physical Processes	
MAC Primary Objective	Determine how magnetosphere-atmosphere coupling processes determine the behavior of both regions.	
MAC Orbital Configuration and Key Measurements	Two spacecraft in the same 400 km circular orbit with 78° inclination and having variable separations varying from 30 seconds to ½-orbit period. Each measures the neutral and charged gas properties and the electromagnetic and particle energy inputs. One satellite in 400 km × 12,000-km orbit with 63.5° inclination images the volume sampled by the lower altitude satellite at high latitudes with spatial resolution better than 100 km.	
MAC Science Objectives	MAC Scientific Merit	MAC Space Weather Relevance
Determine the spatial and temporal scales over which electromagnetic energy is delivered to the atmosphere.	A major feedback in the coupling between the magnetosphere and the AIM.	Critical to development of accurate predictions for satellite drag and ground induced currents.
Establish the critical spatial and temporal scales for ion neutral coupling processes that feedback to the magnetosphere.	The link between spatial scale and temporal persistence determines the heat and momentum transfer.	Applications to the development of ionospheric irregularities and atmospheric heating.
Understand the roles of winds and conductivity in the penetration of magnetospheric drivers to low latitudes.	Established the link between magnetospheric drivers and AIM drivers.	Critical for the assessment of interplanetary influences on the appearance of plasma structures at low latitudes.

TABLE 8.8 Key Parameters to Be Measured by MAC

Notional Instrument	Key Parameters
LEO Thermal Ions (in situ)	Ion velocity vector (±3,000 km/s) Ion temperature 300-10,000 K Major ion composition (H^+, O^+) Total ion concentration
LEO Neutral Gas (remote sensing 100 km to 350 km)	Neutral wind vector profile (±800 m/s) O/N_2 density profile
LEO Fields (in situ)	Magnetic field perturbation vector Electric field vector
LEO Particles (in situ)	Pitch angle/energy distribution Ions and electrons 30 eV to 30 keV
HEO FUV Imaging 1356, LBH-S, LBH-L	Auroral images ~100 km spatial resolution, ~3 minute cadence

craft has propulsion that allows small adjustments in the orbit eccentricity, allowing the satellites to fly in a "string-of-pearls" configuration with controlled temporal separations that vary from 0 seconds to ½-orbit period. Each spacecraft will measure the electromagnetic energy flux and the precipitating energetic particle flux incident to the atmosphere with a temporal resolution of 1 second or better. The same satellites will measure the ion and neutral density and dynamics at altitudes between 100 km and 400 km, thus allowing the response of the atmosphere to the energy inputs to be specified. Spatial and temporal ambiguities will be resolved by identifying spatial features from cross-correlation between the two LEO measurement sets and establishing coherence of the identified features through FUV imaging in the same volume from a high-altitude satellite in a 400 km × 12,000 km orbit with a 63.4° inclination (Table 8.8).

8.5.2 Explorers, Suborbital, and Other Platforms

The relative proximity of the AIM system makes it amenable to observational strategies involving a wide variety of platforms. This attribute is a significant strength in crafting a program that is responsive to budgetary realities and to the changing climate of programmatic risk factors.

8.5.2.1 Explorer Program Enhancement

The NASA Explorer program has long been instrumental in the advancement of AIM science. The Orbiting Geophysical Observatory missions, Atmospheric Explorers, and Dynamic Explorers opened whole new areas of scientific inquiry into the physics and chemistry of the AIM regions. Later missions such as IMAGE (Imager for Magnetopause to Aurora Global Exploration) and Aeronomy of Ice in the Mesosphere have provided system-level information on the behavior of the magnetosphere effective in the upper atmosphere and ionosphere, and the conditions at the boundary between the atmosphere and space, respectively. These notable achievements could be followed by an Explorer mission that investigates the coupling of energy between the regions or the development of large-scale structures and emergent behavior in the system. The ESCAPE and MAC missions are two such examples provided in this chapter.

Budgetary pressures require missions to stay on schedule and respect firm cost caps. These are hallmarks of the Explorer program, whose PI-led missions have historically performed in line with budgetary requirements. Explorers are not "too big to fail," and technical, schedule, and cost risks can be, and have been, identified early. The Explorer office holds reserves at the program level and can thus better control costs and apply fixes in individual cases as they arise on particular missions. All of these aspects of the program support a scientifically broad approach where no mission can impose the risk of cancellation on others because it experiences unforeseen cost increases. A budget line enhancement to return the Heliophysics Explorer program to a level that allows for timely, focused science missions, with a range of scales including the MIDEX, would enable significant new advances toward the AIMI panel's top goals. The panel notes that reorganizing the STP and LWS lines to accommodate a more robust Explorer program may well make the difference between being able to support three versus two heliophysics missions for the decade.

The AIMI panel supports an enhancement of the Heliophysics Explorer budget line to accomplish a broad range of science missions that can address important AIMI science challenges. The mission classes should range from a tiny Explorer that takes advantage of miniaturized sensors and alternative platforms and hosting opportunities, up to a Medium Explorer that could address multiple science challenges for the decade.

8.5.2.2 Constellations of Satellites

The importance of AIM system interactions with neighboring systems above and below dictates the need for a global-scale view of AIM system responses to various drivers. Such views will be enabled by satellite constellations with a range of capabilities in terms of state variable measurements, spatial coverage, and sampling cadence. NASA will benefit from developing innovative technologies; having methods to cost, launch, and implement multisatellite/platform programs; and working with the science community to address the challenges presented by satellite constellation missions.

Small satellites for AIMI science have proved successful and offer a unique perspective on achieving specific AIMI science objectives cheaply (thus the panel's strong support for a tiny Explorer). This path offers low-cost satellites in ad hoc constellations, expendable platforms for exploratory research, opportunities for workforce development, and potential commercial collaborations. NASA and NSF are encouraged to jointly explore opportunities to augment, or to otherwise complement, the NSF program for CubeSat-based science missions for space weather and atmospheric research.

The AIMI panel supports development of the means to effectively and efficiently implement constellation missions, including proactive development of small-satellite capabilities and miniaturized sensors, and pursuit of cost-effective alternatives such as commercial constellations. NASA, in partnership with other agencies where advantageous, is best suited to lead this effort.

8.5.2.3 Suborbital Research

Low-cost access to space via rockets and balloons is a valuable program to the AIM community that has reached a level of maturity enabling sophisticated instrumentation and new science. Present funding for science payloads is inadequate and should be increased to take advantage of the new, near-space-observatory-class capabilities being developed. The sounding rocket program has been particularly fruitful in advancing science, training and educating students, and developing technologies. It is essential that this program be protected from infringement from other programs, including proposed low-cost orbiting platforms. Without a critical mass needed to maintain core capabilities and a small, dedicated cadre of engineers, the efficiencies of scale for maintaining the low-cost sounding rocket program would evaporate and the many key advantages of the program to NASA, including the unique research capabilities it offers to the nation, would be lost.

The AIMI panel supports maintaining a strong suborbital research program by continuing development of observatory-class capabilities such as a high-altitude sounding rocket and long-duration balloons, and expanding funding for science payload development for these platforms.

8.5.2.4 Strategic Hosted Payloads

A simple review of the plethora of robust and useful measurements from NOAA's powerful GOES platform demonstrates the possibilities that continuous observations of select scientific parameters can provide. For example, GOES provides in situ measurements of the magnetic field, energetic particles, and solar emissions. These data are primarily for space weather applications, but they provide important information for research studies as well. In the past decade, a vision for other AIMI instruments in geosynchronous orbit has grown to include imagers of the IT system at wavelengths tuned to the science target. Such observations offer compelling, continuous observations of the IT system over large regions on Earth. The hosting of payloads in these orbits offers a cost-effective way to make a critical measurement for AIMI science that would otherwise be allocated to an Explorer, LWS, or STP mission. Conversely, strategic observations in place reduce the costs of future missions that require the measurements for closure, just as most AIMI missions benefit scientifically from upstream solar wind measurements.

The AIMI panel supports development of a strategic capability to make global-scale AIMI imaging measurements from host spacecraft, notably those in high Earth orbit and geostationary Earth orbit, as is currently done in support of solar (GOES SXT) and magnetospheric (TWINS, GOES, LANL) research.

8.5.3 Ground-Based Facilities

The spaceflight missions discussed in the previous sections will be greatly enhanced by the acquisition of measurements from ground-based and suborbital platforms that leverage the inherent synergy between these different means of accessing the AIM system. Ground-based instruments have an advantage in that all local times are viewed every day. Thus, the physics and evolutionary aspects of waves, electric fields, and plasma structures can be explored over much shorter timescales than from space. Ground-based remote sensing techniques and suborbital platforms are also capable of accessing regions of the atmosphere and space that are not easily probed by orbital vehicles. On the other hand, these types of observations do not provide the global view that is demanded by several of the science questions enumerated above.

AIMI science priorities regarding ground-based facilities are aimed mainly at advancing the knowledge base regarding cross-scale coupling at local and regional scales, since this area of study is fertile ground for scientific discovery, while at the same time addressing aspects of the IT system that hold societal relevance. Thus, very significant contributions to all of the AIMI science goals described in the section titled "Science Goals and Priorities for the 2013-2022 Decade" would be addressed by the imperatives put forth below. Specific areas of contribution are noted.

8.5.3.1 Autonomous American Sector Network

As described in the section "Science Goals and Priorities for the 2013-2022 Decade," AIMI science goal 4 seeks to understand how neutrals and plasmas interact to produce multiscale structures. An example of propagating structures over Japan is shown in Figure 8.16, but such structures also exist over the United States (Figure 8.21). At both locations they propagate in the southeastward direction during daytime until mid-afternoon, switching to southwestward in the late afternoon and evening. A distributed array of ground-based instruments extending from pole to pole and with regional (i.e., continental United States) concentrations would significantly advance understanding of hemispheric variability in these and other anomalies (e.g., the plasma plumes illustrated in Figure 8.10) arising from plasma-neutral interactions in the geospace system. Although a global sensor network is the ultimate vision, focusing the network initially

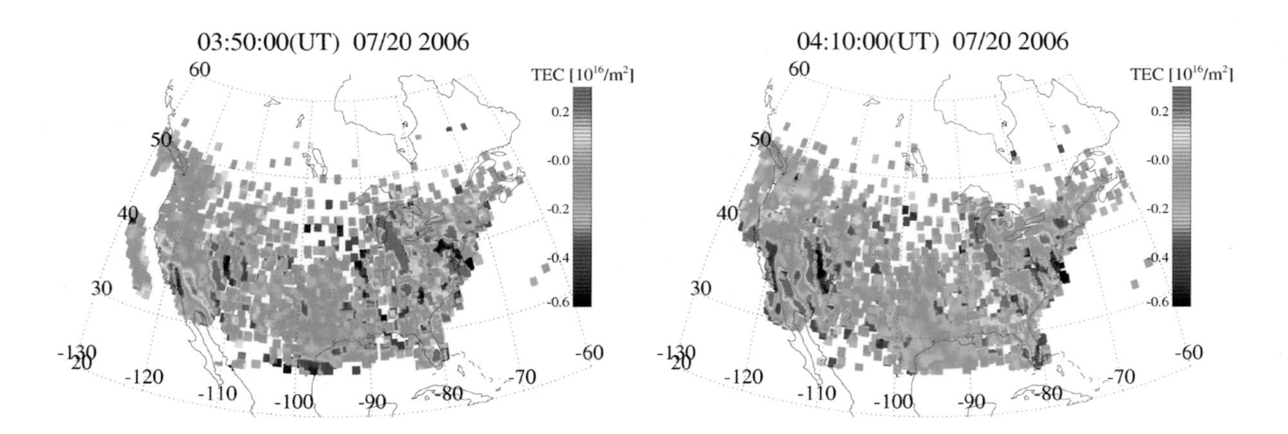

FIGURE 8.21 Two snapshots of total electron content (TEC) perturbations from Global Positioning System measurements, 20 minutes apart, over the continental United States. These are nighttime traveling ionospheric disturbances similar to those depicted in Figure 8.16 and discussed under "Science Goals and Priorities for the 2013-2022 Decade," in the section titled "AIMI Science Goal 4." SOURCE: Adapted from T. Tsugawa, Y. Otsuka, A.J. Coster, and A. Saito, Medium-scale traveling ionospheric disturbances detected with dense and wide TEC maps over North America, *Geophysical Research Letters* 34:L22101, doi:10.1029/2007GL031663, 2007. Copyright 2007 American Geophysical Union. Modified by permission of American Geophysical Union.

in the American longitudinal sector would leverage existing NSF infrastructure investments, providing an achievable goal for the coming decade.

Such a sensor network would provide significant insight into the causes of ionospheric variability relevant to space weather operational needs. One such example is the challenge of predicting and removing errors of ionospheric origin associated with the GPS-based Wide Area Augmentation System (WAAS), which was developed by the Federal Aviation Administration (FAA) to become the primary means of civil air navigation. Of particular interest are the integrity and availability of the WAAS LPV (Localizer Performance with Vertical Guidance) phase of flight that provides vertical guidance to aircraft, enabling descent to 200-250 feet above a runway (LPV approaches are equivalent to the instrument landing systems installed at many runways today). Besides loss of lock on GPS satellites owing to, e.g., signal scintillations due to plasma structures, loss of vertical navigation capability occurs due to lack of knowledge of plasma density gradients and their impact on position accuracy. High-resolution assimilative models can potentially go a long way toward ameliorating the uncertainties underlying such operational problems, but the phenomena responsible for the plasma gradients must be understood so that the correct physics and observational parameters are represented in the model. The network proposed above, and the development of embedded grid and assimilative first-principles models (see below), will represent critical steps forward in this area.

The concept of a globally distributed facility augments the current approach of clustering small, dissimilar instruments around a few large facility-class assets (e.g., ISR facilities, ionospheric heater facilities, rocket launch facilities). There is synergy between this concept and initiatives to deploy AMISR facilities in Antarctica, Argentina, and other locations. The concept also has potential synergy with NSF's emerging small-satellite initiative, which may lead to a complementary network (or constellation) of distributed AIMI sensors in space. The "seamless" assimilation of distributed measurements from ground and from space is at the heart of the heterogeneous facility concept discussed in Appendix C

AIMI Priority: Develop, deploy, and operate a network of 40 or more autonomous observing stations extending from pole to pole through the (North and South) American longitudinal sector. The network nodes should be populated with heterogeneous instrumentation capable of measuring such features as winds, temperatures, emissions, scintillations, and plasma parameters for study of a variety of local and regional ionosphere-thermosphere phenomena over extended latitudinal ranges.

8.5.3.2 Whole-Atmosphere Lidar Observatory

One of the most fundamental and least-understood topics in upper-atmosphere research concerns the vertical evolution of the wave spectrum from the troposphere to the lower and middle thermosphere (ca. 100-200 km) where many of the waves are dissipated. As waves propagate into the more tenuous upper atmosphere, they grow in amplitude exponentially with height; this leads to nonlinear interactions causing energy to cascade between wave scales, and to convective instability resulting in turbulence and mixing of chemical constituents, and deposition of wave momentum into the mean flow (Figure 8.22).

The curious result that the summer mesopause is the coldest region of Earth's atmosphere is due to the downwelling (adiabatic cooling) associated with a global meridional circulation that is gravity wave driven. Recent modeling efforts demonstrate that gravity wave–mean flow and nonlinear interactions can also lead to secondary generation of waves that then propagate to even higher altitudes. How gravity waves are dissipated and drive the mean circulation and thermal structure of the thermosphere remains unclear. In addition, gravity waves interact with longer-period tides and planetary waves and modify their vertical

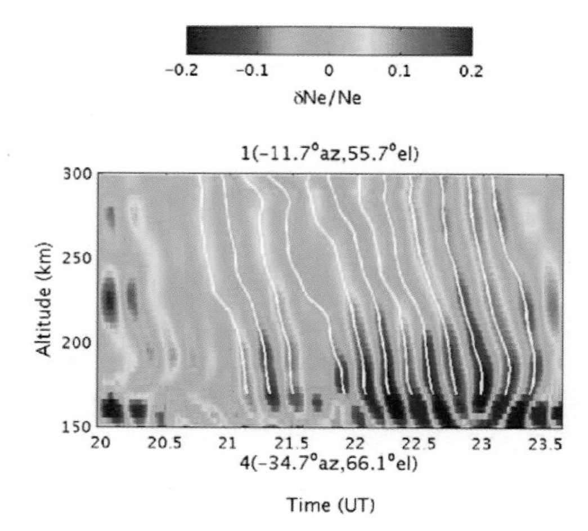

FIGURE 8.22 Gravity wave vertical structures seen in electron densities by the Poker Flat Incoherent Scatter Radar (PFISR) on December 13, 2006. The scale is in terms of percent perturbation relative to the mean; maximum electron density perturbations at the lowest altitudes exceed 20 percent. These authors attribute observed accelerations of the mean thermosphere winds to dissipation of the waves. SOURCE: S.L. Vadas and M. Nicolls, Temporal evolution of neutral, thermospheric winds and plasma response using PFISR measurements of gravity waves, *Journal of Atmospheric and Solar-Terrestrial Physics* 71:740-770, 2009.

propagation characteristics. All of these processes are subgrid processes in general circulation models of the AIM system, and their macroscopic effects need to be parameterized in such models before the influences on the mean state can be determined.

A significant impediment to further progress has been the lack of adequate observations. However, measurements of neutral gas properties from the lower atmosphere to the mid-thermosphere are within the current reach of lidar technologies. The combination of Rayleigh and resonance lidars is currently able to observe winds and temperatures from the ground to 105 km, albeit with signal-to-noise ratio (SNR) that is marginal for advancing the state of knowledge. Large-aperture telescopes and more powerful lasers are the natural remedy for limited SNR in lidar. Previous campaigns using resonance lidar techniques have demonstrated the scientific utility of using large telescope apertures of 3.5 meters at Starfire Optical Range, New Mexico, and the Air Force facility on Haleakala, Hawaii, with correlative passive optics, meteor radars, and rocket payloads to achieve desired resolutions to advance mesosphere and lower thermosphere science. A further demonstration of possibilities using large-aperture telescopes is shown in Figure 8.23, provided by a resonance lidar team working on sodium guide star studies. A 6-meter, zenith-pointing telescope comprising a spinning mercury mirror was coupled to a sodium lidar system and revealed amazing detail in MLT instability structures, identified as Kelvin-Helmholtz billows evident at the base of the sodium layer, at a temporal resolution of 60 milliseconds and a spatial resolution of 15 meters.

The available laser power has also increased exponentially over the years and, when combined with a large-aperture telescope, enables retrieval of winds and temperatures well into the thermosphere using the proven Rayleigh lidar technique. A lidar simulation based on a laser transmitter of 325 watts at 750 pulses per second and an 8-meter telescope can retrieve neutral temperatures at 200 km with 10 percent error at a range resolution of 5 km with 1-hour integration. Obviously the temporal and spatial resolution improves exponentially as altitude decreases, leading to unprecedented measurements of neutral gas properties in

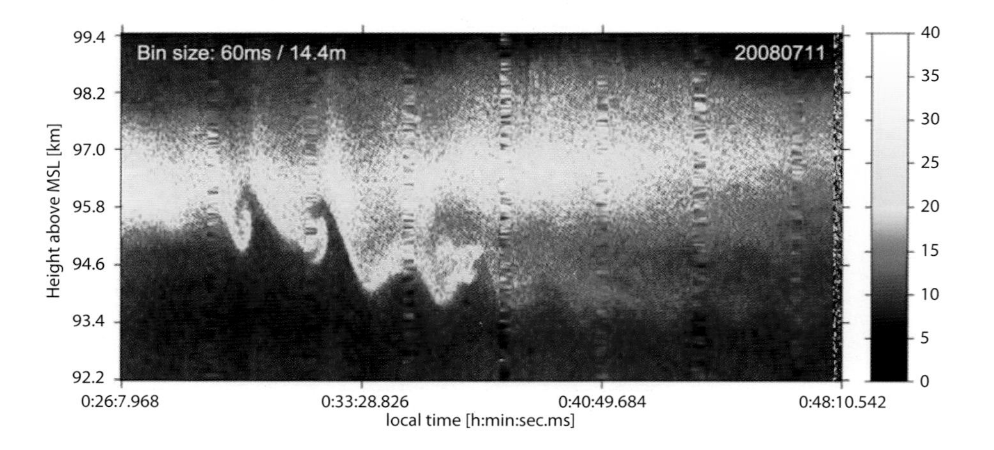

FIGURE 8.23 Large-aperture Na atomic density lidar measurements at 60-millisecond and 15-meter resolution showing detailed Kelvin-Helmholtz structures at the base of the layer. SOURCE: T. Pfrommer, P. Hickson, and C.-Y. She, A large-aperture sodium fluorescence lidar with very high resolution for mesopause dynamics and adaptive optics studies, *Geophysical Research Letters* 36:L15831, doi:10.1029/2009GL038802, 2009. Copyright 2009 American Geophysical Union. Reproduced by permission of American Geophysical Union.

the thermosphere and mesosphere. Recent lidar developments are also providing new possibilities for observations in the thermosphere. A helium resonance lidar is under development to probe the resonance structure of metastable helium in the upper atmosphere. If the lidar demonstrated, wind and temperatures would be derivable from altitudes well above 200 km. The technological advances expected with this program will also help lead to future developments of a lidar system in space for upper-atmosphere research.

AIMI Priority: Create and operate a lidar observatory capable of measuring gravity waves, tides, wave-wave and wave-mean flow interactions, and wave dissipation and vertical coupling processes from the stratosphere to 200 km. Collocation with a research facility such as an incoherent scatter radar (ISR) installation would enable study of a number of local-scale plasma-neutral interactions relevant to space weather.

8.5.3.3 Southern Hemisphere Expansion of Incoherent Scatter Radar (ISR) Network

ISR is an extraordinarily powerful AIMI diagnostic, able to remotely sense the fundamental state parameters of the ionospheric plasma (Ne, Te, Ti, Vi) as a function of range and time. Through the use of ancillary models, higher-order parameters can also be resolved, including conductance, ion composition, Joule heating, electric current systems, and neutral wind fields. The emergence of electronically steerable ISRs in the previous decade has provided a major step forward in AIMI science. The Advanced Modular ISR (AMISR) facilities have demonstrated enormous capabilities to study the ionosphere with unprecedented resolution and precision. One example is provided in Figure 8.22, illustrating the capability of the Poker Flat AMISR (PFISR) to observe the ionospheric signatures of gravity waves in the critical 100- to 300-km-altitude region. A second example is provided in Figure 8.24, illustrating the capability of an AMISR to measure ionospheric flow fields and ion temperatures over small spatial and temporal scales.

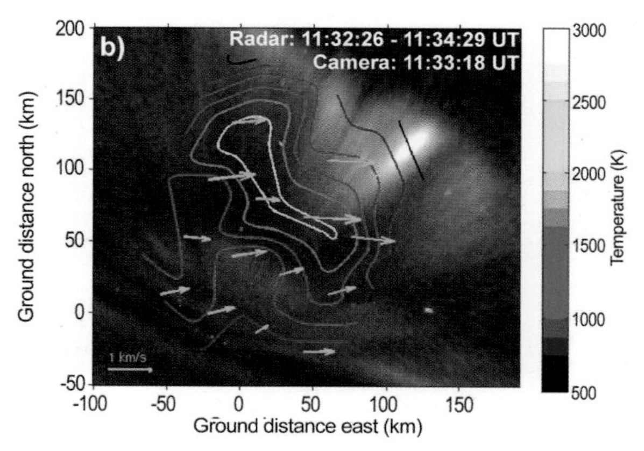

FIGURE 8.24 Composite image showing auroral forms, F-region ion temperature, and F-region ion flows, illustrating the local reduction in electric field in the vicinity of an auroral activation—a consequence of the polarization response of the ionosphere to the increased conductivity produced by the auroral precipitation. SOURCE: J. Semeter, T. W. Butler, M. Zettergren, C.J. Heinselman, and M.J. Nicolls, Composite imaging of auroral forms and convective flows during a substorm cycle, *Journal of Geophysical Research* 115:A08308, doi:10.1029/2009JA014931, 2010. Copyright 2010 American Geophysical Union. Reproduced by permission of American Geophysical Union.

In addition to boasting low operation and maintenance costs, low power requirements, and a highly robust architecture, the AMISR facilities offer an extraordinary degree of experimental flexibility that has not yet been fully realized. For instance, during the 2009 International Polar Year, the PFISR facility was configured to record a vertical profile every 15 minutes for the entire year. This low-duty cycle mode was interleaved seamlessly with other experiments. Figure 8.25 shows an epoch analysis of ionospheric effects caused by corotating interaction regions extracted from these data.

AMISR facilities deployed in the Southern Hemisphere and in the southern polar regions will contribute significantly to understanding of inter-hemispheric variability that serves as the focus of the longitudinal sensor network proposed above. Collocation of the proposed whole-atmosphere lidar with such a deployment will lead to further advances in AIMI science by elucidating wave-plasma and plasma-neutral interactions over a range of scales, as well as contributing to understanding of the spatial and temporal evolution of Joule heating. Thus, with successful AMISR deployments at Poker Flat (PFISR) and at Resolute Bay (RISR), and planned deployments in Argentina and Antarctica, attention over the next decade should turn to developing technologies and strategies to fully exploit the emerging ISR network to address AIMI panel science priorities.

AIMI Priority: Develop and deploy phased-array ISR facilities in the Southern Hemisphere including Antarctica, and develop the technologies and strategies to enable autonomous, extended, and coordinated operation of these facilities.

AIMI Priority: Create a medium-scale research facility program at NSF. The above facilities are candidates for support by the NSF Geospace program and would require that a medium-scale (~$40 million to $50 million) research facility funding program be instituted at NSF to fill the gap between the Major Research Instrumentation (MRI; <$4 million) and Major Research Equipment and Facilities Construction (MREFC; >$100 million) programs.

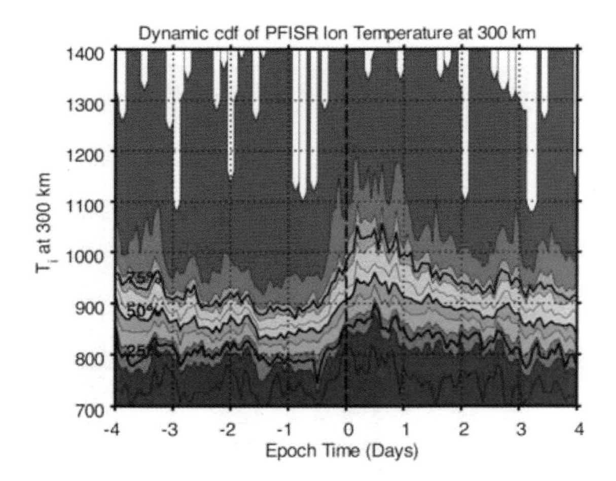

FIGURE 8.25 Epoch analysis of ion temperature affected by the arrival of corotating interaction regions over the course of a year, extracted from AMISR low-duty-cycle measurements. SOURCE: J.J. Sojka, R.L. McPherron, A.P. van Eyken, M.J. Nicolls, C.J. Heinselman, and J.D. Kelly, Observations of ionospheric heating during the passage of solar coronal hole fast streams, *Geophysical Research Letters* 36:L19105, doi:10.1029/2009GL039064, 2009. Copyright 2009 American Geophysical Union. Reproduced by permission of American Geophysical Union.

8.5.3.4 Ionospheric Modification Facilities

Ionospheric modification using high-frequency (HF) radio transmitters, or "heaters," provides a powerful tool for exploring the physics of the upper atmosphere from the ground. Heating facilities treat the ionosphere as a "laboratory without walls," providing insight into complicated plasma physics processes that occur elsewhere in the cosmos but that are difficult or impossible to explore in the laboratory. Ionospheric heaters affect the propagation of radio signals; they generate airglow and radio emissions that can be observed from the ground; they create plasma density irregularities that can be studied using small coherent scatter radars; they provide access to chemical rate constants that are otherwise hard to quantify; they accelerate electrons, mimicking auroral processes; and finally, they modify plasma density and electron and ion temperatures and enhance the plasma and ion lines observed by incoherent scatter.

The DOD operates and maintains the world's largest ionospheric modification facility, HAARP, near Gakona, Alaska. HAARP is not collocated with an incoherent scatter radar, and so its full potential has not been realized since the phenomena it creates cannot be fully diagnosed. Figure 8.26 shows an image of an artificial aurora created at the HAARP facility. Another ionospheric modification facility is under construction at the Arecibo Radio Observatory. While this facility will be modest in power compared to HAARP, its collocation with Arecibo, the world's most sensitive incoherent scatter radar, raises the prospect of discovery science in the areas of artificial and naturally occurring ionospheric phenomena.

The Arecibo heater came about through close collaboration between DOD and NSF. The AIMI panel regards this kind of interagency cooperation as a model to be followed for the utilization of existing ionospheric modification facilities as well as the planning and development of new ones.

AIMI Priority: Fully realize the potential of ionospheric modification techniques through collocation of modern heating facilities with a full complement of diagnostic instruments including incoherent scatter radars. This effort requires coordination between NSF and DOD agencies in the planning and operation of existing and future ionospheric modification facilities.

FIGURE 8.26 Artificial aurora induced by high-power HF radiation from the HAARP heater facility. The rayed structures are about 100 meters in width and are aligned with the geomagnetic field. SOURCE: E. Kendall, R. Marshall, R.T. Parris, A. Bhatt, A. Coster, T. Pedersen, P. Bernhardt, and C. Selcher, Decameter structure in heater-induced airglow at the High-frequency Active Auroral Research Program facility, *Journal of Geophysical Research* 115:A08306, doi:10.1029/2009JA015043, 2010. Copyright 2010 American Geophysical Union. Reproduced by permission of American Geophysical Union.

8.5.4 Theory and Modeling

As noted in numerous examples within this chapter, cross-scale coupling processes are intrinsic to IT system behavior. That is, phenomena highly structured in space and time (e.g., wave dissipation, turbulence, electric field fluctuations) can produce effects (e.g., wind circulation, chemical transport, Joule heating, respectively) over much broader scales. By the same token, larger-scale phenomena create local conditions that can seed development of rapidly changing structures at small spatial scales (e.g., instabilities and turbulence). The current state of affairs is that parameterizations are formulated to approximate the bulk effects of small-scale phenomena in global models whose spatial and temporal resolutions preclude inclusion of physics at smaller scales. Often such parameterizations make ad hoc assumptions about the governing physics and coupling between scales, and are usually artificially tuned to yield results in global-scale models that agree better with observations.

The observational strategies suggested in this report, which place high priority on understanding how local, regional, and global-scale phenomena couple to produce observed responses at all scales, call for complementary development of theory and numerical modeling capabilities that enable self-consistent treatment of cross-scale coupling processes. The fundamental physics of small-scale phenomena needs to be developed and understood, and numerical simulations performed that validate theories and explore parameter space dependencies. These models need to be embedded in regional-scale models so that two-way interactions are self-consistently addressed, and regional-scale models need to be nested at strategic locations within global models to enable cross-scale coupling processes and their implications to be truly understood and emulated.

Finally, researchers know well from terrestrial weather forecasting the concept of assimilating real-time data to nudge the solutions of physics-based models toward the observed state of the system. Global weather models assimilate data of various types over the globe to provide local, regional, and global forecasts as part of our daily lives. A similar path needs to be followed for the IT system to attain a true space weather forecast capability. During the next decade, assimilative models for the IT need to be developed, and such models need to be explored to reveal the types and distributions of measurements that provide optimal characterizations of the system at local, regional, and global scales.

In summary, and as an indication of its priorities for progress in theory and modeling, the AIMI panel notes that:

- Comprehensive models of the AIM system would benefit from the development of embedded grid and/or nested model capabilities, which could be used to understand the interactions between local- and regional-scale phenomena within the context of global AIM system evolution.
- Complementary theoretical work would enhance understanding of the physics of various-scale structures and the self-consistent interactions between them.
- Comprehensive models of the AIM system would benefit from developing assimilative capabilities and would serve as the first genre of space weather prediction models.

8.5.5 Enabling Capabilities

The missions and initiatives outlined above will not be successful if there is not an infrastructure of additional capabilities that enable cheaper and more frequent measurements of the AIM system, that transform measurements into scientific results, that maintain the health of the scientific community, and that serve the needs of 21st-century society. These enabling capabilities (i.e., working group imperatives) fall into the following categories: innovations: technology, instruments, and data systems; theory, modeling, and data exploitation; research to operations and operations to research; and education and workforce, and are detailed below.

8.5.5.1 Innovations: Technology, Instruments, and Data Systems

Key Instrument Development

Principal among the challenges presented to AIM science is a description of the neutral wind over the altitude range from 90 km to 300 km where a transition from a collision-dominated to a magnetized atmosphere occurs. Neutral-wind instrument development for space is a high priority for AIM science. The panel notes that reductions in mass, power, and volume of AIM space sensors would increase the effectiveness of small-satellite missions. Similarly, early action in instrument and technology development reduces the risk of unanticipated cost growth. Performed alongside a revitalized Explorer program, a new heliophysics instrument and technology development program would thus be highly complementary and support rapid scientific advancement.

AIMI Priority: An expanded instrument development program to enable new observational advances that can reduce cost risk and threats when implemented as part of satellite missions.

Data Access

The AIM community uses a wide variety of heterogeneous data sets that sample regions throughout the geospace system and are collected by space- and ground-based instruments, some of them in networks or arrays. The synthesis of data sets within arrays, with other data sets, and with global and regional models to explore frontier science issues is a major challenge for the future. The Solar and Space Physics Information System (SSPIS) was created to enable access to, and digital searching of, the many distributed data resources managed or utilized by NASA. It currently consists of a set of VxOs providing access to distributed data sets with space data model descriptors. New search-and-analysis technologies that make use of these core capabilities are now possible with a potential for significantly enhancing AIM research. In the near future SSPIS will face major challenges in providing data services for manipulation, visualization, and storage of terabyte data sets and model outputs. The development and the implementation of new capabilities are extremely slow given the minimal funding levels of the program (only 1 percent of the NASA heliophysics budget). VxOs have developed to the point that software tools can now be built by end users to support scientific studies that could not conceivably have been performed before. For geospace regions that are populated by heterogeneous sensors, this is an important capability for revealing the processes that drive the development of structure and change in the AIM system.

AIMI Priority: Create enhanced VxOs and interactive data access for system-level understanding.

8.5.5.2 Theory, Modeling, and Data Exploitation

Programmatic Support

High-priority AIMI science described throughout this chapter focuses on multiscale coupling, emergence, nonlinear dynamics, and system-level behaviors. This focus sets new requirements for research programs, computational technologies, and data analysis needs. Discovery science in all these areas requires a means of understanding global connections, sometimes across vast distances or very disparate spatial or temporal scales, while at the same time developing a deep understanding of the individual regions and processes that are elements of these connections. Coupling within the AIM system and with other elements in the Sun-Earth system spans timescales from seconds to centuries and serves to refocus efforts in understanding global change and the role of solar variability in climate. Tackling planetary change and space climate issues is dependent on the availability of historical data sets and continuity in observations of key AIM parameters like atmospheric temperatures, composition and cooling rates, and solar inputs like spectral irradiance and interplanetary magnetic field. Finally, the future of assimilative modeling in space weather prediction rests on a continuing supply of near-real-time observations of the AIM system that provide information on large-scale features like the auroral zone and equatorial electrojet as well as small-scale gradients relevant to the triggering of ionospheric instabilities.

Advances in computer power and speed have reached the point that self-consistent simulations of multiscale coupling in the AIM system are already possible. Existing peta-scale computers have reached 300,000 cores, and powerful mega-core computers are expected in the next decade to provide the computational equivalent of 1 million to 10 million CPU cores. These computational advances will drive a revolution in the realism of simulations and the ability to reproduce the self-consistent global signatures of small-scale processes. These advances in modeling in a very real sense parallel the innovations in observational programs that are high-priority targets in this chapter. An investment in a range of technological capabilities is needed to take full advantage of these powerful computational resources, including new multiscale, multiphysics algorithms (such as adaptive mesh refinement), computational frameworks that couple physics across disparate time and spatial scales, innovative ways to mine, visualize, and analyze massive amounts of data produced by the next generation of multiscale simulations, and data assimilation technologies essential to improve space weather forecasting tools.

These requirements and technological advances compel the following:

AIMI Priority: Establish a re-balanced and expanded Research and Analysis Program with the following elements:

- Solar and space physics (heliophysics) science centers, a new program of interdisciplinary centers (heliophysics scientists with computational experts) that leverages the power of peta-scale computers to create powerful physics-based multiscale models of the AIM system and its coupling to other regions, alongside parallel efforts in data assimilation and data fusion. Similar interdisciplinary theory and modeling efforts in the range of $1.5 million to $5 million per year over 3 to 5 years' duration but not focused specifically on geospace are funded through NSF through its Frontiers in Earth System Dynamics, and AFOSR through its Multidisciplinary Research Program of the University Research Initiative program. NASA funds smaller-scale modeling efforts within the strategic capabilities category in the Living with a Star Targeted Research and Technology program. Given the large costs of these programs, it may work best if NASA, NSF, and AFOSR coordinate their funding of these multidisciplinary programs in order to avoid duplication and ensure that essential projects are funded.
- A strengthened NASA theory program that supports critical-mass groups responding to new theoretical challenges in AIM science using a wide variety of research approaches.
- An enhanced data analysis program (attached to satellite missions and ground-based facilities) that provides a level of support needed to convert new and archived AIM observations into knowledge and understanding.
- An upgraded R&A program that is a reasonable fraction of the overall AIM budget to make sure that expenditures in the program are converted to major advances in science.

AIM Data Environment

The innovative observational and modeling programs described in this chapter will provide essential new information about how the AIM system works. However, this information will be embedded in the relationships *between* data sets as well as within the individual data sets themselves. It will be buried in petabytes of simulation data and in large volumes of heterogeneous data from new and ongoing space missions, from major new ground-based facilities, from suborbital platforms, and from arrays of ground-based all-sky cameras, lidars, radars, magnetometers, GPS receivers, ionosondes, imagers, and other instruments. There must be a parallel effort to develop the tools needed to convert the volumes of data into new knowledge about the AIM system. The challenge is to combine these heterogeneous data sources, housed in archives distributed around the world, into new browse summaries and new data products that contain information about AIM system behaviors in addition to the regional and process information contained in the individual data sets themselves. In turn, this global view will provide needed context for the interpretation of small-scale features and local observations.

Because of these requirements, new efforts in data exploitation and data synthesis are essential ingredients for the future of the AIM research environment. As data sources grow in size and complexity, exploiting the data requires being able to (1) locate specific pieces of information within a large distributed set of worldwide data archives, (2) manipulate and visualize the data while retaining knowledge of version information and all supporting analysis programs, and (3) combine data sources to create new data products while maintaining linkages back to the original data sources. This is where the development of data synthesis capabilities is essential. In the face of constrained budgets, much of this effort can be accomplished by using robust and sustainable commercial off-the-shelf (COTS) technologies that keep pace with new developments. To make heliophysics data "findable" by these commercial technologies requires the development of standard text-based metadata descriptors. Much of the groundwork for this capability has taken place in the last few years with the creation of an array of virtual observatories and the continuing development and implementation of the international Space Physics Archive Search and Extract (SPASE) data model.

The full and complete implementation of SPASE opens the way for the development of an array of shared software tools for essential capabilities in data mining, pattern recognition, statistical analyses, data visualization, and so on, both on the client side and on the server side in the case of large data volumes. Virtual observatories also enable the development of tools that require detailed information about instruments not easily obtained by individual investigators. One example is the calculation of common volumes in which in situ and remote-sensing measurements are made or in which space- and ground-based observations intersect.

AIMI Priority: Develop a data environment that preserves important elements of the current heliophysics data environment, while expanding the capabilities in directions that enhance data exploitation to maximize the scientific value of the data sets.

Data Synthesis

Essential to many of the AIM science frontiers identified in this chapter is the ability to synthesize information from multiple data sets into new knowledge about the AIM system. This includes, for example, mapping between geospace data sets using magnetic fields from continuously running magnetohydrodynamic simulations, browse products that superpose observations along satellite tracks onto global patterns from constellations or imagers, maps that combine information from a large number of individual ground-based instruments into global views, and combinations of ground- and space-based observations that address space-time ambiguities, among others.

AIMI Priority: Explore new data synthesis technologies to leverage the many types of AIM data into new knowledge of the AIM system required for accelerating progress on AIM system science frontiers.

Heterogeneous Data Sets

To accomplish AIM goals in the coming decade, a data environment is needed that draws together new and archived satellite and ground-based geospace data sets from U.S. agencies as well as international partners. This effort is needed to obtain the best possible coverage of the geospace system and describe its evolution over time. This data environment should also provide access to operational space weather data sets, climatological data sets, archived simulation outputs, and the latest technological advances in digital searching, storage, and retrieval and in data mining, fusion, and assimilation, as well as client-side and server-side data manipulation and visualization. These ground and space assets represent a large investment and are vital for the system science goals of the AIM program. The full value of that investment will be realized only if adequate funding is provided. The present Solar and Space Physics Information System (SSPIS) could naturally form the core of this program. The valuable multiagency and international efforts to unify standards and increase interoperability through agreements on data and metadata formats and communication protocols started under the SSPIS program must be an essential component of this effort.

AIMI Priority: Increase investment in the acquisition, archiving, and ease of use of geospace data sets.

Long-Term Data Sets

NASA and NSF should identify essential long-term data sets and pursue maintenance and preservation through funding within the heliophysics data environment activities and related NSF data archives. Some of this data, not in digital formats yet, is in danger of being irretrievably lost. A larger problem is the *continuity* of key long-term data sets, for example, solar spectral irradiance and 30 years of particle precipitation on operational satellites. Mechanisms do not yet exist to continue such data sets except by competing against new satellite missions. New mechanisms must be established to add instrumentation where possible on rides of opportunity and to partner with other agencies to re-establish space environment monitors on key operational satellite payloads.

AIMI Priority: Establish mechanisms for maintenance and continuity of essential long-term data sets.

Laboratory Experiments

Modeling planetary atmospheres and interpreting information in airglow and auroral emissions from Earth's atmosphere, planetary bodies, and comets requires detailed knowledge of differential cross sections and of reaction rates. The extraction of information about planetary environments from these emissions and the design of new and innovative remote-sensing instrumentation are dependent on accurate cross-section information. Laboratory experiments are a primary source of this information and have a long history of advancing the discovery of fundamental processes in all of these environments. Critical information on rates and cross sections for key atomic and molecular processes is still lacking. One particular example where cross-section information is highly uncertain and has impeded progress is oxygen ion precipitation and backsplash during extreme space weather events.

AIMI Priority: Establish a program of laboratory experiments joint between NSF, NASA, and DOE that includes measurements of key cross sections and reaction rates.

8.5.5.3 Research to Operations and Operations to Research

Research is the foundation for future improvements in space weather services. Ionospheric models are just now beginning to reach maturity at a level that can benefit space weather customers. But at the same time, there is a major decline in support for the development and improvement of models, which are critical to operational entities.

Furthermore, model development aimed at improving forecasts requires a dedicated effort focused on validation and verification. Also, in order to quantify a model's performance, there is a need to establish a standard set of metrics for space weather products. Under NASA, NSF, NOAA, and AFOSR sponsorship, a standard set of metrics and a skill score most appropriate for a given space weather application need to be selected that would reflect operational needs. Metrics and skill score should also be used to quantify progress with each new model version.

To reduce "the valley of death" that separates research models and operational systems, the AIMI panel suggests that operational and research agencies fund open-source models. Open-source operational models have the potential to minimize the cost of transitioning research models to operations and of maintaining the models.

AIMI Priority: Improve the effectiveness and level of support for model development, validation, and transition to operation.

Existing and planned assets that are routinely used for purposes other than space weather often have capabilities that could be utilized to return valuable space science measurements. With minimal investment, important data can be obtained from these instruments. One very-low-cost example includes tasking on a daily basis the Altair radar at Kwajalein to measure ionospheric parameters (electron density and plasma irregularities). This can easily be done during short periods interlaced between scheduled radar target acquisitions. These measurements would also benefit the users of the Altair system by providing a method to detect and forecast possible scintillation problems that could affect the radar data.

AIMI Priority: Seek high-leverage opportunities for acquiring new measurements.

It is widely recognized that space environment data provided by NOAA and DOD operational satellites for almost four decades are essential to AIMI science. Data from these satellites are of fundamental importance for space weather forecasting. They are among the most often used data sets for space science research on ionosphere/magnetosphere coupling. In addition, they are essential for studying long-term climatic changes in geospace. The following three-part imperative follows:

AIMI Priority: Leverage national investments in operational satellite data for scientific progress: (1) NOAA and DOD should maintain the space environment sensing capability that has provided data for almost four decades and should continue to acquire observations from low-Earth-orbit satellites of particle precipitation, ion drifts, ion density, and magnetic perturbations similar to those measured from POES and DMSP. (2) Open data policies should be negotiated with DOD, DOE, and other agencies (that also safeguard national security concerns). (3) DOD, NOAA, and NASA should coordinate the archiving of these data sets, making data accessible and creating tools that provide for ease of use. The present Solar and Space Physics Information System could naturally form the core of this program.

8.5.5.4 Education and Workforce

The AIM community recognizes the increasing need to promote education and training in all aspects of space science and space technology. At the same time, few educational institutions have the breadth within their faculty, or the student numbers within a department, to facilitate a complete AIM curriculum. Thus it is important to afford the opportunity for students and faculty to participate in programs that bring together varying expertise from different institutions in intensive training sessions in order to fill this pedagogical gap. NSF, NASA, and DOD should continue to support the development and execution of summer schools and workshops to provide a full spectrum of instruction in geospace science and technology. This effort is considered essential to the proper development of the next generation of space scientists and engineers. NSF should continue its Faculty Development in Space Sciences program, which provides an incentive for universities to hire faculty in geospace research. The federal government should revise export control policies to exempt basic space research from government restrictions such as those mandated under ITAR.

AIMI Priority: As described above, expand and promote education and training opportunities to develop the future generation of AIM scientists and engineers.

9

Report of the Panel on Solar Wind-Magnetosphere Interactions

9.1 SUMMARY OF SWMI SCIENCE PRIORITIES AND IMPERATIVES

The magnetosphere is a central part of the solar and space physics system. Its various regions interact globally in complex, nonlinear ways with each other, with the solar wind, and with the upper atmosphere. These interactions occur by a number of fundamental physical processes that operate throughout the universe. The resulting dynamic variations in the magnetospheric environment also have important practical consequences for a society that is increasingly reliant on ground- and space-based technological systems that are sensitive to Earth's space environment.

Significant progress has been made over the past decade in achieving the science objectives identified in the 2003 decadal survey for solar wind-magnetosphere interactions.[1] This progress has been achieved through a powerful combination of tools, including new data from satellites launched during the decade or just before (e.g., Cluster, IMAGE, THEMIS, TWINS); analysis of data returned from earlier missions; data from instruments flown on non-NASA operational satellites; measurements from suborbital missions and from networks of ground-based observatories; greatly improved numerical simulations; analytical theory; and laboratory work.

For the coming decade, the Panel on Solar Wind-Magnetosphere Interactions (SWMI) identified a set of eight high-priority science goals for research in solar wind-magnetosphere interactions—goals that follow naturally from the progress that has been made and will contribute substantially toward accomplishing the decadal survey key science goals for solar and space physics identified in Chapter 1 of this report. These eight goals (which are not prioritized) are as follows:

> *SWMI Science Goal 1.* Determine how the global and mesoscale structures in the magnetosphere respond to variable solar wind forcing.

[1] National Research Council, *The Sun to the Earth—and Beyond: A Decadal Research Strategy in Solar and Space Physics,* The National Academies Press, Washington, D.C., 2003; and National Research Council, *The Sun to the Earth—and Beyond: Panel Reports,* The National Academies Press, Washington, D.C., 2003.

SWMI Science Goal 2. Identify the controlling factors that determine the dominant sources of magnetospheric plasma.

SWMI Science Goal 3. Understand how plasmas interact within the magnetosphere and at its boundaries.

SWMI Science Goal 4. Establish how energetic particles are accelerated, transported, and lost.

SWMI Science Goal 5. Discover how magnetic reconnection is triggered and modulated.

SWMI Science Goal 6. Understand the origins and effects of turbulence and wave-particle interactions.

SWMI Science Goal 7. Determine how magnetosphere-ionosphere-thermosphere coupling controls system-level dynamics.

SWMI Science Goal 8. Identify the structures, dynamics, and linkages in other planetary magnetospheric systems.

These are ambitious goals, and their accomplishment will require intelligent use of the variety of research tools that are available. Achieving some of them will require new measurements from strategic missions, either already under development (MMS and RBSP[2]) or to be started in the coming decade. Other contributions toward accomplishing these goals will come from community-proposed Explorer missions, suborbital flights, CubeSats, operational satellites, and instruments flown on commercial platforms as rides of opportunity. Studies that combine data from the numerous existing spacecraft will elucidate important aspects of global coupling, while analytical theory, laboratory studies, and especially powerful numerical simulations will complement the various spacecraft measurements in a synergistic attack on the science questions. However, because of budget constraints and as-yet immature technology, a number of critical investigations must be deferred to a later decade, requiring investments in the coming decade in the areas of technology innovation and development, as well as in ways to more effectively couple theory and data.

In this chapter, the SWMI panel advocates a coherent and balanced program of research in solar wind-magnetosphere interactions based on a prioritized set of imperatives that will enable a cost-effective approach to accomplishing the SWMI science goals. These imperatives are sorted into three categories: (1) missions, (2) DRIVE-related initiatives, and (3) space weather. For simplicity, the panel assigned each imperative to a single primary category, although some actually apply to multiple categories. The panel prioritized the imperatives not only within but also across the three categories in order to identify the most important and most cost-effective actions to accomplish its pressing science goals. The five overall SWMI highest-priority imperatives that are developed in this chapter are the following:

SWMI Imperative 1. Enhance the resources dedicated to the Explorer program and broaden the range of cost categories.

SWMI Imperative 2. Complete the strategic missions that are currently in development (MMS, RBSP/BARREL) as cost-effectively as possible.

[2]RBSP—the Radiation Belt Storm Probes mission—was launched successfully on August 30, 2012, shortly after the release of the prepublication version of this report. The mission has since been renamed the Van Allen Probes.

SWMI Imperative 3. Initiate the development of a strategic mission, like MEDICI, to determine how the magnetosphere-ionosphere-thermosphere system is coupled and responds to solar and magnetospheric forcing.

SWMI Imperative 4. Ensure strong continued support for existing satellite assets that can still contribute significantly to high-priority science objectives.

SWMI Imperative 5. Enhance and protect support for theory, modeling, and data analysis, including research and analysis programs and mission-specific funding.

Within the three categories, the panel's full set of prioritized imperatives is as follows. The numbering in each category reflects each priority's relative ranking among all three categories, as shown also in Table 9.4 at the end of this chapter.

9.1.1 Missions

1. Enhance the resources dedicated to the Explorer program and broaden the range of cost categories.
2. Complete the strategic missions that are currently in development (MMS, RBSP/BARREL) as cost-effectively as possible.
3. Initiate the development of a strategic mission, like MEDICI, to determine how the magnetosphere-ionosphere-thermosphere system is coupled and responds to solar and magnetospheric forcing.
4. Ensure strong continued support for existing satellite assets that can still contribute significantly to high-priority science objectives.

9. Develop a mechanism within NASA to support rapid development, deployment, and utilization of science payloads on commercial vehicles and other missions of opportunity. NSF and DOD efforts in this regard are also encouraged.

10. Through partnership between NASA's Heliophysics Division and Planetary Division, ensure that appropriate magnetospheric instrumentation is fielded on missions to other planets. In particular, the SWMI panel's highest priority in planetary magnetospheres is a mission to orbit Uranus.

11. Partner with other space agencies to implement consensus missions, such as a multispacecraft mission to address cross-scale plasma physics.

15. If resources permit, initiate a strategic mission like MISTE to simultaneously measure the inflow of energy to the upper atmosphere and the response of the ionosphere-thermosphere system to this input, in particular the outflow back to the magnetosphere.

9.1.2 DRIVE-Related Actions

5. Enhance and protect support for theory, modeling, and data analysis, including research and analysis programs and mission-specific funding.
6. Ensure continuity of measurements of the upstream solar wind and interplanetary magnetic field.
7. Invest significantly in developing the technologies to enable future high-priority investigations.
8. Ensure strong multiagency support for a broad range of ground-based assets that are a vital part of magnetospheric science.

14. Strengthen workforce, education, and public outreach activities.

16. Create an interagency joint laboratory astrophysics program that addresses issues relevant to space physics.

9.1.3 Space Weather

12. Encourage the creation of a complete architecture for the National Space Weather Program that would coordinate joint research, commercial, and operational space weather observations and define agency roles for producing, distributing, and forecasting space weather products. In addition the SWMI panel encourages all agencies to foster interactions between the research and operational communities and to identify funding for maintaining a healthy research-to-operations/operations-to-research program.
13. Implement a program to determine, based on past observations, the optimum set of measurements that are required to drive high-fidelity predictive models of the environment.

Implementation of these imperatives will enable achievement of the exciting and high-priority science goals laid out in this report, providing a strong foundation for the accomplishment of the long-term actions described earlier in this decadal survey. To summarize, eight overarching SWMI science goals motivate sixteen prioritized, actionable imperatives that are required to enable the goals (these prioritized imperatives and their mapping to decadal categories from Part I are shown in Table 9.4 at the end of this chapter).

9.2 INTRODUCTION TO SWMI SCIENCE

This section gives a brief introduction to the magnetosphere and its interactions with the solar wind and the upper atmosphere. This information provides a context for the subsequent discussion of the past decade's accomplishments and important unanswered questions, leading to the SWMI panel's science goals for the coming decade and the initiatives necessary to accomplish them.

9.2.1 What Is the Magnetosphere?

The magnetosphere (Figure 9.1) is a vast, highly coupled system governed by fundamental physical processes and characterized by complex, nonlinear linkages between its different parts. It is formed by the interaction of the solar wind plasma stream and its embedded magnetic field with Earth's intrinsic magnetic field. Earth with its field is an obstacle in the solar wind flow, carving out a separate plasma domain where Earth's field has dominant control over the motions of the electrically charged particles trapped there. These charged particles come from both the solar wind and Earth's upper atmosphere. This region of dominance, the magnetosphere, extends out to approximately 10 Earth radii on the sunward side of Earth and, in a long "magnetotail," extends to well beyond the Moon on the side away from the Sun. The shape of the magnetosphere is determined by the balance between the pressure exerted by the solar wind plasma and interplanetary magnetic field (IMF) and the pressure of Earth's plasma and magnetic field. Earth is only one of six of the Sun's planets (Mercury, Earth, Jupiter, Saturn, Uranus, and Neptune) that are known to have a magnetosphere by virtue of their intrinsic magnetic fields. Ganymede, one of Jupiter's satellites, also has its own tiny magnetosphere embedded within Jupiter's giant one.

9.2.1.1 Regions

The magnetosphere is made up of regions with different plasma characteristics. As illustrated in Figure 9.1, the shape of the underlying geomagnetic field lines governs the morphology of these various regions. Nearest Earth, there is a relatively cold and dense region called the plasmasphere. The plasmasphere contains plasma that has escaped from the ionosphere, the ionized region of Earth's upper atmosphere. Coincident with the plasmasphere or residing at slightly larger radial distances are higher-energy charged-particle populations called the ring current and radiation belts. Ring-current particles drift azi-

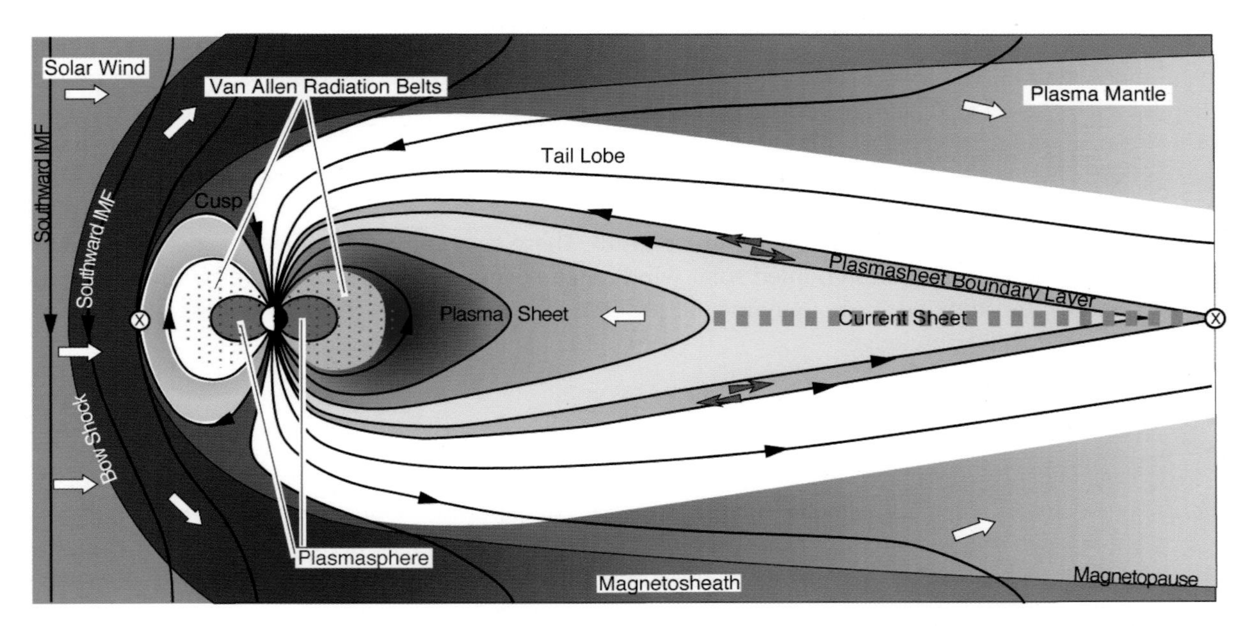

FIGURE 9.1 A schematic diagram of the magnetosphere, with various regions indicated. SOURCE: Adapted from L.A. Weiss, P.H. Reiff, R.V. Hilmer, J.D. Winningham, and G. Lu, Mapping the auroral oval into the magnetotail using dynamics explorer plasma data, *Journal of Geomagnetism and Geoelectricity* 44(12):1121, 1992.

muthally around Earth because of the strong magnetic field gradients, with positively charged ions drifting west and electrons drifting east, producing an electrical current large enough to substantially modify the global magnetic field. At still higher energies, the similarly drifting radiation-belt population possesses tremendous energy per particle, but because of its very low density carries little net current. Beyond the ring current and radiation belt is a low-density, hot plasma called the plasma sheet, extending to large distances into the magnetotail and serving as the reservoir for much of the plasma that ultimately feeds the inner-magnetosphere ring-current and radiation-belt populations. The high-latitude magnetic flux regions that are nearly devoid of plasma and that extend down the magnetotail from the polar caps are called the tail lobes. The magnetopause is the boundary of the magnetosphere, which separates it from the surrounding regions dominated by the solar wind and its magnetic field. Upstream from the magnetopause there is a standing shock wave, the bow shock, at which the supersonic solar wind is slowed and heated, enabling it to flow around the magnetospheric obstacle. The transition region of shocked solar wind between the bow shock and the magnetopause is known as the magnetosheath. A final region, beyond the bow shock, is called the foreshock, in which fast particles move upstream along the IMF and perturb the supersonic flow of the incoming solar wind. Another term that is often used synonymously with the magnetosphere-ionosphere system is "geospace."

9.2.1.2 Physical Processes

Several primary physical processes produce the rich phenomenology of the structured, time-dependent magnetospheric system. "Magnetic reconnection" is the process by which energy stored in magnetic fields is converted to plasma thermal energy, plasma bulk flow energy, and energized particles through a topological magnetic field reconfiguration. Charged particles in general conserve certain quantities of motion called adiabatic invariants, which relate aspects of the particle motion to magnetic field parameters and can lead to reversible energy changes. The term "wave-particle interactions" (WPIs) refers to the general and broad process by which electromagnetic waves and charged particles exchange energy and momentum. "Turbulence" describes the flow state of a fluid (including magnetized fluids) that is chaotic and stochastic. Turbulent media provide a rich opportunity for WPI. These same physical processes, determining the structure and dynamics of Earth's coupled solar wind-magnetosphere system, also govern other planetary magnetospheres; by studying these fundamental processes in our own neighborhood, researchers glean universally applicable knowledge.

9.2.1.3 Coupling to the Ionosphere and Solar Wind

The magnetosphere is physically bounded by the ionosphere and solar wind at its lower and upper extents, respectively. Earth's magnetosphere has no internal plasma sources, and so these boundary regions are the two major sources of magnetospheric plasma. The solar wind is a large source of protons and electrons into the magnetosphere, while the ionosphere contributes not only protons but also heavy ion species like oxygen, helium, and nitrogen, and their accompanying electrons. Furthermore, the solar wind and IMF, through magnetic reconnection and viscous interaction, drive convective flow throughout the magnetosphere. The ionosphere, with its high conductance, regulates and modulates this convective flow. In addition, the neutral gas of the upper atmosphere, known as the thermosphere, can also influence magnetospheric flow through ion-neutral collisions.

9.2.1.4 Space Weather and the Magnetosphere

"Space weather" is the name given to the time-dependent conditions and changes that occur in near-Earth space to the magnetospheric plasmas and fields. These include changes in the plasma density, temperature, and spatial distributions, from the cold plasmasphere to the very energetic radiation belts. In particular, space weather implies changes that have significant impact on technology and society. For example, the variable ionosphere and plasmasphere alter geolocation signals from GPS and transmissions from communication spacecraft; strong magnetospheric currents create geomagnetically induced currents in power distribution systems; energetic particles cause radiation damage to microelectronics and spacefarers; and substorm-related satellite charging causes malfunctions and surface degradation.

9.2.1.5 Magnetospheric Questions That Flow from the Motivations

The motivations underlying the study of solar and space physics[3] apply directly to the study of Earth's magnetosphere and its interaction with the solar wind and upper atmosphere: Earth's magnetosphere (and those of other planets) is a fascinating, complex system, in which fundamental physical processes that operate throughout the universe combine with unique conditions of plasma sources, sinks, and drivers to create dynamic conditions that can affect humans and the technologies they depend on in space and sometimes on the ground.

[3] See the introduction to Part II of this decadal survey report.

- *Motivation 1. Understand our home in the solar system.* The fundamental motivation of wanting to understand the fascinating Sun-Earth system drives the desire to learn what determines the dynamically changing charged-particle environment of the magnetosphere, from the lowest-energy particles emerging from the upper atmosphere, to the hazardous high-energy particles of the radiation belts. What exactly are the relationships between the different magnetospheric regions, and how are these regions coupled to the solar wind and upper atmosphere? How do these linkages determine magnetospheric dynamics? And what new insights do researchers gain from examining the similarities and differences in these regions and processes at other planets?
- *Motivation 2. Predict the changing space environment and its societal impacts.* A thorough understanding of this complex and highly coupled system will make it possible to predict its behavior under a variety of changing conditions, allowing anticipation and recognition of conditions that are adverse to human life and technology. Within the magnetosphere, researchers are particularly interested in how they can evaluate and predict damaging particle populations and electrodynamic fields.
- *Motivation 3. Reveal universal physical processes.* The geospace system constitutes a laboratory for exploring a wide variety of processes that operate throughout the universe. These include the ways in which ionized outflows can be driven from planetary atmospheres; how waves and particles provide coupling between disparate plasma regions; how plasmas interact with neutral materials; how magnetic reconnection occurs and how it energizes particles; how collisionless shock waves work; and how plasma turbulence is generated and dissipated and affects other dynamical processes.

The following sections outline recent progress made toward addressing the fundamental questions raised by these motivations; indicate the outstanding problems where significant progress can be accomplished in the near future, leading to the identification of specific science goals for the coming decade; and lay out the SWMI panel's imperatives for actions that are needed to meet those goals.

9.3 SIGNIFICANT ACCOMPLISHMENTS OF THE PREVIOUS DECADE

9.3.1 Scientific Progression

The science of solar wind-magnetosphere interactions was established near the dawn of the space age, only 50 years ago. Since that time, knowledge of the magnetosphere has grown tremendously, progressing from discovery to understanding.

Early research focused on the discovery and exploration of the different regions that lie above Earth's atmosphere: the radiation belts, the plasmasphere, the plasma sheet, and the solar wind. As researchers discovered these regions and the particle populations that define them, they sought to identify and understand the physical processes that accelerate and transport the particles. They also learned that the system is not static, but rather exhibits substantial variability with identifiable patterns. It became clear that different regions are not independent, but rather are intimately linked, and the variability and processes in one region can be clearly related to those in other regions.

Scientists now appreciate that the magnetosphere is a vast, highly coupled system that involves a wide array of fundamental physical processes and complex linkages between different regions. In the past decade this system's impact on human technology and society has also come to be appreciated: storms and other disturbances in geospace have significantly disrupted and disabled spacecraft and ground-based power grids. Finally, as mankind has reached out to explore the solar system, researchers have discovered that other solar system bodies possess magnetospheres that exhibit many of the same processes found near Earth, but frequently manifested in different ways, and producing very different structure and dynamics.

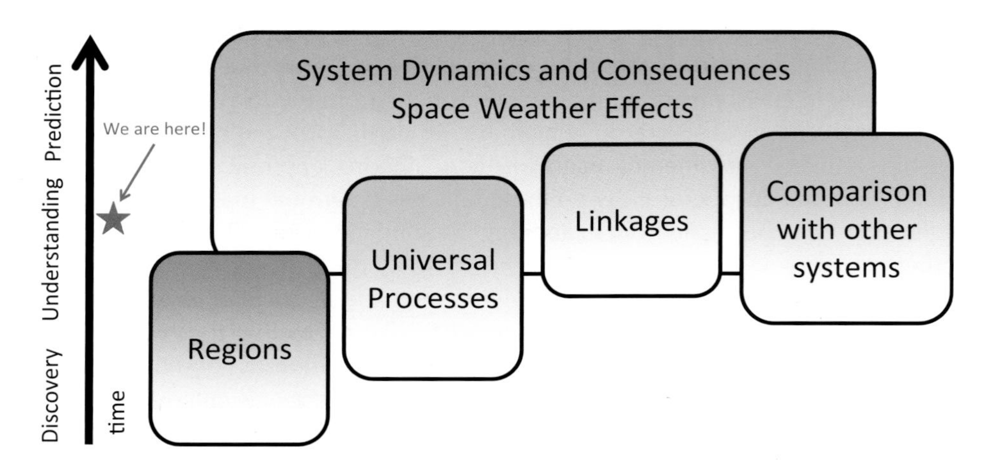

FIGURE 9.2 A schematic of the temporal progression of scientific research into solar wind-magnetosphere interactions, with time increasing upward.

Thanks to observations from new satellite missions, new analyses of data from previous missions, and improved numerical modeling capabilities, substantial progress has been made in the past decade to advance the field along the progression illustrated in Figure 9.2. Here the SWMI panel reviews some of these significant accomplishments. This review is by no means comprehensive, but it serves to illustrate the progress that continues to be made across a broad front of scientific endeavors. These accomplishments lay the foundation for what needs to be done in the future to continue the progression toward a comprehensive and actionable understanding of the SWMI system.

9.3.2 Regions

9.3.2.1 Statistical Description of Magnetospheric Properties

The decades-long reconnaissance of magnetospheric structure and dynamics has culminated in new statistical descriptions of plasma properties in the magnetosphere, particularly their systematic variation in response to solar wind drivers. For example, use of data sets that extend over more than a solar cycle showed the activity-dependent spatial distribution of plasma-sheet fluxes to be well described by a relatively simple convection model coupled with losses. These and similar results have solidified the statistical picture of the structure of the inner magnetosphere, but they do not capture the dynamic evolution of these regions, nor the coupling with the rest of the system.

9.3.2.2 Measuring Invisible Populations

During the past decade, several new techniques were exploited for measuring the plasma density over more extended regions than allowed by single-point, in situ measurements. On the ground, observations of field-line resonances enabled instantaneous determination of the equatorial plasma density over a broad spatial domain. In space, active radio sounding techniques probed the density variation along magnetic field lines, and both radio sounding and extreme ultraviolet (EUV) images revealed not only the dynamics of the boundaries of the plasmasphere, but also actual global plasmaspheric density distributions. These advances made it possible for the first time to observe the global evolution of the plasmasphere in response to variable driving of the magnetosphere.

Energetic neutral atom (ENA) imaging established itself as a valuable tool in determining the global-scale configuration, dynamics, and composition of the ring current (e.g., see Figure 9.6 below). It was found that the peak of the ring-current proton distribution during the main phase of magnetic storms could lie not in the historically expected afternoon location, but in the early morning sector, revealing that the coupling with the ionosphere can very strongly alter the behavior of magnetospheric plasmas. In addition, the hydrogen component of the ring current builds up and decays gradually throughout a magnetic storm, but the oxygen component rises and falls impulsively. These variations in oxygen are often correlated with substorm injections, highlighting the coupling between the magnetosphere and ionosphere. These results represent more than a huge leap forward in system-level knowledge of the ring current; they also offer a tantalizing hint of the dynamic magnetospheric behavior that could be uncovered with higher-resolution, continuous, and global imaging.

9.3.3 Processes

9.3.3.1 Magnetic Reconnection

The recent decade has witnessed substantial progress in understanding how magnetic reconnection works. For example, increased computing power has allowed full-physics simulations describing the essential physics and structure of the diffusion region, the key region where magnetic field lines break and reform. The decoupling of ion and electron motions plays a key role, accelerating the energy release, creating high-speed electron beams, and warping the magnetic field. These predictions have facilitated the first direct detections of the ion diffusion region in the magnetosphere (Figure 9.3) and in the laboratory, as well as glimpses of the much smaller electron diffusion region. The observations in the vicinity of the diffusion region revealed surprisingly that electrons can be accelerated by reconnection to hundreds of kiloelectron volts.

The past decade has also witnessed surprises regarding the triggering and modulation of reconnection: theoretical studies using fully three-dimensional simulations revealed that the added dimension facilitates plasma instabilities that can disrupt the diffusion region, making reconnection highly turbulent. Observationally, reconnection seems to behave differently in different regions. On the dayside magnetopause, as well as recently discovered in the solar wind, magnetic reconnection can be quite steady in time and extended in space. In the magnetotail, however, reconnection is most often patchy and bursty, producing narrow flow burst channels. Using multispacecraft observations, these reconnection-generated flow channels have now been demonstrated to initiate magnetospheric substorms.

9.3.3.2 Wave-Particle Interactions

A delicate balance between acceleration and loss caused by wave-particle interactions controls the variability of radiation belt fluxes during geomagnetic storms. Understanding of magnetospheric plasma waves and their role in radiation belt dynamics has increased significantly during the past decade. Statistical analyses of satellite wave data have led to the development of global models of the wave environment. These have then been used to quantify the rates of energization and scattering loss to the atmosphere. Time-dependent two-dimensional and three-dimensional models for the radiation belts and the ring cur-

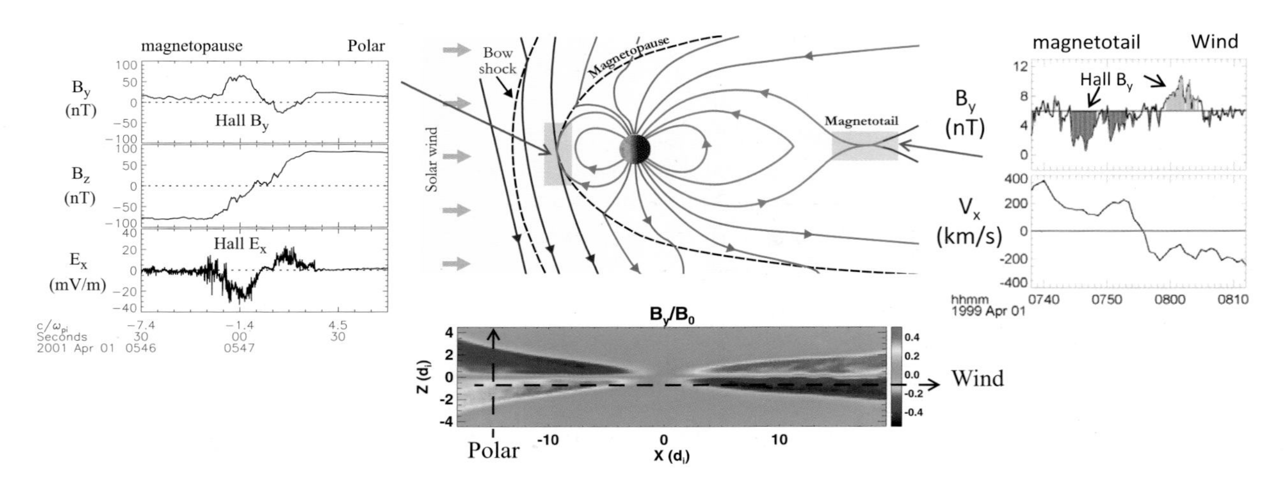

FIGURE 9.3 At the center is a schematic of the magnetosphere indicating the regions where magnetic reconnection usually occurs. The lower panel is a simulation result of the quadrupolar magnetic field topology in the ion diffusion region, and the left and right panels are observations of these field topologies by the Polar and the Wind spacecraft, respectively. For comparison, the relative motions of the two spacecraft through the reconnection region are indicated on the simulation result panel. The Hall field arises from the decoupling of the ions from the electrons. SOURCE: *Left:* Adapted from F.S. Mozer, S.D. Bale, and T.D. Phan, Evidence of diffusion regions at a subsolar magnetopause crossing, *Physical Review Letters* 89:015002, 2002. © 2002 The American Physical Society. *Middle, upper:* C. Day, Spacecraft probes the site of magnetic reconnection in Earth's magnetotail, *Physics Today* 54:10, 2001. *Middle, lower:* Adapted from T.D. Phan, J.F. Drake, M.A. Shay, F.S. Mozer, and J.P. Eastwood, Evidence for an elongated (>60 ion skin depths) electron diffusion region during fast magnetic reconnection, *Physical Review Letters* 99:255002, 2007. © 2007 The American Physical Society. *Right:* Adapted from M. Øieroset, T.D. Phan, M. Fujimoto, R.P. Lin, and R.P. Lepping, In situ detection of collisionless reconnection in the Earth's magnetotail, *Nature* 412:414-417, 2001, doi:10.1038/35086520.

rent, which incorporate the effects of wave-particle scattering, have been developed and are beginning to provide a more realistic picture of storm-time particle dynamics.

Satellite observations of peaks in the radial profiles of radiation-belt electrons have demonstrated that local acceleration due to WPI may at times dominate over traditionally accepted acceleration associated with diffusive radial transport. Diffusive radial transport may actually lead to enhanced losses at the magnetopause, causing a decrease in trapped flux rather than an increase. Particle interactions with "chorus" waves in particular have been shown to provide a major probable source of local acceleration. The source of plasmaspheric hiss, another wave mode known to be responsible for strong losses of radiation belt electrons near the edge of the plasmasphere, has been shown to be discrete chorus emissions generated in the low-density region outside the plasmapause (Figure 9.4). Chorus emissions have also been shown to be the dominant cause of scattering of plasma sheet electrons, leading to their precipitation into the atmosphere, where they produce the diffuse aurora. Intriguingly, recent observations of extremely large-amplitude waves suggest that nonlinear wave-particle physics may play an important role in radiation belt dynamics.

9.3.3.3 Turbulence

In the past decade the presence of large-amplitude MHD fluctuations in the magnetotail plasma sheet has been established by spacecraft measurements, and the spatial scales (~1 R_E) and timescales (~1 minute)

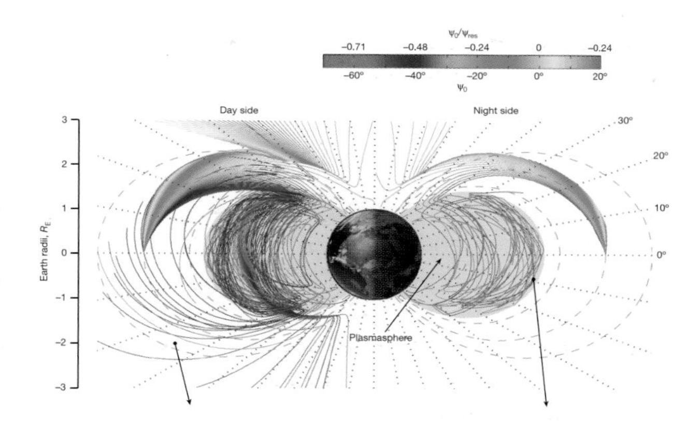

FIGURE 9.4 Ray path calculations showing how discrete whistler-mode chorus emissions generated outside the plasmapause can be refracted into the plasmasphere, become trapped, and eventually merge to form incoherent plasmaspheric hiss. SOURCE: J. Bortnik, R.M. Thorne, and N.P. Meredith, The unexpected origin of plasmaspheric hiss from discrete chorus emissions, *Nature* 452:62-66, 2008, doi:10.1038/nature06741.

---

of the fluctuations have been determined. Global simulations of the solar-wind-driven magnetosphere are reaching high-enough spatial resolutions in the magnetotail to enable them to predict irregular vortical flows at these scales and to statistically match the properties of the vortical flows to observed fluctuations. However, the dynamical nature of these fluctuations, what causes them, and their influence on the behavior of the magnetosphere have not been determined. Fluid turbulence provides one pathway by which energy moves across scale sizes from large to small where energy can be dissipated in the form of heating. When and where turbulent processes play a significant role in magnetospheric dynamics remain unclear.

9.3.4 Linkages

9.3.4.1 Coupling with the Solar Wind

The variable solar wind drives a wide range of variations in magnetospheric behavior. Over the past decade, continuous measurements of the solar wind combined with observations throughout geospace have enabled significant advances in identifying which specific large- and mesoscale solar-wind properties produce different modes of magnetospheric response (e.g., storms, steady magnetospheric convection events, sawtooth events). For example, studies have linked variations in the solar wind dynamic pressure to radiation-belt loss and energization processes. Other studies quantified that the strength of geomagnetic storms depends on both the electrodynamic coupling between the solar wind and the magnetosphere and plasma loading of the magnetosphere, including both ionospheric and solar wind sources. Spacecraft observations and numerical simulations reveal that solar-wind plasma entry into the magnetosphere is surprisingly efficient under "quiescent" conditions of northward interplanetary magnetic field. This plasma in turn participates as a substantial element of the storm-time ring-current development when southward interplanetary magnetic fields couple with and energize the magnetosphere. Additional progress has also been made in delineating the effects of smaller-scale solar wind variations on magnetospheric behavior. While some of the specific processes that mediate this coupling with the solar wind were clarified in the past decade, major questions remain regarding the spatial extent over which they operate and the conditions that control their relative importance (such as small-scale solar wind dynamic pressure variations and how they drive ULF waves regionally and globally in the magnetosphere).

9.3.4.2 Magnetosphere-Ionosphere Coupling

Coupling between the magnetosphere and the ionosphere represents a key linkage in geospace. Over the past decade, combined ground-based and space-based observations, theory, and modeling greatly advanced understanding of this coupling as well as fostering new discoveries and new areas of investigation. Empirical studies of spacecraft data established correlations between solar wind and magnetosphere-ionosphere coupling parameters. For example, solar wind density and dynamic pressure increases lead to enhanced ionospheric outflow. Empirical relationships quantified how electromagnetic energy flux into the ionosphere led to consequent outflow rates. Supporting theory has shown that producing this outflow requires a multistep process involving a combination of WPI and electromagnetic forcing. Researchers have also realized the important consequences this outflowing ionospheric plasma has on the dynamic evolution of the magnetosphere. Observations have shown how this outflow merges with plasmas of solar wind origin in the plasma sheet, creating a multi-species plasma. Theorists have shown how differently reconnection behaves in multispecies plasmas, which in turn substantially modifies its impacts on magnetospheric evolution and topology. Multifluid global-scale simulations have confirmed the major role ionospheric outflow plays in the creation of periodic substorm or so-called sawtooth intervals (Figure 9.5). Although the basic correlations and the fundamental building blocks have been established, the creation of a complete theory of outflow and a detailed understanding of their magnetospheric consequences remains a goal for the next decade.

9.3.5 System Dynamics

The past decade has witnessed a tremendous improvement in understanding how the inner magnetosphere responds to storm-time disturbances as a coherent system of coupled, mutually interacting plasmas. Imaging and global simulations have played a central role by providing quantitative contextual information that ties together single-point observations and gives much-needed global constraints for predictive models. The modern picture that has resulted is one where multiple dynamic linkages are initiated by processes with spatial scales ranging from highly localized to global.

Investigations uncovered key causal relationships between solar wind driving and inner-magnetospheric response. Changes in the north-south component of the IMF were shown to trigger the aurora, ring current injections, and the commencement or cessation of plasmaspheric erosion. Numerical models and global ENA images showed that the ring current is highly asymmetric during the main phase of storms (Figure 9.6). EUV images confirmed the predicted existence of plasmaspheric plumes (see Figure 9.6) and tracked their temporal evolution globally.

The directly driven response of the inner magnetosphere was found to engender electrodynamic coupling among different regions. For example, storm-time ring-current-ionosphere coupling profoundly distorts the inner magnetospheric field and feeds back to the ring current itself, skewing its peak toward dawn (see Figure 9.6). Moreover, subauroral polarization streams (SAPS) were identified as duskside flow channels, arising from ionospheric coupling, that maintain plumes long past the subsidence of solar wind driving. These studies affirmed early theoretical concepts,[4] quantifying just how poorly shielded the innermost magnetosphere can be during rapid changes in magnetospheric convection.

[4] See, for example, R.A. Wolf, M. Harel, R.W. Spiro, G.H. Voigt, P.H. Reiff, and C.-K. Chen, Computer simulation of inner magnetospheric dynamics for the magnetic storm of July 29, 1977, *Journal of Geophysical Research* 87:5949, 1982.

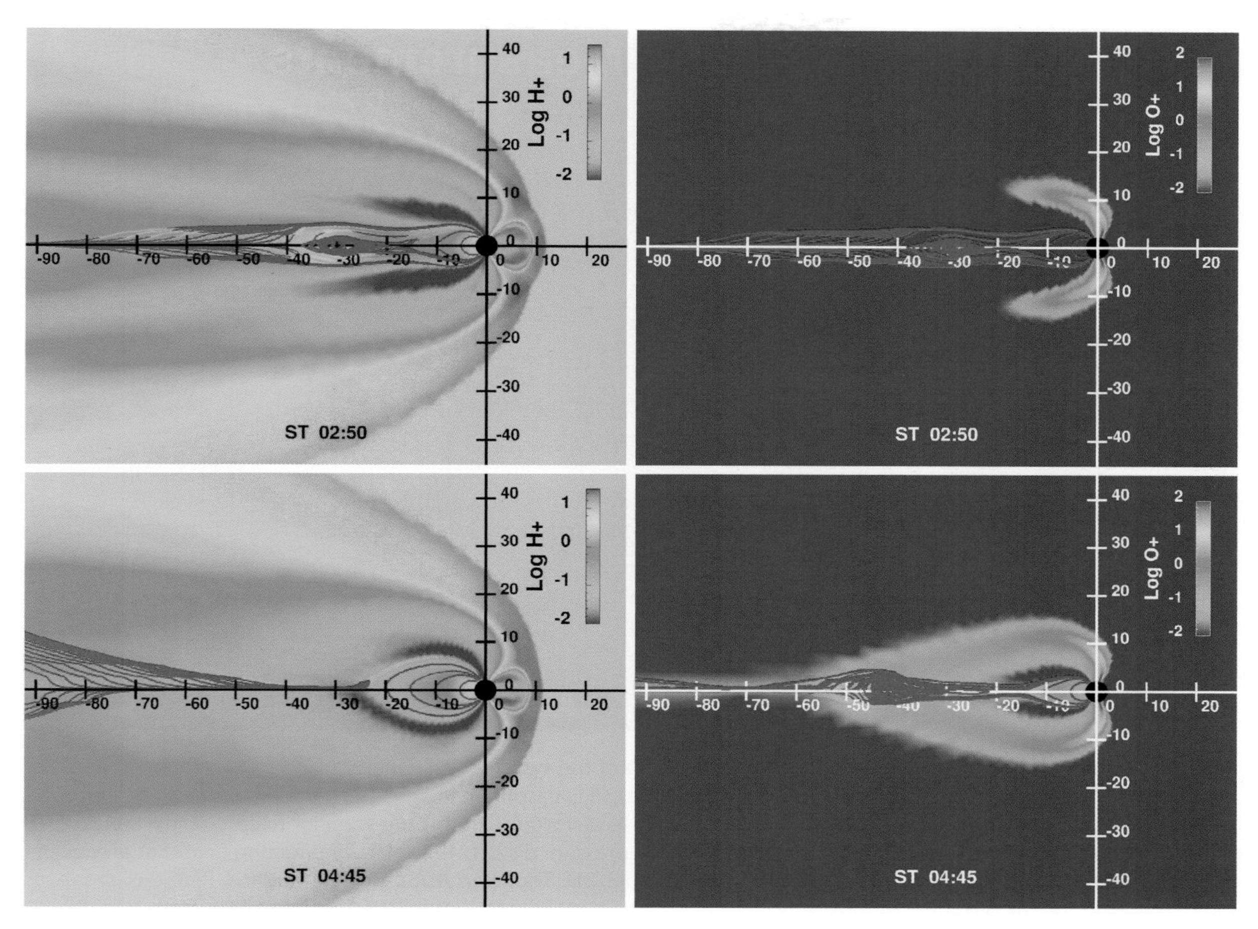

FIGURE 9.5 Multifluid MHD simulation results of substorm initiation without (left-hand panels) and with (right-hand panels) O^+ outflow from the ionosphere. In the upper panels, both simulations show a plasmoid release. In the lower panels (~2 hours later), the magnetosphere has stabilized in the simulation without O^+, while the result with O^+ shows a second plasmoid release. The addition of O^+ as a distinct fluid with a significant contribution to the mass density makes the magnetosphere repetitively unstable. SOURCE: M. Wiltberger, W. Lotko, J.G. Lyon, P. Damiano, and V. Merkin, Influence of cusp O(+) outflow on magnetotail dynamics in a multifluid MHD model of the magnetosphere, *Journal of Geophysical Research-Space Physics* 115:A00J05, 2010. Copyright 2010 American Geophysical Union. Reproduced by permission of American Geophysical Union.

The past decade of research has also identified other ways in which the ionosphere-thermosphere and magnetosphere affect each other: plasmaspheric corotation lag was discovered and interpreted as a consequence of two-way coupling between the magnetosphere-ionosphere-thermosphere regions. Recent work demonstrates that the diffuse aurora is the main source of energy deposition into the ionosphere and that relativistic precipitation can have important effects on atmospheric chemistry, including ozone depletion. Several serendipitous opportunities for imaging of both the northern and the southern aurorae simultaneously provided tests of auroral conjugacy and the dynamical processes thought to drive the aurorae.

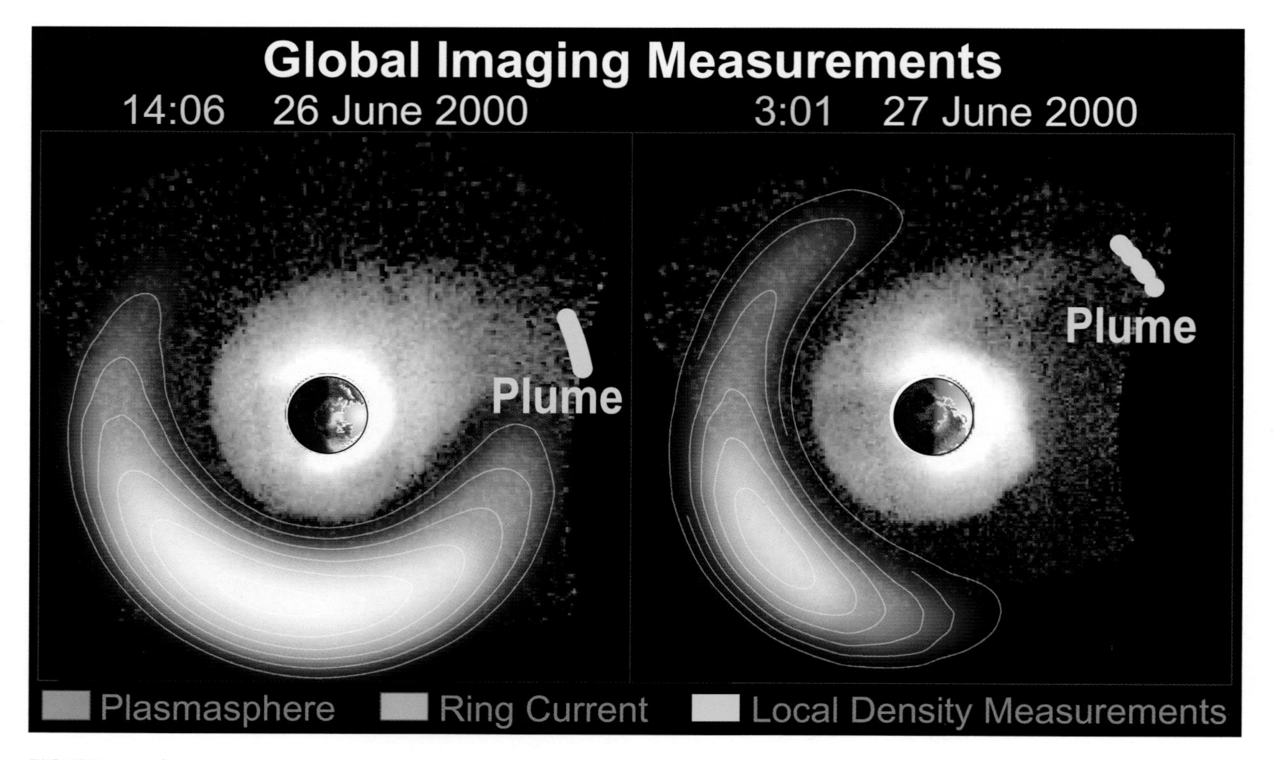

FIGURE 9.6 The past decade produced a new understanding of the system-level dynamics and *mutual* interaction of the plasmasphere (cold plasma) and ring current (warm plasma). Plasmaspheric images from the IMAGE satellite (in green, above) revealed coherent plume structure and plasmapause distortions from time-varying electric fields. IMAGE ENA images (orange, above) show the partial ring current whose pressure distorts the solar wind forcing field that erodes the plasmasphere. SOURCE: Adapted from J. Goldstein, B.R. Sandel, M.F. Thomsen, M. Spasojević, and P.H. Reiff, Simultaneous remote-sensing and in situ observations of plasmaspheric drainage plumes, *Journal of Geophysical Research* 109:A03202, doi:10.1029/2003JA010281, 2004. Copyright 2004 American Geophysical Union. Reproduced by permission of American Geophysical Union.

Coordinated studies employing imaging, remote sensing, modeling, and local measurements have demonstrated the critical importance of hot-cold plasma interactions. Imaging revealed the dynamics of ring-current-plasmasphere overlap (see Figure 9.6) and helped confirm the prediction that growth of ion cyclotron waves in this region can cause scattering losses of both energetic ions and radiation-belt electrons. Theoretical predictions that absence of cold dense plasma facilitates the acceleration of energetic electrons to relativistic energies were confirmed by observations. Studies showed that cold, dense plasma plays a pivotal role in defining the wave environment that controls energetic particle behavior, overturning the decades-old idea of a passive, quiescent plasmasphere in favor of an extremely dynamic, influential one. Other aspects of this influence included partial quenching of dayside magnetopause reconnection by plasmaspheric plumes and the creation of ionospheric density enhancements that refract and scintillate GPS signals, producing ranging errors of tens of meters.

It is clear that much progress has been made in understanding the system-level dynamics of the inner magnetosphere. In the coming decade, this knowledge must be refined and extended to encompass the entire magnetosphere-ionosphere-thermosphere system.

9.3.6 Comparative Magnetospheres

Over the past decade, researchers have made many advances toward understanding the structure, dynamics, and linkages in other planetary magnetospheres or systems with magnetospheric-like aspects.

For the inner rocky planets, there are new results on atmospheric loss at Mars, through numerical modeling of the solar wind interaction with the atmosphere, identification of Venus lightning from high-altitude radio wave measurements, and magnetospheric dynamics at Mercury, with events analogous to substorms at Earth.

In situ data have yielded a better understanding of the dynamics, structure, and linkages of Jupiter's complex magnetosphere. Flux-tube interchange processes transport Io-originating plasma outward through weak, centrifugally driven transport on the dayside. On the evening and nightside, where there is no confinement by the solar wind, this transport occurs through a more explosive centrifugal instability, leading to plasmoid loss. Observations far down the distant magnetotail revealed anti-sunward flows of plasma every few days, as well as bursts of energetic particles accelerated in regions ~200 Jupiter radii down the tail on the dusk flank. Earth-orbiting satellites imaged X-ray emissions from the auroral/polar regions resulting from capture, acceleration, and subsequent atmospheric charge exchange of highly ionized heavy solar wind ions.

There have been advances in theoretical understanding and observational tests of the impact of solar wind dynamic pressure variations on Jovian auroral emissions, and significant progress in understanding magnetospheric interactions with Jupiter's satellites, especially Io. ENA imaging demonstrated that an extensive torus of neutral gas from Europa has a significant impact on Jupiter's magnetosphere.

Extensive measurements have been made of Saturn's highly structured, interconnected, dynamical system. Magnetospheric phenomena reveal two distinct, narrow band modulations near Saturn's rotation period (Figure 9.7a). Plumes of water gas and ice crystals emanate from rifts in the south polar region of Enceladus (Figure 9.7b). Negatively charged hydrocarbon ions were discovered in Titan's ionosphere and may be important to the chemistry of Titan's upper atmosphere. Flux-tube interchange in the middle magnetosphere followed by plasmoid release in the magnetotail was revealed as the primary transport mechanisms for cold Enceladus plasma. Solar wind pressure variations strongly modulate the activity in the outer magnetosphere, including Saturn kilometric radio emission and acceleration of energetic particles in Saturn's ring current (Figure 9.7c). Saturn's rotating ring current results from both relatively symmetric centrifugal acceleration of the sub-corotating cold plasma, and from more asymmetric hot plasma pressure.

9.4 SCIENCE GOALS FOR THE COMING DECADE

Today, researchers stand on the threshold of developing a comprehensive understanding of Earth's magnetosphere, its coupled behavior, and its impacts. This understanding will enable a capability to anticipate, predict, and ameliorate the effects of variable space weather. In this section, the SWMI panel takes stock of where we are in the progression shown in Figure 9.2, and identifies the high-priority science goals that must be pursued in the coming decade. After describing each science goal, the panel discusses how their accomplishment relates to the achievement of the four decadal survey key science goals identified in Chapter 1 (see Box 9.1). Table 9.1 summarizes the expected contributions of the SWMI science goals to the decadal survey key science goals. Table 9.1 demonstrates that the discipline advances through a strategic and thoughtful combination of discovery-class observations promoting new physical models and theories, and the targeted observations needed to differentiate between competing physical theories.

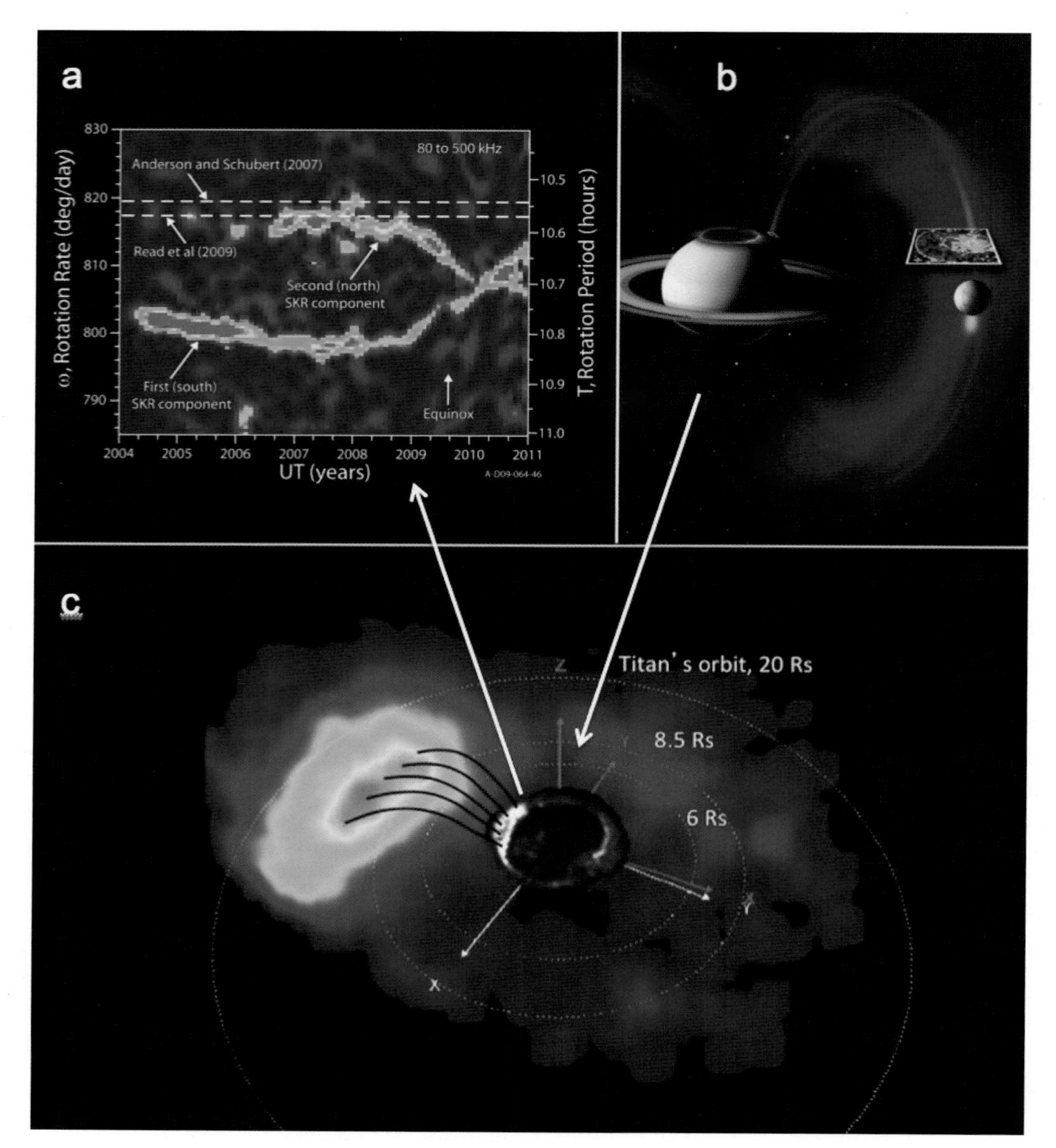

FIGURE 9.7 Saturn's Enceladus-dominated, rotating magnetosphere. (a) Saturnian kilometric radio (SKR) periodicities. (b) Enceladus, its geysers, resulting plasma, and connection to Saturn's ionosphere. (c) Return of energized plasma after tail plasmoid loss. These injections yield bright auroral displays in the same region as the SKR radio emissions. SOURCE: (a) D.A. Gurnett, J.B. Groene, A.M. Persoon, J.D. Menietti, S.-Y. Ye, W.S. Kurth, R.J. MacDowall, and A. Lecacheux, The reversal of the rotational modulation rates of the north and south components of Saturn kilometric radiation near equinox, *Geophysical Research Letters* 37:L24101, doi:10.1029/2010GL045796, 2010. Copyright 2010 American Geophysical Union. Reproduced by permission of American Geophysical Union. (b) JHUAPL/NASA/JPL/University of Colorado/Central Arizona College/SSI. (c) Adapted from D.G. Mitchell, S.M. Krimigis, C. Paranicas, P.C. Brandt, J.F. Carbary, E.C. Roelof, W.S. Kurth, D.A. Gurnett, J.T. Clarke, J.D. Nichols, J.-C. Gérard, et al., Recurrent energization of plasma in the midnight-to-dawn quadrant of Saturn's magnetosphere, and its relationship to auroral UV and radio emissions, *Planetary and Space Science* 57(14-15):1732-1742, doi:10.1016/j.pss.2009.04.002, 2009.

BOX 9.1 DECADAL SURVEY KEY SCIENCE GOALS

1. Determine the origins of the Sun's activity and predict the variations in the space environment.
2. Understand the dynamics and coupling of Earth's magnetosphere, ionosphere, and atmosphere and their response to solar and terrestrial inputs.
3. Determine the interaction of the Sun with the solar system and the interstellar medium.
4. Discover and characterize fundamental processes that occur both within the heliosphere and throughout the universe.

9.4.1 Regions

Observation from instruments on space platforms have provided researchers with a global view of the different plasma regions found in the magnetosphere and enabled a general understanding of their statistical structure and shape. During the past decade, space missions have delivered pathfinder global observations of some of the inner magnetospheric regions. These observations, from the IMAGE and TWINS satellites, were revolutionary in their global perspective but were unfortunately characterized by relatively low spatial and temporal resolution. Also during the past decade, from THEMIS and from serendipitous alignments of Heliophysics Systems Observatory satellites, researchers acquired pathfinder one-dimensional simultaneous in situ observations of the outer magnetosphere, but still have no unambiguous observations of its two-dimensional or three-dimensional structure and evolution. In sum, scientists do not know the instantaneous global and mesoscale structure of each of the various regions, nor how it evolves with time and solar wind driving.

To understand how the system as a whole behaves in response to variations in the solar wind driver requires a better view of the simultaneous evolution of the various parts of the system, leading to the first SWMI science goal for the coming decade.

9.4.1.1 SWMI Science Goal 1. Determine How the Global and Mesoscale Structures in the Magnetosphere Respond to Variable Solar Wind Forcing

Digging Deeper

Investigation into the global and mesoscale magnetospheric reaction to the solar wind is a challenging problem. Much like meteorology, the plasmas of geospace interact in a highly complex, nonlinear way. Actions and reactions feed back on each other. For example, merging of the magnetic fields of the solar wind and Earth may impose up to a few hundreds of thousands of volts across the entire magnetosphere, activating an enormous, global convection cycle that strips away tons of near-Earth plasma and drags Earthward the plasma-loaded magnetic field lines of the distant nightside magnetosphere. In response, geospace creates its own cross-scale network of intricately interconnected electrical currents and fields whose effect is anything but uniform. Partial and temporary shielding occurs in some regions, while amplification of solar wind driving occurs in other regions, although exactly where and on what timescales are poorly known. Internal feedback profoundly modifies the whole system and can outlast by hours the cessation of solar wind forcing.

Predicting the behavior of this highly coupled, self-modifying system will require powerful models that include many physical processes operating over a wide range of spatial and temporal scales. The development and validation of such models will require a strong foundation in observations of global

TABLE 9.1 Expected Level of Contribution of Each SWMI High-Priority Science Goal to Accomplishing the Decadal Survey Key Science Goals of Chapter 1

	Decadal Key Science Goals			
SWMI Science Goals	Determine the origins of the Sun's activity and predict the variations in the space environment	Understand the dynamics and coupling of Earth's magnetosphere, ionosphere, and atmosphere and their response to solar and terrestrial inputs	Determine the interaction of the Sun with the solar system and the interstellar medium	Discover and characterize fundamental processes that occur both within the heliosphere and throughout the universe
Determine how the global and mesoscale structures in the magnetosphere respond to variable solar wind forcing	Significant	Major	Minimal	Some
Identify the controlling factors that determine the dominant sources of magnetospheric plasma	Significant	Major	Minimal	Some
Understand how plasmas interact within the magnetosphere and at its boundaries	Significant	Major	Minimal	Some
Establish how energetic particles are accelerated, transported, and lost	Significant	Significant	Some	Major
Discover how magnetic reconnection is triggered and modulated	Large	Significant	Some	Major
Understand the origins and effects of turbulence and wave-particle interactions	Significant	Significant	Some	Major
Determine how magnetosphere-ionosphere-thermosphere coupling controls system-level dynamics	Significant	Major	Minimal	Some
Identify the structures, dynamics, and linkages in other planetary magnetospheric systems	Significant	Significant	Large	Major

Contribution to Action: Major (●), Large (◕), Significant (◒), Some (◔), Minimal (⊖)

response. This foundation does not yet exist; to provide it will require instant-to-instant determination of the state of the system, at both global and mesoscales. Progress in the coming decade will thus require a comprehensive set of observations that connect global-scale changes to the mesoscale currents, flows, fields, heating, and particle acceleration that modify that global response. For example, continuous, global auroral imaging would enable researchers to follow rapid storm- and substorm-driven changes down to the scale of individual auroral arcs. Nonstop global plasma imaging could resolve the minute-to-minute development of cross-scale, cross-region plasma energization and transport, erosion of the plasmasphere, and development of internal structure. Uninterrupted radar measurements could follow ionospheric flows and fields. Additional observations that would help relate global and mesoscale magnetospheric evolution to solar wind driving would include continuous global distributions of field-aligned currents linking the magnetosphere and ionosphere, detailed measurements of ionospheric ion outflow, and observations of plasma-sheet and cusp plasma conditions and composition. The key to significant progress in this area is to have these global and mesoscale observations available simultaneously.

While touching on decadal survey key science goals 1, 3, and 4, this thrust is aimed squarely at decadal survey key science goal 2. This problem is one of the most challenging scientific problems remaining in the realm of geospace, and one of the most important to solve toward the goal of providing the capability to predict the effects of solar variability on the environment and on society. The geospace community has moved ever closer to that goal over the past several decades; with the proper focus and implementation, the coming decade can more fully realize the benefits to society of this endeavor.

Linkages

Researchers know that key regions are coupled and understand the general nature of the linkages. Some involve transfer of charged particles from one region to another, while others involve electrodynamic connections, and still others involve plasma instabilities and waves. Pathfinder observations have been collected of ionospheric plasma outflows, auroral and radiation-belt precipitation into the ionosphere, solar-wind plasma entry into the magnetosphere, and signatures of nonlinear feedback in the electrodynamic coupling between the solar wind, magnetosphere, and ionosphere. However, still lacking is a quantitative understanding of the linkages, including their critically important nonlinear, feedback aspects. In addition, researchers do not understand the dependence of these linkages on the conditions within the regions nor on the variability of the driver. Furthermore, there is compelling evidence that preconditioning, system memory, and prior history of the solar wind driver confound the coupling. Since history in the solar wind often does not repeat itself (except for recurring patterns such as stream interaction regions), this presents a challenge for general understanding.

To understand how the magnetospheric system as a whole behaves in response to variations in the solar wind driver, researchers need a quantitative understanding of these linkages, including what and how conditions control them and what the feedback processes are and how they work. These needs motivate the two SWMI science goals discussed next.

9.4.1.2 SWMI Science Goal 2. Identify the Controlling Factors That Determine the Dominant Sources of Magnetospheric Plasma

Digging Deeper

There are two primary sources for magnetospheric plasma: the solar wind and the ionosphere. Solar wind plasma enters predominantly via reconnection between interplanetary and magnetospheric magnetic fields; diffusive entry constitutes a smaller contribution. Ionospheric plasma enters by flowing outward into the magnetosphere, as the result of being heated and/or directly accelerated through auroral processes and solar-wind influences. Solar wind and ionospheric plasma sources both show enhancements during geomagnetically active times, but through different causal pathways. Understanding the relative importance of these two sources as a function of time, space, and driving conditions is critical. For example, the ionospheric source is low-charge-state and heavy-ion rich, which affects magnetospheric dynamics differently than does the high-charge-state, proton-rich source, especially in storm-time ring current evolution, reconnection rates, plasma wave excitation and interactions, and instability thresholds.

Multiple observations at the magnetopause have demonstrated that reconnection between the geomagnetic field and the IMF controls solar wind entry into the magnetosphere. However, fundamental questions on this process remain. Even if given the IMF and solar wind conditions, the location and rate of reconnection still cannot be predicted. In addition, the relative importance of diffusive entry is still largely unknown even though assumed to be small most of the time. While diffusive processes are often assumed in cases where reconnection cannot explain observed entry, the conditions under which they occur have not been established and remain a mystery.

As illustrated by Figure 9.8, ionospheric outflow is a multistep process in which electromagnetic and particle inputs, driven by solar wind-magnetosphere interactions, heat the ionosphere. Waves, also driven

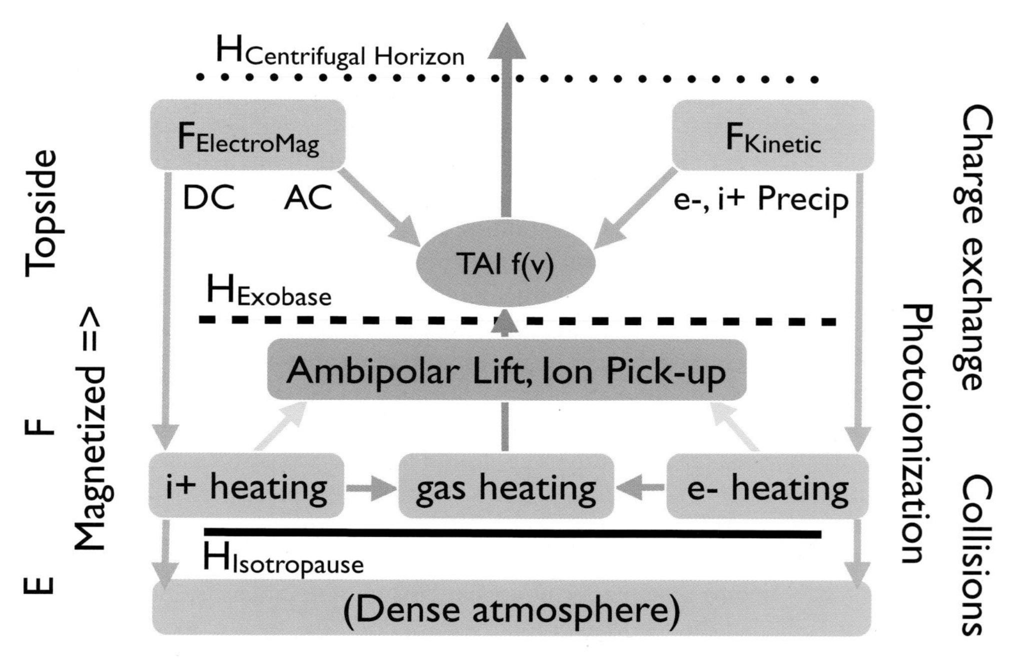

FIGURE 9.8 Schematic illustration of the outflow process. SOURCE: T.E. Moore, L. Andersson, C.R. Chappell, G.I. Ganguli, T.I. Gombosi, G.V. Khazanov, L.M. Kistler, D.J. Knudsen, M.R. Lessard, M.W. Liemohn, J.P. McFadden, et al., Mechanisms of Energetic Mass Ejection (MEME), white paper submitted to the Committee on a Decadal Strategy for Solar and Space Physics (Heliophysics). Adapted from R.J. Strangeway, R.E. Ergun, Y.-J. Su, C.W. Carlson, and R.C. Elphic, Factors controlling ionospheric outflows as observed at intermediate altitudes, *Journal of Geophysical Research* 110:A03221, 2005, doi:10.1029/ 2004JA010829, as adapted by T.E. Moore and G.V. Khazanov, Mechanisms of ionospheric mass escape, *Journal of Geophysical Research* 115:A00J13, 2010, doi:10.1029/2009JA014905. Copyright 2010 American Geophysical Union. Reproduced by permission of American Geophysical Union.

by solar wind-magnetosphere interaction, further accelerate the ions with a component that drives them outward along geomagnetic field lines. While the basic components controlling outflow are understood, quantitative relationships between solar wind conditions and energy inputs, and between energy inputs and resulting outflow, have not been established.

For example, researchers do not know how the amount of electromagnetic energy entering the ionosphere depends on the specific solar wind conditions. In recent years, the spatial and temporal distribution of Poynting flux has received much attention. Driven by both dayside and nightside reconnection and associated fast flows, the associated Poynting flux drives not only convection but also ionospheric outflows and is now recognized as an important term in the ionospheric energy balance. The spatial and temporal variability of precipitating electron flux is also not well established, and the altitude and locations where these inputs deposit their energy are not yet clear. In addition, the atmospheric and ionospheric responses are poorly quantified.

Current understanding is limited by poor knowledge of the wave environment generated by the energy input; which waves accelerate and heat the plasma; and the impact of any feedback and saturation processes. These issues must be resolved in order to develop a predictive understanding of the ionospheric response to solar wind forcing; their resolution also has strong links to decadal survey key science goals 2 and 4. Identifying the factors that control solar wind and ionospheric contributions to magnetospheric populations is fundamental to determining the dynamics of the magnetosphere, ionosphere, and atmosphere, their coupling, and the response to solar wind variability.

SWMI science goal 3 is the panel's second critical goal related to linkages between different regions and populations.

9.4.1.3 SWMI Science Goal 3. Understand How Plasmas Interact Within the Magnetosphere and at Its Boundaries

Digging Deeper

The interaction between the solar wind and magnetosphere results in energy and mass transfer across the magnetic fields at their interface. One key to understanding magnetospheric processes lies at the magnetopause. Past observations by single spacecraft or closely spaced spacecraft established that processes such as magnetic reconnection, diffusive entry, and Kelvin-Helmholtz instability operate at the magnetopause and lead to solar wind entry across the magnetopause These processes, especially reconnection, and their consequences depend on the IMF orientation, solar wind convection electric field, and solar wind pressure. Even within the magnetotail itself, significant ion densities are known to exist in the lobes, yet only a portion eventually enters into the plasma sheet; the remaining fraction escapes down the tail through the distant side of the near-Earth or mid-tail reconnection. This is another critical aspect of solar wind entry: solar wind plasma has not been at least quasi-trapped into the magnetosphere until it crossed the nightside reconnection separatrix. Due to the lack of large-scale observations of the outer magnetosphere, significant questions remain concerning the global consequences of these processes and their relative importance under different solar wind conditions.

Another crucial interface within the magnetosphere is that between the magnetotail and the inner magnetosphere. In response to solar-wind energy input, magnetotail processes produce narrow flow bursts, transporting plasmas and magnetic flux from the distant magnetotail to the inner magnetosphere. While the past decade revealed the size of individual flow channels and their temporal evolution, much remains unknown about their global occurrence properties. This limits the ability to understand the transport of magnetotail particles into the inner magnetosphere, where they interact with other particle populations. The current inability to observe the three-dimensional, time-dependent magnetotail limits the ability to understand transport processes at the inner edge of the magnetotail and their response to time-varying solar wind.

Exquisite regions of overlap between the cold plasmasphere plasma and the hotter populations of the plasma sheet, ring current, and radiation belt define conditions in the inner magnetosphere. In the regions of overlap, these disparate plasmas interact with one another, largely through intermediary electromagnetic waves, leading to important dynamical consequences: enhanced particle precipitation, large-scale instabilities, particle energization, and enhanced transport. Because systems-level measurements have been insufficient, a quantitative understanding is lacking of how these particle interactions are controlled by external driving parameters and how important they are under various conditions. To predict how the system behaves in response to variations in the solar wind requires quantitative understanding of the nature and significance of the fundamental processes driven by these overlapping, disparate plasma populations. Progress in the coming decade requires a program incorporating coordinated multipoint and/or remote-sensing global measurements, along with global numerical simulations and local theory.

This goal directly addresses decadal survey key science goal 2, with significant contributions to understanding fundamental processes (key science goal 4) and enabling prediction of magnetospheric variability (key science goal 1).

9.4.2 Universal Processes

Scientists have identified a range of important physical processes operating in different magnetospheric regions, and understand what their consequences are, locally and in some cases globally. This understanding stems from pathfinder observations of some of these processes and of the solar wind conditions under which they appear to operate. In addition, scientists eagerly anticipate information about reconnection and particle energization from the Magnetospheric Multiscale (MMS) and Radiation Belt Storm Probes (RBSP) missions. However, even with these additions to the Heliophysics Systems Observatory, adequate insight into the nature and consequences of important processes like turbulence and wave-particle interactions will still be lacking. Moreover, scientists do not know the relative effectiveness of these various processes under different conditions, nor how they turn on and off. To understand how the system as a whole behaves in response to variations in the solar wind driver, improved knowledge is needed of how various conditions control these processes, including their own nonlinear feedback. Thus, for the coming decade, including the goals for MMS and RBSP, there are three SWMI science goals (4-6 below) relating to fundamental physical processes in solar wind-magnetosphere interactions.

9.4.2.1 SWMI Science Goal 4. Establish How Energetic Particles Are Accelerated, Transported, and Lost

Digging Deeper

The flux of energetic particles in Earth's radiation belts exhibits extreme variability, with timescales ranging from a few minutes to days. However, current understanding of the underlying physical mechanisms for this variability remains incomplete. The observed variability, particularly for energetic electrons in the outer radiation belt, has been associated with pronounced changes in either the acceleration or the loss processes, both of which are enhanced during geomagnetic activity associated with solar disturbances. The current difficulty in modeling radiation belt dynamics is due to the inability to adequately quantify the variability of the dominant source and loss processes under different levels of geomagnetic activity.

In the essentially collisionless magnetosphere, energetic particles tend to behave adiabatically in the absence of perturbing influences, that is, preserving certain characteristic invariant properties of their gyro, bounce, and drift motions when the magnetic fields they move within evolve slowly relative to the timescales of these motions. Changes in the adiabatic trapped particle motion are primarily due to interactions with various plasma waves and shock fronts, which cause a violation of one or more of the adiabatic particle invariants, leading to loss to the atmosphere, exchange of energy between waves and particles, or radial transport. Waves responsible for such processes either are generated naturally during the injection of medium-energy particles into the inner magnetosphere, or are excited by macroscopic changes in the system caused by solar wind variations, interplanetary shocks, or substorm activity. Accurate modeling of the energetic particle source and loss processes thus requires a global understanding of all important waves, or shock characteristics, and the variability of their power spectra. Moreover, several of these important waves attain amplitudes at which nonlinear scattering occurs, but it is not yet known how pervasive such conditions are, nor how to incorporate them in global radiation belt models.

The radiation belt population is strongly coupled to changes in the medium-energy ring current and plasma-sheet populations, which provide a reservoir of both source particles and energy needed to accelerate a fraction of the lower-energy particles to radiation-belt energies. In this regard, RBSP stands to benefit from existing missions of the Heliophysics Systems Observatory. Electrons in the outer radiation belt are undoubtedly seeded by lower energy particles injected from beyond geostationary orbit. Missions such as Geotail, Cluster, and THEMIS provide these higher-altitude measurements needed to augment RBSP's local studies of acceleration and heating in the radiation belts. Improvements in understanding how the radiation belts respond to changes in the solar wind will require the development of numerical codes capable of simulating the development of these populations and the magnetic field distortions they produce, the incorporation of physically realistic particle scattering and diffusion into such codes, and further detailed in situ observations to establish the exact nature of the scattering. Improved measurements of the spatial distribution of the radiation belt population are needed to discriminate between wave-driven energization processes that occur locally and energization that occurs via spatial diffusion. Further observations of the prompt acceleration of electrons and ions by interplanetary shocks penetrating into the magnetosphere are also required to establish the significance of this process.

Understanding the energization of the radiation belts was one of the top-level science objectives identified in the 2003 decadal survey[5] and led not only to the impressive advances described in Section 9.3 but also to NASA's RBSP. It is expected that RBSP will provide definitive answers to many of the outstanding questions in this area, but since it has not yet launched, the SWMI panel reiterates the enduring importance of those questions and endorses anew the science objectives of the RBSP mission.

Predicting the variability of the highly energetic and thus hazardous populations of our space environment is a central part of decadal survey key science goal 1. Since this variability is explicitly determined by processes operating within the magnetosphere in response to solar wind input, this goal also enables a significant portion of decadal survey key science goal 2. Those processes, as described above, are fundamental ones that presumably operate throughout the universe, including in other planetary magnetospheres, and so accomplishing this goal will also contribute significantly to decadal survey key science goal 4.

The second critical SWMI science goal related to universal physical processes is goal 5.

9.4.2.2 SWMI Science Goal 5. Discover How Magnetic Reconnection Is Triggered and Modulated

Digging Deeper

Magnetic reconnection is a ubiquitous process in plasmas in which magnetic field lines break and reform, causing an explosion powered by magnetic field annihilation. Examples of its fundamental role include releasing the energy that drives solar flares and coronal mass ejections, coupling the solar wind

---

[5] National Research Council, *The Sun to the Earth—and Beyond: A Decadal Research Strategy in Solar and Space Physics,* The National Academies Press, Washington, D.C., 2003; and National Research Council, *The Sun to the Earth—and Beyond: Panel Reports,* The National Academies Press, Washington, D.C., 2003.

to Earth's magnetosphere, and coupling plasma in the heliosphere to that in the interstellar medium. Magnetic reconnection is a multiscale process linking global physics with microphysics, and as such is a grand challenge problem. Geospace provides a unique opportunity to study magnetic reconnection because it is one of the few locations allowing in situ measurements of the reconnection process. These measurements facilitated great progress in the past decade, but critical questions remain unsolved.

How is magnetic reconnection triggered? A fundamental question remains as to exactly how the explosion is allowed to happen in magnetic reconnection. This occurs at very tiny scales and has eluded direct measurement. What turns on reconnection or suppresses it? On the dayside magnetosphere, magnetic reconnection often happens gradually over long time periods, whereas on the nightside, it suddenly turns on, explosively releasing large amounts of energy during a substorm. Why do these two regions behave so differently? Does inflowing turbulence inhibit or actually encourage reconnection? Questions similar to these were called out as one of the highest-priority science objectives of the 2003 decadal survey and are the focus of the upcoming Magnetospheric Multiscale Mission (MMS), now in development. The present survey finds that these questions continue to be of very high importance, and the SWMI panel anticipates major progress from MMS.

In addition to the microphysics questions to be addressed by MMS, there is a second fundamental issue regarding the reconnection process: How is magnetic reconnection modulated? A critical unsolved aspect of reconnection modulation is the relative role of global conditions (boundary conditions) versus microscale physics in determining the properties of reconnection. Addressing this question will require observations at both microscale and macroscales simultaneously. Specific questions regarding modulation are: What determines the temporal characteristics of reconnection? Does reconnection spontaneously create bubbles of magnetic field periodically? Or are these bubbles due to changing inflow conditions? How do changes in boundary conditions modify reconnection? How does shear flow affect the occurrence and efficiency of reconnection? How do different asymmetric inflow or outflow conditions affect the structure and dynamics of reconnection? How do multiple ion species affect reconnection? How does inflowing turbulence modulate magnetic reconnection?

Finally, the third frontier in reconnection physics is understanding its behavior in more realistic geometries. Current understanding of magnetic reconnection is limited primarily to a simplistic two-dimensional picture where magnetic field lines break at simple lines called x-lines. However, the solar wind flow around the tear-drop-shaped magnetosphere can break these simple x-lines into points called null points joined by magnetic separator lines. Does magnetic reconnection happen primarily at null points or separator lines? How fast does reconnection happen in these more complex three-dimensional geometries?

The questions outlined above, while specifically addressing decadal survey key science goal 4, also contribute to decadal survey key science goal 2.

The third SWMI science goal relating to fundamental physical processes is goal 6.

9.4.2.3 SWMI Science Goal 6. Understand the Origins and Effects of Turbulence and Wave-Particle Interactions

Digging Deeper

Turbulence has been detected in the collisionless plasma of Earth's magnetotail, but the nature of the turbulence and its impacts on the dynamics of the magnetosphere are not known. Three major outstanding questions about the nature of the turbulence focus on its dynamics (What is the nature of the turbulent fluctuations and how do they interact to transfer energy?), its driving (What process supplies power to the turbulence?), and its dissipation (What physical processes extract energy from the turbulence and where does the power go?). Three outstanding questions about the impacts deal with three universal properties of turbulence: mixing (Is there evidence for mixing?), energy cascade (How do the turbulent fluctuations interact to transfer energy?), and heating (What is the plasma heating rate and is the heating important?). It is not known whether turbulence affects transport and entropy conservation, and it is not known whether turbulence leads to reconnection events in the magnetotail or alters large-scale dynamics there.

The magnetosheath flow around the magnetosphere is also turbulent, and the same three questions about the nature of the turbulence apply: dynamics, driving, and dissipation. There is an additional question for the turbulent magnetosheath flow: Is eddy viscosity important for the coupling of the flow to the magnetosphere?

The turbulence inside and around the magnetosphere has potential impact on the dynamics of the magnetosphere and its response to solar input (decadal survey key science goal 2). Turbulence in both the magnetosheath and the magnetotail provides unique opportunities to discover and characterize fundamental processes that occur here and throughout the universe (decadal survey key science goal 4).

Wave-particle interactions are ubiquitous in the magnetosphere. Waves clearly play an important role in both energization and loss of ring current and radiation belt particles, but it is necessary to establish which WPIs are most effective. In the inner magnetosphere, current knowledge is limited to the statistical distributions of waves, and so moving forward will require a better characterization of the spatial-temporal structure of the waves and how they are produced. Researchers must understand how wave production is modulated by macroscale plasma properties and how high-frequency waves are modulated by lower-frequency, large-scale fluctuations. They must determine how particle populations are modified by wave-particle interactions and understand how that feeds back on wave generation. Finally, a better understanding is needed of the importance and detailed physics of nonlinear interactions with the large-amplitude waves that are observed.

The instabilities that occur in our solar system are expected to also occur throughout the universe, affecting particle populations in many different environments. Advancing understanding of the waves and their interaction with particles at Earth thus enables decadal survey key science goal 4. Moreover, determining the roles played by waves in energization and loss of magnetospheric particles will be necessary for advancing understanding to the point of predictive capability for the near-Earth environment, enabling decadal survey key science goals 1 and 2.

9.4.3 System Dynamics

Serendipitous multipoint measurements by the Heliophysics Systems Observatory plus the global perspective from the IMAGE and TWINS missions have provided a greater appreciation for the degree to which the various regions and processes interact. Still, researchers have only a rudimentary understanding of how all the pieces fit together to determine the global magnetospheric response to solar wind variability. In particular, we lack a full understanding of how nonlinear feedback between the ionosphere and magnetosphere regulates magnetospheric dynamics. This motivates SWMI science goal 7.

9.4.3.1 SWMI Science Goal 7. Determine How Magnetosphere-Ionosphere-Thermosphere Coupling Controls System-Level Dynamics

Digging Deeper

In previous studies of the magnetosphere-ionosphere-thermosphere system, signs of nonlinear feedback were observed, and that the coupling affects the global transport of plasma is known. This nonlinear coupling involves both electrodynamic communication and mass exchange between these regions, but what controls the nonlinearities is not known. At a fundamental level, the mapping between a location in the magnetosphere to the ionosphere can vary among models by more than 100 km, so ionospheric signatures cannot be properly attributed to the associated magnetospheric regions and processes. Since strong and variable currents significantly distort the magnetospheric magnetic field by different amounts under different conditions, determining this mapping is a challenging but important problem.

There are numerous examples of how magnetosphere-ionosphere-thermosphere coupling plays out in the system-level response, both through electrodynamics and through mass coupling.

On the electrodynamic side, the strength of the convective motion in the ionosphere saturates for high levels of solar wind driving. The exact physical mechanisms for this saturation effect remain unclear, but ionospheric conductance certainly plays a key role. As another example, the electrodynamic coupling with the magnetosphere extends down into the thermosphere, where the ionospheric convection can drive neutral-particle motions that in turn influence ionospheric drifts after the solar wind driving is reduced. These plasma drifts then are communicated back into the magnetosphere, where they affect the shape and evolution of the plasmasphere. Because the plasmasphere dominates the inner magnetospheric mass content and therefore the plasma wave properties, the location of the plasmaspheric edge dramatically influences the acceleration and loss processes of the ring current and radiation belts. Furthermore, it has been shown that strong field-aligned currents closing the partial ring current near the plasmasphere boundary can lead to the formation of SAPS, which have significant impacts throughout the system, including disturbances that propagate in the thermosphere down to equatorial latitudes. Finally, "slippage" between the intrinsic field of the rotating Earth and the magnetosphere is of fundamental interest but still limited basic understanding, perhaps with applications to concepts in coronal physics of interchange reconnection.

On the mass coupling side, it is known that ionospheric plasma is a vital component of the magnetosphere. Observations have shown that the ionosphere can be a dominant mass source for the energetic ring current during magnetospheric storms. Theoretical results indicate that this plasma will alter the wave processes as well as the rates of magnetic reconnection. How this plays out in the macroscopic evolution of the system, and whether it helps determine the global mode of magnetospheric behavior, are not fully understood.

Understanding the dynamic behavior of the coupled system is central to completing decadal survey key science goal 2 and undergirds the ability to make quantitative predictions about the space environment, making this goal also relevant to decadal survey key science goal 1. Furthermore, understanding system-level dynamics will help advance knowledge of the fundamental processes that couple the regions of geospace, providing a connection to decadal survey key science goal 4 as well.

9.4.4 Comparative Magnetospheres

Fifty years of robotic exploration of the solar system have led to flybys of all the major planets and orbital insertion around all but Uranus and Neptune.[6] Researchers have found intrinsic magnetospheres at Mercury, Jupiter, Saturn, Uranus, Neptune, and Ganymede; induced magnetospheres at Venus and comets; and mini-intrinsic magnetospheres on Mars. Pathfinder magnetospheric observations have been made for all of these planets, but comprehensive measurements have been made only at Jupiter and Saturn and presently at Mercury by the MESSENGER mission. Even for Jupiter and Saturn, fundamental magnetospheric questions remain unanswered, including the degree of solar wind influence on the structure and dynamics. For the other planets, especially the outer ice giants, nothing is known about the dynamics or variability. A comprehensive understanding of magnetospheric physics requires measurements from a complete suite of magnetospheric instruments from satellites in orbit around other planets, ideally with

[6]Pluto is no longer considered a major planet; a flyby of Pluto is anticipated in January 2015 by NASA's New Horizons spacecraft.

simultaneous information about upstream solar wind conditions. In recognition of the relatively immature current understanding of other planetary magnetospheres, the corresponding critical science goal for the coming decade is SWMI science goal 8.

9.4.4.1 SWMI Science Goal 8. Identify the Structures, Dynamics, and Linkages in Other Planetary Magnetospheric Systems

Digging Deeper

Six planets in our solar system have strong internal magnetic fields and associated magnetospheres: Mercury, Earth, Jupiter, Saturn, Uranus, and Neptune. The sizes of these magnetospheres vary considerably (from ~1.5 to ~100 planetary radii), and their plasma and field structures are quite different, for reasons scientists have just begun to discover and explore.

At Earth the main plasma sources are solar wind capture and ionization of the upper atmosphere, and magnetospheric dynamics are driven largely by interactions with the solar wind. In other planetary magnetospheres there is considerable variety in both plasma sources and drivers. With no atmosphere or moons, Mercury's small magnetosphere has no significant internal plasma source (aside from a small population of sputtered material from Mercury's surface), and interaction with the solar wind is probably the primary source of energization and structure of the magnetospheric plasma. Giant-planet magnetospheres such as those of Jupiter and Saturn have dramatically different plasma sources and dynamics. Neutral gases produced by volcanic activity on Jupiter's moon Io and water geysers on Saturn's moon Enceladus provide the main internal plasma sources. The rapidly rotating planetary magnetic fields pick up and accelerate new plasma to form dense, spinning plasma disks whose outer edges are flung outward by centrifugal force and replaced by hotter, more tenuous bubbles of plasma that are drawn inward to fill the void. In these rotationally dominated giant magnetospheres, major, fundamental mysteries remain concerning time-variable magnetospheric rotation rates, poorly understood periodicities, plasma transport, and magnetosphere-ionosphere coupling.

Similar effects are expected to occur at Uranus and Neptune, although currently very little is known about the internal structure of the magnetospheres of these planets because of the limited data from the Voyager 2 flybys. Though their small icy moons are not believed to be primary plasma sources, thorough exploration of these magnetospheres will undoubtedly yield major new discoveries. The magnetospheres of Uranus and Neptune are fundamentally different from others in our solar system, in that their magnetic dipole axes are tilted by 59° and 47° relative to their rotational axes—with Uranus's rotation axis lying uniquely very close to the orbital plane of the planet. These large tilts are believed to cause enormous rotationally induced variations and create complicated and unusually dynamic internal plasma structures—which may change profoundly in less than 1 planetary day. So far, the only existing measurements near these planets, from the single flyby of Voyager 2, were too limited to adequately resolve these complex rotational variations.

The space environments of other planets offer natural laboratories for exploration of a wider range of structures and dynamic linkages than are found at Earth. Scientific investigation of other planets enriches understanding of how fundamental processes manifest themselves in real space environments. Discoveries of unique features of other planetary magnetospheres—such as moons with volcanoes and geysers, and drastically tilted magnetic and rotational axes—hint at the variety of new phenomena and challenges to be found as human society expands into the larger universe. These discoveries are universal, applicable to all four decadal survey key science goals, but in particular, comparative magnetospheric studies document the variety of ways that the Sun interacts with the solar system (3) and enable exploration of fundamental processes over a broad range of conditions inaccessible at Earth (4).

9.4.5 Summary

Today, researchers are poised to acquire a comprehensive understanding of the magnetospheric system (Figure 9.9). The SWMI panel has identified eight critical science goals for research in solar wind-magnetosphere interactions that follow naturally from progress over the past decade. Achieving these eight SWMI science goals will contribute significantly to accomplishing the decadal survey key science goals (see Table 9.1). Reaching these goals will lead to a much-improved understanding of the coupled behavior of the magnetosphere and ionosphere. It will shed light on some of the most important physical processes operating in the universe, and it will provide a strong foundation for predicting magnetospheric behavior with societal impact.

Two of the SWMI science goals (4 and 5) will see major contributions from the RBSP and MMS missions already in development as a result of the 2003 decadal survey. Continued analysis of existing data sets can make contributions to all of the goals. However, to fully achieve these goals, new measurements are needed as well as complementary and important contributions from existing missions that provide

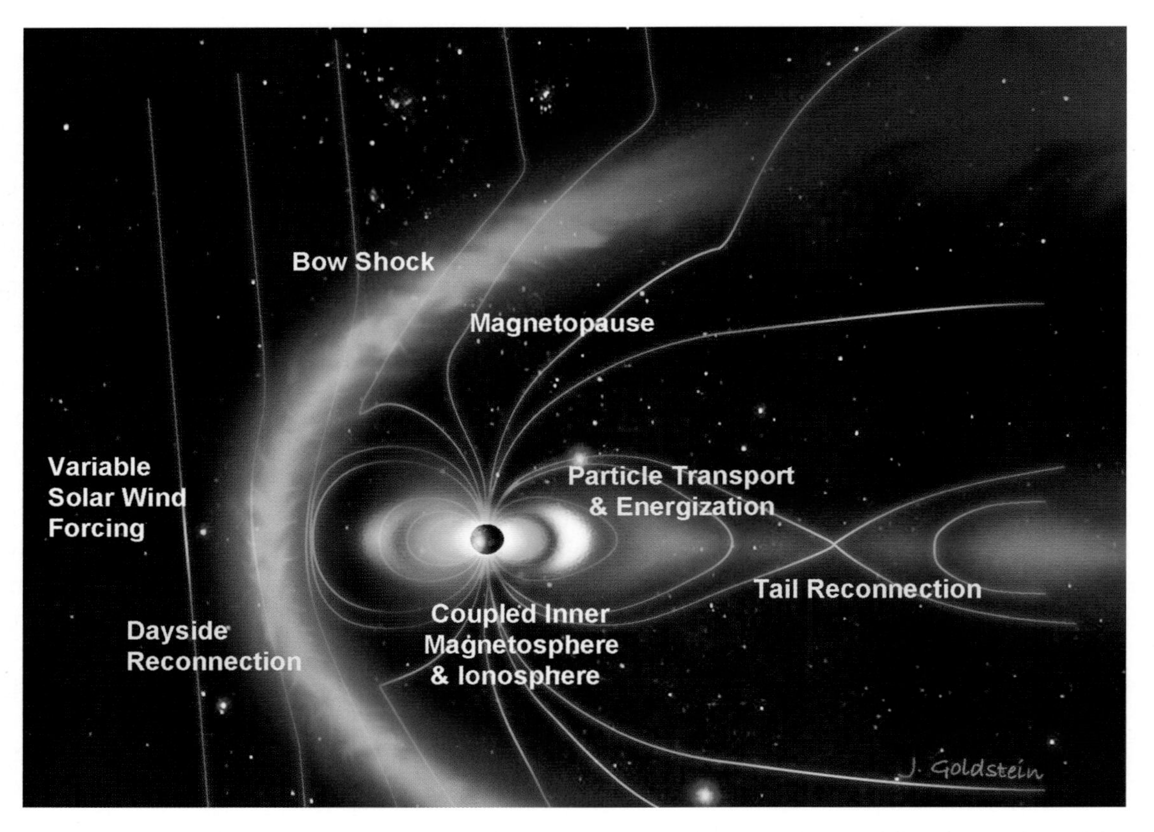

FIGURE 9.9 Illustration of the critical processes that drive the magnetosphere. To achieve a full understanding of the complex, coupled, and dynamic magnetosphere, it is essential to understand, by using a combination of imaging and in situ measurements, how global and mesoscale structures in the magnetosphere respond to variable solar wind forcing, and how plasmas and processes interact within the magnetosphere and at its outer and inner boundaries. SOURCE: Courtesy of Jerry Goldstein, Southwest Research Institute.

information about the external boundary conditions in geospace. As noted before, MMS and RBSP science objectives are extended when pursued in combination with other ongoing magnetospheric missions (Cluster, Geotail, THEMIS). Some will require strategic missions, while others can be acquired with smaller platforms, including Explorers, CubeSats, and commercial satellites. In the next section the initiatives necessary to accomplish the SWMI critical science goals are outlined.

9.5 PRIORITIZED IMPERATIVES

9.5.1 Introduction

The science goals described in Section 9.4 are ambitious, and their accomplishment will require intelligent use of the diverse research tools that are available:

Strategic missions. Strategic missions are the centerpiece of research in solar and space physics, requiring the largest budget and enabling researchers to address major portions of high-priority science that are not accessible by other tools.

Explorers. In science per dollar, NASA's Explorer program has been extremely successful. Small and medium-class missions such as SAMPEX, IMAGE, TWINS, and THEMIS have significantly advanced solar and space physics, and these missions have been productive beyond their prime mission, producing even more results than originally proposed.

Suborbital program. The suborbital program includes rocket and balloon investigations, which do targeted science and can augment strategic missions, increasing total science return. For example, the upcoming BARREL balloon experiment will enhance RBSP by providing measurements of relativistic electron precipitation.

Small platforms. In the coming decade, even smaller platforms will continue to advance technology and will also perform focused science investigations. Three magnetospheric missions are currently under development as part of NSF's new CubeSat program: CINEMA will image the ring current and measure particle precipitation; the CSSWE experiment measures electron and ion precipitation using a miniaturized RBSP REPT instrument; and the FIREBIRD mission will measure the spatial scale of relativistic electron microbursts that are part of bursty radiation belt losses.

Hosted payloads and commercial alternatives. In recent years, there has been increasing awareness of the potential to use alternative spacecraft and procurement methods to lower the cost and increase the frequency of opportunity of solar and space physics missions. These include hosted payloads on commercial satellites, the purchase of "commercial clone" spacecraft, and "data buy" (also called service-level agreement), whereby partial mission costs are paid upfront and the data are bought back after the spacecraft is successfully on orbit. Hosted payloads have already been successfully implemented. For example, the TWINS mission carries out stereo ENA imaging from hosted experiments on non-NASA, U.S.-government spacecraft. Such hosted payloads have the potential for supporting multipoint measurements at a significantly reduced cost, enabling the vision for an ever-more-capable Heliophysics Systems Observatory.

Heliophysics Systems Observatory. As amply demonstrated by the sampling of scientific accomplishments from the past decade presented above, major progress toward the SWMI panel science goals can be expected from analysis of data from the NASA and other agency spacecraft already in orbit. Together with new missions as they come on line, these assets constitute the space-based elements of the Heliophysics Systems Observatory, providing multipoint simultaneous measurements essential to advancing the SWMI science goals.

Ground-based observations. Ground-based measurements are commonly combined with spacecraft measurements to understand magnetospheric processes. In some cases, it has proven beneficial to install additional ground-based instruments in support of a specific spacecraft mission. This is exemplified most recently by the THEMIS mission, which included a 20-camera all-sky imaging array and 21 magnetometers installed in the northern United States and Canada.

Theory and modeling. The complexity of the solar wind-magnetosphere-ionosphere system, combined with its vastness, guarantees that the system will be greatly undersampled regardless of the number of spacecraft placed in orbit. While global remote-sensing capabilities help, they introduce complexities of their own, such as the fact that they often view an optically thin medium. Thus, to fully exploit the variety of available measurements and to put them into the proper interpretive physical perspective, theory and modeling are crucial. The enormous improvements in numerical modeling capabilities over the past decades have led to major progress in physical understanding of the system; with appropriate investment, these improvements will continue through the next decade.

Laboratory experiments. With spacecraft measurements, scientists observe whatever behavior nature provides. There is no way to repeat an experiment or to isolate a single parameter for investigation in the classical model of physical experimentation. This can be remedied by supplementing satellite observations with appropriately designed laboratory experiments, which can reveal significant aspects of the physical processes operating in space. Such experiments also provide a touchstone for testing numerical models.

Grants programs. Finally, the heartbeat of space physics is the set of grants programs administered by NASA and NSF, which underlie essentially all the progress that emerges from the various data sources. These relatively small, investigator-led studies enable researchers to exploit the data from missions and connect them to relevant theoretical frameworks.

The SWMI panel strongly supports the survey committee's conclusion that the key science goals identified in this decadal survey can be most effectively accomplished with a well-balanced program that uses the full spectrum of implementation options. This same approach is needed for the achievement of the SWMI panel goals discussed in this chapter. The optimum balance is of course challenging to identify, but in the course of this decadal survey, the SWMI panel identified a number of areas where new attention or enhanced resources could significantly increase the ability to deliver the important science goals outlined above. Seeking and protecting the appropriate balance of these capabilities is the overarching theme of the imperatives the SWMI panel believes are required to successfully address the critical SWMI science goals outlined above. These imperatives fall into three categories:

- Missions,
- DRIVE-related initiatives (see Chapter 4 for DRIVE description and discussion), and
- Space weather.

The remainder of this section addresses the SWMI panel's imperatives in each of these categories.

9.5.2 Missions

One of the central elements in a robust program of scientific research in solar wind-magnetosphere interactions must be an ongoing sequence of strategic missions obtaining in-space measurements targeted at the highest-priority science objectives. Identification of new mission concepts to address the SWMI science objectives is an important element of this decadal survey. These new missions and the science they address must be considered against the backdrop of accomplishments from past missions, as well as the science that will be addressed by missions that are currently in development and missions that are still operating. Below, the SWMI panel reviews expectations from missions currently under development or still operating and considers the central role played by Explorers, PI-led missions that afford the maximum agility in identifying new and exciting science. The panel then presents the new strategic mission concepts that it found particularly compelling. Finally, some promising concepts are mentioned that will require significant development efforts in the coming decade to enable their application to the outstanding SWMI science problems of the future.

9.5.2.1 Missions Under Development

MMS

The Magnetospheric Multiscale Mission currently under development is designed to address key parts of SWMI critical science goal 5, to "discover how magnetic reconnection is triggered and modulated." Earth's magnetosphere provides one of the few opportunities to observe magnetic reconnection with in situ measurements. Specifically, MMS will focus on the kinetic microphysics of magnetic reconnection that occurs in tiny boundary layers at electron scales. Measuring these boundary layers requires multiple spacecraft with unprecedented spatial and temporal resolution. MMS is poised to provide the first glimpse of these electron layers and continues to be a high science priority for the coming decade.

RBSP/BARREL

The Radiation Belt Storm Probes mission currently in development directly addresses this decade's SWMI critical science goal 4, to "establish how energetic particles are accelerated, transported, and lost." To do so, RBSP's mission objectives are threefold: (1) to discover the relative importance of various candidate mechanisms and when and where they act to accelerate and transport electrons and ions; (2) to quantify the balance between competing acceleration and loss mechanisms that determine time-varying radiation belt intensity and spatial distribution; and (3) to understand—to the point of predictability—how the radiation belts change dynamically. The instruments on the two RBSP spacecraft will provide the measurements needed to characterize and quantify the processes that produce relativistic ions and electrons. They will measure the properties of charged particles that comprise Earth's radiation belts and the plasma waves that interact with them, the large-scale electric fields that transport them, and the magnetic field that guides them, all of which are required to develop a predictive understanding of radiation belt dynamics. The mission promises to make breakthrough discoveries and to rewrite the textbook on universal planetary radiation belt physics.

Over the past decade, the importance of relativistic electron losses from Earth's radiation belts has also become increasingly clear. There is evidence that the radiation belts are emptied during the main phase of geomagnetic storms before being refilled by newly accelerated particles. However, the relative importance of atmospheric precipitation losses versus magnetopause losses is not known, and the processes that scatter relativistic particles into the atmosphere have not been experimentally validated. BARREL (Balloon Array for RBSP Relativistic Electron Losses) will augment RBSP by providing a low-altitude component to quantify electron precipitation. BARREL consists of several balloon campaigns during which an array of

X-ray detectors will be launched on stratospheric balloons to measure electron precipitation in conjunction with RBSP. These measurements will quantify the precipitation loss rate, will probe the global spatial structure of energetic precipitation for the first time, and, when combined with RBSP measurements, will allow for quantitative tests of wave-particle interaction theories.

The critical importance of these two missions for achieving some of the panel's high-priority science objectives for the coming decade leads to the following SWMI imperative:

SWMI Imperative: Complete the strategic missions that are currently in development (MMS, RBSP/BARREL) as cost-effectively as possible.

Solar Probe Plus and Solar Orbiter

These missions, while aimed directly at solar and heliospheric science objectives, are also likely to shed light on fundamental physical processes that are high on the list of science objectives for SWMI, namely, particle acceleration, reconnection, turbulence, and wave-particle interactions. Insights gained about how these processes work near the Sun can potentially help advance understanding of how they operate in near-Earth space, thereby helping address SWMI critical science goals 4 through 6.

9.5.2.2 Heliophysics Systems Observatory

The globally coupled nature of the solar wind-magnetosphere-ionosphere system demands simultaneous measurements of related phenomena in widely spaced locations. Thus, crucial information comes from combining observations from existing operating satellites, both from NASA (ACE, TWINS, SAMPEX, THEMIS, Cluster, planetary missions, and so on) and from other agencies and other nations (e.g., GOES, LANL, DMSP, POES). The demonstrated value of these existing assets far surpasses their original intended use, and it is extremely important that they continue to be supported for the contributions they can make to the evolving science objectives.

In the area of comparative magnetospheres, Juno will enter its prime mission phase when it arrives at Jupiter in 2016, while Cassini at Saturn is approved for a final mission extension to 2017, and Messenger will complete its prime mission early in the decade. Past and current missions continue to provide deep insights into general solar wind magnetosphere interactions. For example, Ganymede's Alfvén wings have led to modern theories of Earth's own polar cap potential saturation mechanism; Saturn's explosive energy releases have much in common with substorm injections at Earth; and Jupiter's interchange motions enabling convection under Io's mass loading have led to similar theories pertaining to inward penetration of fast reconnection flows. As is the case for Earth-orbiting satellites, extended missions for planetary missions that continue to return valuable science data are strongly encouraged.

SWMI Imperative: A high priority of the panel is to ensure strong continued support for existing satellite assets that can still contribute significantly to high-priority science objectives. For example, careful optimization of existing assets to address RBSP and MMS objectives will result in a significant gain in the science return from these missions.

9.5.2.3 Explorers and Alternate Platforms

Since the inception of the space age, the Explorer program has been a mainstay of science return, including significant contributions to the progress outlined in Section 9.3. The Explorer missions, with science objectives and implementations identified competitively, provide scientific agility and the ability to tap into the creativity of the science community. Thus, the SWMI panel supports a significant enhancement of resources dedicated to the Explorer program. This includes increases in the flight cadence of missions in this category, as well as launches of opportunity. The panel also supports the extension of the program limits to include so-called tiny Explorers (TEX) (~$50 million) and enhanced MIDEX (~$400 million) options, but emphasizes the importance of maintaining a high mission cadence.

SWMI Imperative: Enhance the resources dedicated to the Explorer program and broaden the range of cost categories.

The past decades have further demonstrated that suborbital missions and, more recently, CubeSats (a constellation of individual CubeSats might constitute a TEX mission) can make significant and very cost-effective contributions to addressing pressing science questions. They also serve as an instrument development platform and as a training ground for the future workforce. The SWMI panel encourages continued support for science investigations conducted on these small, cost-effective platforms. Moreover, the panel also calls on NASA and other agencies to develop an effective mechanism for supporting payloads on commercial vehicles, consistent with U.S. national space policy. Ideally, such a mechanism would allow the rapid development and deployment that are often required by these types of missions. The SWMI panel supports continuation of a cost-effective management program that applies risk/benefit tools appropriate to this cost category.

SWMI Imperative: Develop a mechanism within NASA to support rapid development, deployment, and utilization of science payloads on commercial vehicles and other missions of opportunity. NSF and DOD efforts in this regard are also encouraged.

9.5.2.4 New Strategic Missions

Analysis of the rich set of data derived from ongoing missions and missions in development, combined with new approaches that will emerge from the Explorer and other smaller mission lines, will provide a firm foundation for accomplishing the SWMI critical science goals outlined in this chapter for the coming decade. However, attaining those goals will also require crucial new observations that have previously been impossible. To obtain those essential observations will require investing in new strategic missions, targeted at the highest-priority science.

In the early stages of this decadal survey, the scientific community submitted a large number of mission concepts addressing different outstanding science questions. In evaluating these concepts, the panel considered the degree to which they address the SWMI science goals. The panel also weighed feasibility issues such as technology readiness and probable cost. Below, the science objectives and implementation strategies for the three top-ranked mission concepts considered by the SWMI panel are summarized. The panel emphasizes that, although it assigned them names for ease of discussion, these are only notional missions, devised to address the panel's science objectives, with feasible technology and within a reasonable budget for a strategic mission. For the purposes of this study, it is the science objectives for the missions, not their specific implementations, that are most important.

Magnetosphere Energetics, Dynamics, and Ionospheric Coupling (MEDICI)

MEDICI (Figure 9.10) is the SWMI panel's highest-ranked new strategic mission concept and the one it most strongly urges as a new start in the coming decade.

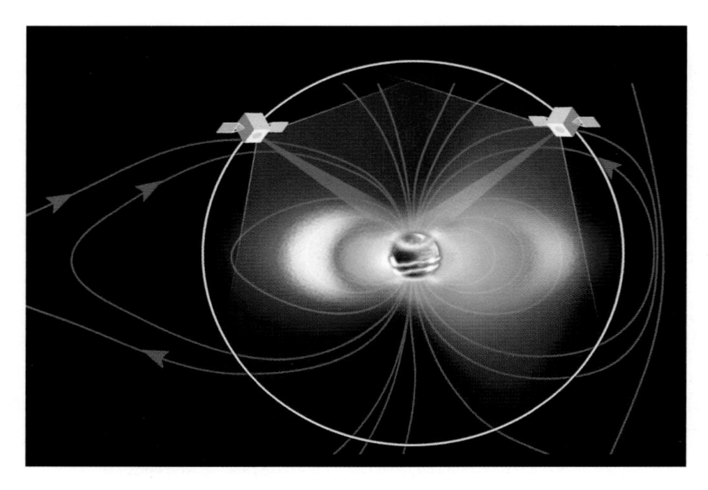

FIGURE 9.10 MEDICI targets complex, coupled, and interconnected multiscale behavior of the magnetosphere-ionosphere system by providing high-resolution, global, continuous three-dimensional images of the ring current (orange), plasmasphere (green), aurora, and ionospheric-thermospheric dynamics and flows as well as multipoint in situ measurements. SOURCE: Courtesy of Jerry Goldstein, Southwest Research Institute.

Science Objectives. Geospace is intrinsically interconnected over diverse scales of space and time. Plasma and fields in the ionosphere and magnetosphere interact, and multiple processes compete simultaneously. Observation of the relationships among components is critical to understand and characterize collective behavior of this complex system across a broad range of spatial scales. MEDICI's science questions address how the magnetosphere-ionosphere-thermosphere system is coupled and responds to solar and magnetospheric forcing. In particular, this mission concept directly addresses SWMI critical science goals 1 and 4, with very significant contributions as well to science goals 2 and 3 (Table 9.2).

MEDICI provides definitive, comprehensive answers to two fundamental science questions that have been outstanding for decades: (1) How are magnetospheric and ionospheric plasma transported and accelerated by solar wind forcing and magnetosphere-ionosphere coupling? (2) How do magnetospheric and ionospheric plasma pressure and currents drive cross-scale electric and magnetic fields, and how do these fields in turn govern the plasma dynamics? Each of these two linked questions focuses on a crucial aspect of the coupled dynamics of geospace. The first question looks at plasma transport: How are the cross-scale, dynamic, three-dimensional plasma structures of the ring current, plasmasphere, and aurora reshaped by acceleration and transport, what controls when and where ionospheric outflow occurs, and what are the cross-scale effects on the system? The second question targets the electrodynamics of magnetosphere-ionosphere coupling: What are the cross-scale, interhemispheric structure and timing of currents and fields that mediate magnetosphere-ionosphere coupling, and how do these MI coupling electromagnetic fields feed back into the system to affect the plasmas that generated them?

Mission Concept. MEDICI is a cross-scale science mission concept that uses both high-resolution stereo imaging and multipoint in situ measurements; as described in Part I of the report, MEDICI is an STP-class strategic mission, but PI-led, following the Planetary Division's Discovery mission class. It also incorporates an array of contemporaneously existing ground-based and orbiting observatories. MEDICI employs two spacecraft that share a high circular orbit (see Figure 9.10), each hosting multispectral imagers,

TABLE 9.2 Level of MEDICI Contributions Toward Achieving SWMI High-Priority Science Goals

	MEDICI Contribution
Goal 1: Determine how the global and mesoscale structures in the magnetosphere respond to variable solar wind forcing.	Major
Goal 2: Identify the controlling factors that determine the dominant sources of magnetospheric plasma.	Large
Goal 3: Understand how plasmas interact within the magnetosphere and at its boundaries.	Large
Goal 4: Determine how magnetosphere-ionosphere-thermosphere coupling controls system-level dynamics.	Large
Goal 5: Establish how energetic particles are accelerated, transported, and lost.	Significant
Goal 6: Discover how magnetic reconnection is triggered and modulated.	Minimal
Goal 7: Understand the origins and effects of turbulence and wave-particle interactions.	Some
Goal 8: Identify the structures, dynamics, and linkages in other planetary magnetospheric systems.	Minimal

Contribution to Goal: Major, Large, Significant, Some, Minimal

magnetometers, and particle instruments. This science payload captures the dynamics of the ring current, plasmasphere, aurora, and ionospheric-thermospheric plasma redistribution (Figure 9.11). In combination with ground-based and low-Earth-orbit data that yield detailed information on field-aligned currents, ionospheric electron densities, temperatures, and flows, MEDICI's imagers and onboard in situ instruments will provide the means to link the global scale magnetospheric state with detailed ionospheric conditions and will yield data for global model validation.

MEDICI takes a major step forward in geospace imaging, by combining and improving crucial elements from several prior missions. The first multispectral, stereo geospace plasma imaging will reveal the three-dimensional structure of cardinal geospace plasmas and provide conjugate views of the northern and southern aurorae. Simultaneous two-point in situ measurements are closely coordinated with plasma imaging from state-of-the-art instruments to uncover transport and electrodynamic connections at different spatial scales throughout the magnetosphere-ionosphere-thermosphere system, enabling major new insights into cross-scale dynamics and complexity in geospace.

The MEDICI mission uses two identical nadir-viewing spacecraft in a shared 8-R_E circular polar orbit with adjustable orbital phase separation (between 60° and 180° separation along track) to enable global stereo, multispectral imaging and simultaneous in situ observations. Circular orbits that avoid the most intense radiation environments provide continuous imaging and in situ measurements and enable a long-duration (up to 10-year) lifetime well beyond the required 2-year mission. MEDICI requires no new technology development; all instruments have high heritage.

Each of MEDICI's key measurement goals contributes essential information about cross-scale geospace dynamics.

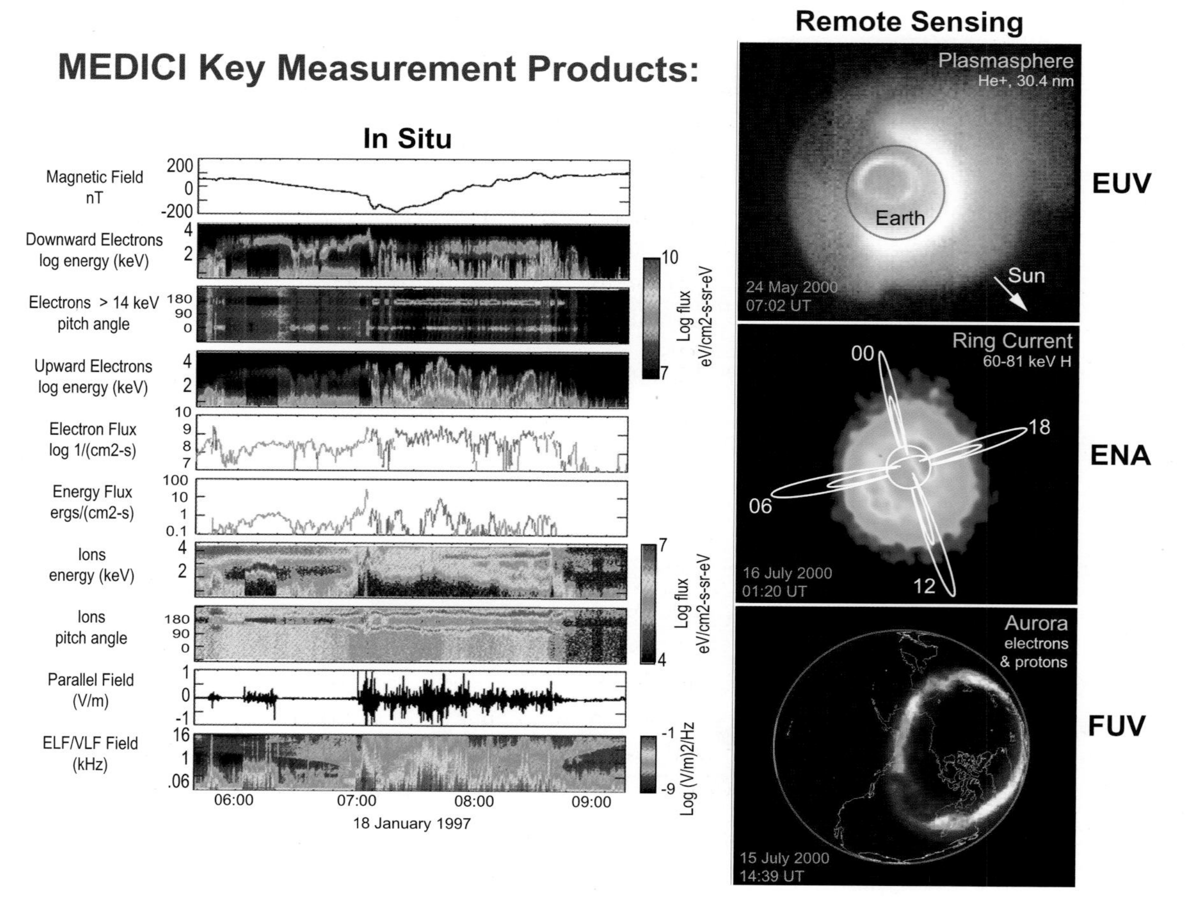

FIGURE 9.11 Examples of pathfinder observations for key MEDICI measurements. (*Right*) Multispectral, stereo, global imaging of hot and cold plasmas and auroral precipitation from IMAGE will be combined with (*left*) in situ observations of critical near-Earth plasmas and fields from the FAST satellite [Carlson et al., 1998] to understand the multiscale processes at work. SOURCE: *Left:* Adapted from C.W. Carlson, J. P. McFadden, R. E. Ergun, M. Temerin, W. Peria, F. S. Mozer, D. M. Klumpar, E. G. Shelley, W. K. Peterson, E. Moebius, R. Elphic, et al., FAST observations in the downward auroral current region: Energetic upgoing electron beams, parallel potential drops, and ion heating, *Geophysical Research Letters* 25(12):2017-2020, doi:10.1029/98GL00851, 1998. Copyright 1998 American Geophysical Union. Reproduced by permission of American Geophysical Union. *Right:* Courtesy of D. Mitchell, Johns Hopkins University, Applied Physics Laboratory.

The first goal is to continuously image the three-dimensional distribution of two critical inner magnetospheric plasmas. Using ENA imaging, the ring current and near-Earth plasma sheet are captured with sufficient temporal and spatial resolution (1 minute, 0.5 R_E) to retrieve the electrical current system that distorts the magnetic field and that connects through the ionosphere producing the electric field. EUV imaging at 30.4 nm captures the plasmasphere with sufficient temporal/spatial resolution (1 minute, 0.05 R_E) to retrieve cross-scale density structures and the global-to-local electric fields that drive the formation and evolution of these structures. State-of-the-art stereo imaging of the optically thin ENA and EUV signals enables determination of the three-dimensional structure of pressure, pitch angle, and density, revealing energization, losses, and plasma sources and resolving cross-scale currents, fields, and flows that bind the system together.

The second goal is to image and measure the ionosphere-thermosphere system using multiple wavelengths of far ultraviolet (FUV): LBH long and short cameras, and spectrographic imaging (SI) at 121.6 nm and 135.6 nm. FUV imaging provides estimates of multiple geophysical quantities: precipitating particle flux, ionospheric electron density and conductivity, and thermospheric O/N_2 ratio. Dynamical features in the ionosphere (e.g., polar cap ionization patches, positive and negative storm effects, neutral atmospheric responses to auroral heating, ionospheric scintillations and plasma bubbles) are tracked with global imaging at 5- to 10-km resolution. Variable phasing of the two circular-orbiting spacecraft provides simultaneous conjugate views of both northern and southern auroral emissions, uncovering the little-understood role of inter-hemispheric asymmetry on the global system behavior.

The third goal is to measure, in situ, the critical near-Earth plasmas and magnetic field in the cusp and near-Earth plasma sheet plasma. Onboard each of the two MEDICI spacecraft, plasma composition, electron plasma conditions, and magnetic field measurements characterize the plasma conditions, field distortions, and storm/substorm activity, and in combination with imaging allow researchers to follow the flow of ionospheric plasma and energy between the ionosphere and magnetosphere.

Complementing and augmenting the high-altitude observations, MEDICI includes funded participation for significant low-altitude components: measurements from a large range of resources, including DMSP or its follow-on Defense Weather Satellite System, IRIDIUM/AMPERE current maps, radar arrays from high to mid latitudes (SuperDARN, Millstone Hill, AMISR), GPS TEC maps, and magnetometer and ground-based auroral all-sky camera arrays. The result will be global specification of the ionospheric electric field and electric current patterns in both hemispheres, essentially completing the electrodynamic picture at low altitude. In principle, MEDICI measurements tackle a broader comparative planetary question, namely, the extent to which magnetospheres can act as shields against atmosphere erosion by the solar wind: Does ion outflow escape or remain in the magnetosphere to be recycled?

Contributions to the Heliophysics Systems Observatory. MEDICI will both benefit from and enhance the science return from almost any geospace mission that flies contemporaneously, such as upstream solar wind monitors, geostationary satellites, and low Earth orbit missions. In particular, by providing global context and quantitative estimates for magnetospheric-ionospheric plasma and energy exchange, MEDICI has significant value for missions investigating ionospheric conditions, outflow of ionospheric plasma into the magnetosphere, energy input from the magnetosphere into the ionosphere, and AIM coupling in general. Thus it will add value to a host of possible ionospheric strategic missions, Explorers, and rocket and balloon campaigns. A mission providing continuous imaging and in situ observations from two separate platforms also plays an important role in providing to geospace predictive models the indispensable validating observations of system-level interactions and processes. The likely long duration of the MEDICI mission will allow it to provide a transformative framework into which additional future science missions can naturally fit.

Table 9.2 summarizes MEDICI's expected level of contribution to the SWMI science goals. In recognition of the need for crucial new observations to enable the accomplishment of these science objectives, the panel identified the following SWMI imperative:

SWMI Imperative: Initiate the development of a strategic mission, like MEDICI, to determine how the magnetosphere-ionosphere-thermosphere system is coupled and responds to solar and magnetospheric forcing.

Magnetosphere-Ionosphere Source Term Energetics (MISTE)

The SWMI panel's second-highest-rated new strategic mission concept, the Magnetosphere-Ionosphere Source Term Energetics (MISTE) mission, seeks to resolve how ionospheric plasma escapes into the magnetosphere (Figure 9.12); as described in Part I of this decadal survey report, MISTE is an STP-class strategic mission, but PI-led, following the Planetary Division's Discovery mission class. This concept directly addresses SWMI critical science goals 2 and 4, with additional contributions to goals 3 and 7. Although the science addressed by this mission is of high merit, the panel is aware that budget realities will likely prevent its initiation as a strategic mission in the coming decade. However, a valuable subset of the science described here may well be within reach of an Explorer mission.

Science Objectives. Certain magnetospheric processes strongly depend on the local concentration of heavy ions, which ultimately originate in Earth's atmosphere. However, this concentration cannot be predicted with much accuracy because researchers cannot yet quantify the amount of outflow from a given input of electromagnetic energy or energetic particle flux into the high-latitude upper atmosphere. Thus, a comprehensive set of combined in situ and remote-sensing measurements is needed to determine how energy input to the high-latitude ionosphere leads to particle outflow and mass loading of the magnetosphere. The MISTE concept has been designed to simultaneously measure the inflow of energy to the upper atmosphere

FIGURE 9.12 The notional MISTE orbit configuration, with 5,000-km apogee and perigee ranging from 200- to 500-km altitude. Phasing shifts will yield magnetic alignments at a range of spacecraft altitude offsets, and precession will provide alignments at all latitudes and local times. SOURCE: Courtesy of NASA and the European Space Agency.

and the response of the ionosphere-thermosphere system to this input, in particular the outflow back to the magnetosphere.

Mission Implementation and Strategy. The fundamental design of the MISTE mission concept is two identical spacecraft in highly inclined elliptical orbits with apogees 180° out of phase, as shown in Figure 9.12. The two satellites can then be positioned on their orbits to obtain magnetic conjunctions so that the energy input, the ionospheric heating and acceleration, and the outflow can be measured simultaneously. With apogee at 5,000-km altitude, the satellites will pass through the auroral acceleration region and observe the energization of the precipitating electrons and the wave heating and acceleration of the ionospheric ions. Perigee will vary from the topside ionosphere (500-km altitude) to below the F-layer peak (200-km altitude) in order to measure the range of energy deposition processes capable of causing ionospheric outflow. With an 80° orbital inclination, apogee will precess, allowing for this magnetic alignment to cover the entire high-latitude region. By varying the positions of the satellites along the orbit tracks, a span of intersatellite distances during these alignments can be sampled.

The notional mission has two dual-spinner spacecraft (spin axis perpendicular to the orbit), with a despun portion of each satellite with faces maintained in the ram and nadir directions and the other portion at a high spin rate of 15 rpm. The despun portion provides high-cadence measurements of cold particle populations and high-cadence auroral imaging. The faster-spinning portion enables the measurement of three-dimensional velocity-space distributions of ions and electrons. It also accommodates long spin-plane booms for high-resolution electric field measurements.

MISTE is designed to make a comprehensive suite of measurements to understand high-latitude energy input and ionospheric outflow. See Figure 9.13 for an example of the breadth of measurements from MISTE. To measure the low-altitude heating, the despun portion includes in situ sensors for measuring the neutral density, neutral winds, ion and neutral mass composition, and ion velocity. The input electromagnetic energy and electromagnetic waves that drive particle acceleration will be measured with axial electric field booms, four wire booms, and a magnetometer. The axial booms will be mounted to the despun platform, with one protruding through the center hole in the spinning section, while the wire booms are on the spinning platform.

The thermal and suprathermal ion and electron measurements, which provide both the input precipitating particle energy and the response to this input, are made by a combination of sensors. Sensors located at the ends of booms measure the very-low-energy ionospheric electron and ion three-dimensional velocity distributions in the 0.1- to ~20-eV range, outside the spacecraft's sheath. Body-mounted sensors will cover ions and electrons over the plasma-sheet energy range from 20 eV to ~50 keV, including ion composition. A final in situ detector will measure the precipitating flux of 50-keV to 5-MeV electrons.

To give a broader context to the in situ measurements, two remote sensing instruments are included. An ultraviolet auroral imager will provide the regional context of the energy input to the upper atmosphere. An active sounder experiment will remotely sense the plasma density along the magnetic field line below the spacecraft. Together, these two instruments will provide a three-dimensional view of the region in which the in situ measurements are made.

The primary objective of the MISTE mission concept is to quantify the amount of ionospheric outflow given a particular intensity of electromagnetic or particle energy input to the upper atmosphere. Therefore, the primary instrumentation of MISTE is the in situ thermal and suprathermal particle detectors and the electric and magnetic field instruments. These observations could be made from a simpler mission design of twin single-spin spacecraft. That is, a critical subset of the MISTE objectives could be obtained with a dramatically reduced payload and mission concept. Ground-based observations of the neutral atmosphere

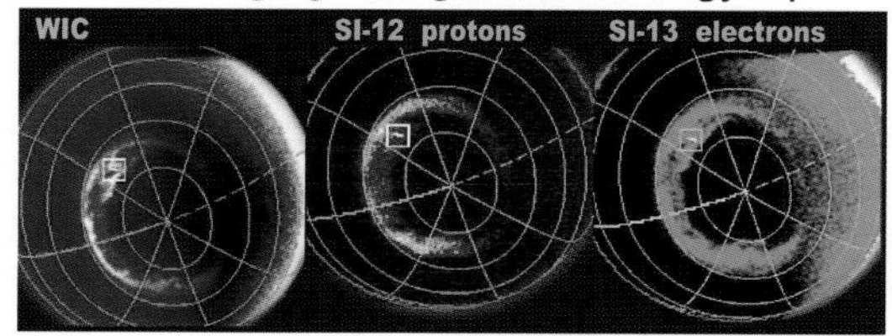

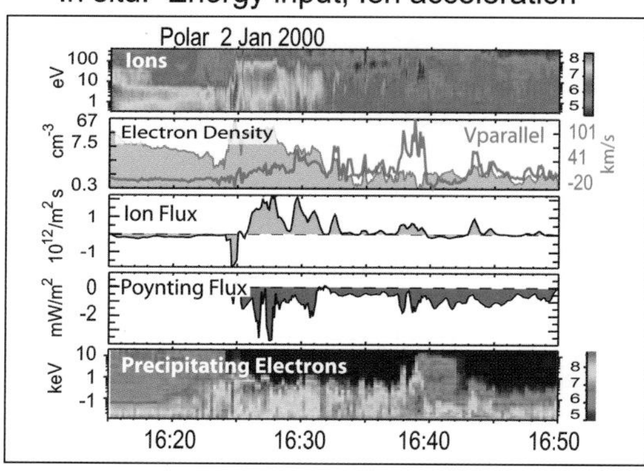

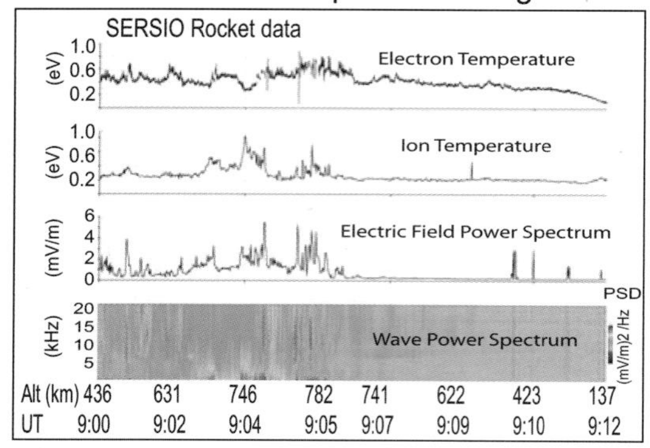

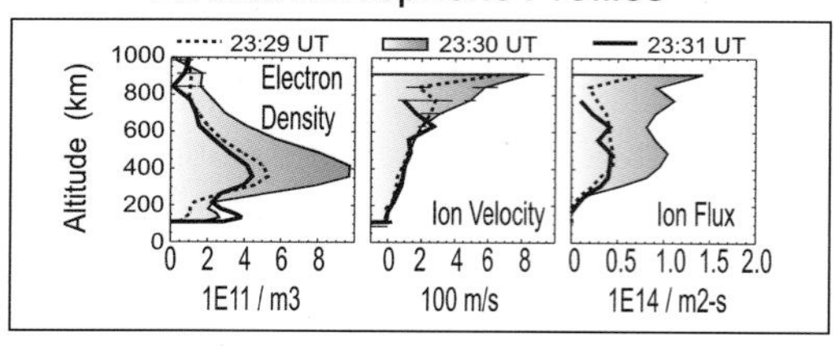

FIGURE 9.13 MISTE will explore the relationship between energy input from the magnetosphere to the ionosphere and the resulting ionospheric outflow. At magnetic alignment, the upper spacecraft (left column of panels) will provide detailed information of the magnetospheric energy inflow (both particle and electromagnetic), while the lower spacecraft (right column of panels) will observe the ionospheric heating and energization. SOURCE: *Upper left:* S.B. Mende, C.W. Carlson, H.U. Frey, T.J. Immel, and J.-C. Gérard, IMAGE FUV and in situ FAST particle observations of substorm aurorae, *Journal of Geophysical Research* 108(A4):8010, doi:10.1029/2002JA009413, 2003. *Upper right and lower left:* Adapted from K.M. Frederick-Frost, K.A. Lynch, P.M. Kintner, Jr., E. Klatt, D. Lorentzen, J. Moen, Y. Ogawa, and M. Widholm, SERSIO: Svalbard EISCAT rocket study of ion outflows, *Journal of Geophysical Research* 112:A08307, 2007. *Lower right:* W. Lotko, The magnetosphere ionosphere system from the perspective of plasma circulation: A tutorial, *Journal of Atmospheric and Solar-Terrestrial Physics* 69(3):191-211, doi:10.1016/j.jastp.2006.08.011, 2007.

---

and regional context could be used to complement a reduced payload, in particular to help understand why certain energy inputs result in particular outflow rates.

Table 9.3 summarizes the contributions MISTE would make to the SWMI critical science goals. The SWMI panel believes that these contributions would enable great progress toward the applicable objectives, but in recognition of probable budget constraints over the coming decade, the panel developed the following SWMI imperative:

SWMI Imperative: If resources permit, initiate a strategic mission like MISTE to simultaneously measure the inflow of energy to the upper atmosphere and the response of the ionosphere-thermosphere system to this input, in particular the outflow back to the magnetosphere.

TABLE 9.3 Level of MISTE Contributions Toward Achieving SWMI High-Priority Science Goals

	MISTE Contribution
Goal 1: Determine how the global and mesoscale structures in the magnetosphere respond to variable solar wind forcing.	◔
Goal 2: Identify the controlling factors that determine the dominant sources of magnetospheric plasma.	●
Goal 3: Understand how plasmas interact within the magnetosphere and at its boundaries.	◒
Goal 4: Establish how energetic particles are accelerated, transported, and lost.	◔
Goal 5: Discover how magnetic reconnection is triggered and modulated.	○
Goal 6: Understand the origins and effects of turbulence and wave-particle interactions.	◒
Goal 7: Determine how magnetosphere-ionosphere-thermosphere coupling controls system-level dynamics.	●
Goal 8: Identify the structures, dynamics, and linkages in other planetary magnetospheric systems.	○

Contribution to Goal	Major	Large	Significant	Some	Minimal
	●	◕	◒	◔	○

Uranus Orbiter

Because of the importance of understanding the range of processes operating in the universe, as well as their operation under different environmental conditions, continued progress in comparative magnetospheres is a key objective for the coming decade. Thus, it seems essential that NASA's Heliophysics Division partner with the Planetary Division to ensure that appropriate magnetospheric instrumentation be fielded on missions to other planets. In particular, the SWMI panel's highest priority in planetary magnetospheres is a mission to orbit Uranus. With a strongly tilted dipole and a rotational axis near the ecliptic plane, Uranus offers an example of solar wind/magnetosphere interactions under strongly changing orientations over diurnal timescales (Figure 9.14).

The complexity of the interactions of Uranus's magnetosphere with the solar wind provides an ideal testbed of the most sophisticated models and theories. Indeed, one could argue that Uranus is too complex a system to study effectively without supporting data; however, the potential discoveries from its dynamo generation and its variability stand to open new chapters in comparative planetary magnetospheres and interiors. A Uranus orbiter is the third-ranked outer planets mission of the 2011 planetary decadal survey[7] and has received extensive study. Key magnetospheric measurements for a Uranus mission would include magnetic field, plasma waves, plasma, energetic particles, dust and neutral mass spectra, and global images

[7] National Research Council, *Vision and Voyages for Planetary Science in the Decade 2013-2022*, The National Academies Press, Washington, D.C., 2011.

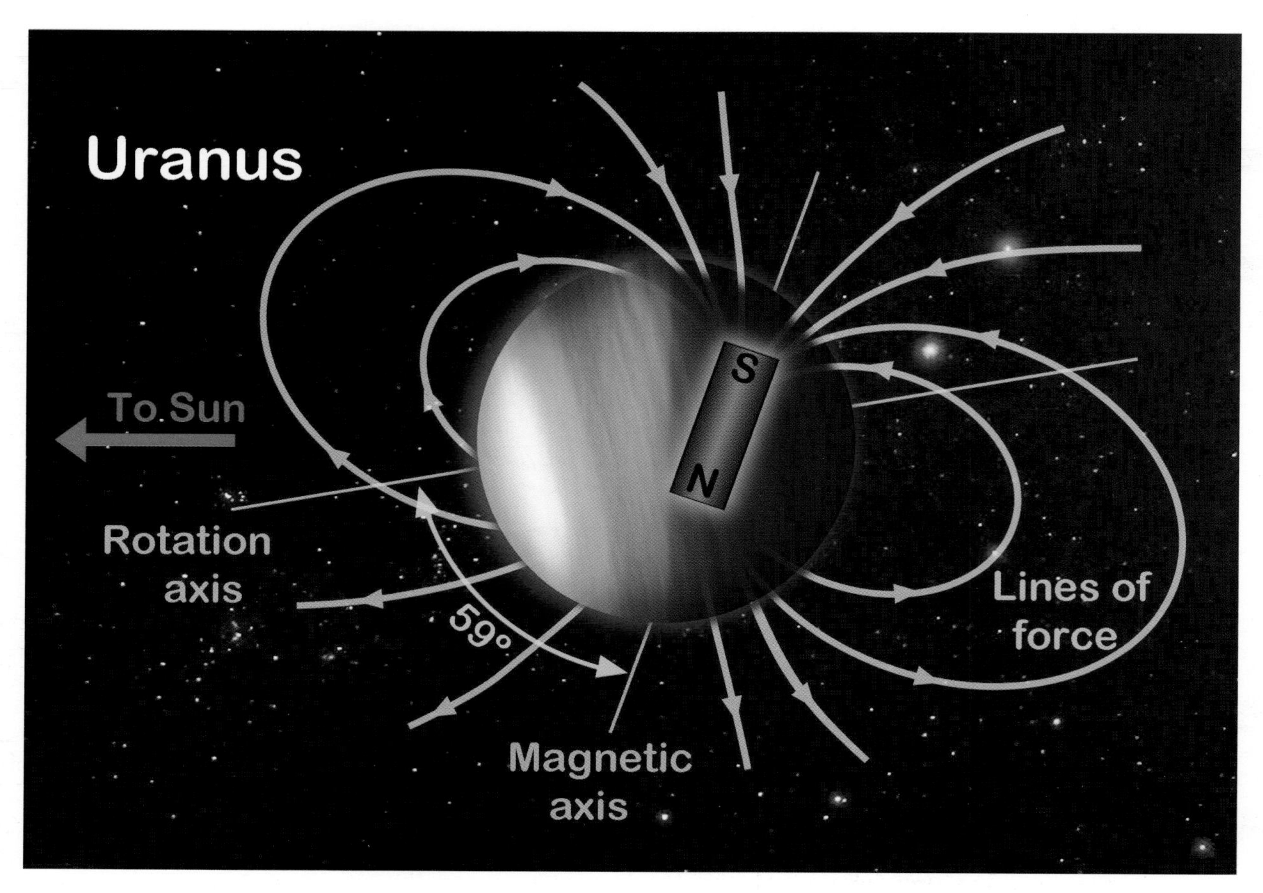

FIGURE 9.14 The magnetic dipole axis of Uranus is strongly tilted with respect to its rotational axis, which in turn lies near the orbital plane. Thus, depending on the season, the effects of solar wind-magnetosphere interaction vary dramatically over the course of each day. Uranus significantly expands the parameter range over which scientists can study magnetospheric structure and dynamics. SOURCE: Courtesy of Jerry Goldstein, Southwest Research Institute.

in UV, IR, and ENA. To make continued exciting progress toward understanding other magnetospheres, the SWMI panel concluded with the following SWMI imperative:

SWMI Imperative: Through partnership between NASA's Heliophysics Division and Planetary Division, ensure that appropriate magnetospheric instrumentation is fielded on missions to other planets. In particular, the SWMI panel's highest priority in planetary magnetospheres is a mission to orbit Uranus.

9.5.2.5 Future Strategic Missions

Determining how mesoscale and global structures in the magnetosphere respond to variable solar wind forcing and understanding how plasmas interact within the magnetosphere and at its boundaries both require observations that match these scales. This effort essentially necessitates global-scale imaging and multisatellite constellation missions to obtain measurements of the physical parameters necessary to accomplish these science objectives. A MEDICI-like mission will address these objectives through global imaging, which is technically feasible primarily for the inner and middle magnetosphere. For the outer magnetosphere, which forms the primary solar-wind entry and energy-storage region, the highly tenuous plasma makes "imaging" by means of local measurements on many spatially separated satellites the best approach to resolving global and mesoscale structure. Below, some exciting mission concepts are highlighted that can address these challenging objectives but will require technology developments in the coming decade to achieve feasible cost and readiness levels.

Magnetospheric Constellation Mission (MagCon)

Science Goals. Understanding the mass and energy transport at global and mesoscales in Earth's magnetospheric plasma sheet and reconnection regions of the near-Earth magnetotail, plus the dayside and flanks of the magnetopause and bow shock regions, can be implemented using a multisatellite in situ mission such as the MagCon mission. The prime overarching objective of the mission is to determine how the magnetosphere stores, processes, and releases energy in the magnetotail and accelerates particles that supply the inner magnetosphere's radiation belts. It would track the spatial-temporal plasma structures and flows associated with the solar wind plasma entry across the magnetopause and transport within and through the magnetotail. On the dayside and flanks the constellation would provide multipoint measurements of the upstream solar wind input, and the response throughout the magnetosphere, enabling determination of how the entire system responds to variable solar wind driving. In the magnetotail it would provide a map of the global plasma flows and field configurations, leading to determination of whether they are internally developed or externally triggered. Throughout the mission, MagCon would provide a global "picture" of these otherwise invisible regions of the magnetosphere.

Mission Concept. MagCon uses many satellites separated by mesoscale distances (~1-2 R_E) that make magnetic field plus plasma and energetic particle distribution function measurements at multiple points simultaneously with relatively rapid cadence. The mission requires a significant number of spacecraft, 36 in the concept the SWMI panel evaluated, to achieve mesoscale spacing while filling a significant fraction of the near-Earth space using orbits with perigees in the 7-8 R_E range and apogees dispersed uniformly up to 25 R_E with low inclination. The satellites would be simple ~30-kg-class spin-stabilized vehicles with their spin axes perpendicular to the ecliptic. Each spacecraft would carry a boom-mounted fluxgate magnetometer, a three-dimensional ion-electron plasma analyzer, and simple energetic ion-electron particle telescopes.

To implement such a constellation requires development of small satellite systems and instruments that can be more cheaply manufactured and tested in a reasonable time frame (2-3 years) with acceptable reliability levels, plus a better match between launch vehicle capabilities and constellation mission needs.

Magnetospheric Constellation and Tomography (MagCat)

Science Goals. With science objectives similar in many respects to those of MagCon, MagCat would address some of the most critical processes in Sun-Earth connections: plasma entry into the magnetosphere, plasma-sheet formation and dynamics, and investigation of bow-shock structure, plasmaspheric plumes, and other mesoscale structures that form in response to solar-wind variability. To achieve this objective requires observations with a minimum spatial resolution of 0.5 R_E at a minimum time cadence of 15 s. MagCat could provide those required measurements.

Mission Concept. MagCat is a 20-spacecraft mission that would provide a combination of two-dimensional images of the equatorial outer magnetosphere and multipoint in situ observations made concurrently and in the same imaged region. The spacecraft would be in two coplanar orbits that pass through critical regions in the magnetotail, flank, and subsolar magnetosphere. Each satellite would transmit radio waves to all others, obtaining 190 line-of-sight densities, enabling tomographic images of plasma density over large regions with an average spatial resolution of 0.32 R_E at 12 s cadence. Each satellite would carry a suite of plasma and field instruments that provide complementary in situ data throughout the imaged area for ground truth as well as revealing the detailed plasma processes in the region. The nominal payload would include a 3-axis fluxgate magnetometer, electrostatic analyzers that measure three-dimensional ion and electron distributions, a relaxation sounder to determine the ambient density, and a radio tomography instrument.

As for MagCon, the pre-CATE estimate for the MagCat mission was deemed beyond the scope of the budget in the coming decade. Thus, there is a similar need in the coming decade to develop cost-effective and efficient manufacturing procedures to mass produce a large number of spacecraft and instruments.

9.5.2.6 International Partnerships

International partnerships involving a consortium of individual space agencies can be an effective way to pool limited resources to achieve an outstanding science goal whose importance is agreed upon by a consensus of these agencies. For example, determining the cross-scale coupling physics involved in key plasma processes is believed to be crucial for complete understanding of the causes and consequences of these processes. None of the past, current (e.g., Cluster and THEMIS), or planned missions (e.g., MMS) are designed to address the cross-scale aspects of these processes. However, a mission concept has been developed in Japan, Canada, and Europe that involves a fleet of spacecraft performing simultaneous in situ measurements at electron, ion, and fluid scales. Such a mission can investigate how turbulence transports and dissipates energy over multiple scales and how kinetic microscale instabilities are modulated by macroscale properties of the plasma, as well as the relative role of global conditions versus microscale physics in determining the structure and dynamics of magnetic reconnection. These are all important aspects of SWMI critical science goals 6 and 7. This and other international, cross-agency partnerships should be pursued when available and possible.

SWMI Imperative: Partner with other space agencies to implement consensus missions, such as a multispacecraft mission to address cross-scale plasma physics.

9.5.3 DRIVE-Related Actions

In this section, the SWMI panel expands on a number of issues that have a material impact on the national ability to conduct an effective and productive solar and space physics research effort.

9.5.3.1 Solar Wind Monitor

Knowledge of upstream solar wind conditions, the interplanetary magnetic field, and solar energetic particles is required in essentially all of the programs that would address the SWMI science objectives. Currently, instruments on the ACE[8] spacecraft, which orbits around the L1 libration point approximately 1.5 million km from Earth, provide these data. Follow-ons to ACE, which could also be instrumented with a solar coronagraph to view Earth-bound coronal mass ejections, are needed both to satisfy SWMI science goals and as part of a space weather forecasting system (see Chapter 7). The SWMI panel does not have

[8]Information about the Advanced Composition Explorer (ACE) is available at http://www.srl.caltech.edu/ACE/.

a preference for which agency conducts this mission, but these data must be available on a continuous basis throughout the coming decade.

SWMI Imperative: Ensure continuity of measurements of the upstream solar wind.

9.5.3.2 Theory, Modeling, and Data Analysis

Efforts related to advancing theoretical and modeling studies of geospace are another key component of the solar and space physics research effort. The future of data analysis is becoming ever more closely linked with modeling, and tools need to be developed to enable a broad segment of the community to access and combine observations and model results. Currently, a great deal of the science return from NASA missions and NSF ground-based measurements occurs through theory, modeling, and data analysis supported in a portfolio of both NASA and NSF grants programs. The high productivity and cost-effectiveness of these programs argue strongly that major increases in science output can be achieved with modest funding increases. The SWMI panel therefore strongly encourages the agencies to enhance the funding of R&A programs. In a constrained funding environment, this is one of the most cost-effective ways to ensure that the high-priority science objectives outlined in this decadal survey will be accomplished.

In addition, the panel supports funding enhancements to enable the creation of a new program within NASA's research portfolio to bring critical-mass multidisciplinary science teams together to address grand challenge problems. However, these proposed Heliophysics Science Centers must not be created at the expense of the current research programs, e.g., the Heliophysics Theory program. They must also be competed regularly and be focused on addressing a key science question.

The SWMI panel also strongly affirms the vital role that theory and modeling supported by mission funding play in advancing the science objectives of the missions. The panel believes that this role should be protected by requiring external reviews of the impacts of any theory-funding reductions on the ability of a mission to fulfill its science objectives.

SWMI Imperative: Enhance and protect support for theory, modeling, and data analysis, including research and analysis programs and mission-specific funding.

9.5.3.3 Innovation and Technology

In reviewing the white papers submitted from the community, the SWMI panel found that many relied on tried-and-true measurement techniques. However, while much can be accomplished with new applications of state-of-the-art technology, it is also clear that to accomplish the most challenging science objectives of the future, the development of new, innovative concepts will be necessary. Indeed, in an observation-driven field such as solar and space physics, it is often new observations that point the way to the science questions of the future.

Advanced instrumentation for constellation missions is an area of particular need for ultimate realization of the science goals the SWMI panel has outlined. Targeted research on how to acquire, analyze, and display the simultaneous measurements from many spacecraft is also needed, along with development of new ways to integrate global modeling and global observations.

Another area where technology development efforts could have a big scientific payoff is global imaging systems, particularly systems that can stand off, well outside the magnetosphere's bow shock, and image its boundary structures using a range of techniques such as neutral atom imaging (ENA), scattered light, and X-ray emission. New instrumentation with increased sensitivity and resolution is needed to achieve spatial resolutions (~0.5 R_E) and image cadences (~15 min) sufficient to image the boundaries and visualize their motions and structure. New techniques for reducing background noise are also needed.

A number of the science goals outlined in Section 9.4 involve connecting phenomena and signatures that occur in the ionosphere with their corresponding phenomena and signatures in the magnetosphere. Because this connection occurs primarily via the geomagnetic field, which is strongly distorted and highly variable due to currents flowing within the magnetosphere itself, an accurate mapping between the ionosphere and magnetosphere for all relevant conditions is lacking. Current efforts to relate magnetospheric and ionospheric physics thus typically rely on empirical models based on statistical analysis of large data sets acquired over long time periods and under a variety of conditions. Techniques to definitively establish the instantaneous mapping are thus urgently needed.

It is the SWMI panel's view that the current limited investment in technology development has discouraged the development of new instrument concepts. Therefore, the panel strongly supports the creation of a robust Heliophysics Instrument and Technology Development Program (HITDP) in the context of the NASA ROSES. In the panel's opinion, this program should be funded sufficiently to support several grants at levels significantly larger than is possible at present through supporting research and technology.

Moreover, the missions of the future will require new satellite and systems technologies. Such technologies are most appropriately pursued by the Office of the Chief Technologist, in close consultation with the Heliophysics Division so that real mission needs are addressed. Possible technology development needs noted by the panel include, but are not limited to, techniques that make it feasible to produce and deploy large numbers of identical spacecraft at an affordable price and with high radiation tolerance; solar sails; advanced propulsion and power; low-cost launch vehicles; mass-production techniques; component miniaturization; and wireless communications within a satellite.

Another area where near-term investments will help reduce the risk and long-term cost of new instruments is in raising the technology readiness level (TRL) of the instruments of the future, particularly in the area of providing in-space operating experience. The panel believes that suborbital flights are cost-effective ways to mature instrument technologies for future science applications, even if no immediate science can be obtained from flying suborbitally. Thus, while the panel applauds and supports the science output of the suborbital program, it also encourages the utilization of suborbital flights for increasing the TRL of instruments with long-term science applications, but not necessarily science output from the flight itself.

SWMI Imperative: Invest significantly in developing the technologies to enable future high-priority investigations.

9.5.3.4 Ground-Based Instrumentation

Since the science objectives established by this panel emphasize a global view of the coupled magnetosphere-ionosphere-thermosphere system the SWMI panel strongly supports networks of ground-based instruments. For example, the MEDICI mission will be greatly enhanced by the data from magnetometer arrays and the SuperDARN network, and both MISTE and MEDICI benefit from the data provided by incoherent scatter radars (ISRs) and networks of all-sky cameras. The panel concluded that existing facilities, e.g., ISRs, SuperDARN, and magnetometer arrays, should continue to be supported, upgraded, and perhaps enhanced. Moreover, research drawing on increasing access to data from these instrument suites via the Internet would greatly benefit from the development of standards for data collection and access.

The SWMI panel takes note of successful NASA contributions in the past decade to ground-based assets that directly support their space missions, e.g., the THEMIS all-sky camera network, for which an example image is shown in Figure 9.15. The panel encourages continuation and expansion of these efforts. The

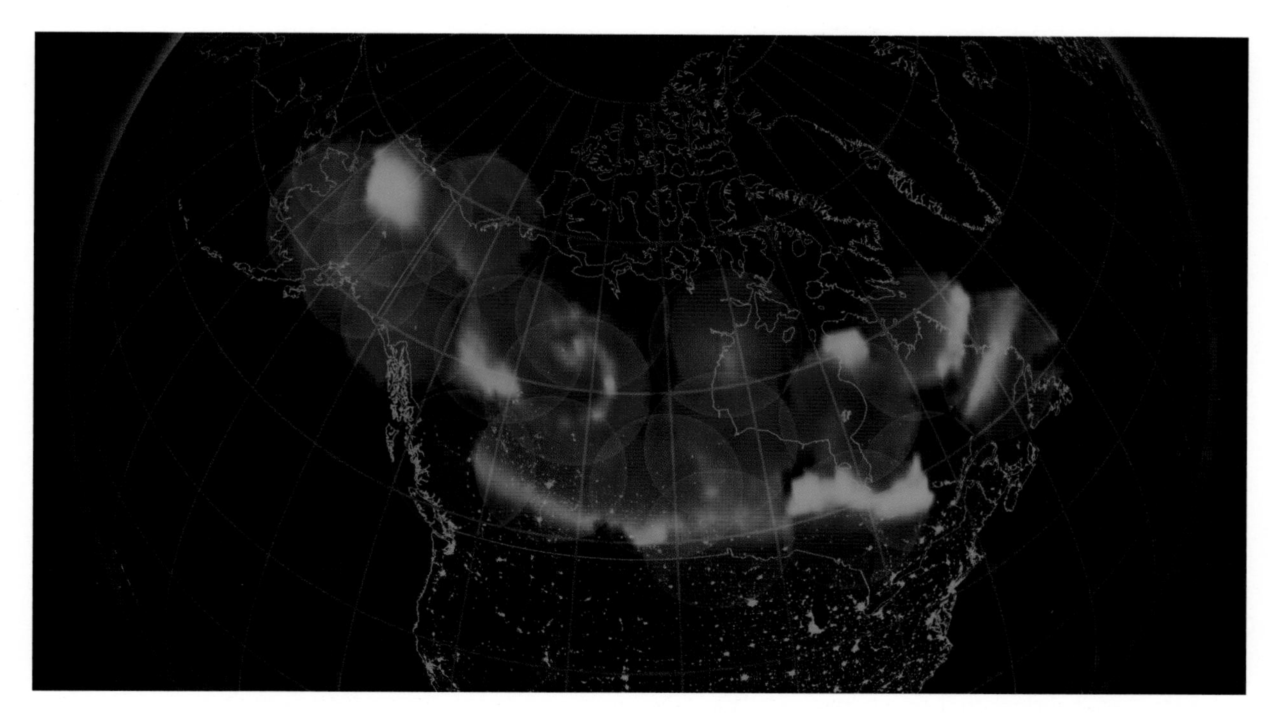

FIGURE 9.15 Composite auroral image constructed from the THEMIS All-sky Imager network. SOURCE: Courtesy of NASA/Goddard Space Flight Center Scientific Visualization Studio; available at http://www.youtube.com/watch?v=GEebRsRnwm0.

panel also notes that DOD supports a number of ground-based observatories that contribute significantly to the science objectives outlined above and encourages continued sponsorship of those facilities as well. The panel further supports negotiation of international agreements to enable coordination and collaboration with non-U.S. ground-based capabilities.

SWMI Imperative: Ensure strong multiagency support for a broad range of ground-based assets that are a vital part of magnetospheric science.

9.5.3.5 Laboratory Studies

The SWMI panel endorses the recommendation in the plasma science decadal survey that the Department of Energy be the prime steward for laboratory plasma science. However, to enable research on basic plasma physics that will be of greater utility to SWMI science objectives, the panel supports the creation of an interagency joint laboratory astrophysics program. This program should be competed on a regular basis, include selection criteria that focus on issues relevant to space physics, contain a mechanism for outside investigators to have access to supported facilities, and be open to proposals from any institution.

SWMI Imperative: Create an interagency joint laboratory astrophysics program that addresses issues relevant to space physics.

9.5.3.6 Education and Workforce

At the most basic level, the field of solar and space physics needs a robust, well-trained, and talented workforce to accomplish its science goals, as well as an educated populace that recognizes the value of addressing these exciting scientific challenges. The SWMI panel therefore endorses curriculum development efforts across all academic levels, as well as faculty development programs. The panel notes in particular the success of the NSF Faculty Development in Space Science program and strongly supports its continuation and enhancement throughout the coming decade. The panel further endorses funded training opportunities for both undergraduate and graduate students, especially for participation in the development of flight hardware. Such opportunities could be provided by science grant augmentations, stand-alone education and public outreach grants, and mission-related funding.

SWMI Imperative: Strengthen workforce, education, and public outreach activities.

9.5.4 Space Weather

As suggested by its title, the long-term goal of this decadal survey is to have the knowledge to ensure the well-being of a society dependent on space. Actionable knowledge of space environment effects involves the ability to characterize conditions anywhere in the system at any time in the past, as well as to predict future conditions with good fidelity. This capability requires an understanding of the full, coupled solar-terrestrial system that encompasses all the regions, processes, and coupling described above, across spatial scales from meters to hundreds of Earth radii. It includes understanding the fundamental microscopic physics as well as the global system behavior in response to variable driving. The ultimate objective in the study of solar wind-magnetosphere interactions is to know how solar and solar-wind input at various spatial and temporal scales determines the nature and behavior of magnetospheric populations, structures, and processes and to be able to predict those that have significant space weather impacts.

There are three aspects of accomplishing this long-term goal:

1. Establishment of the foundation of comprehensive scientific understanding;
2. Development of sound, validated space environment models; and
3. Fielding of the optimum operational assets to drive those models.

The scientific program presented above will put in place some of the tools essential to achieving this vision, particularly by defining outstanding questions that still inhibit a comprehensive scientific understanding.

The development of sound, validated space environment models requires a healthy research-to-operations/operations-to-research program. This in turn clearly necessitates communication and coordination between research-oriented agencies and operational agencies with end-use requirements so that a robust and adequately funded process exists for transitioning scientifically sound and operationally useful models between the two emphases.

Observations are critical for an effective space weather program because they support research and development of models and they drive models in their operational phase. Space weather observations are available from government research and some operational programs as well as from the commercial sector. Currently, each group in isolation develops observational requirements and observing systems to fulfill those requirements. To make more effective use of limited resources, these observations should be nationally coordinated, allowing research groups to provide input and possibly additional payloads to operational or commercial endeavors. Similarly, coordination would allow operational and commercial entities to guide and use research observations. It is highly desirable that the National Space Weather Program (NSWP) be augmented to provide a mechanism for coordinating space weather observations, allowing for input to mission definition and payloads from all groups. The SWMI panel encourages the placement and coordination of space-weather-related observing systems on research, commercial, and government platforms to the greatest extent possible. The panel also supports development of mechanisms for real-time data acquisition and distribution from *all* available platforms.

More generally, to make the most of limited resources, an effective space weather program must be accomplished through coordinated activities within the government as well as with the commercial sectors. Currently several commercial and government groups provide various forms of space weather products with some limited coordination, yet many functions overlap. The panel urges that the NSWP be enhanced and clarified in order to delineate the roles for each agency, e.g., NASA, NOAA, NSF, the U.S. Geological Survey, DOD, and DOE, in producing and distributing data and models providing forecasting and space-weather-related products. The panel strongly endorses the undertaking of a high-level study to design a more complete overarching architecture for the NSWP that would define agency roles and coordinate observations.

SWMI Imperative: Encourage the creation of a complete architecture for the National Space Weather Program that would coordinate joint research, commercial, and operational space weather observations and define agency roles for producing, distributing, and forecasting space weather products. In addition the SWMI panel encourages all agencies to foster interactions between the research and operational communities and to identify funding for maintaining a healthy research-to-operations and operations-to-research program.

Achieving the key science goals of this decadal survey requires identifying the optimum set of operational observations to drive models that will enable specification and prediction of the environment throughout the magnetosphere. This effort may ultimately require an operational "great observatory" of satellites in appropriate orbits for monitoring crucial aspects of the input from and response to solar wind variability. In addition to providing the input necessary for high-fidelity environmental specifications, these measurements would provide routine context information for future targeted science experiments, much as magnetospheric activity indices are used today. Potential elements of such a space weather observatory could include a solar wind monitor (including IMF, energetic particle, and potentially coronagraphic measurements); high-altitude synoptic imaging of the aurorae, ring current, plasmasphere, and of the outer magnetospheric boundary; constellation observations of plasma entry and global tail structure; low-altitude, DMSP-like satellites to observe the magnetospheric input into the ionosphere and its response; multiple geosynchronous measurements of plasma, energetic particles, and magnetic field; RBSP-like monitors of the inner magnetospheric radiation environment; and a fine mesh of appropriate ground-based measurements. One important near-term investment is to determine, based on past observations of this nature, the optimum set of measurements that are required to drive high-fidelity predictive models of the environment.

SWMI Imperative: Implement a program to determine, based on past observations, the optimum set of measurements that are required to drive high-fidelity predictive models of the environment.

9.5.5 Prioritization

The SWMI imperatives presented above would all greatly enhance the ability to accomplish the science goals the panel has outlined, thereby providing the foundation needed for addressing the decadal survey's

TABLE 9.4 Summary of SWMI Prioritized Imperatives

Rank	Imperative	Missions	DRIVE	Space Weather
1	Enhance the resources dedicated to the Explorer program and broaden the range of cost categories.			
2	Complete the strategic missions that are currently in development (MMS and RBSP/BARREL) as cost-effectively as possible.			
3	Initiate the development of a strategic mission, like MEDICI, to determine how the magnetosphere-ionosphere-thermosphere system is coupled and responds to solar and magnetospheric forcing.			
4	Ensure strong continued support for extended missions that can still contribute significantly to high-priority science objectives.			
5	Enhance and protect support for theory, modeling, and data analysis, including research and analysis programs and mission-specific funding.			
6	Ensure continuity of measurements of upstream solar wind.			
7	Invest significantly in developing the technologies to enable future high-priority investigations.			
8	Ensure strong multiagency support for a broad range of ground-based assets that are a vital part of magnetospheric science.			
9	Develop a mechanism within NASA to support rapid development, deployment, and utilization of science payloads on commercial vehicles and other missions of opportunity. NSF and DOD efforts in this regard are also encouraged.			
10	Through partnership between NASA's Heliophysics Division and Planetary Division, ensure that appropriate magnetospheric instrumentation is fielded on missions to other planets. In particular, the SWMI panel's highest priority in planetary magnetospheres is a mission to orbit Uranus.			
11	Partner with other space agencies to implement consensus missions such as a multispacecraft mission to address cross-scale plasma physics.			
12	Encourage the creation of a complete architecture for the National Space Weather Program that would coordinate joint research, commercial, and operational space weather observations and define agency roles for producing, distributing, and forecasting space weather products. In addition the SWMI panel encourages all agencies to foster interactions between the research and operational communities and to identify funding for maintaining a healthy research-to-operations and operations-to-research program.			
13	Implement a program to determine, based on past observations, the optimum set of measurements that are required to drive high-fidelity predictive models of the environment, and to put in place a plan to ensure that the optimum set of observational capabilities is maintained.			
14	Strengthen workforce, education, and public outreach activities.			
15	If resources permit, initiate a strategic mission like MISTE to simultaneously measure the inflow of energy to the upper atmosphere and the response of the ionosphere-thermosphere system to this input, in particular the outflow back to the magnetosphere.			
16	Create an interagency joint laboratory physics program that addresses issues relevant to space physics.			

key science goals. However, in times of constrained budgets, it may not be possible to enact all of these imperatives, and the panel has, therefore, undertaken a prioritization process to help identify which of these are the most important. The panel's prioritization—which informed the survey committee but does not carry its imprimatur—is based on the overarching objective to identify initiatives that will most cost-effectively enable the science of the future. Accordingly, the SWMI panel adopted the following criteria:

- *General considerations for prioritization*
  —Focus on elements that are directly relevant to the enumerated SWMI science goals.
  —Consider historical evidence about contributions of various elements to scientific progress.
  —Consider the cost of an imperative versus its likely return.
- *For mission-related imperatives*
  —How well does the mission address the SWMI science goals?
  —Is it highly focused or does it address a broader range of the goals?
  —How feasible is it?
  —Will it fit within the expected budget?
  —What is the science return for the cost?
  —Does it require technology development?
  —What is the broader impact (for example, for the science objectives of the other panels and for the development of a forecasting system for space weather)?
- *For other capabilities*
  —How central is the capability to the accomplishment of the SWMI science goals?
  —Is it currently in danger?
  —Could it make a much bigger contribution with a modest enhancement?

Using these criteria, the SWMI panel prioritized its imperatives not only within but also across the three categories discussed above—missions, DRIVE initiative, and space weather—in order to identify the most cost-effective approach to accomplishing its science goals. The full set of prioritized SWMI imperatives, with their mapping to the three categories, is presented in Table 9.4.

9.6 CONNECTIONS TO OTHER DISCIPLINES

9.6.1 Solar and Heliospheric Physics

It is clear from the foregoing discussion that the variable solar wind is the dominant driver of magnetospheric dynamics. To fully understand the ways in which long-term solar variations as well as short-term eruptive events such as solar flares or coronal mass ejections can produce dramatic effects of importance to humans at Earth, solar wind measurements upstream of Earth's magnetosphere are essential. At present a nearly continuous record of measurements of the solar wind flow velocity, density, and temperature is available from the very early days of the space program up to the present time. Any interruption in the continuity of these measurements would have serious consequences for the ability to study the effects that solar variations have on Earth's magnetosphere, ionosphere, and atmosphere, and for the ability to forecast significant societal impacts.

9.6.2 Atmosphere and Ionosphere

The past decade has significantly reinforced appreciation of the key influences the ionosphere has on magnetospheric behavior and of the importance of magnetospheric driving for ionospheric behavior. It is thus impossible to understand one without taking into account the two-way coupling with the other. While this decadal survey separates the two regions for ease of discussion, it is clear that they must be understood as a coupled system. During the coming decade one of the most important emphases of space physics research will be to clarify quantitatively the coupling of these elements in order to enable progress toward a predictive understanding of both.

9.6.3 Planetary Science

The magnetospheres of other planets display not only certain close similarities, such as the formation of bow shocks and radiation belts, but also many processes that are markedly different, such as the source of charged particles within the radiation belts, which can vary from a mixture of the solar wind and ionosphere at Earth, to lava volcanoes and water geysers on small moons at Jupiter and Saturn. This rich diversity has provided surprising discoveries of the breadth of expression of fundamental physical processes that play important roles in the acceleration of charged particles and the generation of magnetic fields in planetary magnetospheres. Both planetary and magnetospheric understanding is thus enriched by the comparative study of magnetospheres. With orbital missions at many of the solar system's other planets, comparative magnetospheric studies should increasingly provide insights into the diversity of magnetospheric processes and how they couple to the planets themselves.

9.6.4 Physics and Astrophysics

Many of the processes and phenomena that determine magnetospheric behavior act throughout the universe in settings as diverse as laboratory plasmas and supernova shock waves. The ability to observe and diagnose these processes in situ, with no interference from walls, is a powerful tool to enable basic understanding with broad applicability. Examples of such processes include magnetic reconnection, shock acceleration of charged particles, and the physics of rapidly rotating magnetized bodies. Moreover, the study of comparative magnetospheres not only elucidates the behavior of magnetized planets in our solar system, but also points the way to the potential breadth of behavior of magnetized bodies throughout the universe.

9.6.5 Complex Nonlinear System Studies

Eruptive solar processes such as solar flares and spontaneously occurring episodic dynamical variations in planetary magnetospheres known as magnetic substorms involve highly nonlinear systems. Similarly, planetary radiation belts evolve through a complicated and nonlinear combination of wave-particle processes that accelerate particles to very high energies, transport them throughout the system, and cause their loss. Although many of these phenomena are at best poorly understood, their study serves as an important driver for furthering understanding of the mathematical analysis of complex nonlinear systems. Predicting the behavior of the vast, coupled, multiscale, and nonlinear Sun-magnetosphere-atmosphere system presents a major challenge to mathematical and computational methods.

10

Report of the Panel on Solar and Heliospheric Physics

To grasp immediately the nature of the field of solar and heliospheric physics, one need only look at the striking image of the 2010 solar eclipse shown in Figure 10.1. The solar atmosphere extends outward from the Sun's surface, apparently without end. In fact, this extended atmosphere—the heliosphere—does end at roughly 140 AU, where the solar wind runs into the galaxy at the heliopause. Our Earth, its upper atmosphere, and its magnetosphere—and the other planets of the solar system—are all embedded deep inside this extended stellar atmosphere. Hence, the vast physical domain stretching from the center of the Sun out to the heliopause defines our home in space. The heliopause is where our home ends and the rest of the universe begins.

10.1 PHYSICS OF THE SUN AND HELIOSPHERE—MAJOR SCIENCE GOALS

To achieve the key science goals set forth by this decadal survey in Chapter 1, it is necessary to understand the physical processes that create and sustain connections within the heliosphere. Figure 10.1 demonstrates, however, that gaining such an understanding requires overcoming major fundamental physics challenges. The most striking feature of the image is its beautifully intricate structure—obviously due to the Sun's magnetic field—that has its origins deep below the solar surface. Understanding the processes that produce the solar field is a prerequisite for achieving the goals of the survey, but the recent, unexpected deep solar minimum demonstrates that understanding of the Sun's dynamo is still rudimentary. Consequently, the Panel on Solar and Heliospheric Physics (SHP) identified as the first of four SHP science goals for the upcoming decade, SHP1, to **determine how the Sun generates the quasi-cyclical variable magnetic field that extends throughout the heliosphere.**

The extended structure of Figure 10.1 shows that the magnetic field provides a coupling across the solar and heliospheric domain, from the solar interior to the galaxy. In fact, the field not only couples the solar and heliospheric domain but also is essential for its creation. All current theories for heating the corona and accelerating the wind postulate the magnetic field as the central agent. Although this is not immediately obvious from Figure 10.1, because it is only a snapshot, the magnetic coupling of the Sun and heliosphere is always dynamic. The field at the solar surface and low corona is observed to vary dra-

FIGURE 10.1 White-light image of the solar corona out to 4 solar radii during the solar eclipse of November 7, 2010. SOURCE: Courtesy of M. Druckmüller, Brno University of Technology, M. Dietzel, Graz University of Technology, S. Habbal, Institute for Astronomy, and V. Rušin, Slovak Academy of Sciences.

matically on temporal scales ranging from seconds to decades. In situ measurements of the heliosphere indicate that that variability creates a broad range of temporal scales, from subseconds for discontinuities produced by explosive events to decades for the solar cycle evolution. Furthermore, the heliosphere is inherently turbulent, and so it produces internal dynamics down to kinetic dissipation scales. All those heliospheric dynamics couple directly to Earth's magnetosphere and upper atmosphere, again by the magnetic field, producing the myriad forms of space weather that we must understand and mitigate. The SHP panel therefore identified as its second major science goal for the coming decade, SHP2, to **determine how the Sun's magnetism creates its dynamic atmosphere.**

In addition to its ceaseless dynamics, a defining feature of our home in space is that it is filled with high-energy radiation, both electromagnetic—ultraviolet (UV) rays, X rays, and gamma rays—and particulate. High-energy particle radiation, especially protons, is the greatest threat to human exploration of the solar system and to deep-space missions. The most dangerous bursts of energetic particles are due to the giant explosions of the solar magnetic field and matter known as coronal mass ejections/eruptive flares. Those events also produce the most destructive space weather at Earth, with damaging consequences ranging from communication and GPS blackouts to the loss of high-orbiting satellites and power transformers on the ground.

In recent years, scientists have made substantial progress in understanding the Sun's magnetic explosions. Researchers know where on the Sun they are likely to occur but are far from understanding when and how large the explosions will be and, in particular, how so much of their energy produces particle radiation. It should be emphasized that the processes of magnetic energy storage and release are commonly observed in all laboratory and cosmic plasmas; therefore, understanding them in the Sun and heliosphere will be a fundamental advance for all physics. Thus, the SHP panel's third major science goal, SHP3, is to **determine how magnetic energy is stored and explosively released.**

As stated above, the heliopause is where the Sun's extended atmosphere ends and the galactic medium begins. That interface where two regions collide is generally rich in unexplored and unique physics. For example, the dominant energy form in the region is due not to the solar wind plasma or magnetic field but rather to interstellar neutrals that stream freely into the heliosphere and charge exchange with the solar wind ions. The magnetic structure of the region is also unique in that the field is wound into such a tight spiral due to solar rotation that it is almost cylindrical rather than radial. It should also be noted that the structure of the outer heliosphere has important effects close to home: it determines the penetration of high-energy galactic cosmic rays into near-Earth space. In situ measurements by the Voyager spacecraft and new methods to globally image the outer boundary of the heliosphere are spurring a revolution in understanding of this region. During the next decade, the Voyager spacecraft should pass the heliopause and enter interstellar space. Extending their presence robotically, humans will have left their home in space for the first time and entered the universe—a truly historic event. The coming decade will be critical for gaining an understanding not only of how our space environment is created and driven, but also of how it ends. That is the SHP panel's final major science goal for the decade, SHP4, to **discover how the Sun interacts with the local galactic medium and protects Earth.**

Associated with each of the SHP panel's four science goals are several SHP actions that, if carried out, promise substantial progress in achieving the goals. Box 10.1 summarizes the SHP panel's four major science goals and 14 associated actions.

10.2 SOLAR AND HELIOSPHERIC PHYSICS IMPERATIVES

To achieve the four major science goals listed in Box 10.1, the SHP panel developed a strategy that consists of a set of imperatives for the federal agencies involved in solar and heliospheric research. The imperatives—actions that are essential for future progress—are listed briefly below according to the relevant agency (or agencies) and discussed in detail below in this chapter.

10.2.1 Prioritized Imperatives for NASA

1. Complete the development and launch of the Interface Region Imaging Spectrograph (IRIS) and Solar Probe Plus (SPP) missions, and deliver U.S. contributions to the European Space Agency–National Aeronautics and Space Administration (ESA–NASA) Solar Orbiter mission. The measurements from those missions are central to the SHP panel's strategy for addressing SHP science goals 1, 2, and 3 in the next decade (§10.5.3.1).

2. Augment the heliophysics Explorer budget to expand launch opportunities and add new cost-effective mid-size launch vehicles (§10.5.3.2).

BOX 10.1 SOLAR AND HELIOSPHERIC PHYSICS PANEL'S MAJOR SCIENCE GOALS AND ASSOCIATED ACTIONS

SHP1. Determine how the Sun generates the quasi-cyclical variable magnetic field that extends throughout the heliosphere.

a. Measure and model the near-surface polar mass flows and magnetic fields that seed variations in the solar cycle.
b. Measure and model the deep mass flows in the convection zone and tachocline that are believed to drive the solar dynamo.
c. Determine the role of small-scale magnetic fields in driving global-scale irradiance variability and activity in the solar atmosphere.

SHP2. Determine how the Sun's magnetism creates its dynamic atmosphere.

a. Determine whether chromospheric dynamics is the origin of heat and mass fluxes into the corona and solar wind.
b. Determine how magnetic free energy is transmitted from the photosphere to the corona.
c. Discover how the thermal structure of the closed-field corona is determined.
d. Discover the origin of the solar wind's dynamics and structure.

SHP3. Determine how magnetic energy is stored and explosively released.

a. Determine how the sudden release of magnetic energy enables both flares and coronal mass ejections to accelerate particles to high energies efficiently.
b. Identify the locations and mechanisms that operate in impulsive solar energetic-particle sites, and determine whether particle acceleration plays a role in coronal heating.
c. Determine the origin and variability of suprathermal electrons, protons, and heavy ions on timescales of minutes to hours.
d. Develop advanced methods for forecasting and nowcasting of solar eruptive events and space weather.

SHP4. Discover how the Sun interacts with the local galactic medium and protects Earth.

a. Determine the spatial-temporal evolution of heliospheric boundaries and their interactions.
b. Discover where and how anomalous cosmic rays are accelerated.
c. Explore the properties of the heliopause and surrounding interstellar medium.

---

3. Substantially enhance the following key elements of the NASA heliophysics research and analysis programs so that the science community can use the new measurements to attain the SHP science goals outlined above (§10.5.3.4):

- Create new heliophysics science centers composed of teams of theorists, numerical modelers, and data experts to tackle major science problems.
- Implement a graduated increase in individual principal investigator grant programs to 20 percent of Heliophysics Division funding, primarily by increasing the size of individual grants.
- Augment the mission operations and data analysis program to facilitate systems science using the Heliophysics Systems Observatory.
- Ensure that all newly confirmed missions have Phase E budgets adequate to ensure mission success and that they sponsor a guest-investigator program as part of their success criteria.

- Form a consolidated heliophysics instrument and technology development program for innovative instrument concepts.
- Provide flight and training opportunities through increased funding of the Low Cost Access to Space program for science-payload development and data analysis.

4. To attack SHP science goals 3 and 4, develop the Interstellar Mapping and Acceleration Probe (IMAP), a mission to observe the interaction of the heliosphere with the interstellar medium, and measure suprathermal ion populations with unprecedented resolution (§10.5.2.2). IMAP is the SHP panel's highest-priority mid-size mission.
5. To attack SHP science goals 2 and 3, support U.S. participation in the Japanese-led Solar-C mission that will measure the magnetic coupling of the lower atmosphere into the corona (§10.5.2.3).
6. To maximize the science returns on limited resources, extend the Explorer mission development model to mid-size strategic missions (§10.5.3.3).
7. To attack SHP science goal 3, develop the Solar Eruptive Events (SEE) mission to image electron and ion acceleration in SEEs with unprecedented resolution. This is the SHP panel's highest-priority mission concept for an LWS-class strategic mission (§10.5.2.4).
8. Carry out advanced planning for the Solar Polar Imager and Interstellar Probe missions (§10.5.2.6, §10.5.2.7).
9. To develop solar-sail propulsion for future Heliophysics Division missions, invest about $50 million as "seed money" in a full-scale solar sail demonstration mission by partnering with the Office of the Chief Technologist's Technology Demonstration Missions program (§10.5.2.8).

10.2.2 Prioritized Imperatives for NSF

1. Provide base funding sufficient for efficient and scientifically productive operation of the Advanced Solar Technology Telescope, which is also central to achieving the SHP panel's science goals (§10.5.4.1).
2. Establish a midscale projects funding program similar to NASA's Explorer model (§10.5.4.2).
3. Fund the development and operation of the Frequency-Agile Solar Radiotelescope (FASR) to produce three-dimensional images of the solar atmosphere with high temporal and spatial resolution (§10.5.4.3).
4. Fund the development and operation of the Coronal Solar Magnetism Observatory (COSMO), a large-aperture coronagraph, chromosphere, and prominence magnetometer (§10.5.4.4).
5. Double the size of the National Science Foundation (NSF) small-grants programs to support the effort required for analyzing and modeling the new data sets (§10.5.4.5).
6. Broaden the definition of NSF's Solar-Terrestrial Research Program to include outer-heliosphere research (§10.5.4.6).

10.2.3 Prioritized Multiagency Imperatives

1. Ensure continued support of current ground-based observations of the Sun, especially the line-of-sight and vector measurements of the solar magnetic field that are being used routinely for space weather operations (§10.5.5.1).
2. Ensure continued support of laboratory facilities for accurate measurement of atomic properties and for instrument calibrations (§10.5.5.2).
3. Develop real-time and near-real-time instrumentation and data streams from future missions and ground-based facilities. In collaboration with other agencies, fly missions devoted to space weather objectives. Establish an interagency clearinghouse and archive for space weather data (§10.5.5.3).
4. Ensure continuity of real-time solar wind measurements from L1 (§10.5.5.4).

5. Identify key long-term solar and heliospheric data sets and recommend approaches to ensure that they are continued and archived (§10.5.5.5).

6. Continue support of laboratory plasma physics at current or higher levels to complement spacecraft measurements in understanding basic heliophysical processes (§10.5.5.6).

7. NASA, the National Oceanic and Atmospheric Administration (NOAA), and the Department of Defense are encouraged to develop a plan for a mission at the L5 Lagrangian point to conduct high-priority helioseismology studies and develop advanced capabilities to forecast space weather (§10.5.5.7).

10.3 SIGNIFICANT ACCOMPLISHMENTS DURING THE PREVIOUS DECADE

There has been considerable progress along with many surprising discoveries in the disciplines that constitute solar and heliospheric physics since publication in 2003 of the National Research Council (NRC) decadal survey, *The Sun to the Earth—and Beyond: A Decadal Research Strategy in Solar and Space Physics.*[1] In the following four subsections, organized by the four SHP science goals outlined in Section 10.1, the panel describes a small sampling of recent developments. Accomplishments or goals that address the three decadal survey guiding motivations (M1-M3) are also noted.[2]

10.3.1 Determining How the Sun Generates the Quasi-cyclical Variable Magnetic Field That Extends Throughout the Heliosphere

As emphasized in the 2003 decadal survey,[3] an enduring major science goal is to determine how the Sun generates its quasi-cyclical variable magnetic field. The practical goal of such research is to learn enough to be able to help to predict the changing space environment and its societal impact (motivation M2). Here the SHP panel sketches four of the many notable accomplishments of the past decade toward meeting these goals.

With a wide array of ground- and space-based sensors, solar activity in its many forms was observed to fall in 2008-2009 to low levels not seen for nearly a century. Solar activity is driven by the solar magnetic field, and the recent decline in the polar magnetic field flux is shown in Figure 10.2. The sharp decline was not generally expected. However, researchers noted that before the activity minimum measurements by ground-based instruments and space-based instruments on the Solar and Heliospheric Observatory (SOHO) showed unusually small amounts of magnetic flux near the poles of the Sun. Considered by some to be a useful precursor of the strength of an activity cycle, the low flux levels led to predictions that the current solar cycle maximum would be the lowest since polar flux measurements became available—as appears to be the case. Other venerable solar activity precursors that suggested a high level of activity proved to be spectacularly unreliable. Vital observational work in this research is continuing with ground-based and space-based assets, particularly because there is a chance that the Sun is entering a sustained period of low activity.

The effects of record low solar activity were observed throughout the heliosphere. Among the prominent effects with societal consequences were these: cosmic ray fluxes near Earth reached the highest levels on record, and reduced heating of Earth's upper atmosphere by solar UV radiation led to less drag on satellites.

---

[1] National Research Council (NRC), *The Sun to the Earth—and Beyond: A Decadal Research Strategy in Solar and Space Physics,* The National Academies Press, Washington, D.C., 2003; and NRC, *The Sun to the Earth—and Beyond: Panel Reports,* The National Academies Press, Washington, D.C., 2003.

[2] The motivations referred to in this section are those outlined in the introduction to Part II of this report: M1, Understand our home in the solar system; M2, Predict the changing space environment and its societal impact; M3, Explore space to reveal universal physical processes.

[3] NRC, *The Sun to the Earth—and Beyond: Panel Reports,* 2003, p. 12.

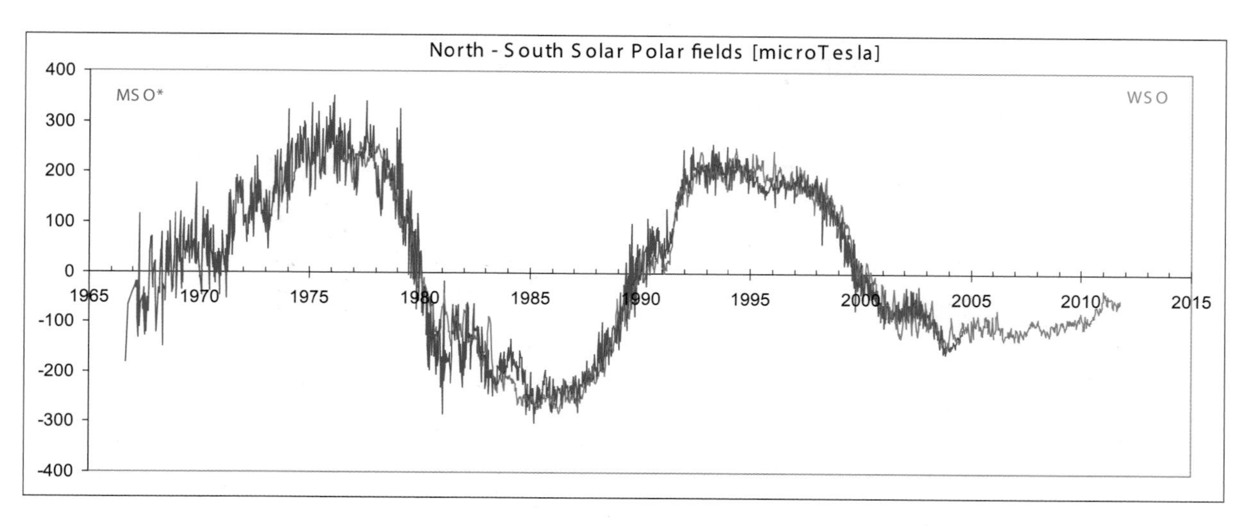

FIGURE 10.2 Average value of the magnetic flux density at the Sun's polar regions during the last four cycles. The polar flux density peaks at the times of maximum. The recent cycle minimum had the weakest polar flux ever recorded. SOURCE: Courtesy of Leif Svalgaard, Stanford University. Red, data from Wilcox Solar Observatory; blue, data from Mt. Wilson Observatory scaled to match.

Those surprising events and the rapid development of flux-transport dynamo models have led to intense recent work in solar dynamo modeling. Two key issues that involve the transport of magnetic flux have emerged: the speed with which meridional flows move poloidal magnetic flux to the solar poles and the extent to which this moving flux stays near the surface or diffuses inward on its journey to the poles. Because of the importance of those issues, there is a critical need for measurements of meridional flow and its variability. The measurements are difficult, and a consensus view on how to proceed has yet to emerge.

Observational spatial resolution has greatly improved through image-processing techniques applied to ground-based observations made with the 1.6-m New Solar Telescope (NST) at the Big Bear Solar Observatory and other 1-m-class apertures and direct observations made with the 0.5-m Hinode satellite. But it is the combination of these state-of-the-art observations with advanced numerical modeling that has, after 400 years of speculation, revealed the main physical processes at work in sunspot penumbral filaments, bright umbral dots, bright faculae, and small-scale magnetic elements. Figure 10.3 shows a numerical simulation of a sunspot and an actual photograph. Good observational and modeling progress has also been made with respect to the more complicated upper layers of the solar atmosphere.

Continuing and improved helioseismic measurements of the solar interior from instruments including those on the SOHO and SDO spacecraft, launched in 1995 and 2010, respectively, and from the GONG ground-based network, have shown cycle-related changes in large-scale internal zonal and meridional flows. The varying flows may play a driving role rather than just being consequences of the solar activity cycle, and they narrow the wide range of realistic solar dynamo models. Helioseismic observations of active regions have shown subsurface helical flows whose strength is closely related to flare activity. In addition, a new analytic method shows surprisingly deep-seated effects of large active regions well before they emerge at the surface (Figure 10.4). Either or both of those findings may develop into useful forecasts of strong solar activity with societal significance.

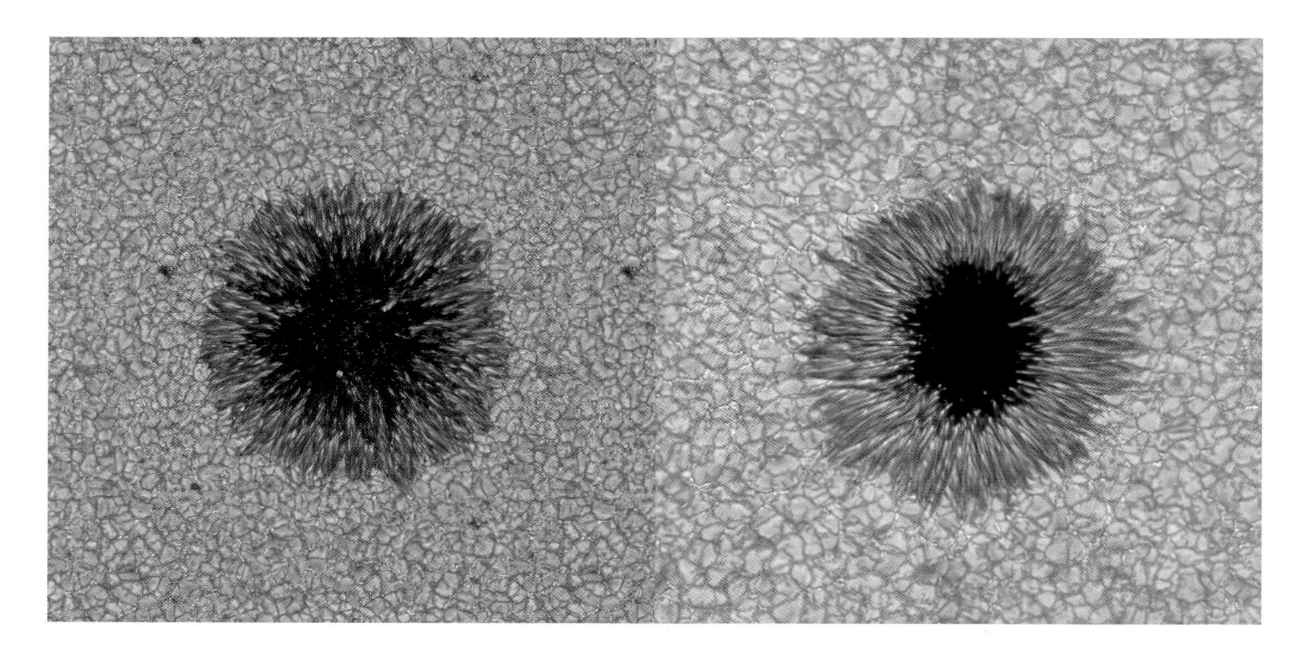

FIGURE 10.3 Numerical simulation of a sunspot (*left*) and a very-high-resolution image (*right*) from the New Solar Telescope at Big Bear Observatory, which is operated by the New Jersey Institute of Technology. Detailed comparisons have elucidated the physics of such solar features. SOURCE: *Left*, courtesy of M. Rempel, High Altitude Observatory. *Right*, courtesy of Big Bear Solar Observatory. For further information on the simulation, see M. Rempel, Numerical sunspot models: Robustness of photospheric velocity and magnetic field structure, *Astrophysical Journal* 750(1):62, 2012.

The Sun sustains life on Earth by its nearly steady output of light. Measurements of total solar irradiance (TSI) with different space-based instruments have consistently shown a cycle variation on the order of 0.1 percent but give conflicting results about the absolute value of the TSI. The conflicts have recently been resolved in favor of a lower value that improves agreement with climate research results. The original conflicting results and new results are shown in the two panels of Figure 10.5. The new results indicate that the TSI at the recent cycle minimum was less than at previous minima. It has also been suggested that this level is about as low as can be expected even if a long cessation of solar activity occurs (as during the Maunder minimum).

10.3.2 Determining How the Sun's Magnetism Creates Its Dynamic Atmosphere

Major advances have been made during the past decade in understanding the dynamic and coupled nature of the Sun's extended atmosphere, which reaches from the chromosphere and corona to the end of the heliosphere (motivations M1-M3). The advances highlight the need for direct sampling of the solar wind in the deep inner heliosphere as close to the Sun as possible, with high-resolution measurements of plasma and electromagnetic fluctuations; high-spatial- and high-temporal-resolution observations of the solar atmosphere from the photosphere into the corona; accurate magnetic field measurements in the low-beta chromosphere and corona; and advanced theoretical models that can couple closely to the upcoming observations.

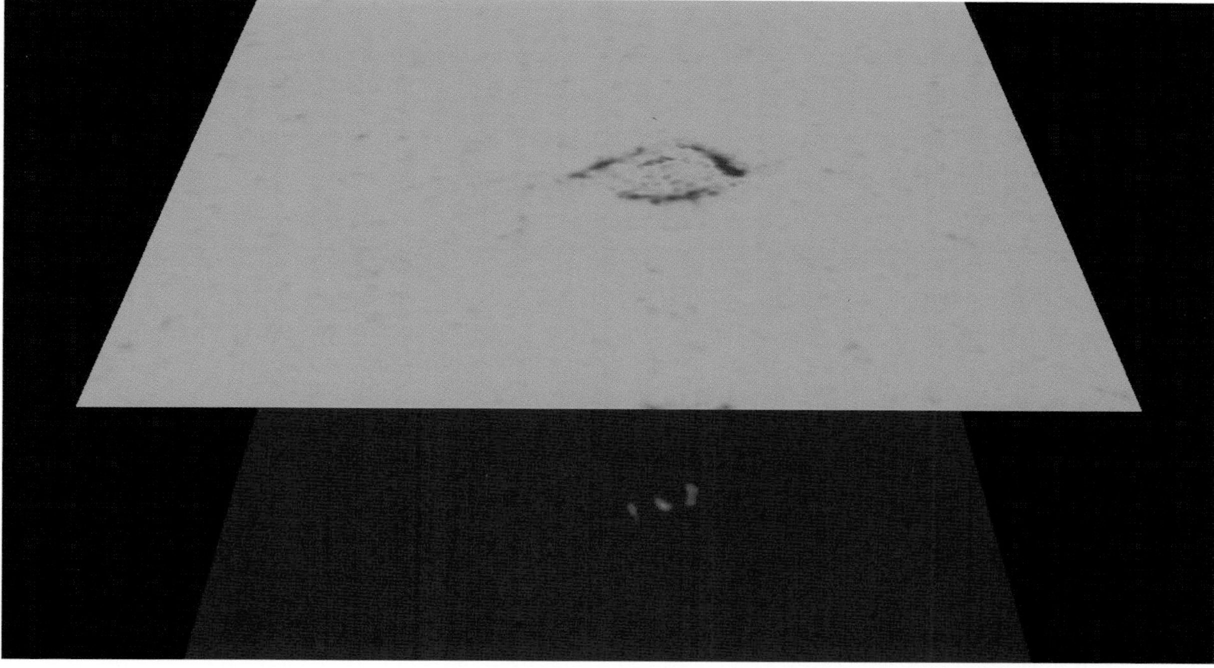

FIGURE 10.4 Upper panel shows the surface magnetic field (grey) with a travel-time perturbation map constructed at a depth of 42-75 million meters (blue) about a day before an active region appeared at the surface as seen in the lower panel. SOURCE: Adapted from S. Ilonidis, J. Zhao, and A. Kosovichev, Detection of emerging sunspot regions in the solar interior, *Science* 333:993-996, 2011. Reprinted with permission from AAAS.

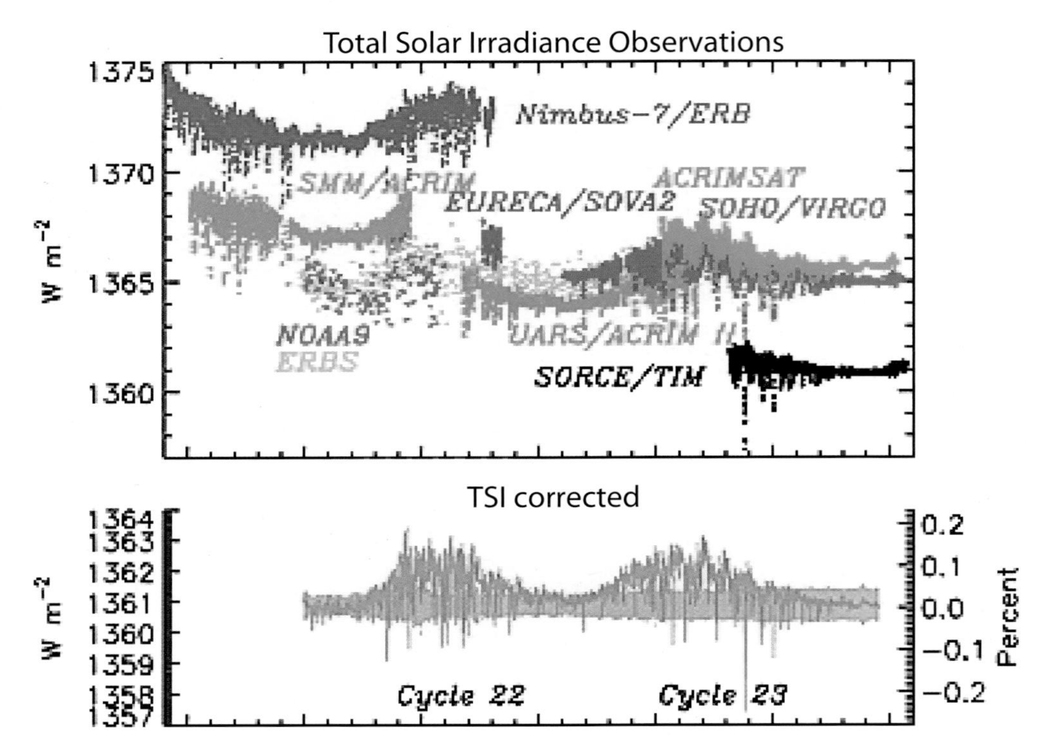

FIGURE 10.5 Recent improvements in the accuracy and precision of measurements of total solar irradiance have resolved several basic issues. The upper panel shows the original discordant measurements before recalibration. SOURCE: G. Kopp and J. Lean, A new, lower value of total solar irradiance: Evidence and climate significance, *Geophysical Research Letters* 38(1):L01706, doi:10.1029/2010GL04577, 2011. Copyright 2011 American Geophysical Union. Reproduced by permission of American Geophysical Union.

In situ measurements of the solar wind plasma velocity distribution functions and electromagnetic fluctuations permit us to identify the dominant kinetic physics responsible for the generation and dissipation of Alfvénic fluctuations and other types of solar wind turbulence (motivations M1 and M3). Progress has been made in measuring and understanding the types of turbulent fluctuations in the solar wind and the likely dissipation mechanisms, but important questions remain. Observations from spacecraft—such as Cluster, Wind, and ACE—and archival data from Helios have been used to probe the power in fluctuations perpendicular and parallel to the local magnetic field as a function of scale and have demonstrated that these fluctuations are highly anisotropic. Higher time-resolution measurements have allowed researchers to trace the evolution of the turbulent cascade and dissipation past ion kinetic scales toward electron scales. More work is needed to determine the relative fraction of fluctuations dissipated by ions and electrons and to distinguish between different mechanisms for dissipation, such as ion-cyclotron resonance, kinetic Alfvén waves, or the formation of small-scale current sheets. It is also unclear where broadband Alfvénic fluctuations and other types of solar wind turbulence, such as 1/f noise, originate, either in the corona or during propagation in the inner heliosphere. Further theoretical work and joint analysis of solar wind plasma and electromagnetic fields will allow a determination of whether nonlinear plasma physics can explain the highly nonadiabatic expansion of the solar wind.

In the past decade, substantial progress has also been made in understanding the development of kinetic instabilities in the solar wind due to nonthermal ion and electron distribution functions; in particular, how temperature anisotropies are limited by the mirror, firehose, and cyclotron instabilities (motivations M1 and M3). A surprising result is that the mirror instability appears to play a stronger role than the cyclotron instability in limiting temperature anisotropy. Instabilities have also been proposed to explain the observed restriction of differential ion flow to the local Alfvén speed, but observational confirmation of these theories has remained elusive.

In situ measurements of the solar wind are also a powerful diagnostic of the evolving connection between the corona and interplanetary space, and measurements in the past decade have revealed new features in the solar wind and related them to the structure and dynamics of the inner heliosphere and evolving Sun. The anomalously low levels of solar activity in the recent solar minimum were associated with substantial decreases in the density and pressure of the solar wind. Charge state and composition allow changes in the coronal sources of the solar wind to be tracked independently of changes in the solar wind speed due to the evolution of the plasma as it expands into interplanetary space. It is now understood that the three distinct forms of wind—fast, slow, and transient (associated with ICMEs)—can be identified clearly by their ionic charge-state signatures (using O^{7+}/O^{6+}) without assumptions about the dynamic evolution of the wind.

Figure 10.6 shows the fractions of the three solar wind components for the decade 1998-2008. The compositions of the three winds provide vital clues to their sources in the Sun and the mechanisms of their formation. The panel notes the particular importance of furthering understanding of the slow wind, as its source and origin have constituted one of the outstanding problems in solar and heliospheric physics. With new insights into the slow wind's origin and with the upcoming Solar Probe Plus and Solar Orbiter missions, researchers are poised to solve this problem definitively in the coming decade.

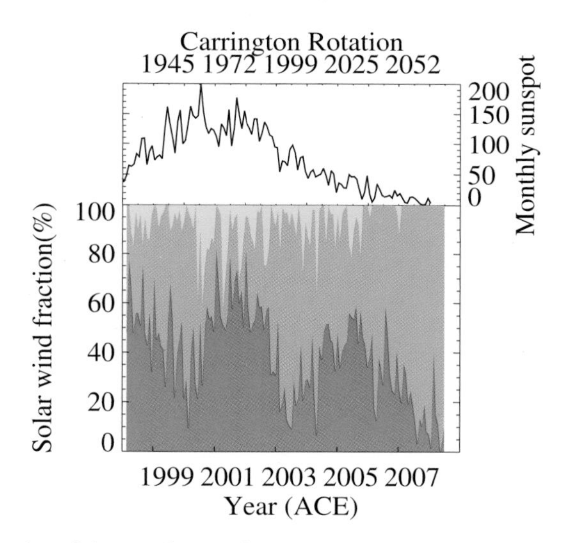

FIGURE 10.6 Sunspot number (*top*) and three solar wind components (*bottom*) during 1998-2008: interplanetary coronal mass ejections (yellow), coronal hole wind (green), and noncoronal hole wind (orange). SOURCE: L. Zhao, T.H. Zurbuchen, and L.A. Fisk, Global distribution of the solar wind during solar cycle 23: ACE observations, *Geophysical Research Letters* 36:L14104, doi:10.1029/2009GL039181, 2009. Copyright 2009 American Geophysical Union. Reproduced by permission of American Geophysical Union.

An unexpected discovery has been the observation of ubiquitous reconnection events in the solar wind, often on a very large scale of about 1 million kilometers. The reconnection events constitute a major puzzle in that they do not appear to accelerate particles to speeds greater than the local Alfvén speed. Again, observations nearer to the Sun will be critical for understanding the physical properties of this type of reconnection and its role in driving solar wind dynamics. Figure 10.7 details the structure of a giant reconnection event in the solar wind. Is there evidence of remnants of plumes in the inner heliosphere? Are there remnants or observable consequences of type II spicules? Do nanoflares indicate a heating process that continues throughout the solar atmosphere? Are small-scale nanoflares and turbulent current sheets (reconnection sites) physically related to solar wind discontinuities, and do they represent a form of turbulence heating mechanism that may heat the corona from the solar wind acceleration region to the outer heliosphere? The need for high-resolution observations can be seen in all levels of the solar atmosphere. The Solar Optical Telescope of the Hinode satellite has provided high-resolution images of the solar chromosphere in the Ca II and Hα spectral lines. At the solar limb, these images reveal a very rich, relentlessly dynamic, and highly structured environment. A new type of spicule has been discovered that may play a crucial role in providing mass and energy transfer to the corona. Figure 10.8 shows images of these solar spicules.

The solar atmosphere and wind are both created and structured through the medium of the Sun's powerful magnetic field; consequently, accurate measurements of the field in the corona and wind are

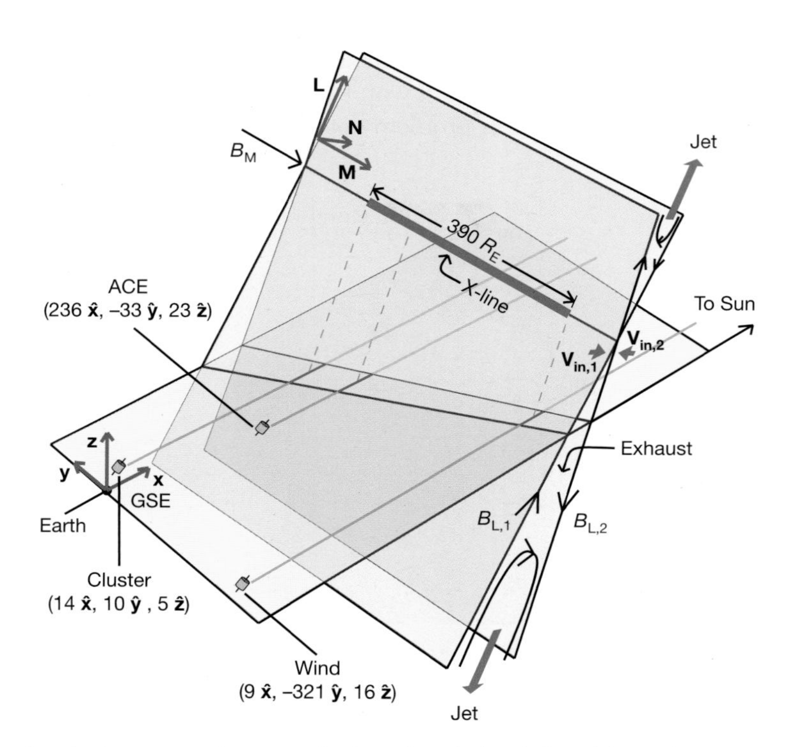

FIGURE 10.7 Structure of a giant reconnection event in the solar wind as inferred from the combination of Cluster, Wind, and ACE data. SOURCE: Reprinted by permission from Macmillan Publishers Ltd.: *Nature*, T.D. Phan, J.T. Gosling, M.S. Davis, R.M. Skoug, M. Øieroset, R.P. Lin, R.P. Lepping, D.J. McComas, C.W. Smith, H. Reme, and A. Balogh, A magnetic reconnection X-line extending more than 390 Earth radii in the solar wind, *Nature* 439:175-178, 2006.

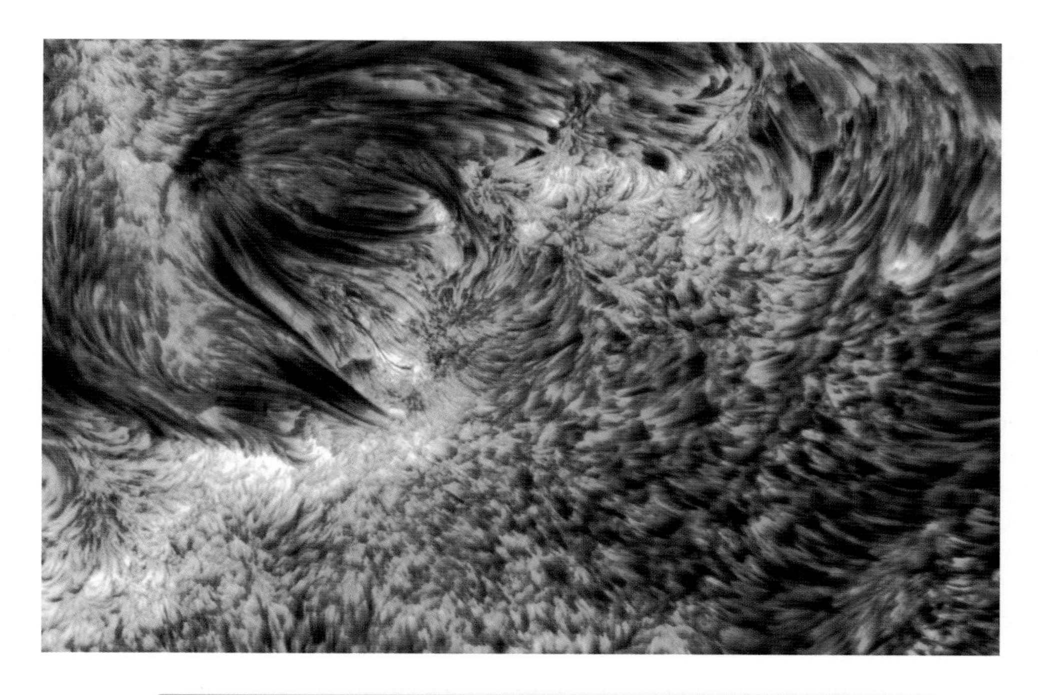

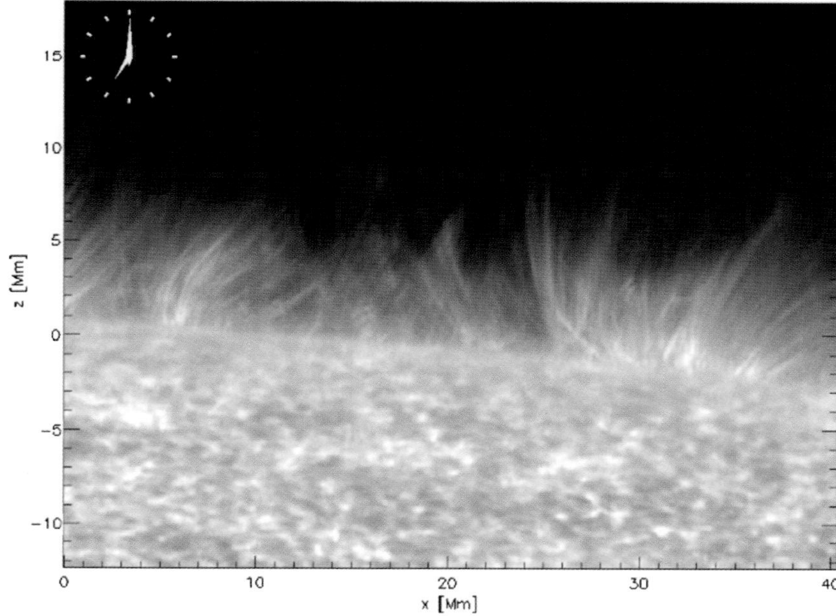

FIGURE 10.8 Images of solar spicules on the disk (*upper*) made with the Swedish Solar Telescope, La Palma, Spain, and on the limb (*lower*) made with the Solar Optical Telescope on Hinode. SOURCE: *Upper*, Courtesy of Bart De Pontieu, Lockheed Martin Solar and Astrophysics Laboratory. *Lower*, Hinode is a Japanese mission developed and launched by ISAS/JAXA, with NAOJ as domestic partner and NASA and STFC (United Kingdom) as international partners. It is operated by these agencies in cooperation with the European Space Agency and NSC (Norway).

essential for understanding our home in space. The lack of such measurements of the corona is one of the greatest obstacles to advancing solar and heliospheric science. Hinode and SDO can determine the full vector field accurately in the photosphere, but the plasma beta is high there, and so it is not possible to extrapolate the magnetic field into the low-beta corona reliably. Two major advances during the past decade promise to overcome the magnetic-field measurement obstacle: the first observations of the full chromospheric vector field on the disk, and the first maps of the coronal field above the solar limb.

The prevalence of high-resolution extreme UV (EUV) narrowband images from instruments on the TRACE (Transition Region and Coronal Explorer) and SDO (Solar Dynamics Observatory) spacecraft have revealed that warm (1 million Kelvin) coronal loops are up to 3 orders of magnitude overdense and hence cannot be in steady state as previously believed (Figure 10.9). Observations and theoretical models of the structures imply that they are still unresolved. How coronal structures are heated is still unknown, but it has been suggested that including the coupling between the chromosphere and corona will prove essential in determining the heating mechanism.

Great progress has also been made in the past decade in achieving closure between observations and predictions from theory or models. The first three-dimensional magnetohydrodynamic (MHD) ab initio numerical simulation of the corona was successfully performed. The chromosphere now stands as the modeling frontier with qualitatively more challenging simulations requiring radiation coupled to MHD (R-MHD) and no assumption of local thermodynamic equilibrium. Fully three-dimensional R-MHD numerical simulations spanning the upper convection zone through the corona—treated as a coherent system—are starting to appear but cannot yet address all the salient physical ingredients on scales larger than a few granules or one supergranule. This state of affairs should improve in the coming decade given the expected rapid progress in numerical hardware and software.

10.3.3 Determining How Magnetic Energy Is Stored and Explosively Released

Major solar flares and associated fast coronal mass ejections (CMEs) are the most powerful explosions and particle accelerators in the solar system, and they produce the most extreme space weather. In the past decade, substantial progress was made in understanding how magnetic energy is stored and explosively released on both large and small scales (motivation M1).

RHESSI hard X-ray (HXR) imaging-spectroscopy measurements have shown that accelerated electrons often contain about 50 percent of the solar-flare energy release and provided strong evidence that energy release-electron acceleration is associated with magnetic reconnection. In one occulted flare, HXR and microwave measurements from the acceleration region, high above the thermal soft-X-ray loop tops (Figure 10.10), showed that essentially all electrons in this region were accelerated to over about 15 keV with no detectable thermal plasma. The accelerated-electron and magnetic-field energy densities were comparable. In large flares, the energy in ions over 1 MeV and in electrons over 20 keV appears comparable. RHESSI gamma-ray imaging of flare-accelerated ions of about 30 MeV shows emission from two small footpoints rather than an extended region. In the largest flare, the footpoints straddled the flare-loop arcade (Figure 10.11), indicating that ion acceleration is also related to magnetic reconnection.

For the first time, flares were detected in total solar irradiance (TSI) with the SORCE/TIM instrument. For the X17 October 28, 2003, flare, the TSI showed both impulsive (HXR-like) and gradual (soft-X-ray-like) components with a peak increase of about 100 ppm. The total radiated energy (over about 10^{32} ergs) and associated-CME kinetic energy (about 10^{33} ergs) were comparable. In addition, the SDO/EVE instrument has discovered an EUV late phase in flares, a second enhancement delayed many minutes after the X-ray peak. In addition, global EUV observations with SDO/AIA and STEREO/EUVI have revealed long-distance "sympathetic" interactions between magnetic fields in flares, eruptions, and CMEs during August 1-2, 2010,

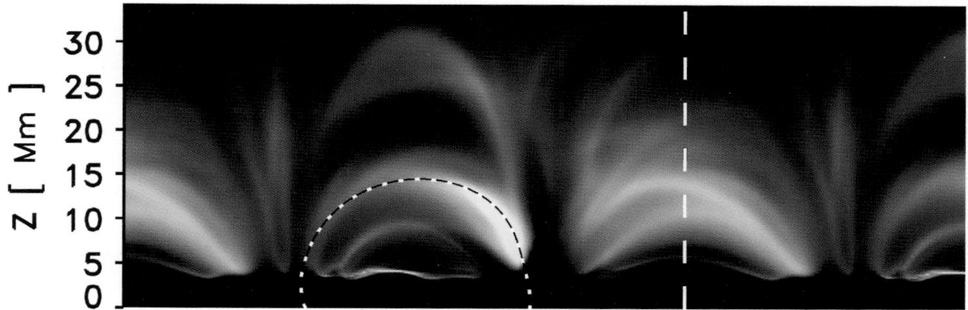

FIGURE 10.9 *Top:* Solar Dynamics Observatory (SDO) observation of 1 million-kelvin active region loops on February 23, 2011. *Bottom:* First three-dimensional magnetohydrodynamic ab initio model for coronal loops. SOURCE: *Top*, Courtesy of NASA/SDO and the AIA science team. *Bottom*, B.V. Gudiksen and A. Nordlund, An ab initio approach to solar coronal loops, *Astrophysical Journal* 618(2):1031-1038, 2005, reproduced by permission of the AAS.

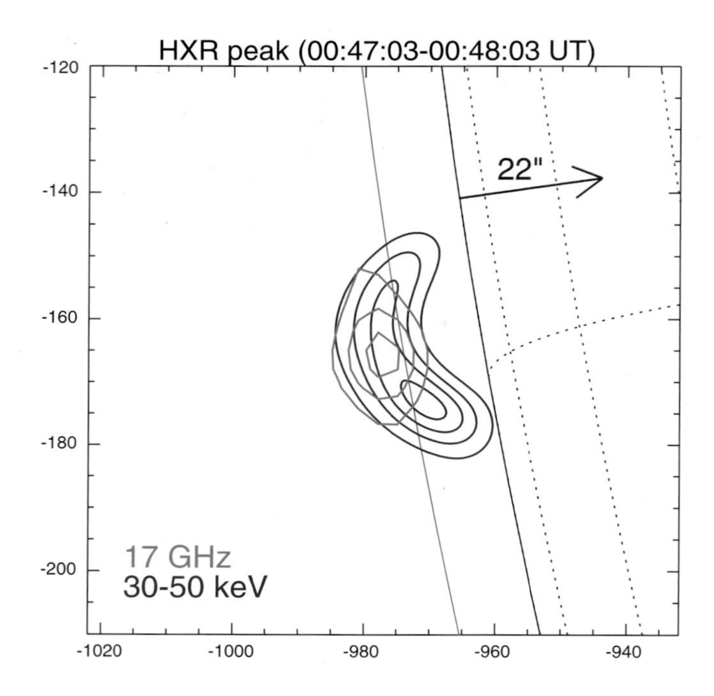

FIGURE 10.10 Hard X-ray (HXR; blue) and microwave (magenta) images of the electron-acceleration/energy-release region in the December 31, 2007, solar flare, in which the bright HXR footpoints were occulted (with HXR and microwave limbs indicated). SOURCE: Courtesy of S. Krucker, University of California, Berkeley, "Electron Acceleration in Solar Flares: RHESSI Hard X-ray Observations," presentation at the Interdisciplinary Workshop on Magnetic Reconnection, February 11, 2010, Southwest Research Institute, San Antonio, Tex.; modified after S. Krucker, H.S. Hudson, L. Glesener, S.M. White, S. Masuda, J.-P. Wuelser, and R.P. Lin, Measurements of the coronal acceleration region of a solar flare, *Astronomical Journal* 714(2):1108-1119, 2010. Reproduced by permission of the AAS.

probably due to perturbations of the magnetic-field topology, indicating that ion acceleration is also related to the magnetic reconnection responsible for the flare.

The acceleration profiles of CMEs below about 4 solar radii (R_S) are closely synchronized with flare-HXR energy releases, implying that the fast CME expansion creates a flare current sheet below the CME, where reconnection results in the particle acceleration responsible for the hard X rays (Figure 10.12). Theory and remote-sensing observations agree that many ejected CMEs contain a magnetic flux-rope structure. The shocks produced by fast CMEs (presumably the main solar energetic-particle [SEP] accelerators) can now often be identified in SOHO/STEREO coronagraph images. Indeed, STEREO/HI images CMEs from the deep corona to 1 AU, revealing their solar origins, propagation to Earth, and space weather impacts.

In a surprising discovery, STEREO A&B detected a burst of 1.6- to 15-MeV energetic neutral hydrogen atoms (ENAs) arriving from the Sun hours before SEPs arrived from the E79 December 5, 2006, solar event. The ENAs were produced by either flare or CME-shock-accelerated protons charge-exchanging with coronal ions. Thus, ENAs provide a new probe of ion acceleration near the Sun.

The largest SEP events are generally thought to be due to CME-driven shock acceleration that can quickly reach energies of 0.1-1 GeV near the Sun. In the past decade, accelerated ions and enhanced resonant waves predicted by diffusive shock-acceleration theory were detected in a few events. Although large

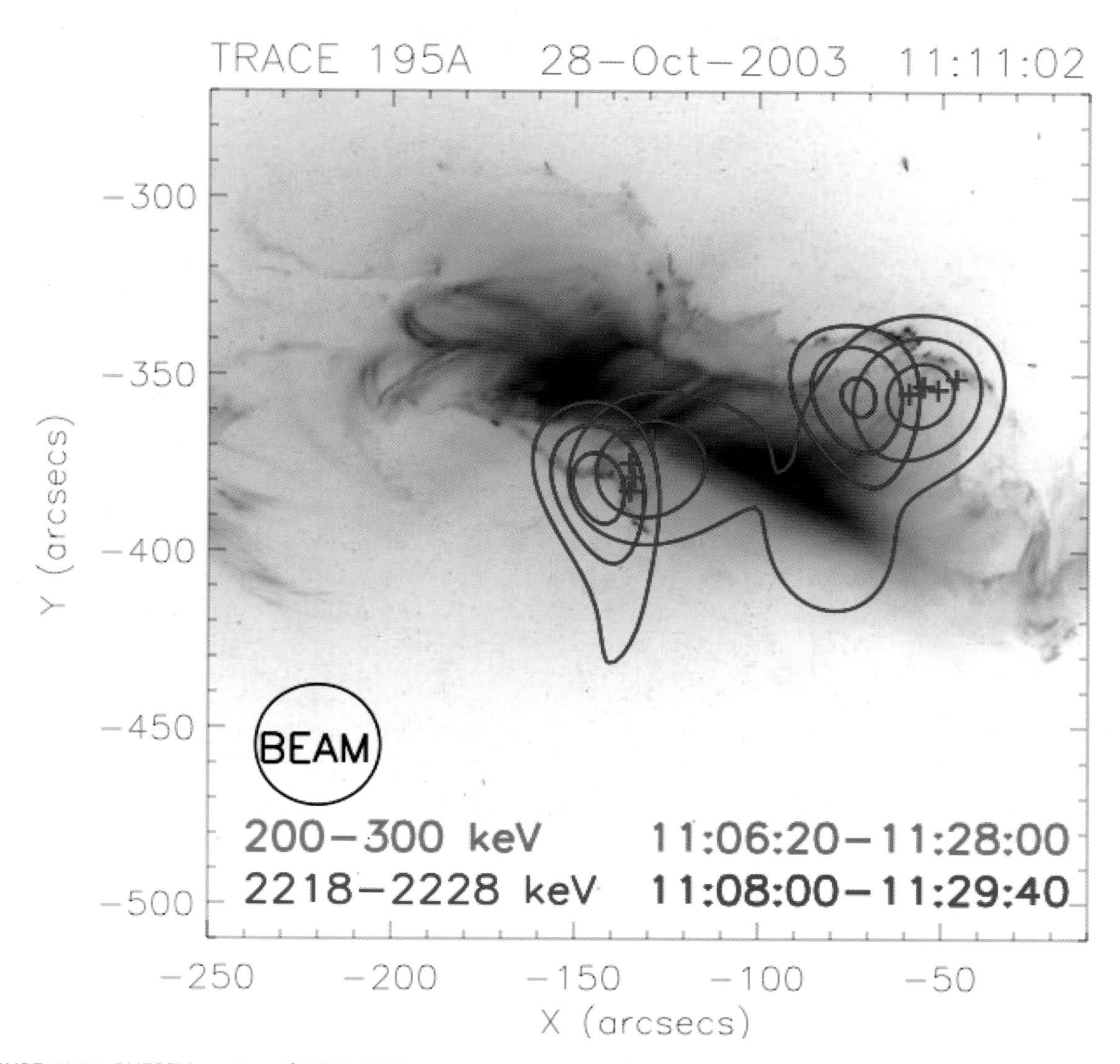

FIGURE 10.11 RHESSI imaging of 2.223-MeV neutron-capture γ-ray line emission (produced by ions over about 30 MeV) and 200- to 300-keV HXR emission (produced by electrons over about 200 keV) for the October 28, 2003, flare, superimposed on a TRACE image of the flare loop arcade. The ions over 30 MeV are confined to two footpoints straddling the loop arcade, similar to the electrons over 200 keV except for a significant (roughly 10^4 km) displacement between the ion and electron footpoints. SOURCE: G.J. Hurford, S. Krucker, R.P. Lin, R.A. Schwartz, G.H. Share, and D.M. Smith, Gamma-ray imaging of the 2003 October/November solar flares, *Astrophysical Journal* 644:L93-L96, doi:10.1086/505329, 2006. Reproduced by permission of the AAS.

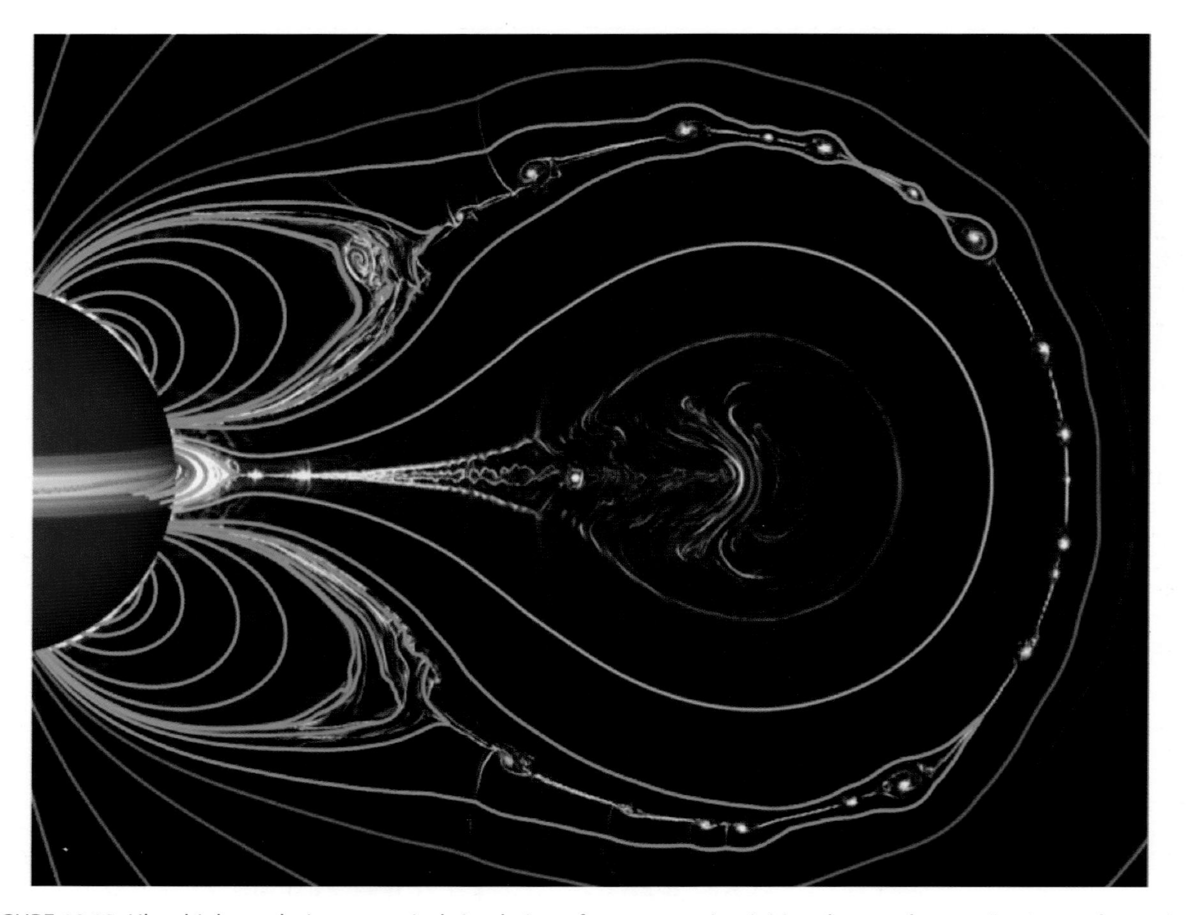

FIGURE 10.12 Ultra-high-resolution numerical simulation of a reconnection-initiated coronal mass ejection and eruptive flare. White contours indicate high current densities. Note the vertical flare current sheet below the erupting plasmoid. The plasmoid undergoes sudden acceleration coincident with the onset of the flare (reconnection in this sheet). A similar result is obtained in ideally driven models. SOURCE: J.T. Karpen, S.K. Antiochos, and C.R. DeVore, The mechanisms for the onset and explosive eruption of coronal mass ejections and eruptive flares, *Astrophysical Journal* 760(1):81, 2012. Reproduced by permission of the AAS.

SEP events are usually associated with strong shocks at 1 AU, there is seldom a one-to-one correspondence between observations and theory, and this shows the need for near-Sun measurements.

Solar-cycle 23 produced about 100 large SEP events, including the largest ground-level event (observed with neutron monitors) since 1956. New instrumentation on ACE, SOHO, and Wind permitted important advances in measuring SEP elemental, isotopic, and ionic-charge-state composition from H to Ni ($1 \leq Z \leq 28$). SEP composition and spectral variations depend on ionic charge-to-mass (Q/M) ratios, probably because of interactions with excited waves and variable shock geometry, but testing acceleration and transport models will require near-Sun measurements where most of the acceleration occurs. SEPs remain a major radiation hazard for spacecraft and astronauts, but the discovery that relativistic electrons provide about a 1-hour warning of arriving SEP ions has provided a new forecast tool (motivation M2).

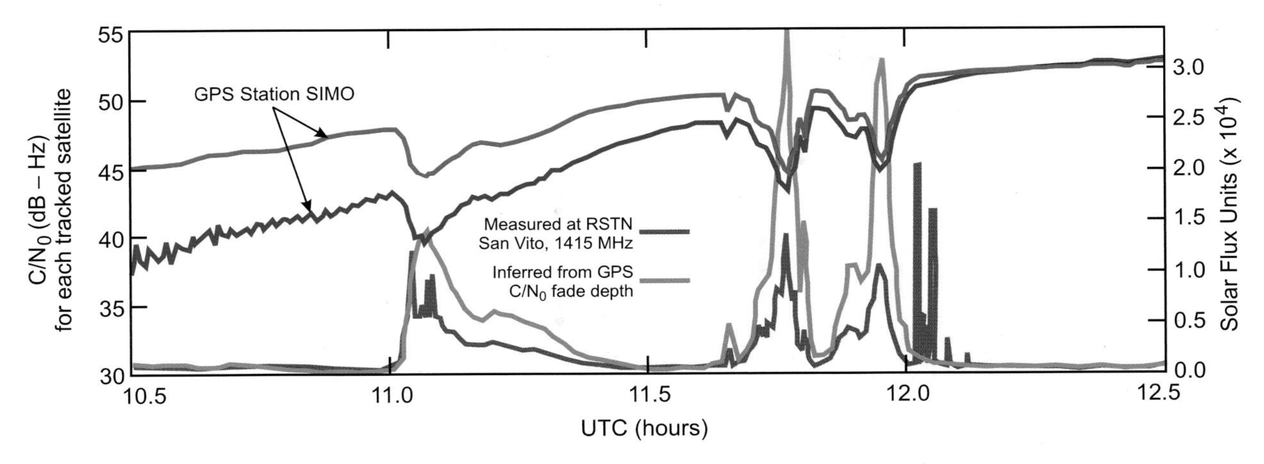

FIGURE 10.13 The effect of "radio noise" from the powerful October 28, 2003, flare on GPS carrier signal-to-noise strength (C/N_0). The red and blue traces show GPS C/N_0 for signals at two receiving stations. The purple trace shows the radio flux density at 1,415 MHz measured by the U.S. Air Force Radio Solar Telescope Network station at San Vito, Italy. The green curve shows the radio flux density inferred from the GPS L1 carrier fade (1,575 MHz). Note the inverse correlation between the C/N_0 fade and the flare's radio flux density. SOURCE: Adapted from P.M. Kintner, Jr., B. O'Hanlon, D.E. Gary, and P.M. Kintner, Global Positioning System and solar radio burst forensics, *Radio Science* 44:RS0A08, doi: 10.1029/2008rs004039, 2009. Copyright 2009 American Geophysical Union. Reproduced by permission of American Geophysical Union.

Radio bursts from major flares interfere with wireless communications and GPS (Figure 10.13), and the "Halloween" solar storm events of 2003 disrupted aircraft navigation systems. Strong magnetic fields in CME-driven disturbances can generate powerful geomagnetic storms that accelerate radiation-belt "killer" electrons and induce ground-level currents that disrupt electric-power grids. The past decade has seen substantial progress in modeling CMEs, shocks, and SEPs from solar eruptions.[4] One system uses CME and real-time solar wind data to drive models that forecast effects on power grids (motivation M2). Furthermore, uncertainties in forecasting CME Earth-arrival times have been reduced from ±12 hours to ±3 hours by using STEREO observations more than 1 day in advance.

The most common SEP events, with 10^4 events per year near solar maximum, are small "impulsive" events associated with coronal jets that are enriched in ^{3}He and heavy ions up to $Z \approx 80$ by amounts that depend on mass-to-charge ratios. ACE and SOHO observations indicate Fe charge states substantially higher than ambient values, most likely because of electron stripping during acceleration in the low corona. ^{3}He and Fe are also enriched in many large SEP events; this indicates that remnant suprathermal particles from previous impulsive flares are an important source of seed particles for CME-shock acceleration.

The past decade has seen a new appreciation of the frequency of occurrence of the halo solar wind (HSW) and its importance in local dynamics. The HSW, a nonthermal tail extending far beyond the thermal ion and electron distributions, makes an important contribution to the local pressure even if the relative density of the halo is small compared with the rest of the solar wind. The HSW and suprathermal tails provide information on nascent particle acceleration in local sites and transport mechanisms for remotely accelerated particles. Distribution functions of locally accelerated suprathermal tails in the heliosphere

[4]See, for example, links to models on the home page of the Community Coordinated Modeling Center (CCMC) at http://ccmc.gsfc.nasa.gov/.

and heliosheath commonly have a power-law index and gradual rollover at higher speeds. Observations reveal that local acceleration occurs in solar wind compression regions and that interplanetary transport modifies spectra of remotely accelerated particles. The variability of the power law and its implications for acceleration models have been topics of focused study.

10.3.4 Discovering How the Sun Interacts with the Local Galactic Medium and Protects Earth

The past decade produced one of the most dramatic advances in space physics. As the Voyager spacecraft approached the termination shock (TS) and entered the heliosheath, a series of groundbreaking discoveries were made. These in situ measurements, combined with all-sky heliospheric images by the Interstellar Boundary Explorer (IBEX) and Cassini mission, led to major advances in our understanding of how the solar system interacts with the interstellar medium (SHP science goal 4; decadal survey key science goal 3).

Voyager 1 (V1) crossed the TS in December 2004 at 94 AU in the Northern Hemisphere, and Voyager 2 (V2) crossed in the Southern Hemisphere in August 2007 at 84 AU. On their locations on the shock, neither Voyager found any evidence of the acceleration of the higher-energy anomalous cosmic rays (ACRs) observed in interplanetary space. That puzzle inspired several new ideas of where and how particles may be accelerated (motivation M3).

In addition, the solar wind did not get heated at the TS nearly as much as expected (Figure 10.14); apparently, 80 percent of the supersonic-flow energy went into suprathermal particles (Voyager measures only thermal plasma). Earlier calculations suggested that acceleration of pickup ions may be the primary dissipation mechanism at the TS. However, that had not been explicitly incorporated into heliospheric models, and so the V2 observation that heating of solar wind protons accounted for only about 20 percent of the dissipation was generally surprising. Remarkably, suprathermal-tail energy spectra have remained power laws with index around −1.5, which is consistent with suprathermal neutral-atom distributions observed in Cassini/INCA images.

Those discoveries affect other disciplines, including astrophysics and plasma physics; the heliosphere has become a test bed for other astrospheres and plasma sheaths. Moreover, the heliosheath plasma environment is unlike any other plasma in the heliosphere.

Beginning about April 2010, V1 observed the heliosheath plasma to become nearly stagnant (zero speed). The V1 plasma instrument is not functional, but the Low-Energy Charged-Particle (LECP) instrument can determine two components of the local plasma velocity by observing its effect on energetic ions. Voyager scientists reported that the flow velocity near the ecliptic plane was near zero, and the north-south component was also near zero on the basis of higher-energy observations. There is no consensus interpretation of those unexpected observations, but one idea is that V1 entered a "transition region," possibly a precursor of the heliopause, starting in April 2010.

To determine the plasma-flow direction at the Voyager locations more precisely, both spacecraft were commanded to execute a series of "roll" maneuvers, allowing the LECP instruments to determine the plasma velocity normal to the ecliptic plane. The rolls have occurred, but results are not yet available. Such maneuvers may also be used after the Voyagers cross the heliopause.

Another major surprise came from global energetic neutral-atom (ENA) maps by IBEX and Cassini. IBEX discovered a completely unpredicted, narrow (about 20°) ribbon of ENA emissions from the outer heliosphere, apparently ordered by the local interstellar magnetic field (ISMF), as indicated by comparison with global MHD models. More than a half-dozen theories have tried to explain the ribbon origin. The hypothesized physical mechanisms operate in disparate regions from the TS to beyond the heliopause and out to the local bubble. CASSINI found a similar, but much broader, feature at higher energies (Figure 10.15).

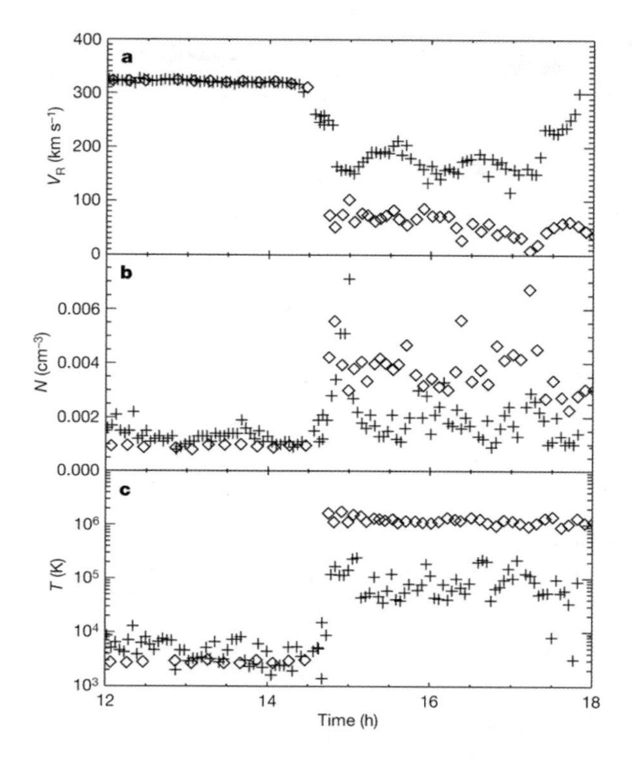

FIGURE 10.14 The radial speed (a), density (b), and temperature (c) in the heliosheath (red) are compared with that expected (black) on the basis of the crossing of Neptune (x-axis scale in hours). Note that the velocity drop and density increase were much lower than expected, and the temperature increase was less than 10 percent of that expected. SOURCE: Reprinted by permission from Macmillan Publisher Ltd.: *Nature:* J.D. Richardson, J.C. Kasper, C. Wang, J.W. Belcher, and A.J. Lazarus, Cool heliosheath plasma and deceleration of the upstream solar wind at the termination shock, *Nature* 454:63-66, 2008, doi:10.1038/nature07024.

The IBEX ribbon appears to evolve on timescales as short as 6 months, demonstrating that the heliosphere-interstellar-medium interaction is more dynamic than expected.

It was not known to what extent the ISMF would play a role in shaping the outer heliosphere. The two Voyagers crossed the TS at different distances, indicating a distinct north-south (and east-west) asymmetry of the global heliosphere. That effect was attributed to a tilt of the ISMF related to the velocity vector of the solar system relative to the interstellar cloud (Figure 10.16). Earlier measurements by SOHO/SWAN also indicated an ISMF influence. The IBEX and Cassini images are apparently organized by the ISMF as well (see Figure 10.16). The ISMF orientation and magnitude are poorly constrained. Models indicate that the ISMF may be strong and provide most of the pressure in the local cloud. The orientation of the local ISMF differs from that of the large-scale interstellar field, but the exact orientation is still uncertain.

In 2009, the galactic cosmic-ray (GCR) intensity at Earth reached the highest level of the space age (Figure 10.17), owing mainly to the reduced interplanetary magnetic field and an extended period of low solar activity. The extended solar minimum, reduced sunspot number, and record cosmic-ray intensity have led to suggestions that we may be entering an extended period of minimum activity such as was

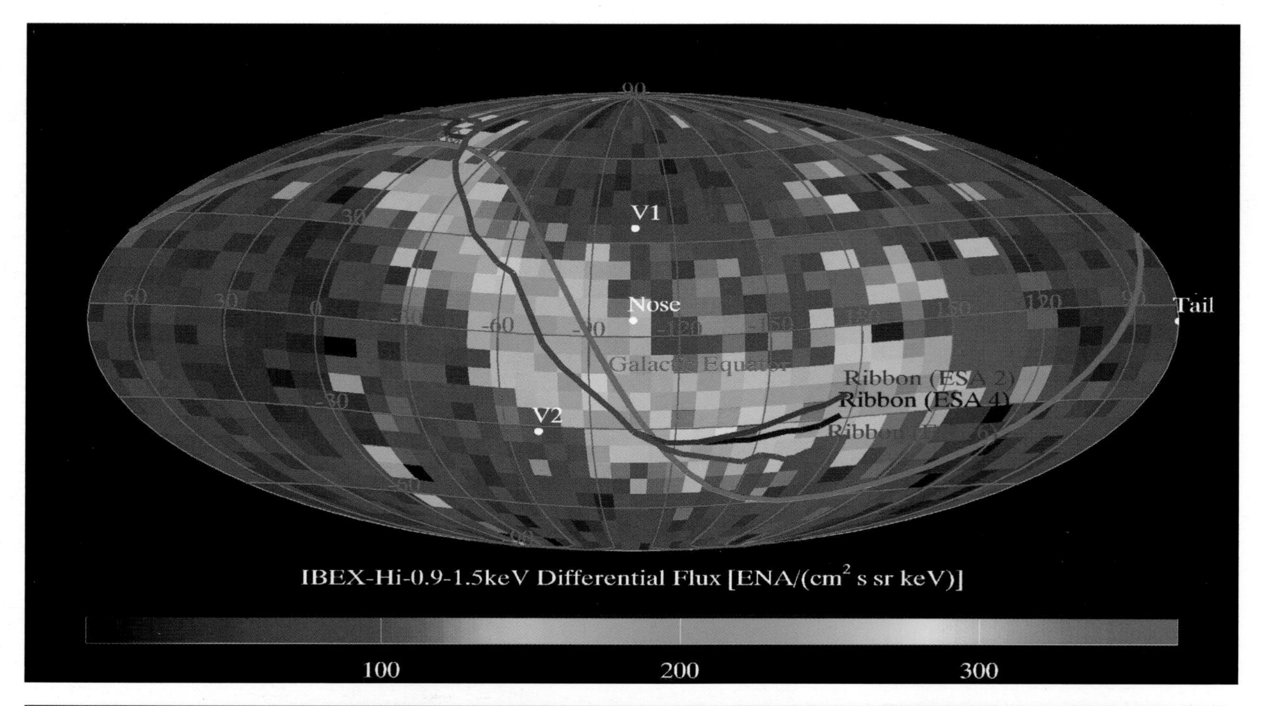

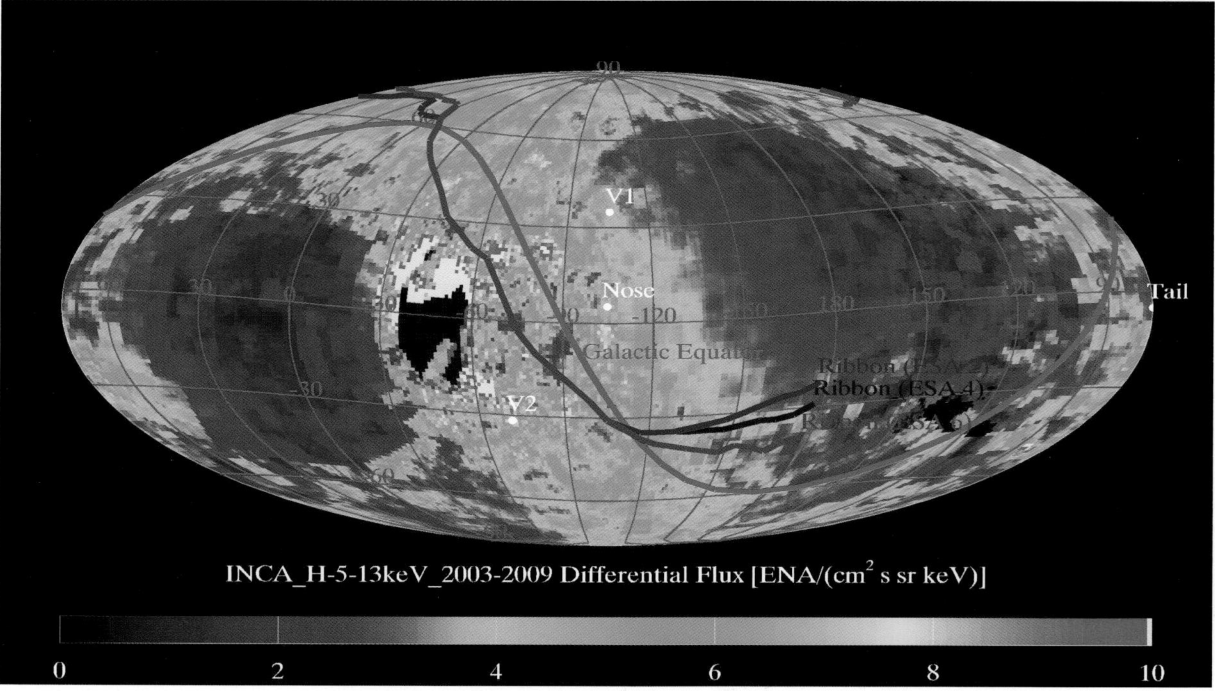

FIGURE 10.15 The unexpected ribbon seen in 0.9- to 1.5-keV energetic neutral atoms (ENAs) with IBEX and the 5- to 13-keV INCA belt. These maps depict integrated line-of-sight global maps of ENAs. Previous models, based on ENA production in the heliosheath, predicted concentrated, uniform emission near the nose. None of the earlier models predicted the ribbon or belt. SOURCE: Interstellar Boundary Explorer Mission Team.

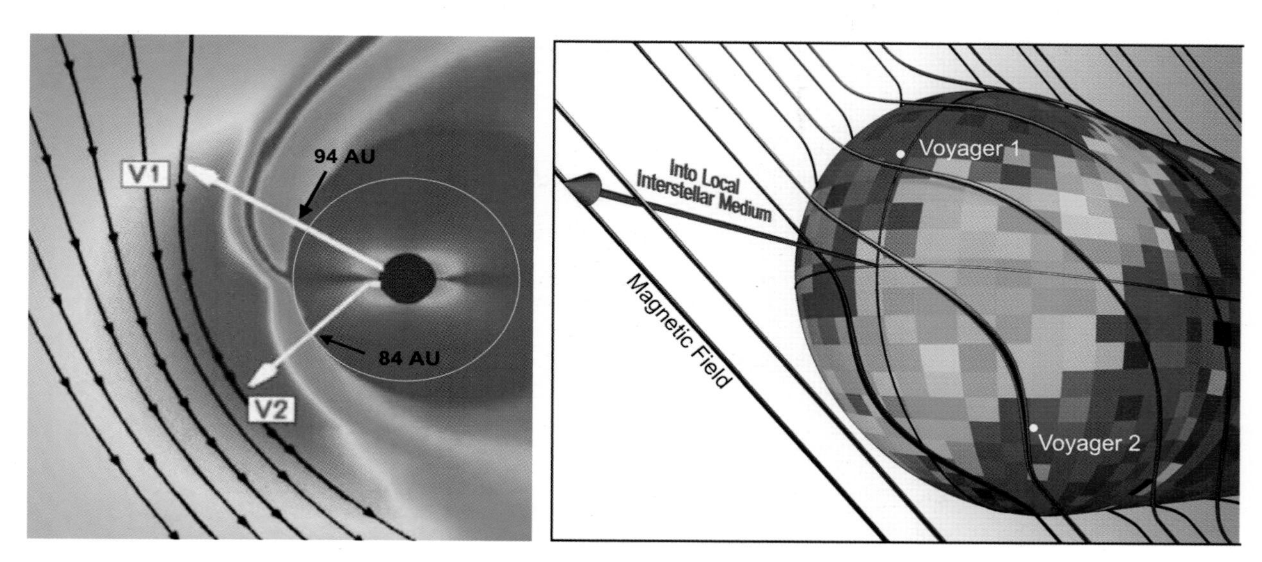

FIGURE 10.16 Crossing of the termination shock by Voyager 1 (V1) and Voyager 2 (V2) at different locations (*left*) indicated a strong influence of the interstellar magnetic field in distorting the heliosphere (Opher et al. 2006). *Left panel* shows a side view of the heliosphere where color contours indicate the magnetic-field intensity. Related results were obtained by IBEX (*right panel*), where the ribbon location appears to be influenced by the interstellar magnetic field that shapes the heliosphere. SOURCE: *Left*, M. Opher, E.C. Stone, and P.C. Liewer, The effects of a local intersteller magnetic field on Voyager 1 and 2 observations, *Astrophysical Journal* 640:L71, 2006. Reproduced by permission of the AAS. *Right*, D.J. McComas, F. Allegrini, P. Bochsler, M. Bzowski, E.R. Christian, G.B. Crew, R. DeMajistre, H. Fahr, H. Fichtner, P.C. Frisch, H.O. Funsten, et al., Global observations of the interstellar interaction from the Interstellar Boundary Explorer (IBEX), *Science* 326:959-962, doi:10.1126/science.1180906, 2009.

observed (in sunspot, ^{14}C, and ^{10}Be data) during the Dalton minimum (1800-1820) or the Maunder minimum (1645-1715).

10.4 SOLAR AND HELIOSPHERIC OBJECTIVES FOR THE COMING DECADE

In the following sections, the panel outlines steps that will continue the recent pace of progress while offering opportunities for significant breakthroughs. The first four subsections of this section summarize opportunities related to each of the SHP panel's four major science goals (§10.2). The panel then outlines new opportunities provided by the continuation of existing space-based and ground-based programs (§10.4.6.4 and §10.4.7.2) and by the completion of programs that are now in development (§10.4.6.1-§10.4.6.3 and §10.4.7.1). Goals addressed by proposed new programs are described in Section 10.5.

10.4.1 Determine How the Sun Generates the Quasi-cyclical Variable Magnetic Field That Extends Throughout the Heliosphere

The variable magnetic field that creates the heliosphere and produces space weather events results from processes in the solar interior and at the surface. Consequently, probing the solar interior and surface to determine the origins of the Sun's magnetic activity is a major goal for the next decade (SHP actions

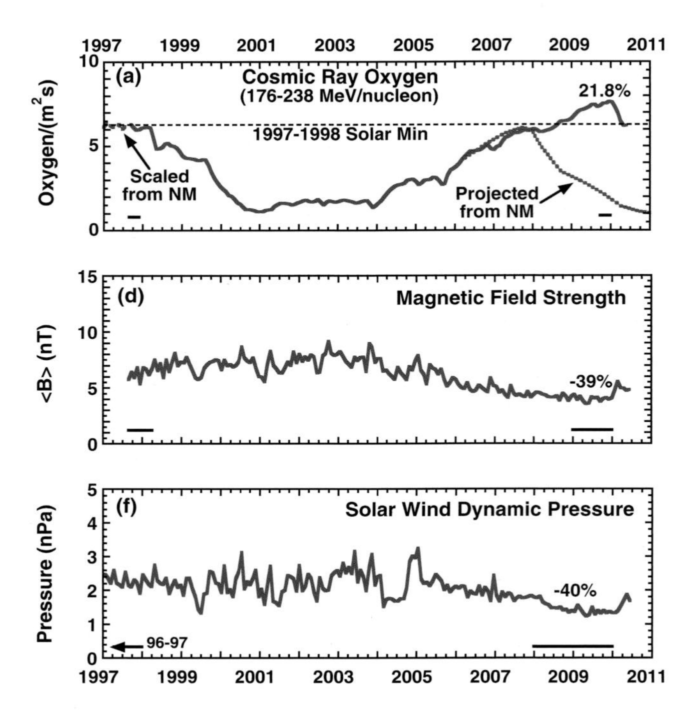

FIGURE 10.17 The galactic cosmic-ray intensity in 2009 was a record for the space age (Mewaldt et al., 2010). At the same time, the interplanetary magnetic field strength (E.J. Smith and A. Balogh, *Geophysical Research Letters* 35:L22103, 2008) and solar wind dynamic pressure (D.J. McComas et al., *Geophysical Research Letters* 35:L18103, 2008) were about 40 percent lower than during the previous solar minimum. Note that the cosmic-ray maximum was about 2 years later than projected. SOURCE: R.A. Mewaldt, A.J. Davis, K.A. Lave, R.A. Leske, E.C. Stone, M.E. Wiedenbeck, W.R. Binns, E.R. Christian, A.C. Cummings, G.A. de Nolfo, M.H. Israel, et al., Record-setting cosmic-ray intensities in 2009 and 2010, *Astrophysical Journal Letters* 723:L1-L6, 2010. Reproduced by permission of the AAS.

1a and 1b; motivation M1). The processes are not yet well understood, mainly because of the challenges of measuring the solar interior and surface in crucial locations and on timescales of multiple solar cycles.

Helioseismology is revealing surprising properties of much of the solar interior. Currently, models of the Sun that agree with helioseismology measurements are based on element abundances that disagree with recent values obtained by spectral line analysis. The discrepancy challenges the foundations of astrophysics and must be solved in the next decade. It is widely thought that the polar regions and tachocline play large roles in the solar cycle, and these locations are difficult to observe with helioseismology from a single near-Earth location. A goal for the next decade is to start to probe these regions in sufficient detail to define whether and how they affect the course of solar activity. Attaining that goal involves simultaneous surface velocity measurements from near Earth and a vantage point separated by a sizable fraction of an astronomical unit (stereohelioseismology).

The ESA-led Solar Orbiter will furnish brief, pioneering observations at moderate inclinations relative to the ecliptic plane during the 2020s. The science return from this mission will be greatly enhanced if NASA supports both additional telemetry coverage and U.S. investigations with non-U.S. instruments. White papers offered other paths to obtain in-ecliptic stereohelioseismic observations, for example, L5 (§10.5.2.5) and Safari. Sustained observation of the polar regions requires a high-inclination solar orbit. Such orbits are not easily reached, but a successful investment in this decade in solar-sail technology (§10.5.2.8) will enable effective long-term use of high-inclination orbits in the 2020s to probe portions of the solar interior that are important for the solar cycle.

Magnetic fields at the visible surface of the Sun—the photosphere—have been studied for decades. The well-observed photosphere links the obscured internal sources of activity with the faint corona and sparsely sampled heliosphere. Conditions measured in this narrow layer provide scientists' main insights into the origin of the solar cycle and sources of space weather. One long-standing goal is to understand how magnetically active regions erupt and disperse at the surface. There are major uncertainties about the influence of small spatial-scale magnetic fields on the solar cycle and on TSI. Open issues include how much small-scale fields contribute to the magnetic and TSI budgets of the Sun and how their properties change over the course of a solar cycle (SHP action 1c). Attacking that frontier is a major goal of the 4-meter-aperture Advanced Technology Solar Telescope (ATST).

An important societal goal is to improve the quality of real-time models of the solar magnetic field to assist in forecasting space weather events and heliospheric conditions (motivation M2). Those models use observed magnetic boundary conditions as the basis of outward extrapolation of the magnetic field. Improving the quality and extending the measurements above the surface with optical and radio methods are goals that will lead to better understanding and improved space weather forecasts. At present, limited measurements of the solar magnetic field are made by a few ground-based observatories and with the Hinode and SDO space missions. Augmenting those measurements with ground-based data from the proposed FASR and COSMO facilities, along with the space-based data from the proposed JAXA-led Solar-C mission, would greatly contribute to meeting the goal.

Great strides in understanding stellar interiors and activity cycles have been made with new high-precision measurements of oscillations and variability. A goal for the next decade is to use such observations to help solve fundamental questions about the dynamo process, internal structure and dynamics, rotation, and activity cycles of stars similar to the Sun. Information from a broad array of stars will sharpen understanding of the Sun's physics (motivation M3).

10.4.2 Determine How the Sun's Magnetism Creates Its Dynamic Atmosphere

Section 10.3.2 describes a few of the impressive advances of the past decade, but researchers are still far from understanding how the Sun's variable magnetic field structures and powers an atmosphere that extends from the bottom of the chromosphere out to the distant boundaries of the heliosphere. For example, how the corona couples dynamically to the solar wind is uncertain. The chromosphere clearly plays an important role in the injection of energy into the corona, but it is observed only in a narrow way and is poorly understood. The mechanisms that heat the corona and accelerate the wind constitute one of the central problems in all space science. It is expected that in the coming decade the revolutionary new observations from the missions and projects discussed in this survey, combined with next-generation theory and models, will resolve many of these outstanding problems.

To make progress in understanding the solar wind's dynamics and structure, it is necessary to go as close to the Sun as possible to measure the properties of the wind at its origin (motivations M1-M3). That is the goal of Solar Probe Plus and Solar Orbiter, new missions that will each make unique observations of the structure and evolution of the connection between the corona and interplanetary space. SPP will measure solar wind characteristics well within the Alfvén radius where the solar magnetic field still controls the dynamics of the wind. The probe will truly be a discovery mission in that it will explore a region of the heliosphere that has never been visited before. Solar Orbiter will bring a suite of instruments designed for coordinated in situ and remote imaging both close to the Sun and out of the ecliptic plane.

The chromosphere is another key region of our home in space that has yet to be understood. Major goals for the upcoming decade are to measure the structure and dynamics of the chromosphere accurately and to understand their role in the origin of the heat and mass fluxes into the corona and wind (motivation M1). Attacking those problems requires simultaneous observation of emission from the photosphere to the corona at high spatial, temporal, and spectral resolution. That is the motivating strategy for Solar-C and for the IRIS Explorer mission. IRIS will deliver pioneering observations of chromospheric dynamics in preparation for Solar-C, which will observe the full, coupled solar atmosphere, including all detailed plasma and magnetic measures, with spatial resolution never before achieved—about 0.1 arc-seconds. With its vast improvement in resolution and coverage, Solar-C, like SPP, will be a discovery mission.

The defining feature of the photosphere-corona system is its magnetic coupling; understanding this coupling requires accurate measurement of the magnetic field in the corona, especially the full vector field so that the free energy can be measured. Measurement of the coronal field has challenged solar and heliospheric physics for decades, but with the recent advances in both ground-based and space-based instrumentation, researchers are finally poised to meet this challenge. Solar-C will determine the vector field in the chromosphere with high resolution and over extended duration, thereby permitting reliable extrapolation of the field into the corona. At the same time, ATST, FASR, and COSMO would measure the coronal magnetic field directly from the ground. With those revolutionary capabilities, it will be possible to follow the buildup and release of magnetic energy in the corona and address many of the most fundamental questions in solar and heliospheric science and space weather, such as the physical processes that produce flares and CMEs.

One of the fundamental questions is the origin of the thermal structure of the closed-field corona. Researchers have long observed coronal loops but have never definitively seen the heating process, which is expected to occur at scales well below that provided in present images. The heating process is expected to have clear signatures in the internal structure of coronal loops. Solar-C is designed specifically to have the spatial, temporal, and spectral resolution required to reveal this internal structure and dynamics. Those observations will be pioneering. Simultaneously with observing the plasma structure, Solar-C, ATST, and FASR would constrain the properties of electric currents in the corona and thereby probe the heating mechanism itself. With those missions and projects in the coming decade, enormous progress will be made toward achieving one of the central goals in solar and heliospheric science: understanding how the Sun produces the hot closed corona.

10.4.3 Determine How Magnetic Energy Is Stored and Explosively Released

Solar cycle 23 was the best-observed cycle of the space era,[5] and scientists now have greatly improved understanding of basic processes in large solar eruptions as well as more sophisticated models of these events. However, key questions remain. Expected progress toward SHP science goal 3 is outlined below for associated SHP actions 3a-d:

- *Determine how the sudden release of magnetic energy enables both flares and CMEs to accelerate particles to high energies efficiently.* There are rather complete models of particle acceleration and transport

[5]Since the last solar maximum, Hinode, STEREO, Fermi, and SDO have joined SOHO and RHESSI to provide solar imaging over 360° with much greater spatial, temporal, and spectral resolution. In addition, in situ instruments now encircle the Sun. These unprecedented observatories promise exciting new observations of CME and flare eruptions as solar activity increases. Expected to come on line in the coming decade or shortly are ATST, which will measure coronal magnetic fields; SPP and Solar Orbiter, which explore SEP, CME, and interplanetary properties near the Sun; and IMAP, a spacecraft to be placed at L1 to observe ENAs from the heliospheric boundary region that also requires background measurements of the solar wind.

at CME-driven shocks, but key questions remain about conditions near the Sun: Why does SEP acceleration efficiency vary so greatly from event to event, and how do preceding CMEs apparently improve acceleration efficiency? SPP and Solar Orbiter will directly measure the seed populations and physical conditions necessary for particle acceleration, investigating the roles of shocks, reconnection, waves, and turbulence. The near-Sun measurements, backed by 1-AU spacecraft, will relate acceleration-region conditions with 1-AU intensities, spectra, and composition to discover why and how SEP acceleration varies and how particles are transported in radius and longitude (motivation M1). High-resolution CME and shock images from FASR will aid these studies. Figure 10.18 shows a stage of the June 13, 2010, coronal wave with the approximate position of the wavefront that forms a weak shock and the outline of a solar eruption.

- *Identify the locations and mechanisms that operate in impulsive SEP sites, and determine whether particle acceleration plays a role in coronal heating.* During solar-active periods, low-coronal reconnection activity causes thousands of impulsive SEP events each year. While close to the Sun, SPP and Solar Orbiter will improve the statistical accuracy and temporal resolution of measured intensity and composition variations by 1-2 orders of magnitude over 1-AU data, enabling improved correlations with images of coronal jets and other reconnection sites, tests of models for acceleration and ion fractionation, and searches for quiet-time coronal emission. In addition, the NuSTAR Explorer X-ray mission will search,

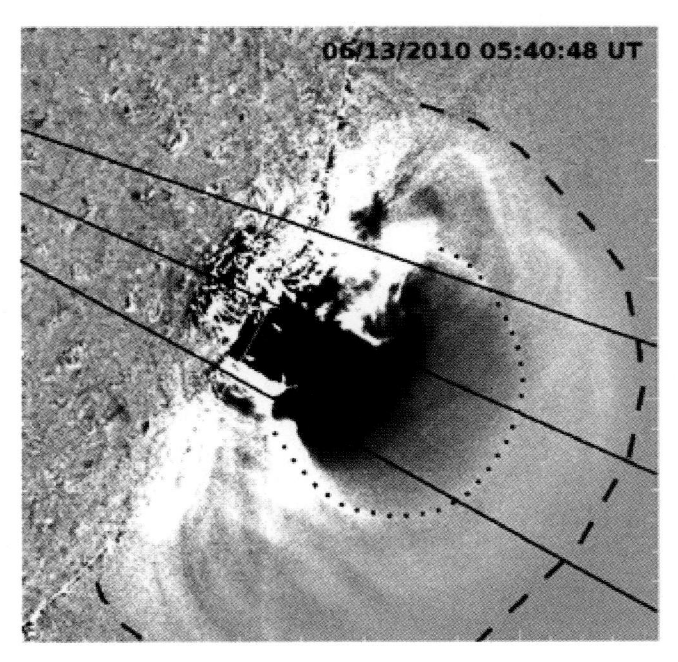

FIGURE 10.18 The formation of shocks low in the corona is a critical missing piece in understanding sudden solar energetic particle onsets. This Solar Dynamics Observatory/Atmospheric Imaging Assembly image shows a stage of the June 13, 2010, coronal wave with the approximate position of the wavefront (dashed black curve) that forms a weak shock and the outline of a solar eruption (dotted curve). The shock was formed at about 1.2 R_S and observed here at 1.4 R_S. SOURCE: K.A. Kozarev, K.E. Korreck, V.V. Lobzin, M.A. Weber, and N.A. Schwadron, Off-limb solar coronal wavefronts from SDO/AIA extreme-ultraviolet observations—Implications for particle production, *Astrophysical Journal Letters* 733: L257, 2011. Reproduced by permission of the AAS.

with more than 100 times the sensitivity of earlier studies, for microflares and nanoflares that may heat the corona. IMAP high-resolution composition and charge-state measurements will trace 1-AU impulsive events while remote-sensing and near-Sun observations provide spatial and temporal structures of solar and interplanetary acceleration regions.

- *Determine the origin and variability of suprathermal electrons, protons, and heavy ions on timescales of minutes to hours.* Discovering suprathermal-ion production mechanisms is one key to understanding particle acceleration. Pioneering observations by Ulysses, ACE, Wind, and STEREO revealed the importance of suprathermal ions and raised questions about their origin, but these studies had limited time resolution and statistical accuracy. SPP and Solar Orbiter will measure the evolution of ion and electron halo solar wind and suprathermal tails close to the Sun, providing improved opportunities to isolate solar and interplanetary contributions and to test theoretical models. Comprehensive measurements of suprathermal-ion composition, spectra, and intensity fluctuations by IMAP will relate energetic-particle populations to interplanetary structures and physical processes.
- *Develop advanced methods for forecasting and nowcasting of solar eruptive events and space weather.* Combining SPP and Solar Orbiter in situ observations with 1-AU imaging and in situ data will provide ground truth for SEP-acceleration models and thereby improve SEP forecasts (motivation M2). The data may also reveal how monitoring critical near-Sun conditions (for example, active-region, shock, and suprathermal-seed properties) can aid forecasting. Multipoint measurements of SEP radial and longitudinal distributions will clarify current environmental-model uncertainties.

It is critical that forecasters develop predictive capabilities for space weather events while maintaining comprehensive measurements for nowcasting solar wind and energetic-particle inputs into geospace (motivation M2). IMAP, like ACE before it, will be a keystone of the Heliophysics Systems Observatory (HSO) by providing comprehensive solar wind data; diagnostics of suprathermal-ion and electron sources; solar wind and energetic-particle inputs into geospace; and evolving interplanetary magnetic-field properties. IMAP will also provide unprecedented measurements of suprathermal-tail variability and determine how seed populations are related to higher-energy particles accelerated by shocks, waves, and disturbances.

FASR would provide many new observations important with respect to space weather, including observations of coronal magnetic fields in active regions and their evolution before, during, and after flares and CMEs; "real-time" observations of coronal-shock locations and properties; measurements of the spectral evolution of electron energy-distribution functions; and radio flux-density spectra in communication bands.

10.4.4 Discover How the Sun Interacts with the Local Galactic Medium and Protects Earth

The coming decade offers unique opportunities for additional breakthroughs in understanding how the heliosphere and local galactic medium interact. Those opportunities address associated SHP actions 4a-c:

- *Determine the spatial-temporal evolution of heliospheric boundaries and their interactions.* Solar-cycle changes in the solar wind dynamic pressure affect the structure of heliospheric boundaries. In the next few years, Voyager and IBEX observations will determine how the global heliosphere responds to increased solar activity. Complementary solar wind, pickup-ion, and anomalous and galactic cosmic-ray observations by ACE, Wind, and STEREO will provide context for evolving solar conditions and measure the cosmic-ray response to global-heliosphere changes. IMAP, with its unprecedented roughly 80 times greater sensitivity and duty cycle in ENA maps (and 10 times higher angular resolution), will deliver definitive measurements of the fine structure and detailed evolution of the global heliosphere. By combining those breakthrough observations with expected advances in theory and modeling,[6] researchers will understand the structure of the heliosheath in detail, including how time-dependent effects propagate in the heliosheath and affect heliospheric structures. IMAP will make it possible to solve the mystery of the IBEX ribbon and INCA belt and to discover the implications of these vast structures for the heliosphere and the local galactic medium (motivation M1).

- *Discover whether particles are accelerated in the heliosheath.* When Voyager 1 crossed the TS, particles up to about 1 MeV/nucleon peaked at the shock, but, surprisingly, the intensity of higher-energy ACRs continued to rise well after the shock. There are several theories about the location and mechanism of acceleration of these high-energy ACRs. One theory suggests that the highest-energy ACRs originate at the flanks of the heliosphere rather than at the nose because of the blunt TS shape. Another idea invokes random "hot spots" along the TS due to large-scale turbulence. It is also proposed that acceleration occurs in the heliosheath as particles move within random plasma compressions. Still another theory suggests that particle acceleration arises from contracting magnetic islands that result from magnetic reconnection in the heliosheath. The last two ideas predict that the main acceleration occurs near the heliopause.

Distinguishing among those theories requires Voyager energetic-particle, solar wind, and magnetic-field measurements through the heliosheath to the heliopause. In addition, current global-MHD models need to be expanded to include turbulence, magnetic reconnection, and feedback from suprathermal particles. Hybrid and kinetic codes are needed to complement global-MHD studies. Voyager data will test quantitative predictions from the models.

The Voyagers are exploring a new region, the heliosheath, which, unlike the region inside the TS, is not dominated by supersonic solar wind. Compressive magnetic structures and possibly turbulence or magnetic reconnection dominate this region. Heliosheath physics is not yet well understood, but new measurements and models will spark new advances. Understanding heliosheath physics is also important for interpreting IBEX, Cassini, STEREO, and IMAP ENA observations. Finally, understanding the heliosheath is important because of the role it plays in modulating the intensity of galactic cosmic rays that penetrate into the inner solar system and reach Earth. Figure 10.19 shows a schematic of the heliosphere, the heliosheath, and the interstellar medium.

- *Explore the properties of the heliopause and surrounding interstellar medium.* The Voyagers are expected to cross the heliopause into the local interstellar medium (LISM) within the next decade and will provide the first measurements of interstellar magnetic-field strength and orientation and, it is hoped, measurements of interstellar plasma properties (motivation M3). They will also measure interstellar cosmic-ray spectra, which may distinguish among acceleration-transport models or reveal contributions from nearby sources. Interstellar cosmic-ray measurements have other consequences: they represent the maximum cosmic-ray intensity that Earth has experienced (important for interpreting ^{10}Be archives) and they establish the maximum intensity of the radiation to which future space travelers can be exposed (constraining interplanetary environmental models). Simultaneous near-Earth measurements will establish the absolute intensity drop from interstellar space to Earth.

The Voyagers will provide the first measurements of the structure of the heliopause. What is the role of instabilities and reconnection near the heliopause? How are cosmic rays modulated? How thick is the heliopause? Heliopause models exist, but surprises are inevitable, such as when the TS was crossed. In addition, if the Voyagers enter interstellar space roughly coincidentally with increased solar activity, plasma-wave data may constrain LISM kinetic properties (such as the turbulence level). IBEX and IMAP

---

[6] J.D. Richardson et al., The Heliospheric Interaction with the LISM: Observations and Models, white paper submitted to the Decadal Strategy for Solar and Space Physics (Heliophysics), Paper 227; V. Florinski et al., The Outer Heliosphere-Solar System's Final Frontier, white paper submitted to the Decadal Strategy for Solar and Space Physics (Heliophysics), Paper 76.

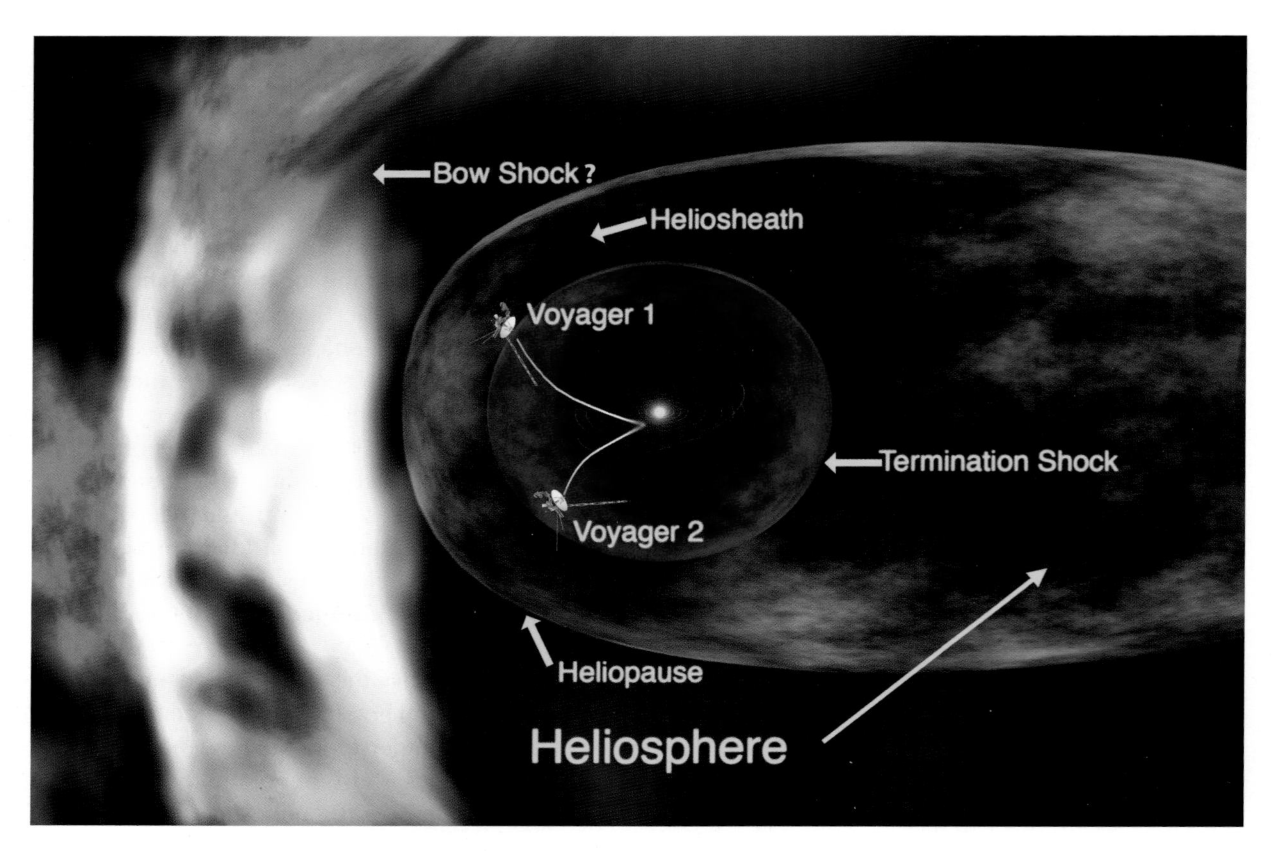

FIGURE 10-19 A schematic image of the regions of the heliosphere, including the approximate locations of Voyagers 1 and 2. Voyager 1 crossed the termination shock at 94 AU in December 2004, and Voyager 2 crossed at 84 AU in August 2007. For details, see http://science1.nasa.gov/missions/voyager/. SOURCE: Courtesy of NASA/JPL/Walt Feimer. NOTE: This figure was prepared several years ago when scientists expected Voyager to discover a bow shock. New results from the Interstellar Boundary Explorer (IBEX) have shown that instead of a shock, there is a softer interaction—a bow wave, similar to how water piles up ahead of a moving boat. For IBEX results, see D.J. McComas, D. Alexashov, M. Bzowski, H. Fahr, J. Heerikhuisen, V. Izmodenov, M.A. Lee, E. Moebius, N. Pogorelov, N.A. Schwadron, and G.P. Zank, The heliosphere's interstellar interaction: No bow shock, *Science* 336:1291, doi: 10.1126/science.1221054, May 2012.

will complement those observations and provide context with ENA maps of the global response of the heliosheath to changing solar activity.

IBEX made the first in situ detections of interstellar H and O, including secondary O (neutral oxygen produced when interstellar O^+ charge-exchanges in the outer heliosheath), which tracks deflected interstellar flow around the heliosphere. IBEX studies of interstellar H, O, He, and Ne are forcing re-examination of LISM characteristics. Resolving LISM properties (for example, whether the solar system is still in the local interstellar cloud or in transition between clouds) requires interstellar-flow observations with high sensitivity and angular resolution. The detailed neutral and high-precision pickup-ion observations with IMAP will provide abundances of He, N, O, and Ne and the flow vector and temperature of He, O, and Ne, which will strongly constrain models of the ionization state and radiation environment of the interstellar medium.

Key isotope ratios (D/H, $^{3}He/^{4}He$, $^{22}Ne/^{20}Ne$) will place strong constraints on big bang cosmology and the evolution of matter in the galaxy. From high-precision observations of secondary O and He, the strength and orientation of the local interstellar magnetic field and the structure of the outer heliosheath can be independently deduced. Extended IBEX observations will provide reconnaissance, and IMAP will have all the necessary capabilities for decisive measurements. The Voyagers will provide ground truth through and beyond the heliopause, and theory and modeling will have the challenge of reconciling these observations.

10.4.5 Contributions of the SHP Panel's Program to Achieving the Decadal Survey's Key Science Goals

This section summarizes the impact that the SHP panel's proposed program can have on realizing the key science goals of this decadal survey (see Chapter 1).

The top section of Table 10.1 illustrates how the space missions and ground-based facilities described in this chapter will address the SHP panel's science goals and associated actions outlined in Box 10.1. Note that implementation of the entire SHP program would make major contributions to all four SHP goals and all 14 SHP actions (for details, see §10.4.1-10.4.4 and descriptions of the individual actions in §10.4.6-10.5).

The bottom section of Table 10.1 illustrates how the SHP panel's program can help realize decadal survey key science goals 1-4 during the coming decade and beyond. As has already been noted, several space missions and ground-based facilities promise to make transformative contributions to decadal survey key science goal 1 with unique remote-sensing measurements that will discover how the Sun's magnetic activity transfers matter and energy from the photosphere through the atmosphere and out into the heliosphere. The exploratory Solar Orbiter and SPP missions, including multiple passes through the solar corona by SPP, will address decadal survey key science goal 1 by providing the first in situ observations of the inner heliosphere close to the Sun. These truly transformative observations will reveal new details of solar and interplanetary processes and greatly improve researchers' ability to understand, model, and predict variability in the space environment. The activities also support decadal survey key science goal 2 in that they will lead to a better understanding of the interplanetary plasma and energetic-particle environments that impinge on the magnetosphere.

Similarly, IMAP will image temporal and spatial variations of the outer heliosphere with about 80 times better sensitivity while providing comprehensive measurements of solar wind and energetic-particle inputs to geospace. Many of the measurements will also reveal details of basic solar and heliospheric processes that are at work in astrophysical systems throughout the galaxy, fulfilling decadal survey key science goal 3. Finally, the planned observations, spanning a broad array of plasma environments, will permit the study and understanding of the fundamental physical processes that act within the heliosphere and in astrophysical plasmas throughout the universe (decadal survey key science goal 4).

10.4.6 Goals for the Ongoing Program and Missions in Development

10.4.6.1 Science Goals for Solar Probe Plus

The space age has revolutionized understanding of the Sun and heliosphere, but after more than a half-century two of the most fundamental questions in heliophysics remain unanswered: Why is the solar corona so much hotter than the photosphere? How is the solar wind accelerated? Those questions are critical for understanding our solar system, determining the effects of solar inputs on the dynamic heliosphere, and developing predictive space weather capabilities. Achieving closure requires direct samples of electromagnetic fields, plasma, and energetic particles in the solar corona and across the Alfvénic transition between the corona and solar wind. Recognizing the urgency of such observations, the 2003

TABLE 10.1 Illustration of Contributions of Planned and Proposed Missions and Ground-Based Facilities to SHP Panel Goals and Actions and How the Proposed SHP Panel Program Addresses the Four Decadal Survey Science Goals Discussed in Chapter 1

		Missions and Their Contributions									
		Strategic Missions			Opportunity	In Development			Ground-Based		
SHP Panel Science Goals and Actions[a]		SEE	IMAP	L5	Solar-C	IRIS	SPP	Solar Orbiter	ATST	FASR	COSMO
Dynamo and Activity Cycle	1a. Polar mass flows, solar cycle	o	o	+	o			++	o	o	+
	1b. Deep mass flows, dynamo			++	o			++	o	o	o
	1c. Small-scale fields, Large-scale, variability	+		+	+	+	o	o	++	++	++
Sun's Magnetic Field to Dynamic Atmosphere	2a. Chromosphere dynamics	+	o		++	++	o	o	++	++	++
	2b. Magnetic free energy	o	o	+	++	+	+	o	++	++	++
	2c. Corona thermal structure	+		+	++	+	++	++	+	++	+
	2d. Solar wind origin dynamics and structure	+	+	+	+	o	++	+	+	++	+
Magnetic Energy Storage and Release	3a. Eruptive events energy/acceleration	++	++	++	++	o	++	++	+	++	+
	3b. Impulsive SEP mechanisms, role	++	+	++	+		++	++	o	++	o
	3c. Suprathermal particle origin, variability	+	++	+	+		++	++		+	
	3d. Solar events forecasting	++	o	++	++	+	+	o	o	+	+
Interaction with Galactic Neighborhood	4a. Heliospheric boundary evolution	o	++	o				o			
	4b. Anomalous cosmic-ray origin	o	++	o							
	4c. Heliopause and interstellar medium		++								
Decadal Survey Key Science Goals											
1. Determine the origins of the Sun's magnetic activity and predict the variations in the space environment		++	o	++	++	++	++	++	++	++	++
2. Understand the dynamics and coupling of Earth's magnetosphere, ionosphere, and atmosphere and their response to solar and terrestrial inputs		o	+	o	o	o	o	o	o	o	o
3. Determine the interaction of the Sun with the solar system and the interstellar medium		o	++	o	o	o	o	o	o	o	o
4. Discover and characterize fundamental processes that occur both within the heliosphere and throughout the universe		++	++	++	++	++	++	++	++	++	++

KEY: o, supporting role; +, significant contribution; ++, unique to transformative.

[a] Note that the SHP panel goals and actions are represented in shorthand form with key words. For the complete missions, see Box 10.1.

decadal survey recommended implementation of a solar probe as a large-class NASA mission.[7] SPP, started in 2009, will begin its voyage of discovery in 2018 and serve as a keystone of the strategy for solar and heliospheric science in the coming decade.

The goals of SPP are to determine the structure and dynamics of the Sun's coronal magnetic field, to understand how the solar corona and wind are heated and accelerated, and to determine what mechanisms accelerate and transport energetic particles. To accomplish those goals, SPP is equipped with a tailored payload for the first near-Sun in situ measurements of solar wind ion and electron thermal plasma, suprathermal and energetic particles, and DC to high-frequency electromagnetic fields. Remote observations include a large-field-of-view white-light imager to provide global context and a directional radio receiver to locate and track flares and shocks. Multiple Venus encounters will gradually lower perihelion from 35 R_S to 9.5 R_S, producing more than 1,000 hours inside 20 R_S, including substantial time within the Alfvén critical point and providing samples of all solar wind types.

SPP will trace the flow of energy that heats and accelerates the solar corona and solar wind (SHP action 2d). Observations of magnetic-reconnection exhausts, jets, shocks, and plasma properties—including wave-particle coupling, heat flux, and mass flux—will directly indicate how the Sun's convective motion and magnetic field create its dynamic atmosphere and how magnetic free energy is transmitted from the photosphere to the corona. Measurements will determine the energy budget of the solar wind as it evolves from the corona and thereby detect signatures of heating and dissipation responsible for the high temperature of the outer corona and extended heating of the solar wind. As described in Section 10.3.2, composition measurements that determine ionic charge states would facilitate this study by identifying the various types of wind and constraining their solar origin.

SPP will determine the structure and dynamics of the plasma and magnetic fields at the sources of solar wind. Measurements will reveal the steady-state mapping between photospheric sources and coronal structures and emerging solar wind. Full-sky maps of the suprathermal-electron strahl and pitch-angle distribution will unambiguously identify when the spacecraft is on closed magnetic-field lines rooted at both ends in the corona.

The Sun accelerates high-energy particles in solar flares and at CME-driven shocks, where suprathermal particles from multiple sources are the seed particles. Testing particle-acceleration models at 1 AU is hampered by lack of knowledge of source conditions and by acceleration and transport ambiguities. By closing within the Alfvén point, SPP will survey plasma-field and seed-particle properties in the prime-acceleration region of CME-driven shocks, providing ground truth for acceleration models and revealing the causes of energetic-particle variability needed to improve SEP forecasts (motivation M2 and SHP actions 3a and 3d). SPP is expected to observe directly about 10 strong CME-driven shocks within 20 R_S, providing comprehensive shock, turbulence, and seed-particle properties for comparison with accelerated-particle spectra, composition, and pitch-angle distributions.

Near the Sun, impulsive SEP events associated with flares appear as sharp spikes, enabling subminute timing comparisons with flares, jets, coronal waves, and radio bursts. Flare-particle studies will test acceleration and charge- and mass-dependent fractionation models, survey neutron-decay protons and electrons, and discover how flare-accelerated particles escape and are transported in longitude (SHP action 3b). Measuring near-Sun suprathermal-particle properties will provide breakthroughs in understanding of the relative importance of local acceleration and solar sources (SHP action 3c).

Sending a spacecraft into the last unexplored region of the heliosphere will produce transformative results throughout the field of solar and space physics and will illuminate fundamental physical processes that occur in stellar atmospheres and energetic astrophysical objects across the universe (motivation M3).

[7] NRC, *The Sun to the Earth—and Beyond: A Decadal Research Strategy in Solar and Space Physics*, 2003.

The SHP panel gives its unqualified endorsement to the SPP mission as long as it satisfies cost and schedule guidelines.[8]

10.4.6.2 Science Goals for Solar Orbiter

In October 2011, ESA's Science Programme Committee unanimously selected Solar Orbiter as the Cosmic Vision M1 mission, and it was scheduled for launch in 2017. Solar Orbiter will investigate links between the solar surface, corona, and inner heliosphere from as close as 62 R_S by using a comprehensive payload that combines remote-sensing and in situ measurements. NASA's LWS program will provide the launch vehicle and critical science instruments and investigations. Two instruments that had been descoped by NASA were restored with ESA funding, and so the complete payload will fly as originally planned. When close to the Sun, Solar Orbiter will observe emissions, solar wind, and energetic particles from a single area for much longer than is possible from 1 AU and will provide improved insight into the evolution of sunspots, active regions, coronal holes, and other solar features and phenomena. Solar Orbiter's high spatial and time-resolution observations close to the Sun and its long observations during near corotation with the Sun will probe key questions for understanding the formation of the heliosphere and the generation of space weather events:

- How and where do the solar wind plasma and magnetic field originate in the corona (SHP action 2d)?
- How do solar transients drive heliospheric variability (SHP actions 2d and 3a)?
- How do solar eruptions produce energetic-particle radiation that fills the heliosphere (SHP actions 3a-c)?
- How does the solar dynamo work and drive connections between the Sun and heliosphere (SHP action 1b)?

A unique aspect of the mission occurs when the spacecraft's orbital plane is increased to about 35° solar latitude, permitting definitive measurements of polar magnetic fields and high-quality observations of solar oscillations in the polar region and thereby supplying a missing link in observations of solar global-circulation patterns (SHP action 1a). Solar Orbiter and SPP observations will overlap in time, permitting many opportunities for coordinated inner-heliosphere measurements that will greatly increase the science return from both missions. Solar Orbiter out-of-ecliptic measurements contemporaneous with near-ecliptic measurements will provide unprecedented insights into the evolving three-dimensional inner heliosphere and outer corona. For example, the remotely observed polar magnetic field combined with in situ observations by SPP will provide tests of the magnetic flux transport model (SHP action 1a).

Data return using U.S. tracking assets to provide enhanced temporal coverage for helioseismology and other uses would provide improved measurements of the solar interior, including such aspects as variations near the bottom of the convection zone and meridional flows, which are important for understanding the generation of solar activity. The scientific return of Solar Orbiter would also be greatly enhanced by providing postlaunch funding opportunities for investigations that would not directly support U.S. instruments but would have important involvement of U.S. investigators.

The SHP panel strongly endorses NASA's highly leveraged participation in this mission.

---

[8]Solar Probe Plus successfully completed its mission design review in November 2011 and proceeded into preliminary design. As this report went to press, it was scheduled for launch in July 2018.

10.4.6.3 Science Goals for the Interface Region Imaging Spectrograph

IRIS is a small Explorer scheduled for launch in June 2013 for a 2-year prime mission. The IRIS science goals focus on three themes of broad importance in solar and plasma physics, space weather, and astrophysics, aiming to understand how internal convective flows power atmospheric activity:

- Which types of nonthermal energy dominate in the chromosphere and beyond (SHP action 2a)?
- How does the chromosphere regulate mass and energy supply to the corona and heliosphere (SHP actions 2a and 2b)?
- How do magnetic flux and matter rise through the lower atmosphere, and what role does flux emergence play in flares and mass ejections (SHP actions 2a, 2b, and 3a)?

10.4.6.4 Goals for the Heliophysics Systems Observatory

The Heliophysics Systems Observatory (Table 10.2) will provide unique opportunities to observe solar and interplanetary activity and heliosphere-interstellar interactions evolve over a new solar cycle.[9] Since February 2011 and continuing through 2019, STEREO and near-Earth observatories (SDO, Hinode, SOHO, and ground-based facilities) have combined to provide a 360° view of the Sun. This total-Sun view can observe far-side active regions emerge and develop and can observe far-side flares and CMEs that often cause near-Earth solar-particle events. The solar origins of Earth-directed CMEs and flares will be observed with ultra-high precision by SDO, which delivers continuous measurements of the photospheric vector magnetic field and of coronal XUV structure. In addition, STEREO and near-Earth spacecraft (ACE, Wind, SOHO, and GOES) now provide broad longitudinal coverage of solar wind, pickup ions, CMEs, shocks, SEPs, and radio bursts, enabling greatly improved nowcasts and forecasts of the interplanetary environment. Simultaneously, the Voyagers traverse the heliosheath, and IBEX maps the outer heliosphere and samples interstellar flows.

An important goal enabled by the HSO is the comprehensive study of solar activity and explosive events. SOHO, ACE, and GOES studies show that about 10 percent of the CME kinetic energy often goes into accelerated particles. Uncertainties arise from limitations of knowledge of CME geometries and single-point sampling of particle intensities. STEREO and near-Earth spacecraft have enabled multipoint observations of cycle-24 SEP events, which indicate surprisingly broad longitudinal SEP distributions. Combined multipoint in situ and stereoscopic solar and CME studies by STEREO, SOHO, and SDO (to be augmented with radial coverage by Solar Orbiter and SPP) will provide more precise measurements of CME evolution, shock-acceleration efficiency, and its controlling properties (SHP actions 3a-c).

On a related note, recent SDO observations show near-simultaneous closely connected "sympathetic" eruptions over broad regions of the Sun (§10.3.3), which complicate eruption forecasts from active regions that may, somehow, be triggered remotely. Coordinated observations by SDO, STEREO, and other 1-AU resources during 2013-2019 (with 360° solar viewing) may reveal the nature of those connections.

The launch of the Fermi astrophysical observatory supplements RHESSI solar gamma-ray observations of eruptive events with extended temporal and energy coverage (to about 300 MeV). Combining those with STEREO and near-Earth SEP coverage can determine the escape efficiency of flare-accelerated particles from large solar eruptions (SHP action 3b).

---

[9] J.G. Luhmann et al., Extended Missions: Engines of Heliophysics System Science, white paper submitted to the Decadal Strategy for Solar and Space Physics (Heliophysics), Paper 167; C.W. Smith et al., The Case for Continued, Multi-Point Measurements in Space Science, white paper submitted to the Decadal Strategy for Solar and Space Physics (Heliophysics), Paper 246; M.H. Israel et al., The Effect of the Heliosphere on Galactic and Anomalous Cosmic Rays, white paper submitted to the Decadal Strategy for Solar and Space Physics (Heliophysics), Paper 116.

TABLE 10.2 Solar and Heliospheric Space Missions

Mission	Description	SHP Major Science Goals
Advanced Composition Explorer (ACE)	Studies elemental and isotopic composition of solar wind, solar energetic particles, and cosmic rays; provides real-time solar wind and magnetic-field data from L1	2, 3, 4
Geostationary Operational Environmental Satellites (GOES)	National Oceanic and Atmospheric Administration meteorologic satellites that provide real-time solar X-ray, solar energetic-particle, magnetic-field, and X-ray imaging data	3
Hinode	JAXA-led mission that measures the full solar vector magnetic field and coordinated optical, X-ray, and EUV images	1, 2, 3
Interstellar Boundary Explorer (IBEX)	Provides ENA all-sky images of heliospheric boundary and measures interstellar H, He, O, and Ne neutral gas at 1 AU	4
Ramaty High Energy Solar Spectrographic Imager (RHESSI)	Explorer mission that provides spatial and time-resolved X-ray and gamma-ray images of solar flares	2, 3
Solar Dynamics Observatory (SDO)	Provides full-disk Dopplergrams, vector magnetography, UV and EUV images, and EUV irradiance and spectra at high cadence	1, 2, 3
Solar Mass Ejection Imager (SMEI)	Multiagency mission led by the U.S. Air Force with an all-sky camera that images the corona and CMEs out to more than 1 AU	2, 3
Solar and Heliospheric Observatory (SOHO)	ESA-NASA mission providing solar wind and solar energetic-particle data from L1; can act as a backup for magnetograms and UV coronal images	1, 2, 3
Solar Terrestrial Relations Observatory (STEREO)	Provides coronagraph images, EUV, solar wind, interplanetary magnetic field, radio, and solar-particle coverage at increasing longitudinal separation from Earth	2, 3, 4
Wind	Provides solar wind, magnetic-field, plasma-wave, radio-burst, solar-particle, and anomalous cosmic-ray data from L1	3, 4
Voyager Interstellar Mission	The Voyagers provide magnetic-field, plasma, radio, suprathermal, anomalous, and galactic cosmic-ray data in the heliosheath; one or both may cross the heliopause	4

Another key objective for the HSO is the study of the prolonged solar minimum and the variable heliosphere. During the long solar minimum of 2008-2009, many solar and interplanetary measures reached extremes for the space age, including a record-low interplanetary magnetic field (IMF) strength and solar wind dynamic pressure, with reduced solar wind He/H ratios and freeze-in temperatures. The weakened solar wind and IMF can be related to continuing high cosmic-ray intensities (see Figure 10.17) and to a smaller heliosphere.

At the same time, the Sun's polar magnetic field has declined substantially, and the solar dipole is less pronounced. Sunspots are apparently weakening, and average CME mass is reduced. Those and other observations have prompted suggestions that we may be entering another Maunder minimum or at least a lower solar-activity level than seen for about 100 years. Researchers thus have a unique opportunity to track solar and interplanetary phenomena (SHP actions 1a-c) with the most powerful instrumentation of the space age while the Sun is apparently undergoing dramatic changes and providing clues to its past and future behavior. In addition, with 1-AU spacecraft currently spread in longitude, and by 2019 spread to near the heliopause, there is a unique opportunity to study how the heliosphere shields against cosmic rays in response to the enigmatic behavior of the Sun (SHP actions 4a and c). Can we afford to wait for opportunities like this to return?

10.4.7 Goals for Ground-Based Facilities

10.4.7.1 Science Goals for the Advanced Technology Solar Telescope

The Advanced Technology Solar Telescope (ATST) is a ground-based, 4-m-aperture solar telescope whose first observations are expected in 2016. By far the largest optical solar telescope in the world, ATST will provide revolutionary observational resolution as small as 30 km. Comparing observations on this scale with equally resolved numerical models will transform physical understanding of many solar features from informed speculation to hard science. ATST will furnish enough intensity to feed a large array of sensitive and powerful instruments. As a general-purpose community facility, ATST will address a wide variety of ever-changing science goals over its decades-long lifetime.[10]

Within the photosphere, nearly all the Sun's magnetic flux is in the form of small, dynamic elements. Constantly energized by convection, this magnetic sea directs upward flows of mass and energy to create the chromosphere, corona, and solar wind. Studying that fundamental process is a major initial ATST science goal. The key observations will include dynamic magnetic and velocity field measurements of small magnetic features at several heights in the solar atmosphere. Small magnetic features may contribute to the total solar irradiance. ATST observations will be used to define the spectral emission characteristics of a sample of those features as an important contribution to understanding TSI variations. Revolutionary observations will be made of larger magnetic features, such as sunspots and active regions, and of transient drivers of space weather, such as flares and erupting prominences.

ATST will also provide high-resolution observations in the infrared portion of the solar spectrum where molecular signatures appear. That capability will be used to study indications of extraordinarily cool molecular clouds in the upper photosphere. In addition, ATST can be operated as a coronagraph. The primary goal of early coronal observations will be to characterize the magnetic and changing velocity fields of coronal features above both active and quiet regions on time and spatial scales that have heretofore been beyond reach. The relative importance of heating of the corona by dissipation of wave motions excited from below will be a specific research target.

10.4.7.2 Science Goals for Ground-Based Solar Research

Important science results of the past decade were accomplished at ground-based facilities.[11] A summary of the principal U.S. ground-based observatories and their observational emphasis is given in Table 10.3. Compared with space missions, these facilities can be far larger, more flexible and exploratory, and longer-lived. Accordingly, emphasis is on achieving high spatial resolution (as with the NST and the ATST), making unique measurements of physical processes at long wavelengths (as with E-OVSA and FASR), and collecting sufficient light flux to make high-time-resolution, high-precision measurements of

---

[10]S.L. Keil et al., Science and Operation of the Advanced Technology Solar Telescope, white paper submitted to the Decadal Strategy for Solar and Space Physics (Heliophysics), Paper 130; S.L.Keil et al., Generation, Evolution, and Destruction of Solar Magnetic Fields, white paper submitted to the Decadal Strategy for Solar and Space Physics (Heliophysics), Paper 131; T. Ayers and D. Longcope, Ground-based Solar Physics in the Era of Space Astronomy, white paper submitted to the Decadal Strategy for Solar and Space Physics (Heliophysics), Paper 3.

[11]T. Ayres and D. Longcope, Ground-based Solar Physics in the Era of Space Astronomy, white paper submitted to the Decadal Strategy for Solar and Space Physics (Heliophysics), Paper 3; A.A. Pevtsov, Current and Future State of Ground-based Solar Physics in the U.S., white paper submitted to the Decadal Strategy for Solar and Space Physics (Heliophysics), Paper 218; K.P. Reardon et al., Approaches to Optimize Scientific Productivity of Ground-Based Solar Telescopes, white paper submitted to the Decadal Strategy for Solar and Space Physics (Heliophysics), Paper 224; S. McIntosh et al., The Solar Chromosphere: The Inner Frontier of the Heliospheric System, white paper submitted to the Decadal Strategy for Solar and Space Physics (Heliophysics), Paper 193.

TABLE 10.3 Dedicated Ground-Based Daily and General Research Solar Facilities

Data Type	AFRL	AFWA	BBSO	MLSO	MSO	MWO	NSO	OVSA	SFO	WSO
Helioseismic far-side images							G			
Photospheric images	S	G		S			G		S	
Photospheric longitudinal B fields						S	G			S
Photospheric vector B fields							S			
Photospheric spectral line profiles							S			
Photospheric sunspot B fields						S				
Chromospheric images	S	G	G	S			G			
Chromospheric longitudinal B fields						S	S			
Chromospheric spectral line profiles							S			
Chromospheric radio emission		G								
Coronal Thomson scatter images				S						
Coronal emission line images	S			S						
Coronal radio emission		G						Y		
General solar research			Y		Y		Y	Y		

NOTE: G, global network; S, single instrument; Y, yes; AFRL, Air Force Research Laboratory; AFWA, Air Force Weather Agency; BBSO, Big Bear Solar Observatory; MLSO, Mauna Loa Solar Observatory; MSO, Mees Solar Observatory; MWO, Mt. Wilson Observatory; NSO, National Solar Observatory; OVSA, Owens Valley Solar Array; SFO, San Fernando Observatory; WSO, Wilcox Solar Observatory.

magnetic and velocity fields; long-term synoptic observations; and novel frontier observations of various kinds (as with COSMO).

An important science goal is to continue comparison of highly resolved observations with numerical models to define the physical nature of photospheric features, such as sunspots, faculae, and cool molecular clouds. Another goal is to understand the physical behavior of magnetic fields in the chromosphere and corona—on both small and large scales—to support study of the flow, storage, and eruption of energy and mass in these poorly understood regions (motivation M1). The recent, mostly unexpected, behavior of the solar cycle and the inability to predict its course make synoptic studies of the magnetic field and of surface and interior mass flows that are related to the solar dynamo a high-priority goal (motivation M3). Ground-based observations are increasingly used in near-real-time data-driven models of the heliosphere and space weather. Two goals are to improve the quality of the measurements and to extend them upward into the chromosphere and corona (motivation M2).

A new science goal is to develop strategies and instrumentation to observe the earliest stages of an emerging active region at high spatial and temporal resolution. That cannot be done now except as a matter of blind-targeting good luck.

10.5 IMPERATIVES FOR THE HEALTH AND PROGRESS OF SOLAR AND HELIOSPHERIC PHYSICS

This section describes the imperatives proposed by the SHP panel for three groups: NASA, NSF, and multiagency. All imperatives are based in part on white-paper input. Of 288 white papers submitted by the community (for a list of titles, see Appendix I), about 150 were relevant to solar and heliospheric physics. Each was reviewed by at least two SHP panel members and many were also brought to the attention of the interdisciplinary working groups. The SHP panel is grateful to the community for its hard work, creative ideas and suggestions, and insight into issues that affect the health of the solar and heliospheric physics discipline.

10.5.1 NASA Missions in Development

The IRIS, SPP, and Solar Orbiter missions are in development. The new NASA mission concepts and other initiatives proposed here are predicated on the assumption that NASA will complete the development and launch of those three missions, each of which will make critical progress toward achieving SHP science goals 1-3 outlined above (§10.1).

10.5.2 New Imperatives for NASA

Sections 10.5.2.2-10.5.2.4 review three high-priority mission concepts that the SHP panel recommends for consideration during the coming decade. Two of the concepts—IMAP (§10.5.2.2) and SEE (§10.5.2.4)—underwent an independent cost and technical evaluation (CATE) that is described in Appendix E and briefly summarized in §10.5.2.1). Solar-C (§10.5.2.3) is an opportunity for NASA to participate in an international mission developed under the leadership of Japan. It is discussed below, but as an international mission it could not be reviewed in the same manner as other concepts and, in particular, it could not undergo an independent cost and technical review.

Sections 10.5.2.5-10.5.2.7 discuss one high-priority concept (L5) that should be considered for the following decade and two concepts—the Solar Polar Imager and the Interstellar Probe—that address high-priority goals but require new propulsion technology and possibly other new technologies.

10.5.2.1 Solar and Heliospheric Physics Panel Participation in the Cost and Technical Evaluation Process

At the SHP panel's first meeting in November 2010, it reviewed 30 white papers (for a list of titles, see Appendix I) describing future mission concepts, including potential strategic and Explorer missions for this decade and beyond. The SHP panel's second meeting included 19 invited talks on mission concepts, ground-based facilities, theory, modeling and data centers, and new technology. The panel selected four candidates for consideration by the steering committee for submission to the CATE process (SEE, IMAP, L5, and Reconnection and Microscale [RAM]) and reviewed two concepts for Solar-C.

The survey committee decided to include all four SHP-endorsed concepts in the CATE process; it also gave the SHP panel and the other panels uniform guidance that included a request to focus mission concepts around key objectives so as to minimize the required payload and cost. "Captains" later documented science objectives, measurement requirements, and mission and instrument concepts. Those "pre-CATE" activities resulted in schematic spacecraft and mission designs, refined science objectives, identification of potential risks, and estimated costs. Using that information, the survey committee selected IMAP and SEE for the full CATE process. The panels and the survey committee used the CATE results for final priority setting.

The IMAP, SEE, and L5 concepts are described below, as is an opportunity for NASA to participate in the Solar-C mission, which addresses objectives similar to those of the RAM concept.

10.5.2.2 Interstellar Mapping and Acceleration Probe

Our heliosphere, its history, and its future in the galaxy are key to understanding conditions on our evolving planet and its habitability over time. By exploring our global heliosphere and interactions, we develop key physical knowledge of the interstellar interactions that influence our home system in its current state, the history and destiny of our solar system, and the habitability of exoplanetary star systems.

Outer heliospheric science is an exciting, rapidly developing field because of groundbreaking all-sky images of the heliospheric boundaries based on energetic neutral atoms (ENAs) from the IBEX mission and Cassini-INCA in concert with dual in situ heliosheath observations from the Voyagers and IBEX measurements of interstellar neutral H, He, O, and Ne flow.

The surprising ENA "ribbon" (§10.3.4) demonstrates the importance of the interstellar magnetic field in the interaction of the heliosphere with our galactic neighborhood. The physical processes that form ENA spectra and the ribbon are hotly debated because of complex interactions between solar wind, pickup ions (PUIs), and suprathermal particles. The big picture provided by IBEX, complemented by Voyager observations, shows that the asymmetry of the heliosphere (§10.3.4) is shaped by the surrounding galactic magnetic field and that the physical processes that control the interaction exist on relatively small spatial and temporal scales (months). IMAP provides the next "quantum leap" forward in understanding the heliosphere through substantial improvements in spatial, temporal, and energy resolution[12] and much broader energy coverage than that of IBEX.

Observations from many spacecraft in the HSO contribute dramatically to understanding of SEP events, of the importance of suprathermal ions for efficient energization (§10.3.3), of the sources and evolution of solar wind (§10.3.2), of solar wind and SEP inputs into geospace, and of the evolution of the solar-heliospheric magnetic field (§10.3.1). Those observable phenomena are controlled by myriad complex and poorly understood physical effects that act on distinct particle populations (Figure 10.20). IMAP combines highly sensitive PUI and suprathermal-ion sensors to provide the species, spectral coverage, and temporal resolution to associate emerging suprathermal tails with interplanetary structures and physical processes (SHP action 3c).

IMAP orbits the inner Lagrangian point (L1) with comprehensive and highly sophisticated instruments to make the key observations that answer these fundamental questions:

- What is the spatiotemporal evolution of heliospheric boundary interactions?
- What is the nature of the heliopause and the interaction of the solar and interstellar magnetic fields?
- What are the composition and physical properties of the surrounding interstellar medium?
- How are particles injected into acceleration, and what mechanisms energize them throughout the heliosphere and heliosheath?
- What are the time-varying physical inputs at L1 into the Earth system?

The mission's heliospheric focus highlights the importance of making IMAP ENA maps and maps of ACR and CGR particles concurrently with in situ Voyager measurements of the heliospheric boundary region (motivation M3). IMAP enables understanding of particle acceleration through unprecedented collection power and time-resolved measurements of suprathermal ions that originate in the solar wind, interstellar medium, and inner heliosphere; enables environmental monitoring that is critical for effective background evaluation and removal from ENA maps and interpretation of PUI distributions; enables comprehensive interplanetary monitoring in support of geospace interaction studies; and enables space weather observations at the ideal location, L1 (SHP action 3d and motivation M2).

Answering the fundamental IMAP questions requires:

- High-resolution mapping and time evolution of heliospheric boundaries;

[12] D.J. McComas et al., Interstellar Mapping Probe (IMAP) Mission Concept: Illuminating the Dark Boundaries at the Edge of Our Solar System, white paper submitted to the Decadal Strategy for Solar and Space Physics (Heliophysics), Paper 188.

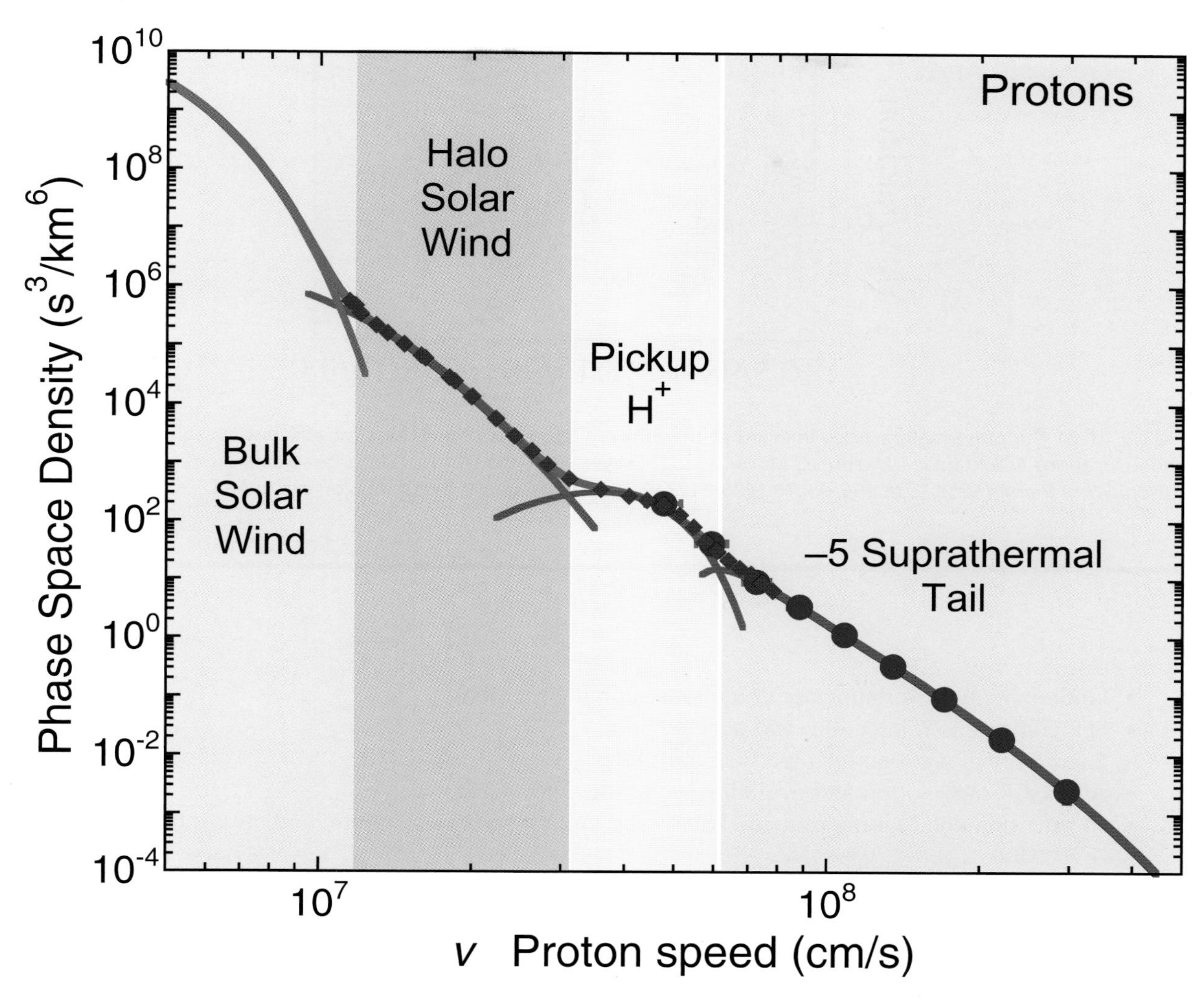

FIGURE 10.20 Four distinguishable components in the measured solar wind distribution (from Gloeckler et al., 2012): bulk solar wind (red); much hotter halo solar wind (blue); interstellar pickup ion H+, observable at 1 AU during the deep solar minimum (green); and suprathermal tail with a spectrum approximated by a power law with an exponential rollover. (A –5 slope in phase space density corresponds to a differential intensity with a slope of –1.5; see Section 10.3.4.) SOURCE: Adapted from G. Gloeckler, L.A. Fisk, G.M. Mason, E.C. Roelof, and E.C. Stone, Analysis of suprathermal tails using hourly averaged proton velocity distributions at 1 AU, pp. 136-143 in *Physics of the Heliosphere: A 10 Year Retrospective*, Proceedings of the 10th Annual International Astrophysics Conference (J. Heerikhuisen, G. Li, N. Pogorelov, and G. Zank, eds.), Volume 1436, American Institute of Physics, Melville, N.Y., doi:10.1063/1.4723601, 2012.

- Properties of interstellar neutral gas flow and its composition for H (including isotopes), He, O, and Ne (also to address big bang cosmology with the first in situ D/H observations) and properties of the outer heliosheath;
- PUI composition (implications for big bang cosmology and nucleosynthesis with a dedicated PUI instrument: He^3/He^4 and Ne^{22}/Ne^{20} with better than 5 percent accuracy);
- Seed populations of energetic particles with high time resolution (several minutes) (Figure 10.21);

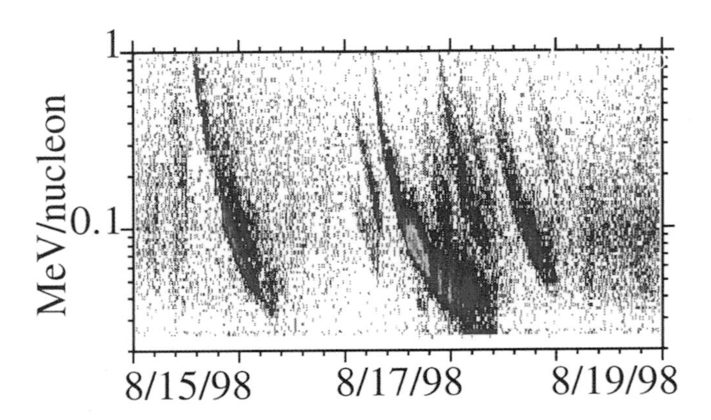

FIGURE 10.21 Suprathermal particles injected at the Sun provide seed populations for efficient energization at widely varied locations. SOURCE: G.M. Mason, J.E. Mazur, and J.R. Dwyer, ^{3}He enhancements in large solar energetic particle events, *Astrophysical Journal* 525:L133-L136, doi:10.1086/312349, 1999. Reproduced by permission of the AAS.

- Underlying time variations of ubiquitous suprathermal ions;
- SEP composition, injection, and acceleration;
- Suprathermal and energetic-particle transport;
- ACR/GCR modulation and evolution with time; and
- L1 environmental monitoring and solar wind input for magnetospheric and atmospheric science.

IMAP spacecraft and instrument implementation is based largely on ACE with ENA imaging infused from IBEX. IMAP is a Sun-pointed spinner, with spin-axis readjustment every few days to provide all-sky maps every 6 months. The L1 placement avoids magnetospheric ENA backgrounds and allows continuous interplanetary observations. Mission goals are achieved with a 2-year baseline, including transit to L1, with possible extension to longer operation. IMAP combines the following measurement capabilities, for which no further development effort is required:

- *High-resolution ENA maps.* Two ENA cameras produce new ENA observations of the heliospheric boundary over an extended energy range (0.3-20 keV, 3-200 keV) with substantially improved sensitivity (≈80 times the combined sensitivity and duty cycle of IBEX-Hi and CASSINI/INCA), spatial (10 times the IBEX-Hi angular resolution), and energy and time resolution (for example, oversampling of polar regions with few-day time resolution) compared with prior observations.
- *High-resolution and high-sensitivity ISM flow collection.* An ISM neutral atom camera and the first dedicated PUI sensor will take coordinated high-sensitivity observations of the interstellar gas flow through the inner solar system. The ISM neutral camera provides ISM flow observations of H, D, He, O, and Ne at 5-1,000 eV with pointing knowledge of 0.05° and over 10 times the combined sensitivity or duty cycle of IBEX-Lo, also extending the ENA maps to <0.3 keV. The PUI sensor provides distributions of interstellar H^+, $^{3}He^+$, $^{4}He^+$, N^+, O^+, $^{20}Ne^+$, $^{22}Ne^+$, and Ar^+ and inner-source C^+, O^+, Mg^+, and Si^+ over 100 eV to 100 keV/e with a combined sensitivity or duty cycle 100 times that of SWICS, also providing solar wind heavy-ion composition.

- *High-cadence suprathermal-ion observations.* Overlapping with the PUI sensor, a suprathermal-ion sensor provides composition (0.03-5 MeV/nuc) and charge state (0.03-1 MeV/e) for H through ultra-heavy ions (1-min cadence for H and He).
- *Solar wind and interplanetary monitoring suite.* This suite mitigates backgrounds for high-sensitivity ENA observations and provides societally important real-time solar wind and cosmic-ray monitoring. It measures solar wind ions (0.1-20 keV/e) and electrons (0.005-2 keV) every 15 s, the magnetic field at 16 Hz, and SEP, anomalous cosmic-ray, and galactic cosmic-ray electrons and ions (H-Fe) over 2- to 200-MeV/nuc.

10.5.2.3 Solar-C

Solar-C is a Japan-led mission expected to include substantial contributions from the United States and Europe.[13] It builds on the highly successful Yohkoh and Hinode collaborations with the United States' most reliable partner. As with Yohkoh and Hinode, Japan will provide the satellite and launch. Almost all NASA funding would go to the U.S. science community for state-of-the-art instrumentation and data analysis. Hence, Solar-C presents an important opportunity to leverage NASA science funding.

The science objectives of Solar-C are to determine:

- How the energy that sustains the Sun's atmosphere is created on small scales and transported into the large-scale corona and solar wind;
- How magnetic energy is dissipated in astrophysical plasmas; and
- How small-scale physical processes initiate large-scale dynamic phenomena, such as CMEs and flares, which drive space weather.

Achieving those objectives is a prerequisite for meeting SHP panel science goals 2 and 3. Solar-C is central to the science strategy for the next decade; therefore, the panel strongly endorses U.S. participation in the mission. As with Hinode, the data should be open to the full U.S. science community. Furthermore, a competitive Solar-C guest-investigator program, overseen by NASA, that follows the guidelines of the general guest-investigator program initiative would achieve maximum science benefit (§10.5.3.4).

To meet its three objectives, Solar-C will obtain highly precise spectroscopic and polarimetric measurements designed to determine the full-vector magnetic field accurately, especially in the chromosphere, and high-throughput measurements designed to resolve the plasma dynamics. Furthermore, spectroscopic measurements that seamlessly cover each temperature domain of the solar atmosphere—the photosphere, lower chromosphere, upper chromosphere, transition region, inner corona, and high-temperature flare—will be obtained to improve understanding of the entire chain of energy transport and dissipation. Finally, high-spatial-resolution measurements will be obtained for resolving elementary physical processes.

Solar-C can meet its measurement strategy with three strawman instruments designed to deliver an order-of-magnitude improvement over present measurement capabilities:

- A Solar UV-visible-IR telescope that will resolve and measure magnetic fields and gas dynamics in the lower atmosphere—from the photosphere through the upper chromosphere—with a diffraction-limited telescope that has an aperture 1.5 m in diameter.
- An EUV/FUV high-throughput spectrometer that will measure spectral lines in the FUV-EUV region from plasma in the upper chromosphere, transition region, lower corona, and flares simultaneously to

[13] G. Doschek et al., The High-Resolution Solar-C International Collaboration: Probing the Coupled Dynamics of the Solar Atmosphere, white paper submitted to the Decadal Strategy for Solar and Space Physics (Heliophysics), Paper 60.

trace energy flow throughout the solar atmosphere and follow the energy released by such processes as magnetic reconnection and instabilities.

- An X-ray imaging (spectroscopic) telescope that will resolve and measure the plasma in the hot corona to improve our understanding of its elemental structure, origins, and dynamics.

The strawman instruments outlined above are only for the purposes of planning and costing the mission. Concrete plans for the instruments and for the roles of the international partners are urgently needed; consequently, it is a *high* priority of the SHP panel that, as with Hinode, NASA and its partners form a Science and Technology Definition Team for Solar-C as soon as possible. Although the NASA contribution has yet to be decided, the panel expects that NASA contributions would involve the most technically challenging elements, such as the focal-plane packages (cameras, detectors, and so on), which would afford the U.S. science community an opportunity to make critical advances in remote-sensing capabilities. The total cost to NASA through Phase E should be capped at $250 million.

Solar-C presents a unique opportunity for solar and space physics to make flagship-level science advances for the cost of an Explorer.

10.5.2.4 Solar Eruptive Events Mission

Major solar eruptive events, consisting of both large flares and fast massive CMEs, are the most powerful explosions and particle accelerators in the solar system (Figure 10.22). They produce the most extreme space weather, generating SEPs that pose a major radiation hazard for spacecraft and humans, intense photon emissions that disrupt GPS and communications (see Figure 10.13), and storms in the magnetosphere that can cause power blackouts and disable satellites. Thus, understanding the fundamental physics of solar eruptive events is one of the most important goals of heliophysics.

Observations indicate that the flare energy-release particle-acceleration region is high above the flare X-ray loops (§10.3.3). The acceleration of fast CMEs is synchronized with the flare energy release, and this suggests that magnetic reconnection in the current sheet behind the CME both generates the flare and accelerates the CME. The CME-driven shock then accelerates SEPs. Despite much progress in developing this picture, fundamental physics questions remain:

- How is magnetic energy suddenly released to produce both a flare and a CME?
- How are CMEs accelerated to high speeds?
- How can flares accelerate electrons and ions so efficiently?
- How are escaping SEPs accelerated to such high energies?
- How can the magnetic energy for major solar eruptive events be accumulated in the corona?

To make major breakthroughs, a single-spacecraft SEE[14] mission in low Earth orbit, with powerful new instruments and a roughly 10-m boom for optics and occulter (Figure 10.23), will provide, for the first time, the following detailed measurements of accelerated electrons and ions plus ambient plasma conditions in the energy-release particle-acceleration regions (SHP action 3a):

- *Focusing Optics X-ray Spectroscopic Imager (FOXSI).* Provides HXR (about 2 to over about 80 keV) imaging (about 7 arcsec) spectroscopy (less than about 1-keV full-width half-maximum [FWHM]) of accelerated electrons and hot thermal plasmas in the high-coronal-energy-release particle-acceleration region

[14] R.P. Lin et al., Solar Eruptive Events (SEE) 2020 Mission Concept, white paper submitted to the Decadal Strategy for Solar and Space Physics (Heliophysics), Paper 162.

FIGURE 10.22 Images and energy budget of the major components of a SEE mission. This comprehensive picture required input from instruments on five spacecraft. SOURCE: A.G. Emslie, H. Kucharek, B.R. Dennis, N. Gopalswamy, G.D. Holman, G.H. Share, A. Vourlidas, T.G. Forbes, P.T. Gallagher, G.M. Mason, T.R. Metcalf, R.A. Mewaldt, R.J. Murphy, R.A. Schwartz, and T.H. Zurbuchen, Energy partition in two solar flare/CME events, *Journal of Geophysical Research* 109:A10104, doi:10.1029/2004JA010571, 2004. Copyright 2004 American Geophysical Union. Reproduced by permission of American Geophysical Union.

---

for most flares, even in the presence of intense footpoint emission, with about 30-100 times RHESSI's dynamic range and sensitivity. FOXSI will also detect accelerated electrons in impulsive SEP events and type III radio bursts in the corona, as well as nanoflares.

- *Energetic Neutral Atom Spectroscopic Imager (ENASI).* Provides ENA imaging (about 0.1 R) spectroscopy of about 4- keV to about 30-MeV SEPs accelerated by CME-driven shocks at about 1.5-10 R_S

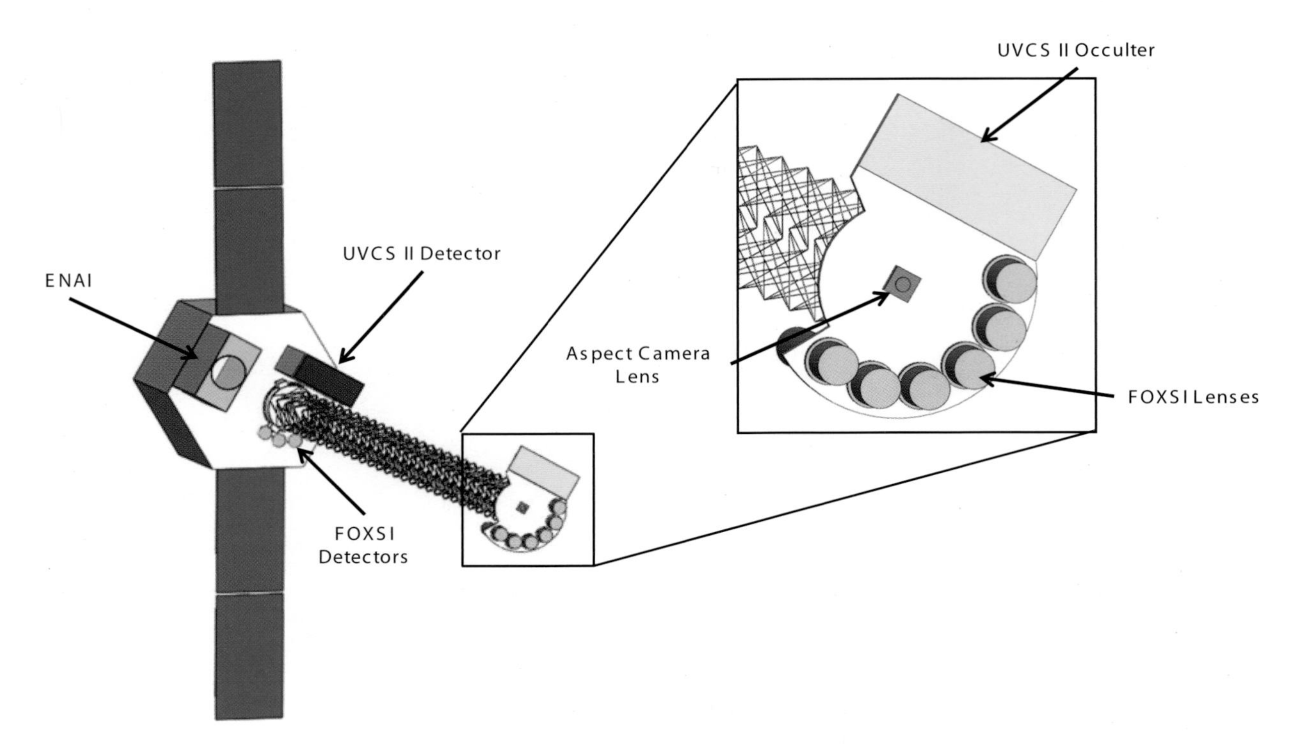

FIGURE 10.23 Schematic spacecraft and instrument accommodation for the minimum SEE mission SOURCE: Courtesy of the Aerospace Corporation.

altitudes, including the SEP seed population down to about 4 keV, with about 1,000 times STEREO's sensitivity. ENASI should also detect ENAs from ions accelerated in impulsive SEP events in the corona.

- *Gamma-Ray Imaging Spectrometer (GRIS).* Provides imaging of a few to hundreds of millions of electron-volt flare-accelerated ions through their gamma-ray line emissions with sufficient spatial resolution (about 7 arcsec), spectral resolution (a few thousand-electron-volt FWHM), and sensitivity to follow the evolution of the ion footpoints, even for normal-size flares.
- *UV-EUV Imaging Spectrometer (EUVIS).* Provides measurements of the ambient density, electron and ion temperatures, ionization states, composition, and flow and turbulent velocities in the flare energy-release particle-acceleration region with high-cadence imaging (less than about 10 arcsec) spectroscopy ($\lambda/\Delta\lambda$ over 3,000) up to about 1.2 R_S. EUVIS should also detect downward-going protons accelerated over 10 keV through their redshifted Lyman alpha emission.
- *UV Coronagraph Spectrometer (UVCS II).* Provides the same measurements as EUVIS but from about 1.2 to 10 R_S.
- *White-Light Coronagraph (WLC).* Provides imaging of CME structure and evolution from 1.5 to 15 R_S.

SEE would operate autonomously in a store-and-dump mode (like RHESSI) with a large onboard memory. Some SEE measurements (such as HXR or ENA of near-Sun SEP intensities) may be good precursors of these major eruptions; near-real-time data could be downlinked for space weather warnings.

ATST, FASR, and COSMO can make crucial measurements of coronal magnetic fields in the energy-release particle-acceleration regions. SPP and Solar Orbiter would provide ideal complementary in situ SEP/CME measurements close to the Sun and solar imaging (coronagraph, heliospheric imager, and HXR).

The minimum SEE mission evaluated by the CATE process included only FOXSI, ENASI, and UVCSII, launched on a Taurus 3210. A EUVIS type of instrument and a WLC are part of Solar-C and Solar Orbiter, respectively. GRIS requires more mass and power than the other instruments combined; flying it on ultra-long-duration balloon flights at solar maximum should be investigated.

A two-instrument SEE mission (FOXSI and ENASI) was deemed by the CATE process to fit in the mid-scale line (less than $480 million), and it provides tremendous new science. One or more of these new instruments, of appropriate size, would also be appropriate in an Explorer mission.

Regarding development status, a FOXSI instrument is scheduled for rocket flight in early 2012 and the balloon-borne GRIPS (Gamma-Ray Imager/Polarimeter for Solar flares) payload is being developed for a 2012 flight. UVCSII is based on proven SOHO UVCS technology. The ENASI instrument uses silicon semiconductor detectors that have flown successfully on STEREO/IMPACT LET and STE instruments and RHESSI modulation-grid imaging methods. EUVIS is based on the Hinode technology, and UVCSII is based on UVCS on SOHO. A 10-m extendable boom will be flown on the NuSTAR SMEX mission (planned for launch in 2012).[15]

10.5.2.5 L5 Mission Concept

The L5 mission concept would place a spacecraft carrying imaging and in situ instruments in about a 1-AU orbit near the L5 Lagrangian point (Figure 10.24).[16] From that location, the mission could make major advances in helioseismology by probing for longitudinal variations in tachocline magnetic fields, observe emerging active regions before they affect Earth, study CME evolution with stereoimaging and in situ data, and make major advances in space weather forecasting. The spacecraft and payload could rely on STEREO heritage. A Doppler magnetograph and UV spectrograph would be added.

An important science objective for L5 is to address the question, How does the solar dynamo drive magnetic activity on the surface? Global helioseismology has had remarkable success in revealing the Sun's internal radial structure and internal rotation. However, global helioseismology does not resolve longitudinal variations of the solar interior. Local helioseismic techniques (such as time-distance helioseismology) provide longitudinal information, but only in the upper third of the convection zone if viewed from a single vantage point.

Calculations predict longitudinal variations as signatures of the magnetic field at the tachocline. Simultaneous observations (Earth + L5) will detect both ends of long, deep, wave ray paths that penetrate to the tachocline (see Figure 10.24). Combining L5 and near-Earth observations will probe variations over a large longitude range. The relatively stable L5-Earth separation and increased solar-surface coverage also enable improved measurements of rotational and meridional flows. Active longitudes and persistent surface "hotspots" of magnetic activity are probably also associated with hotspots in the tachocline region and require longitudinal resolution.

Another science objective concerns the question, Can helioseismology forecast strong flare activity? Recent studies show that the strength and vorticity of subsurface flows around active regions are closely

[15] This panel report was completed in late 2012. An update in June 2013 to the information above follows: FOXSI was launched successfully in November 2012 from the White Sands Missile Range in New Mexico; GRIPS is now planned to have its first test flight in September 2014; and NASA's Nuclear Spectroscopic Telescope Array (NuSTAR) began its 2-year mission on June 13, 2012, aboard a Pegasus XL rocket launched from Kwajalein Atoll in the Marshall Islands.

[16] A. Vourlidas et al., Mission to the Sun-Earth L5 Lagrangian point: An Optimal Platform for Heliophysics and Space Weather Research, white paper submitted to the Decadal Strategy for Solar and Space Physics (Heliophysics), Paper 273.

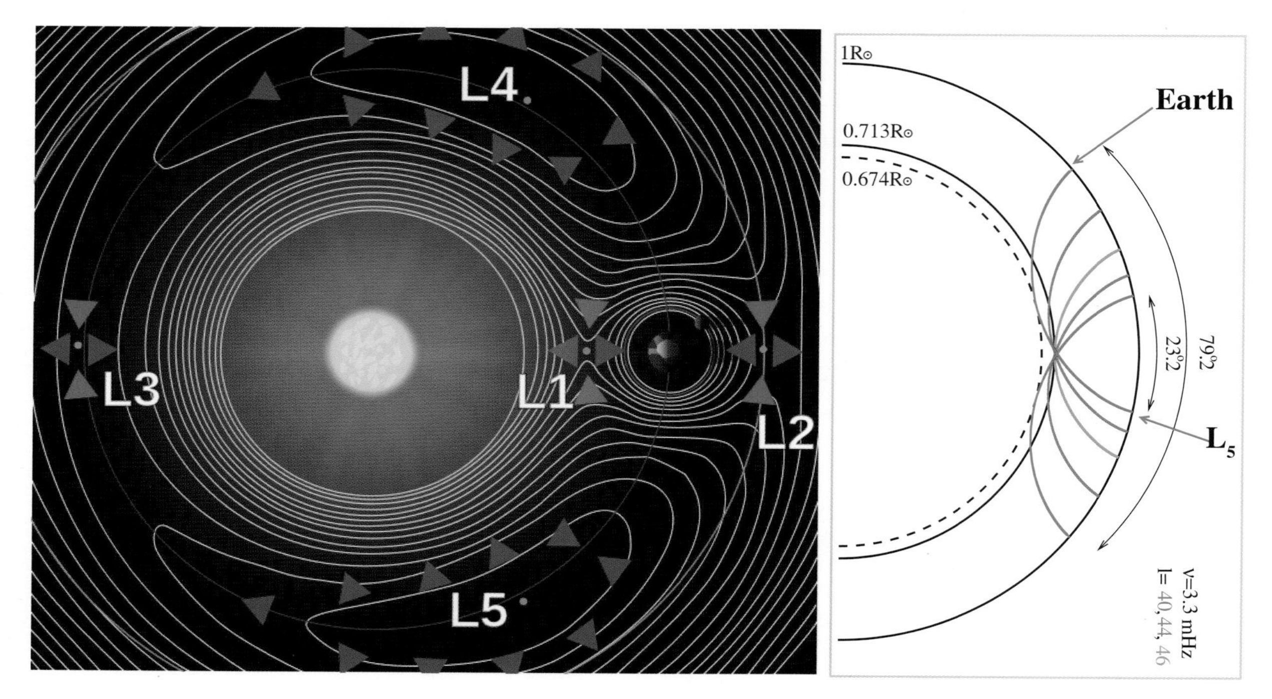

FIGURE 10.24 *Left:* The five Lagrangian points in the Earth-Sun system include L5, a broad gravitational plateau about 60° east of central meridian, where relatively little energy is needed to maintain an orbit extending about 40° to 90° east of the Earth-Sun line. *Right:* If travel times along the indicated ray paths are measured, thermal anomalies and flows in the tachocline can be resolved in longitude for the first time. For stereohelioseismology studies, L5 would be better suited than Solar Orbiter because of its continuous duty cycle at a relatively stable location. SOURCE: *Left*, NASA/WMAP Science Team. *Right*, Courtesy of S.P. Rajaguru, Indian Institute of Astrophysics, Bangalore, India.

related to flare activity and have the potential to become a forecast tool (see §10.3.1). In addition, there are indications that large active regions may be detected enough before they emerge to provide useful forecasts. L5 can be coupled with GONG/SDO to attempt probing vorticity deeper below the surface. Most important, L5's additional longitude coverage will allow active-region evolution studies for a much longer period than is now possible.

Instruments at L5 could also contribute to new advances in space weather forecasting. From L5, it is possible to observe and forecast Earth-directed CMEs with high precision, observe emerging and developed active regions about 4 days earlier than from Earth, forecast corotating interaction regions about 4 days before they cause geomagnetic storms, and measure photospheric magnetic fields over about 60° of additional longitude, improving models of coronal magnetic fields rotating toward Earth.

The pre-CATE L5 concept included the following instruments, all with excellent heritage:

- White-light coronagraph images, 2-225 R_S;
- EUV images, full-disk, 4 wavelengths;
- Doppler-magnetograph, full-disk;
- Off-limb spectroscopy, 8 wavelengths, 2 and 3.5 R_S;

- Hard X-ray imaging spectroscopy, full-disk, 64-150 keV;
- Solar and interplanetary energetic particles, 0.05- to 100-MeV/nuc, and electrons;
- Solar wind ion composition and electrons; and
- Magnetometer.

Two science phases are envisioned: drift to L5 at about 38° per year with continuous collection of science data and orbit around L5, 45°-90° from the Sun-Earth line. A long extended mission is possible.

Goddard Space Flight Center has studied a similar concept called Earth-Affecting Solar Causes Observatory[17] (EASCO), featuring about a 2-year low-thrust trajectory to L5 and using solar-electric propulsion, saving more than 200 kg compared with hydrazine.

In summary, the L5 mission concept promises important breakthroughs in both helioseismology and space weather (motivation M2). It would make major advances toward SHP panel science goals 1 and 3, including, in particular, actions 1b and 3d. The panel encourages NASA, NOAA, and the Department of Defense (DOD) to carry out an interagency study of an L5 mission (see §10.5.5.7).

10.5.2.6 Solar Polar Imager Mission

Current understanding of the Sun, its atmosphere, and the heliosphere is severely limited by a lack of good observations of the Sun's polar regions. The Solar Polar Imager (SPI) mission concept,[18] a NASA vision mission with strong international interest, would go into a 0.48-AU circular orbit with 60° inclination to conduct extended (many days per orbit) observations of the polar regions, enabling the determination of polar flows down to the tachocline, where the solar dynamo is thought to originate. The rapid 4-month orbit, combined with in situ and remote-sensing instrumentation, will enable unprecedented studies of the physical connections between the Sun, the solar wind, and SEPs.

Instrumentation could include a Doppler magnetograph, white-light coronagraph, EUV imager, UV spectrograph, TSI monitor, energetic-particle spectrometer, solar wind composition spectrometer, and magnetometer and could provide studies of the polar magnetic field over the solar cycle, the three-dimensional global structure of the corona, and solar wind, energetic-particle, and TSI variations with latitude. Most important, SPI would measure the temporal evolution of time-varying flows, differential rotation, and polar-region meridional circulation down to the tachocline, addressing SHP science goal 1. Finally, SPI would explore how space weather forecasts can benefit from a polar perspective (SHP science goal 3 and motivation M2).

Solar-sail propulsion is proposed to place SPI into its orbit. Recent advances demonstrate that solar sails are technically feasible and effective for maneuvering in the heliosphere. A technology-readiness plan is outlined in Section 10.5.2.8. The SHP panel strongly encourages NASA to develop the propulsion technology needed to launch SPI during the 2023-2033 decade.

10.5.2.7 Interstellar Probe Mission Concept

Recent in situ measurements by the Voyagers, combined with all-sky heliospheric images from IBEX and Cassini, have made outer-heliospheric science one of the most exciting and fastest-developing fields of heliophysics. The measurements have transformed knowledge of the boundaries of the heliosphere. The

---

[17] N. Gopalswamy et al., Earth-Affecting Solar Causes Observatory (EASCO): A New View from Sun-Earth L5, white paper submitted to the Decadal Strategy for Solar and Space Physics (Heliophysics), Paper 99.

[18] P. Liewer et al., Solar Polar Imager: Observing Solar Activity from a New Perspective, white paper submitted to the Decadal Strategy for Solar and Space Physics (Heliophysics), Paper 156.

Voyagers are now deep in the heliosheath, and one or both may cross the heliopause in the next decade. Although they have performed spectacularly, the Voyager instruments are 1970s-vintage and, for example, are unable to measure suprathermal heavy ions or interstellar-plasma elemental and ionic charge-state composition. The interstellar probe[19] would make comprehensive, state-of-the-art, in situ measurements of plasma and energetic-particle composition, magnetic fields, plasma waves, ionic charge states, energetic neutrals, and dust that are required for understanding the nature of the outer heliosphere and exploring our local galactic environment.

Advanced scientific instrumentation for an interstellar probe does not require new technology, as the principal technical hurdle is propulsion. (Also required are electric power from a low-specific-mass radioactive power source and reliable, sensitive, deep-space Ka-band communications.) Advanced propulsion options, which could be pursued with international cooperation, should aim to reach the heliopause considerably faster than Voyager 1 (3.6 AU/year). Possibilities include solar sails and solar electric propulsion alone or in conjunction with radioisotope electric propulsion.[20,21] The panel did not find either the ballistic or the nuclear electric power approach to currently be credible. In summary, to enable achievement of this decadal survey's key science goals in the coming decades, the SHP panel believes high priority should be given by NASA toward developing the necessary propulsion technology for visionary missions like SPI and interstellar probe.

10.5.2.8 Solar-Sail Propulsion for Heliophysics Missions

Solar sails have long been envisioned as a simple, inexpensive means of propulsion that could provide access to and maintenance of unstable orbits that would otherwise require, if they were possible at all, large and expensive propulsion systems. Solar sails can use solar photons to propel inner-heliosphere spacecraft to high velocities ($\Delta v > 50$ km/s) and can provide low-thrust propulsion to maintain missions in non-Keplerian orbits that are not feasible by other means. Solar sails will enable a number of important heliophysics missions, including the Solar Polar Imager (§10.5.2.6), an interstellar probe (§10.5.2.7), and a solar wind monitor several times farther upstream than L1. All indications are that solar-sail propulsion (SSP) is technically feasible and very effective for maneuvering in the heliosphere.[22]

Recently, the NASA Office of the Chief Technologist (OCT) selected a small sail-technology demonstration mission for implementation in the near future. However, for future missions like the Solar Polar Imager, a critical follow-on step will be flight validation of a full-scale (about 150 × 150-m) SSP system. That could be accomplished by NASA's Heliophysics Division's investing about $50 million as "seed money" in the full-scale SSP development effort over the next decade by partnering with the OCT Technology Demonstration Missions program (or other technology program as appropriate). In addition, investing a modest part of the seed money (about 10 percent) to fund grants to NASA centers and universities for solar-sail mission design, trajectory analysis, and so on, would lead to new mission applications for heliophysics exploration.

---

[19] R. McNutt et al., Interstellar Probe, white paper submitted to the Decadal Strategy for Solar and Space Physics (Heliophysics), Paper 195.

[20] R. McNutt et al., Interstellar Probe, white paper submitted to the Decadal Strategy for Solar and Space Physics (Heliophysics), Paper 195.

[21] L. Johnson et al., Solar Sail Propulsion: Enabling New Capabilities for Heliophysics, white paper submitted to the Decadal Strategy for Solar and Space Physics (Heliophysics), Paper 122.

[22] R.P. Lin et al., Expansion of the Heliophysics Explorer Program, white paper submitted to the Decadal Strategy for Solar and Space Physics (Heliophysics), Paper 160; E. Moebius et al., NASA's Explorer Program as a Vital Element to Further Heliophysics Research, white paper submitted to the Decadal Strategy for Solar and Space Physics (Heliophysics), Paper 205.

Its advocacy for this seed-money allocation emphasizes the high priority that the SHP panel places on developing SSP technologies to enable future expeditions to the inner and outer heliosphere and to the local interstellar medium.

10.5.3 Summary of NASA-Related Imperatives Developed by the Panel on Solar and Heliospheric Physics

10.5.3.1 Completion of the Development and Launch of the Heliophysics Flight Program Now in Development

SHP Imperative: The highest-priority objectives for heliophysics during the coming decade are to complete the development and launch of the IRIS and SPP missions and to fulfill U.S. contributions to the Solar Orbiter mission.

Justification: These new missions each offer unique opportunities for important breakthroughs in achieving SHP goals outlined in Section 10.1 (see §10.4.6.1-§10.4.6.3). IRIS is a small Explorer that will investigate how internal convective flows power solar-atmospheric activity. SPP will make mankind's first visit to the solar corona carrying a payload of in situ and remote-sensing instruments to discover how the corona is heated, how the solar wind is accelerated, and how the Sun accelerates particles to high energy. The ESA-NASA Solar Orbiter mission has a remote-sensing and in situ payload that, after a NASA launch, will investigate links between the solar surface, corona, and inner heliosphere from as close as 62 R_S and as high as 35° solar latitude. The SHP panel gives its unqualified endorsement to the mission concepts under development in October 2011 as long as they satisfy cost and schedule guidelines.

10.5.3.2 Expansion of the Heliophysics Explorer Program

SHP Imperative: In light of the success of heliophysics Explorers in addressing focused, high-priority science targets in all three disciplines, the SHP panel gives high priority to expanding flight opportunities in the heliophysics Explorer program and adding new cost-effective middle-size launchers to the manifest.[23]

Justification: The Explorer line contains a mix of small and middle-size Explorers and a stand-alone mission of opportunity (SALMON) component supporting participation in space programs of other agencies and nations. These missions have been highly innovative, extremely successful, and implemented on time and within budget. There is a large supply of new, cutting-edge ideas for Explorer missions in all heliophysics disciplines. Since 2001, 43 heliophysics-related Explorer missions have been proposed; 15 (35 percent) were rated category 1, but only 3 have been implemented. Ramping up the heliophysics Explorer program to about $150 million per year could enable a launch opportunity (Explorer or SALMON) every year.

Currently available (as of May 2012) launch vehicles are not adequate to support middle-size Explorers. Adding cost-effective low-end Atlas, Minotaur, or Falcon-9 launchers to the Explorer manifest would enable future MidEx mission opportunities.

[23] R.P. Lin et al., Expansion of the Heliophysics Explorer Program, white paper submitted to the Decadal Strategy for Solar and Space Physics (Heliophysics), Paper 160; E. Moebius et al., NASA's Explorer Program as a Vital Element to Further Heliophysics Research, white paper submitted to the Decadal Strategy for Solar and Space Physics (Heliophysics), Paper 205.

10.5.3.3 Development of Strategic Missions in the Explorer Mode

SHP Imperative: Given the Explorer program's excellent record in delivering state-of-the-art science within or below cost and on schedule, the SHP panel gives high priority to extending the Explorer mission-development model to middle-size strategic missions.

Justification: The heliophysics Explorer program has an unmatched record of innovative implementation and high science return on investment while staying within cost and schedule guidelines. Adopting the Explorer model for strategic missions whenever possible implies that missions are competed, led by principal investigators (PIs), and cost-capped. Unlike for Explorers, the science would be restricted. To encourage innovation, considerable latitude should be allowed to achieve the science objectives. That approach should be possible for missions up to about $500 million; the Planetary Division already has PI-led missions of this size. That is consistent with the 2003 decadal survey recommendation that for strategic STP and LWS missions "NASA should (1) place as much responsibility as possible in the hands of the principal investigator, (2) define the mission rules clearly at the beginning, and (3) establish levels of responsibility and mission rules that are tailored to the particular mission and to its scope and complexity."[24]

10.5.3.4 Recovery of an Effective NASA Grants Program

The heliophysics grants program is the foundation of the NASA science enterprise, but its effectiveness has been severely compromised in recent years by budget cuts in both research and analysis and new-missions guest-investigator (GI) programs and by the dearth of opportunities to develop innovative instrument concepts. In all of 2010, there were only 12 advertisements for heliophysics postdoctoral positions (see Appendix D, "Education and Workforce Issues in Solar and Space Physics")—clear evidence that the typical PI grant size is no longer sufficient to support postdoctoral researchers. The imperatives described below are essential for recovering an effective NASA science grants program.

Establishment of Heliophysics Science Centers

Achieving the SHP panel's science goals (§10.1) requires solving a number of major science problems. Many are sufficiently mature that important progress toward achieving closure between theory and observations can be expected. Examples are the following:

- Generation, emergence, and detection of active regions and other subsurface structures;
- Dynamic coupling of the ambient solar corona to the inner heliosphere;
- Magnetic reconnection in the Sun and heliosphere;
- Acceleration and transport of high-energy particles from the Sun;
- Origin and evolution of extreme solar storms and their impact on the geospace environment;
- Structure of the large-scale heliosphere and its interaction with the interstellar medium; and
- Complexity, nonlinearity, and cross-scale coupling through physical processes, such as turbulence, plasma-neutral coupling, and wave-particle interactions.

Making ground-breaking advances on these major problems requires teams that combine different expertise.[25] The SHP panel therefore strongly supports the creation of new heliophysics science centers composed of teams of theorists, numerical modelers, and data experts working collectively to tackle the

[24] NRC, *The Sun to the Earth—and Beyond: A Decadal Research Strategy in Solar and Space Physics*, 2003, pp. 19 and 157-158.

[25] A. Bhattacharjee et al., Advanced Computational Capabilities for Exploration in Heliophysical Science (ACCEHS): A Virtual Space Mission, white paper submitted to the Decadal Strategy for Solar and Space Physics (Heliophysics), Paper 12.

field's most compelling science problems. (See the section "Venture: Venture Forward with Science Centers and Instrument and Technology Development" in Chapter 4.)

NASA Individual-Principal-Investigator Grants Program

To achieve its broad science goals, the NASA science enterprise requires coordinated programs of missions, data analysis, theory and modeling, and technology development. Heliophysics individual-PI grants programs (SR&T, TR&T, and GI programs) are the core support of the division's non-mission-hardware science and are the bedrock for developing new science understanding and mission concepts. Many transformational advances, ranging from the prediction of a solar wind to the development of far-side imaging, are based on research supported by the core grants program.

Although PI-grant programs provide unmatched science return on investment, they have recently suffered severe losses, and their funding is now less than 10 percent of total division funding. That investment is inadequate; fully realizing the returns of the HSO with its enormous data sets from multiple missions demands a higher level of PI-grant funding. Therefore, the SHP panel assigns high priority to the Heliophysics Division's gradually increasing the investment in individual-PI-grant programs to 20 percent of division funding over the next decade.[26]

Furthermore, the SHP panel suggests that the increase be implemented primarily by increasing grant size rather than grant number. The present grant size of $100,000 to $150,000 per year is inadequate to support even individual-PI investigations and restricts the scope and depth of the science that such grants can address too much. It also leads to enormous inefficiencies for the community and NASA in the number of proposals that are written and reviewed. It is an imperative of the SHP panel that NASA work toward doubling the size of SR&T, TR&T, and GI grants to $200,000 to $300,000 per year.

Heliophysics Systems Observatory and MO&DA Support

The HSO relies on a distributed network of missions (implemented for specific science goals) to address strategic objectives that require multipoint or remote-sensing observations on long timescales.[27] The distributed network provides comprehensive measurements of the whole Sun-heliosphere system, including Sun-Earth interactions, spanning a few years up to almost a 22-year Hale cycle. It fuels new system science by extending individual missions beyond their prime phase.

Resource allocation among extended HSO missions is determined through the senior-review process, which evaluates future scientific priorities for each mission. The present 5-year budget requests show flat or declining HSO funding. In addition to supporting existing HSO missions, the budget must accommodate new missions, such as RBSP (renamed the Van Allen Probes) and SDO, that finish their prime mission in or before FY 2015; this will inevitably lead to forced termination of or severe cuts in current HSO missions. As a consequence, key systems-science objectives are endangered, and essential legacy data sets may be foreshortened at a time when solar activity is apparently evolving in unexpected ways. Multipoint observations throughout the heliosphere and from the Sun to geospace regions need to be maintained to enable systems science. The SHP panel assigns high priority to augmenting MO&DA support by annual inflationary increases plus $5 million to $10 million per year to accommodate new missions so that senior-review decisions can be prudently based on strategic evaluations of existing and emerging assets.

---

[26] J.A. Klimchuck, Maximizing NASA's Science Productivity, white paper submitted to the Decadal Strategy for Solar and Space Physics (Heliophysics), Paper 135; C.M.S. Cohen et al., Protecting Science Mission Investment: Balancing the Funding Profile for Data Analysis Programs, white paper submitted to the Decadal Strategy for Solar and Space Physics (Heliophysics), Paper 45.

[27] J.G. Luhmann, Guest Investigator and Participating Scientist Programs, white paper submitted to the Decadal Strategy for Solar and Space Physics (Heliophysics), Paper 166.

Adequate MO&DA Budgets and Mission-Related Guest-Investigator Programs

The science return from missions depends critically on adequate funding to analyze returned data, to develop relationships of observations across the HSO, and to test theories and models that are required to answer the science questions that motivated these missions quantitatively. Confirmed missions need to have adequate Phase E research budgets that ensure it is possible to achieve mission success. In addition, the budget for all missions needs to include an adequate GI program (typically 3 percent of the total mission cost spread over the prime mission) administered by NASA as part of the mission success criteria. The GI programs enable the community at large to exploit new data further and relate them to other fields of heliophysics research.[28]

Heliophysics Instrument Development Program

Heliophysics is now exploring the boundaries of its domain, performing increasingly complex measurements, and preparing for truly predictive capabilities. Progress hinges on growing the capabilities of new instrumentation. High-priority items include the following:

- UV-blind ENA detectors that promise breakthroughs in resolution and efficiency;
- Detectors for MeV ENAs to open new windows on localized acceleration processes; and
- Large-format, high-efficiency array detectors and rapidly switching polarization modulators that can measure heating and acceleration processes in the solar atmosphere.

Achieving such capabilities requires increased support for new instrument concepts (recommended in the 2003 decadal survey[29]). Also needed are opportunities for low-cost rides into space to boost the technology readiness level and means of leveraging technology-development resources in the Office of the Chief Technologist.

Therefore, the SHP panel strongly advocates that current funding for instrument development within SR&T, LWS, and the Low Cost Access to Space (LCAS) program be consolidated into a comprehensive heliophysics instrument development program[30] that includes funding to cross the "valley of death" gap (technology readiness levels 4-6). If the program grows to about 2 percent of the heliophysics flight program budget, or about $6 million in current-year funds, comparable savings can be achieved from flight programs.

Low-Cost Access to Space Program

The NASA Heliophysics Division LCAS program provides opportunities for flying space experiments on sounding rockets (100- to 1,000-km apogee altitude), balloon payloads, and CubeSats to address new science opportunities, develop new instruments and technology, provide complementary science and underflight calibrations for missions, and train the next generation of space scientists and engineers. The SHP panel endorses the NRC's 2010 recommendation to increase funding support of NASA's suborbital programs.[31] In particular, funding for science-payload development, flight, and data analysis is inadequate and needs to be doubled.

---

[28]A.J. Tylka, Heliophysics System Science and Funding for Extended Missions, white paper submitted to the Decadal Strategy for Solar and Space Physics (Heliophysics), Paper 269; J.G. Luhmann et al., Extended Missions: Engines of Heliophysics System Science, white paper submitted to the Decadal Strategy for Solar and Space Physics (Heliophysics), Paper 167.

[29]NRC, *The Sun to the Earth—and Beyond: A Decadal Research Strategy in Solar and Space Physics,* 2003, p. 11.

[30]E.R. Christian, Heliophysics Instrument and Technology Development Program (HITDP), white paper submitted to the Decadal Strategy for Solar and Space Physics (Heliophysics), Paper 34.

[31]NRC, *Revitalizing NASA's Suborbital Program: Advancing Science, Driving Innovation, and Developing a Workforce,* The National Academies Press, Washington, D.C., 2010.

Funding support for NASA's Wallops Flight Facility (and subcontractors) appears adequate to support 20 or more flights per year, double the current rate. The SHP panel encourages continued development of new capabilities within the NASA LCAS program. High-altitude sounding rockets with three or four vehicles can provide longer observing time and carry heavier payloads. Balloons provide the only practical flight opportunities (because of launch-vehicle issues) for very heavy scientific payloads, such as the Sunrise solar experiment and the GRIPS Gamma-Ray Imager/Polarimeter mentioned above. Ultra-long-duration balloons (the first successful flight was in 2009) promise to provide about 100-day flight times in the near future. CubeSats can provide even longer observing times.

Conclusions of the Panel on Solar and Heliospheric Physics Regarding NASA's Grants Program

Achieving the key science goals of this decadal survey requires establishing an effective NASA heliophysics grants program by implementing or augmenting the following key elements:

- Creation of new heliophysics science centers composed of teams of theorists, numerical modelers, and data experts to tackle the most compelling science problems;
- A graduated increase in individual-PI grants programs to about 20 percent of division funding, primarily by increasing the size of individual grants, to realize the science returns from the space missions and develop future ones;
- An augmented heliophysics MO&DA program to enable systems science by using the HSO;
- Adequate Phase E research budgets for confirmed missions to ensure mission success and appropriately sized GI programs as part of mission success criteria to ensure mission science productivity;
- Formation of a consolidated heliophysics instrument development program to facilitate innovative instrument concepts and make future mission implementation cost-effective; and
- Increased funding for science-payload development, flight, and data analysis in the LCAS program to provide urgently needed flight and training opportunities.

10.5.4 NSF-Related Initiatives

The following initiatives are based mainly on more than 30 white papers dealing with current or potential NSF-funded facilities and programs (for a list of titles, see Appendix I).

10.5.4.1 Advanced Technology Solar Telescope Operations

SHP Imperative: The SHP panel attaches high priority to NSF's providing base funding sufficient for efficient and scientifically productive operation of the ATST to realize the scientific benefits of the major national investment in it.

Justification: Starting in 2017, the world-leading ATST will provide the most detailed and most accurate measurements of the Sun's plasma and magnetic field ever obtained.[32] The measurements promise to revolutionize understanding of the Sun and heliosphere. Realizing the abundant scientific payoff from the major national investment in ATST will require adequate sustained funding from NSF for operation, development of advanced instrumentation, and research-grant support for the ATST user community. The National Solar Observatory (NSO) will close all existing facilities except for a synoptic program, but projections indicate that the current NSO base funding will be adequate only to realize a small fraction of

[32] S. Keil et al., Science and Operation of the Advanced Technology Solar Telescope, white paper submitted to the Decadal Strategy for Solar and Space Physics (Heliophysics), Paper 130; K.P. Reardon et al., Approaches to Optimize Scientific Productivity of Ground-Based Solar Telescopes, white paper submitted to the Decadal Strategy for Solar and Space Physics (Heliophysics), Paper 224.

the scientific potential of the ATST. Additional annual funding of around 4-5 percent of the capital cost of the ATST will dramatically enhance its science yield.

10.5.4.2 Need for a National Science Foundation Midscale Projects Line

SHP Imperative: The SHP panel strongly supports the establishment within NSF of a competed funding program for midscale projects.

Justification: NSF does not have a means of funding midscale projects ranging from about $4 million to $135 million in FY 2010 dollars. Several previous bodies, including the 2010 decadal survey of astronomy and astrophysics (Astro2010),[33] have recommended that NSF implement a competed funding program for midscale projects similar to NASA's Explorer model. A midscale funding program would enable NSF to provide a more balanced and flexible response to scientific opportunities that are currently too large to allow funding by the Major Research Instrumentation program and too small to qualify for funding by the Major Research Equipment and Facilities Construction line. A midscale program would offer excellent return by exploiting new techniques and instrumentation more rapidly, by ensuring a broader scientific portfolio of exciting and timely programs, and by offering additional opportunities for training scientists, engineers, and students. Examples of solar facilities that could be funded by this new line are the Frequency-Agile Solar Radiotelescope (FASR) and the Coronal Solar Magnetism Observatory (COSMO). They are described in more detail below.

10.5.4.3 Frequency-Agile Solar Radiotelescope

The SHP panel assigns high priority to NSF's funding of construction and operation of FASR[34] to produce three-dimensional images of the solar atmosphere with high temporal and spatial resolution. Solar radio emission provides uniquely powerful sources of diagnostic information with the potential for transformational insights into solar activity and its terrestrial impacts. That is because a number of distinct emission mechanisms operate at radio wavelengths: thermal free-free emission is relevant to the quiet solar atmosphere, thermal gyroresonance emission is a key mechanism operative in solar active regions, nonthermal gyrosynchrotron emission from electrons with hundreds of kiloelectron volts 10 MeV plays a central role in flares and CMEs, and a variety of coherent emission processes—such as plasma radiation, familiar to aficionados of type II and type III radio bursts—make it possible to trace electron beams and shocks in the solar corona and heliosphere. Radio observations and the diagnostic information that they provide are highly complementary to next-generation observations at optical and infrared wavelengths provided by the ATST and COSMO on the ground, at EUV wavelengths by Solar-C, and at X-ray wavelengths by SEE in space.

FASR is a solar-dedicated radiotelescope that provides a unique combination of superior imaging capability and broad instantaneous frequency coverage and thus exploits the powerful diagnostics available at radio wavelengths. The potential value of such a facility has been recognized in high-priority recommendations of FASR in two previous decadal surveys conducted by the National Research Council. The 2003 decadal survey of solar and space physics recommended FASR as its highest-priority "small project"[35] in recognition of the unique and transformative role that it will play in addressing basic research questions

[33]NRC, *New Worlds, New Horizons in Astronomy and Astrophysics,* The National Academies Press, Washington, D.C., 2010, p. 28.

[34]D.E. Gary et al., The Frequency-Agile Solar Radiotelescope, white paper submitted to the Decadal Strategy for Solar and Space Physics (Heliophysics), Paper 86; D.E. Gary et al., Particle Acceleration and Transport on the Sun: New Perspectives at Radio Wavelengths, white paper submitted to the Decadal Strategy for Solar and Space Physics (Heliophysics), Paper 87.

[35]NRC, *The Sun to the Earth—and Beyond: A Decadal Research Strategy in Solar and Space Physics,* 2003, p. 54.

and its complementary role with respect to other important instruments being developed to address the Sun and heliosphere. The Astro2010 decadal survey characterized FASR as a "compelling" midscale project and recognized it, with ATST, as a core facility in the U.S. ground-based solar portfolio.[36] An independent analysis of cost and technical readiness (CATE analysis) described FASR as "doable today." FASR thus has broad constituencies in solar and space physics and in astronomy and astrophysics.

The major advance offered by FASR over previous solar radio instrumentation is its unique combination of ultra-wide-frequency coverage, high spectral resolution, and high image quality. FASR measures the polarized brightness temperature spectrum over a broad frequency range (50 MHz to 21 GHz, or 1.4-600 cm) along every line of sight to the Sun as a function of time. Radiation in this radio wavelength range probes the solar atmosphere from the middle chromosphere to well into the corona. In essence, FASR images the entire solar atmosphere in three dimensions once every second from the chromosphere through the corona while retaining the capability to image a restricted frequency range with time resolution as small as 20 ms. In so doing, FASR enables fundamentally new, unique observables, including:

- Quantitative measurements of coronal magnetic fields, both on the disk and above the limb, under quiet conditions and during flares;
- Measurements of tracers of energy release and the spatiotemporal evolution of the electron distribution function during flares;
- Imaging CMEs and the associated coronal dimming, "EIT waves,"[37] and coronal shocks; and
- Imaging of thermal emission from the solar atmosphere from chromospheric to coronal heights, including the quiet Sun, coronal holes, active regions, and prominences.

FASR's panoramic view allows the solar atmosphere and physical phenomena therein to be studied as a coupled system. Its powerful and unique capabilities allow it to address such high-priority science questions as these:

- *The nature and evolution of coronal magnetic fields*
  —What is the quantitative distribution of coronal magnetic fields?
  —How do coronal magnetic fields evolve in time and space?
  —How is magnetic energy stored?
- *The physics of flares*
  —What is the physics of magnetic energy release?
  —How and where are electrons accelerated?
  —What are the relevant particle transport processes?
- *The drivers of space weather*
  —How are CMEs initiated and accelerated?
  —What is the origin of coronal shocks?
  —How are solar energetic particles in the heliosphere accelerated?
- *The physics of the quiet sun*
  —How are the solar chromosphere and corona heated?
  —What is the origin of the solar wind?
  —What is the structure of prominences and filaments?

---

[36]NRC, *New Worlds, New Horizons,* 2010, p. 191.
[37]So named because they were first discovered by the Extreme-ultraviolet Imaging Telescope (EIT) on the SOHO spacecraft.

BOX 10.2 FASR SPECIFICATIONS

Angular resolution	$20/\nu_{GHz}$ arcsec
Frequency range	50 MHz to 21 GHz
Number of data channels	2 (dual polarization)
Frequency bandwidth	500 MHz per channel
Frequency resolution	Instrumental: 4,000 channels Scientific: minimum (1%, 5 MHz)
Time resolution	About 1 s (full spectrum sweep) 20 ms (dwell)
Polarization	Full Stokes (intensity and linear polarization and circular polarization)
Number of antennas deployed	A (2-21 GHz): about 100 B (0.3-2.5 GHz): about 70 C (50-350 MHz): about 50
Size of antennas	A (2-21 GHz): 2 m B (0.3-2.5 GHz): 6 m C (50-350 MHz): log-periodic dipole array
Array size	4.25 km east-west x 3.75 km north-south
Absolute positions	1 arcsec
Absolute flux calibration	<10%

FASR comprises three antenna arrays that are designed to observe meter (FASR C), decimeter (FASR B), and centimeter wavelengths (FASR A). The instrument exploits the long heritage of Fourier synthesis imaging developed over several decades in radioastronomy. The techniques are mature, and the technologies robust. High-level instrument specifications are given in Box 10.2.

10.5.4.4 Coronal Solar Magnetism Observatory

The SHP panel's other high-priority candidate for the NSF midscale line is the COSMO large-aperture coronagraph and chromosphere and prominence magnetometer,[38] which would complement the ATST and FASR. Until explosively released, most of the mass and energy that produce space weather events are stored in the coronal magnetic field. Coronal faintness is an observational challenge especially for sensitive magnetic and velocity measurements. Responding to a next-step suggestion of the 2003 decadal survey,[39] COSMO is proposed as a suite of three instruments: a ground-based, 1.5-m-aperture solar coronagraph supported by a 12-cm-aperture white-light K-coronagraph (under construction) and a 20-cm-aperture chromosphere and prominence magnetometer. COSMO will be a facility for use by the solar physics research community. The facility will take continuous daytime synoptic measurements of magnetic fields

[38] S. Tomczyk et al., COSMO, The Coronal Solar Magnetism Observatory, white paper submitted to the Decadal Strategy for Solar and Space Physics (Heliophysics), Paper 268; J. Burkepile et al., The Importance of Ground-Based Observations of the Solar Corona, white paper submitted to the Decadal Strategy for Solar and Space Physics (Heliophysics), Paper 25; S. McIntosh et al., The Solar Chromosphere: The Inner Frontier of the Heliospheric System, white paper submitted to the Decadal Strategy for Solar and Space Physics (Heliophysics), Paper 193.

[39] NRC, *The Sun to the Earth—and Beyond: A Decadal Research Strategy in Solar and Space Physics*, 2003.

in the solar corona and chromosphere to support understanding of solar eruptive events that drive space weather and investigation of long-term coronal phenomena. The large emission-line coronagraph will be the largest refracting telescope in the world.

COSMO complements other facilities and missions by providing 300-days-per-year limb observations of the coronal magnetic field down to 1 Gauss over a 1° field of view. Long-term synoptic observations will directly address how the coronal magnetic field responds to the sunspot cycle and how the switch in polarity of the global field manifests itself in the heliosphere.

On a shorter timescale, COSMO will provide new information on the evolution of and interactions between magnetically closed and open regions that determine the changing structure of the heliospheric magnetic field and provide verification of extrapolations of the photospheric field.

COSMO will detect polarization signatures of nonpotential fields to afford insight into the roles of energy storage, magnetic reconnection, helicity creation and transport, and flux emergence in CME formation. It will also provide white-light images of CMEs in the low corona needed to determine basic properties of CMEs in early stages of formation. Particles accelerated by CME-driven shocks have the highest particle energies and pose the greatest space weather hazards. COSMO can detect compressions and distortions in the field that are due to the formation and passage of a CME and associated waves. Density compressions resulting from shock formation can produce intensity enhancements in white-light coronal images. COSMO has the potential to detect shock and CME locations simultaneously.

COSMO will provide routine observations of prominence, filament, and chromospheric magnetic fields and prominence flows. Those observations will constrain prominence densities and determine how prominence and coronal magnetic fields interact, how and where magnetic energy is stored (for example, by flux and helicity transport), and how it is released (for example, by instabilities, reconnection, or dissipative heating). COSMO will provide the first routine high-time-cadence measurements of coronal magnetic fields in and around flares and associated CMEs as seen above the solar limb. The high-time-cadence observations can help to determine when, where, and how magnetic energy is released by CMEs and by dissipative processes that result in flares.

10.5.4.5 National Science Foundation Small-Grants Program

SHP Imperative: To support the community effort required to analyze, interpret, and model the vast new solar and space physics data sets of the next decade, the SHP panel assigns high priority to NSF's adopting the policy of doubling the size of its small-grants programs.

Justification: The coming decade should see major advances in solar and space physics with new ground-based and space-based instrumentation coming on line, new computational technology becoming available, and new personnel in the nation's universities and other institutions. Funding will be needed to achieve the science promise of the new capabilities and researchers. Core support for new researchers, especially in universities, comes from NSF small-grants programs, which include the CEDAR, GEM, SHINE, and base grants programs and associated postdoctoral and young-faculty programs. The NSF small-grants program is inadequate to cover the science requirements of existing facilities and personnel. Whether measured by publications, data analyzed, or students educated, increased investment in the small-grants program is by far the most effective strategy for NSF to use to achieve its science and education goals.

10.5.4.6 Broadening the Definition of the National Science Foundation's Solar-Terrestrial Research Program to Include Outer-Heliosphere Research

SHP Imperative: New scientific discoveries about the outer heliosphere may have important effects on the history and future of our planet, and this justifies broadening the scope of NSF's Solar-Terrestrial Research program.[40]

Justification: NSF's Solar-Terrestrial Research program is in the Geospace Section of the Atmospheric and Geospace Sciences Division. To be funded, proposals to the program must demonstrate direct relevance to solar-terrestrial science or be transformational in advancing understanding of the outer solar system. Even high-quality proposals in outer-heliospheric research may not be funded if they are not perceived to be directly relevant to changing terrestrial conditions. Outer-heliosphere research is not funded in other NSF programs and is "falling between the cracks" in NSF. Recent discoveries suggest that solar interactions with the local interstellar medium may play a much larger role in influencing terrestrial conditions than previously thought. In addition, this exciting and rapidly emerging field is providing new insights into processes occurring in the Sun, solar wind, planetary magnetospheres and atmospheres, and the galaxy. All that suggests that the definition of solar-terrestrial research in NSF should be broadened to include solar wind interactions with the local interstellar medium.

10.5.5 Multiagency Imperatives

10.5.5.1 Ground-Based Solar Observations

SHP Imperative: The SHP panel gives high priority to NSF's and other agencies' support (where it is appropriate to their missions) of continuation of ground-based observations of the Sun.[41]

Justification: The Sun's scientific and societal importance stimulates a broad array of research approaches. Much of the current understanding of the Sun originates in observations made on the ground. Continuing ground-based observations serve crucial roles in studies of long-term variations of the Sun, in attaining the highest spatial resolution of solar features, in measuring wavelengths from the infrared to the radio, in exploring new ideas, in testing numerical models, in developing new instruments, and in verifying space-based results. National and university-based solar observatories are supported mainly by NSF, and support from mission-oriented agencies is sometimes available when specific mission objectives align with ground-based capabilities. In spite of an outstanding record of scientific accomplishments, support for ground-based observations and for research based on them has eroded. The SHP panel believes that agencies, especially NSF, should work to reverse that trend.

---

[40] G.P. Zank, Role of the National Science Foundation ATM/GEO in Promoting and Supporting Space Physics, white paper submitted to the Decadal Strategy for Solar and Space Physics (Heliophysics), Paper 284.

[41] T. Ayres and D. Longcope, Ground-based Solar Physics in the Era of Space Astronomy, white paper submitted to the Decadal Strategy for Solar and Space Physics (Heliophysics), Paper 3; F. Hill et al., The Need for Synoptic Optical Solar Observations from the Ground, white paper submitted to the Decadal Strategy for Solar and Space Physics (Heliophysics), Paper 108; A. Pevtsov, Current and Future State of Ground-based Solar Physics in the U.S., white paper submitted to the Decadal Strategy for Solar and Space Physics (Heliophysics), Paper 218; A.G. Kosovichev, Solar Dynamo, white paper submitted to the Decadal Strategy for Solar and Space Physics (Heliophysics), Paper 141.

10.5.5.2 Laboratory Astrophysics and Calibration Facilities

SHP Imperative: NASA, the National Oceanic and Atmospheric Administration (NOAA), and NSF are encouraged to continue their support of laboratory facilities for accurate measurement of atomic properties and for instrument calibrations as needed for several research programs and missions.[42]

Justification: Understanding the Sun and heliosphere requires observing them remotely. Diagnosing these observations requires both theoretical knowledge of radiation across the electromagnetic spectrum and well-calibrated instrumentation. Heliophysics research relies heavily on measurements of atomic and molecular collision physics, especially for highly charged ions that dominate X-ray and EUV production. Proper interpretation of observations requires cross sections for calculating excitation, direct and dielectronic recombination, single-charge and multiple-charge exchange, lifetimes, and other values. And calibration facilities (for example, at the National Institute of Standards and Technology) are required for accurate calibration and characterization of NASA and NOAA instrumentation before flight. Without such facilities, the science return from missions and ground-based remote-sensing observations can be substantially limited. The SHP panel assigns high priority to continued support of these facilities by NASA, NOAA, and NSF, preferably through budget line items for laboratory facilities.

10.5.5.3 Enhanced Space Weather Monitoring and Modeling

SHP Imperative: To optimize space weather capabilities, the SHP panel recommends that all future solar and heliospheric space missions and ground-based facilities consider including capabilities for delivering space weather data products. It would be a benefit for NASA to develop multiagency missions whose primary purpose is space weather monitoring. An interagency clearinghouse and archive of space weather data would benefit forecasters, researchers, and model-builders.

Justification: During the past decade, the direct application of solar and heliospheric physics to the protection of life and our high-technology society has demonstrated that the nation's investment in the field is returning tangible benefits. The SHP panel recognizes the great benefit that would be realized if studies of new NASA missions considered adding space weather measurements and real-time data-transmitting capabilities where appropriate. Examples of possible missions include IMAP, SEE, and L5. The technical and financial effects of adding space weather instrumentation could be addressed by partnering with other NASA directorates or other agencies. Furthermore, the SHP panel recognizes a benefit in NASA consideration of partnering with other agencies to fly missions whose primary purpose is space weather operations but that also have science objectives, such as an L5 mission (§10.5.2.5).

Currently, ground-based neutron monitors and solar optical telescopes and radiotelescopes provide unique real-time data. In the future, an expansion of SOLIS to a three-site network and completion of FASR and COSMO would provide key data. The panel encourages ground-based facilities to assess data products that could improve space weather capabilities. Finally, to expand the availability of space weather data to forecasters, researchers, and model-builders, the SHP panel encourages development of an interagency clearinghouse for real-time and near-real-time data and an active national archive of current and past space weather data.

---

[42] F.G. Eparvier, The Need for Consistent Funding of Facilities Required for NASA Missions, white paper submitted to the Decadal Strategy for Solar and Space Physics (Heliophysics), Paper 69; A. Chutjian et al., Laboratory Solar Physics from Molecular to Highly-Charged Ions, Meeting Future Space Observations of the Solar Plasma and Solar Wind, white paper submitted to the Decadal Strategy for Solar and Space Physics (Heliophysics), Paper 39.

10.5.5.4 Continuity of Real-Time Solar Wind Data from L1

SHP Imperative: Maintaining the continuity of real-time solar wind data from L1 is essential for the preservation and improvement of current space weather forecasting capabilities.

Justification: Observations of the solar wind and the Sun from 1 million miles upstream from Earth (orbiting the L1 Lagrangian point) have proved to be extremely valuable. The ACE mission has returned solar wind observations in real time since 1998, and the NOAA Space Weather Prediction Center uses them to drive 5 of its 11 space weather watches and warnings with a lead time of up to 60 minutes. The real-time data are also downloaded about 1 million times a month by about 25,000 unique customers, including deep-sea drilling, surveying, mining, and airline companies. NOAA, NASA, and the U.S. Air Force are refurbishing the DSCOVR spacecraft to provide operational real-time data to partially replace the data supplied by the aging ACE spacecraft. IMAP will be able to replace and expand on the suite of real-time space weather data provided by ACE and DSCOVR.

10.5.5.5 Preserving Key Solar-Heliospheric Data Sets

SHP Imperative: Agencies that engage in solar-heliospheric research or use solar and heliospheric data products (such as NASA, NOAA, NSF, and DOD) are encouraged to initiate an internal or external study that would identify key long-term data records and recommend approaches to ensure that they are continued and archived.[43]

Justification: A number of long-term solar and heliospheric data sets are essential for tracking the evolution of the solar-heliosphere system on timescales ranging from minutes to millennia. Examples include neutron-monitor records of cosmic-ray variations, TSI measurements, and solar wind properties upstream of Earth. Some of the data sets are often in danger of being discontinued because of lack of funding or changes in agency priorities.

10.5.5.6 Laboratory Plasma Physics

SHP Imperative: NASA and NSF are encouraged to continue supporting laboratory plasma-physics research at current or higher levels because it complements space- and ground-based measurements in efforts to understand basic heliophysical processes.[44]

Justification: Facilities that approximate solar and space plasma environments in the laboratory have long been part of the scientific investigation of space. Next-generation experiments have been emerging to provide experimental capabilities to study detailed processes directly relevant to space and solar plasmas. Some fundamental plasma processes are difficult, if not impossible, to measure and confirm with spacecraft but are potentially within reach through laboratory experiments. Examples include studies of magnetic reconnection and associated particle acceleration, solar coronal loops, and large-scale dynamo experiments at the Madison Dynamo Experiment facility. Moderate funding for laboratory heliophysics in

---

[43] J.M. Ryan et al., Ground-Based Measurements of Galactic and Solar Cosmic Rays, white paper submitted to the Decadal Strategy for Solar and Space Physics (Heliophysics), Paper 235; P. Pilewskie, The Total and Spectral Solar Irradiance Sensor: Response to the NAS Decadal Strategy for Solar and Space Physics, white paper submitted to the Decadal Strategy for Solar and Space Physics (Heliophysics), Paper 221; F. Hill et al., The Need for Synoptic Solar Observations from the Ground, white paper submitted to the Decadal Strategy for Solar and Space Physics (Heliophysics), Paper 108.

[44] B. Brown et al., An Experimental Plasma Dynamo Program for Investigations of Fundamental Processes in Heliophysics, white paper submitted to the Decadal Strategy for Solar and Space Physics (Heliophysics), Paper 18; H. Ji et al., Next Generation Experiments for Laboratory Investigations of Magnetic Reconnection Relevant to Heliophysics, white paper submitted to the Decadal Strategy for Solar and Space Physics (Heliophysics), Paper 120.

support of space-plasma and solar research, preferably through budget line items for laboratory facilities, would be beneficial.

10.5.5.7 Interagency Planning for a Future L5 Mission

SHP Imperative: The SHP panel assigns high priority to the development by NASA, NOAA, and DOD of plans for a mission at the L5 Lagrangian point to conduct helioseismology studies and develop advanced capabilities to forecast space weather.

Justification: From L5, it is possible to forecast the arrival of Earth-directed CMEs with high precision, to observe emerging or developed active regions and corotating interaction regions about 4 days earlier than from Earth, and to measure photospheric magnetic fields over about 60° additional longitude, improving magnetic-field models of regions rotating toward Earth.[45] In addition, stereohelioseismology studies combining L1 and near-Earth observations have the potential to enable even longer solar-activity forecasts (§10.3.1 and §10.5.2.5).

The SHP panel encourages NASA, NOAA, and DOD to study means of achieving an L5 orbit for minimal cost and to define the spacecraft, instrument, and mission requirements needed to optimize the scientific and operational opportunities from a future L5 mission.

10.6 CONNECTIONS TO OTHER DISCIPLINES

The SHP panel has interests that are connected to those of the Panel on Atmosphere-Ionosphere-Magnetosphere Interactions (AIMI) and the Panel on Solar Wind-Magnetosphere Interactions (SWMI). Solar photon radiation and energetic particle outputs are primary energy inputs into and thus drivers of dynamic processes in Earth's atmosphere, ionosphere, and magnetosphere and those of other planets. The Sun varies on timescales of minutes (flares, CMEs, and SEPs), days (27-day solar rotation), years (11-year solar-activity cycle), and even longer. There are corresponding responses throughout the heliosphere, including at Earth and other planets. Successful space weather modeling and forecasting of magnetospheric and ionospheric events include continuous tracing through interplanetary space and monitoring of conditions upstream of Earth. Consequently, most SHP panel initiatives also support AIMI and SWMI goals. Robust observational and modeling programs of solar outputs and interactive collaborations among the SHP, AIMI, and SWMI communities are critically important for current and future heliophysics research and space weather operations.

10.6.1 Earth Science and Climate Change

Solar-irradiance variations in the near-ultraviolet, visible, and near-infrared have direct forcing on Earth's climate change because of energy deposition in the lower atmosphere and at the surface and oceans. The solar-energy input and other indirect forcings, such as top-down coupling of solar ultraviolet heating and photochemistry in the stratosphere, and the possible influence of galactic cosmic rays, affect both global climate changes and regional changes that are due to subtle changes in the atmosphere, ocean circulation patterns, and cloud cover. Solar-irradiance observations and reconstruction and modeling of long-term variations, such as improved solar-dynamo models, are important support for the Earth science

[45] A. Vourlidas et al., Mission to the Sun-Earth L5 Lagrangian Point: An Optimal Platform for Heliophysics and Space Weather Research, white paper submitted to the Decadal Strategy for Solar and Space Physics (Heliophysics), Paper 273; N. Gopalswamy et al., Earth-Affecting Solar Causes Observatory (EASCO): A New View from Sun-Earth L5, white paper submitted to the Decadal Strategy for Solar and Space Physics (Heliophysics), Paper 99.

goal of understanding natural influences on climate change. The SHP panel encourages enhanced collaboration among the solar, heliosphere, Earth science, and climate-change communities, in particular between the NASA Heliophysics Living With a Star targeted research and technology and NASA Earth science radiation and stratosphere programs.

10.6.2 Astrophysics

Understanding the Sun as a star, understanding how solar structure and dynamics are related to those of other stars, and understanding relationships between the heliosphere and astrospheres are examples of how solar, stellar, galactic, and heliospheric physics research efforts benefit each other. Basic plasma processes that include particle acceleration, reconnection, and turbulence in the heliosphere also occur in other astrophysical environments. Studies of stellar evolution, through asteroseismology—the study of a star's internal structure by analyzing the frequency spectra of its oscillations—and other means, and improved understanding of the range of variability of Sun-like stars also benefit Sun-climate studies of long-term solar variations, such as the Maunder minimum.[46] It is highly desirable to continue and improve collaborations between heliophysics and astrophysics programs in NSF.

10.6.3 Comparative Planetology and Astrospheres

The magnetosphere represents a boundary in space plasmas created through the interaction of Earth's magnetic field and solar wind. Comparable boundaries exist around planets, their moons, asteroids, comets, the heliosphere, and astrospheres. Detailed study of plasma boundaries mutually benefits understanding of boundaries surrounding Earth, planets (including exoplanets), moons, comets, the heliosphere, and astrospheres surrounding other stars. It is highly desirable to continue and improve collaborations among heliophysics, planetary science, and astrophysics programs to elevate and generalize understanding of planetary and astrophysical boundaries in space.

---

[46]K. Schrijver et al., The Solar Magnetic Dynamo and Its Role in the Formation and Evolution of the Sun, in the Habitability of Its Planets, and in Space Weather around Earth, white paper submitted to the Decadal Strategy for Solar and Space Physics (Heliophysics), Paper 240.

Appendixes

A

Statement of Task and Work Plan

The Space Studies Board shall establish a heliophysics survey committee to develop a comprehensive science and mission strategy for heliophysics research for a 10-year period beginning in approximately 2013. The survey committee, informed by up to five study panels that will also be established by the board, will broadly canvas the field of solar and space physics and:

1. Provide an overview of the science and a broad survey of the current state of knowledge in the field, including a discussion of the relationship between space- and ground-based science research and its connection to other scientific areas;
2. Identify the most compelling science challenges that have arisen from recent advances and accomplishments;
3. Identify—having considered scientific value, urgency, cost category and risk, and technical readiness—the highest priority scientific targets for the interval 2013-2022, recommending science objectives and measurement requirements for each target rather than specific mission or project design/implementation concepts; and
4. Develop an integrated research strategy that will present means to address these targets.

SCOPE

This "decadal survey" follows the National Research Council's (NRC's) previous survey in solar and space physics, *The Sun to the Earth—and Beyond: A Decadal Research Strategy in Solar and Space Physics,* which was completed in 2002 and published in final form in 2003. The scope of the study will include:

- The structure of the Sun and the properties of its outer layers in their static and active states;
- The characteristics and physics of the interplanetary medium from the surface of the Sun to interstellar space beyond the boundary of the heliosphere; and

- The consequences of solar variability on the atmospheres and surfaces of other bodies in the solar system, and the physics associated with the magnetospheres, ionospheres, thermospheres, mesospheres, and upper atmospheres of the Earth and other solar system bodies.

In order to ensure consistency with other advice developed by the NRC for NASA, the following additional scope guidance is provided:

- With the exception of interactions with the atmospheres and magnetospheres of solar system bodies, which are within scope, planetary phenomena are out of scope (these other topics are being addressed by an ongoing decadal survey in planetary science);
- Basic or supporting ground-based laboratory and theoretical research in solar and space physics are within scope, noting that the findings and recommendations in the present survey should be harmonized with those developed and reported by the ongoing astronomy and astrophysics decadal survey; and
- Consistent with the current astronomy and astrophysics decadal survey, recommendations related to ground-based implementations (e.g., ground-based solar observatories) will be directed to the National Science Foundation (NSF).

Without undertaking a detailed analysis of operational space weather user or provider requirements, the survey committee will describe the value of these services to society and examine the role of NASA and NSF research in underpinning and improving these services.

In addition to an integrated review of the current state of scientific knowledge and recommendations for future basic research directions to advance our understanding, the survey will provide implementation recommendations separately for NASA and NSF.

For each science target, the committee will establish criteria on which its recommendations depend and identify developments of sufficient significance that they would warrant an NRC reexamination of the committee's recommendations. The committee will also make recommendations to the agencies on how to rebalance programs within budgetary scenarios upon failure of one or more of the criteria.

WORK PLAN

General Approach

The decadal survey will focus on the research aspects of the broad range of solar and space physics (within NASA, referred to as "heliophysics"). As such, it will address primarily the responsibilities of NASA and NSF and will provide recommendations to these two agencies. However, the survey will also address issues of particular interest to the National Oceanic and Atmospheric Administration and the Department of Defense, including the current state of capabilities and future directions in space weather monitoring and operations. We note that basic research and derived applications share a common knowledge base and community of experts and practitioners.

The Space Studies Board, working through its Committee on Solar and Space Physics, will establish a survey steering committee ("committee") of approximately 16 members. The committee will be responsible for the overall organization and execution of the new study, as well as the production of a final consensus report that will undergo the usual NRC review processes. The final report will represent a comprehensive and authoritative analysis of the subject domain and a broad consensus among research community stakeholders. To do so, it is anticipated that the committee will utilize approximately five specialized study subpanels, with allocation of the domain of study among them to be determined by the committee and the Space Studies Board. The specific structure of the survey will be determined following a planning meeting and workshop to be held in spring 2010. An important role of the panels will be to solicit broad input from the research community about issues of scientific and programmatic priorities in the field.

The work of the study subpanels will not result in separate, independent reports; subpanel conclusions and recommendations will be considered by the committee and used to prepare a single final report.

It is essential that the study activity solicit and aggregate inputs from across the solar and space physics community and the country via town hall meetings, sessions at geographically dispersed professional meetings, solicitation of white papers, and aggressive use of electronic communications and networks. The committee may also convene focused workshops on special topics of interest, such as how to better accomplish the transition from research to operations or the potential for small satellites to advance space science and address ongoing issues related to workforce and training.

It is critically important that the recommendations included in the final report be achievable within the boundaries of anticipated funding. To that end, it is anticipated that NASA and NSF will provide an up-to-date understanding of these limitations at the time of survey initiation. In designing and pricing the study, the NRC will include resources for independent, expert cost analysis support, which will be used when appropriate to improve cost estimations, expose and bound uncertainties, and facilitate cost comparisons among missions with varying heritage and technical maturity.

In addition to a review of the current state of scientific knowledge and recommendations for future directions to advance our understanding, implementation recommendations are sought for NASA and NSF. These three central tasks are described below.

A. Scientific Assessment and Recommendations

The report should provide a clear exposition of the following:

- An overview of heliophysics science—what it is, why it is a compelling undertaking, the relationship between space- and ground-based science research; and its connections to other areas of scientific inquiry, for example, fundamental processes in astrophysical plasmas; another connection is with basic plasma physics;
- A broad survey of the current state of knowledge; and
- An inventory of the top-level scientific questions, prioritized by value, urgency, and technical readiness, that should guide NASA flight mission selections and supporting research programs and NSF's primarily ground-based investigations and supporting research programs.

In order to ensure consistency with other advice developed by the NRC, specific guidelines for the scientific scope of the survey are as follows:

- With the exception of interactions with the atmospheres and magnetospheres of solar system bodies, which are within scope, planetary phenomena are out of scope (these other topics are being addressed by an ongoing decadal survey in planetary science); and
- Basic or supporting ground-based laboratory and theoretical research in solar and space physics are within scope, except those covered by the ongoing astrophysics decadal survey.

B. Implementation Recommendations for NASA

As was done in the previous heliophysics decadal survey, the section of the report that provides recommendations to NASA will reflect the agency's charter to conduct flight mission investigations and, consistent with the current astronomy decadal survey, recommendations related to ground-based implementations (e.g., ground-based solar observatories) will be directed to the NSF. The prioritized science goal inventory from the previous section will form the basis for flight mission prioritization. NASA will provide an axiomatic budget runout for the survey decade, and the committee will allocate these resources into "budget packages." Recommended budget packages are:

- Small flight investigations appropriate for the Explorer class;
- A mix of small, medium, and large strategic flight investigations appropriate to goals of the Solar-Terrestrial Probes program and the Living With a Star program; and
- If budget guidelines permit, flagship flight investigations may also be recommended.

Life-cycle cost classifications defined by NASA currently are as follows: very small/mission of opportunity, less than $250 million; small, $250 million to $500 million; medium, $500 million to $750 million; large, $750 million to $1,000 million; and flagship, over $1 billion. These cost ranges are inclusive of launch vehicles costs.

The survey should address the role of missions of different sizes (small, medium, and large), based on these cost bracket categories. Note that candidates for individual Explorer and lower-cost flight projects, including sub-orbital programs, should not be prioritized; they will be solicited and selected as principal-investigator-led missions on the basis of science merit via NASA's standard Explorer Announcement of Opportunity (AO) and NASA Research Announcements processes. However, the strategic value of these smaller mission programs to the overall science agenda should be appraised and described.

The survey should prioritize "science targets" for the mission implementation opportunities, recommending science objectives and requirements for these missions rather than specific mission design/implementation concepts. However, mission conceptual designs should be used within each prioritized science area to establish feasibility within its notional budget package. This budget feasibility will be demonstrated by the NRC's costing subcontractor on the basis of nominal point designs. Concept mission designs should not be represented in the final, prioritized list; missions to be flown will be solicited and selected by NASA on the basis of science and implementation merit versus risk via NASA's standard AO processes.

The prioritized flight program list recommended in the report should be provided in terms of scientific problems and measurements, with validated feasibility, and their associated "budget packages," not in terms of specific missions. Missions prescribed in *The Sun to the Earth—and Beyond* and other recent NRC reports that are not yet in formulation or development must be reprioritized. In contrast, the Solar Orbiter, Solar Probe Plus, and IRIS missions are in formulation and should not be prioritized. The flight investigations priority list should be supported by a summary of the assumptions underlying the relative rankings. This summary should, to the extent possible, be accompanied by decision rules that could guide NASA in adjusting the queue in the event of major unanticipated scientific developments or technical, cost, or other programmatic changes. It is understood that initiation of missions on this list will depend on actual resource availability and programmatic factors.

For Solar Probe Plus (SPP), cognizant of the fact that the previous decadal survey ranked Solar Probe as their highest priority for a "large-class" mission, the decadal survey will consider how the scientific rationale for a solar probe mission is substantially affected by:

a. Scientific developments since the completion of the 2002 decadal survey; and

b. The projected timeline and instrument complement of the particular implementation of the previous survey's recommendation, Solar Probe Plus.

In the context of the committee's charge to establish criteria on which its recommendations depend and to identify developments of sufficient significance that they would warrant a reexamination of the committee's recommendations, the committee will also make recommendations in order to maintain program scientific balance. The committee will make recommendations to NASA (including for SPP) on programmatic or cost trigger points for consideration of possible rescoping, deferral, cancellation, application of additional resources at the possible expense of other Heliophysics Division programs, or a reassessment by the NRC or by the NASA Advisory Council's Heliophysics Subcommittee.

Other welcome recommendations would include those that address NASA-funded supporting research programs to maximize the science return from the currently operating and pending flight mission investigations and a discussion of strategic technology development needs and opportunities relevant to NASA heliophysics science programs. All such recommendations must be achievable within the constraints of anticipated funding.

The committee report shall reflect an awareness of the science and space mission plans and priorities of potential foreign and U.S. agency partners and identify opportunities for cooperation, as appropriate and achievable within budget constraints. Examples would be the proposed Japanese Solar-C or SCOPE projects. NASA's Science Mission Directorate will provide authoritative, formal NASA policy and heliophysics program documentation, as required.

C. Implementation Recommendations for NSF

For NSF, the survey and report shall encompass all ground-based observational techniques, as well as analysis of data collected and relevant laboratory and theoretical investigations (including modeling and simulation). Thus, the study will assess the NSF-supported infrastructure of the field, including research and analysis support, the educational system, instrumentation and technology development, data distribution, analysis, and archiving, and theory programs. The committee shall also recommend any changes to this infrastructure that it deems necessary to advance the science and to capture the value of facilities in place.

In addition, the survey will examine the NSF's "CubeSat" program, which is developing small-satellite science missions to advance space weather and atmospheric research, as well as providing opportunities to train the next generation of experimental space scientists and aerospace engineers.

The committee shall review relevant programs of other nations and will comment on NSF opportunities for joint ventures and other forms of international cooperation.

Committee Meetings and Workshops

We expect an approximately 16-member survey steering committee to meet approximately six times during the course of the study. The steering committee will be supported by approximately four 12-member study panels that will meet three times during the course of the study. We also may convene topical workshops to explore particular subjects in more depth, e.g., the transition to operations for space weather models and the potential for smaller satellites to conduct research currently performed by larger spacecraft.

B

Instrumentation, Data Systems, and Technology

SUMMARY

This appendix, drawn largely from the analysis of this decadal survey's Innovations: Technology, Instruments, and Data Systems Working Group, summarizes information provided to the survey committee regarding areas where innovations would have a substantial impact in solar and space physics. The most significant advances over the next decade and beyond are most likely to derive from new observational techniques; from innovative approaches to access and to dissemination, maintenance, storage, and use of data; and from more capable observational platforms.

INSTRUMENTATION DEVELOPMENT NEEDS AND EMERGING TECHNOLOGIES

Progress toward meeting the scientific and technical challenges for solar and space physics over the coming decade hinges on improved observational capabilities and novel instrumentation. During times of diminished flight rates, the most promising instrumentation concepts must be brought to spaceflight readiness so that new instruments are selectable. Recent solar and space physics missions have typically flown slightly modified versions of existing instrumentation, due to limited instrument development opportunities and to the risk-averse programmatic environment. Conversely, incremental development of selected sensors has sometimes been carried out during implementation, thus increasing risk, cost, and schedule vulnerability.

A primary reason for this is the scant and fragmented instrument development support in the heliophysics programs Supporting Research and Technology, Living With A Star, and Low Cost Access to Space. Often in competition for limited resources, instrument development proposals are not funded because they cannot promise immediate science closure, even though the developed technology would enable breakthrough measurements in a strategic sense. Also, there is no effective program to pick up after basic development and ensure technical readiness for a mission. Such development comes at a higher cost than can be currently supported. Disturbingly, the scarcity of resources for instrument development has led to a substantially smaller and weaker instrument development cadre in the solar and space physics field than is adequate.

Deliberate investment in new instrument concepts is necessary to acquire the data needed to further solar and space physics science goals, reduce mission risk, and maintain an active and innovative hardware development community. To demonstrate the need for a dedicated NASA-funded instrument and technology development program, this section describes several notional instruments that could carry out science investigations considered for this decadal survey. Several of these cut across disciplines and thus serve a variety of important goals. These are representative examples, and specifics may change over time as new technology becomes available and scientific progress occurs. The dynamic nature of the requirements underscores the need for proactive management. Adequate, strategic resources are needed for a combination of basic development, system integration, and technology readiness level (TRL)-boosting steps, the latter usually being the most costly.

Airglow Imaging in the Visible and Far Ultraviolet (AIMI and SWMI)

Earth's upper atmosphere (30-1,000 km) plays a key role in the interaction with the magnetosphere. The lower regions (<200 km) are virtually inaccessible by in situ probes. Improved instrumentation is needed to effectively study the energy and plasma coupling by revealing the global interaction and providing altitude-specific information. Wind measurements are critical. Composition (ion/neutral) and temperature, as well as the type and energy of precipitating particles, and ionospheric conductivities can be effectively deduced from induced atmospheric emissions. This can be provided by auroral and airglow imaging in the visible and far ultraviolet. Such an instrument requires mirrors with higher-reflectance coatings than are currently available, as well as narrower-band filters and blazed gratings with high ruling densities. "Solar blindness" in the ultraviolet regime needs to be improved substantially, and basic development of such components has been completed. System integration and in-space testing are needed to boost the system TRL.

Visible and Infrared Lidar Probing of the Upper Atmosphere (AIMI and SWMI)

Optical techniques for remote sensing of the upper atmosphere are limited by the inherent line-of-sight integration. Improving lidar technology holds great promise. Like radar, lidar has three-dimensional resolving capability and is increasingly used to measure density distributions, drift speeds, and temperatures of atmospheric constituents and trace elements. Three-dimensional resolution is especially important in characterizing atmospheric waves propagating from the lower atmosphere.

Ground-based resonance lidars currently probe altitudes of 80 to 110 km, and Rayleigh lidars probe from 30 to 80 km with power apertures of 10 to 20 Wm^2. To probe above 110 km requires a substantial increase in power aperture. High-frequency, high-power lasers are now available, and with a 10 by 10 array of 1 m^2 telescopes, more than 1,000 times the current capability is possible. While the components exist, system integration development is needed.

A space-based helium lidar could provide Doppler measurements over the entire 250- to 750-km altitude range. From low Earth orbit, a 10-Wm^2 system is sufficient to provide spatial profiles at 10-km altitude resolution. Technology development is necessary to bring such a system to spaceflight readiness, whether on the International Space Station (ISS) or as a free-flyer.

Magnetosphere-to-Ionosphere Field-Line Tracing Technology (SWMI and AIMI)

Determining how magnetosphere-ionosphere-thermosphere coupling controls system-level dynamics requires accurate observational knowledge of the magnetic connection between magnetospheric and ionospheric phenomena. One technique that is ready for a technological boost involves firing a high-energy electron beam along the magnetic field from a spacecraft in the equatorial magnetosphere and detecting the location of the ionospheric foot point by optically imaging the airglow spot the beam produces.

This straightforward technique faces a number of technical issues. First, finding and imaging the beam spot from the ground in the presence of auroral activity is challenging. Simply increasing the power of the electron beam leads to a second challenge—preventing space charge within the beam and increased beam divergence. The third challenge is perhaps the most severe: extracting negative charge from an ungrounded spacecraft in the tenuous magnetosphere can lead to catastrophic spacecraft charging. A high-current plasma contactor may provide a solution, but electron-beam pointing will degrade. Issues of optimizing power-storage systems, accelerators, and plasma contactors for the spacecraft remain.

High-Resolution High-Cadence Infrared-Ultraviolet Imaging of the Solar Atmosphere (SHP)

Tracing the transformation of magnetic into kinetic and thermal energy in the solar atmosphere is key to understanding processes that control heating and acceleration of flows and particles, as well as large explosive phenomena with substantial impact on our space environment. Present detectors cannot adequately measure these processes, which occur on spatial and temporal scales that push the observation requirements to higher angular resolution (<0.1°) and faster cadence (<10 s). Imaging at the smallest scales with the highest time resolution requires large-format, high-speed, high-efficiency detectors (advanced charge-coupled devices, complementary metal-oxide semiconductors) and fast polarization modulators.

Design concepts exist, but implementation requires compromises between spatial and temporal capabilities. Basic development must be followed by a staged approach with system integration and TRL boosting. Synergism with ground-based development is possible in the visible spectrum.

High-Angular-Resolution Energetic Neutral Atom Imaging (SWMI and SHP)

Key magnetospheric processes associated with solar wind driving and ionospheric coupling occur from global scales down to those of individual flux tubes, requiring both effective remote sensing and detailed in situ measurements. Energetic neutral atom (ENA) imaging provides the global view, but currently with insufficient spatial resolution, which would require ≈1° angular resolution and greater sensitivity. The same advances will be sufficient for improving observations of the heliospheric boundary, for which 2-3° is adequate.

Recent developments promising breakthrough improvements include ultra-thin foils for good statistics at moderately low energies, higher-efficiency electrostatic configurations for multiple-coincidence measurements, sensor designs with intrinsic low sensitivity to visible and ultraviolet light, and improved charged-particle rejection techniques. Important development steps include a system-level imager design for an ultraviolet-blind detector, followed by TRL boosting.

Solar Flare Neutral Energetic Particle Imager (SHP)

Direct observations of neutral atoms at mega-electron-volt energies from solar flare regions provide remote information about acceleration sites at the Sun, including spatial and temporal variations of the acceleration processes (SHP3). Such particles have been observed serendipitously with the STEREO (Solar Terrestrial Relations Observatory) High-Energy Telescope/Low-Energy Telescope, which was designed to measure solar energetic ions. An optimized detector promises much greater sensitivity. This breakthrough observation opens the door to another complementary flare diagnostic technique. Substantial advances can be expected by optimizing proven techniques and applying them within the natural shield of Earth's magnetic field. However, to study flare events in detail requires development of dedicated sensors with intrinsic ion suppression and sufficient angular resolution.

Technologies using electrostatic deflection and collimator-based imaging exist for lower energies. The techniques must be expanded toward higher energies, and new system integration and development are needed. This instrumentation requires basic system development, followed by staged boosting of the system's TRL.

DATA SYSTEMS

Data from NASA's heliophysics missions and many ground-based observatories can be obtained currently online, either directly from individual websites or through central archives such as the Solar Data Analysis Center, the Space Physics Data Facility, or the National Space Science Data Center. These data archives are also accessible through virtual observatories (VxOs), whose goals are to provide one-stop access to validated science data from many observatories, along with the necessary tools for cross-mission analysis and visualization. Access to sophisticated modeling tools is provided by repositories such as the Community Coordinated Modeling Center (CCMC). Such agency-sponsored facilities host physics-based or empirical models developed by the user community and allow users to perform their own simulations.

Current Status

Significant progress has been made over the past decade in defining the fundamental components of the data environment (virtual observatories, archives, etc.) and in starting to build and integrate them. However there continues to be a dearth of tools for using and analyzing data. However, projected data requirements for new projects are not as demanding as the leap from the Solar and Heliospheric Observatory to the Solar Dynamics Observatory (SDO). New requirements can probably be met with existing technologies and software. For instance, daily generation of Advanced Technology Solar Telescope data in 2018 is estimated to be ~4 TB, about the same as the current SDO export rate. It is also noted that some segments of the research community still suffer from the lack of effective data policies enforced by sponsoring agencies.

Data systems supporting heliophysics research over the past decade have evolved from stand-alone, custom-built "stove-pipes" to distributed, interacting systems that leverage software and technologies developed by the community. Much of this welcome development has come through NASA's Heliophysics Data Environment (HPDE) enhancement and the National Science Foundation's (NSF's) Directorate for Computer and Information Science and Engineering and Office of Cyberinfrastructure. Many heliophysics data sets and models are hosted at multiple data archives and modeling centers, each with different architectures and formats. And much of the work on data systems infrastructure is funded through individual principal investigator teams. This results in uncoordinated software development, unpredictable support life cycle, and data analysis tools with limited scope. Such activity also draws funds and focus away from scientific research and analysis activities, since investigators are obliged to provide data sets and analysis tools as deliverables. Unfortunately, many of the existing archives, modeling centers, and VxOs are not inter-compatible, despite significant overlap in content or access.

The current lack of coordination among data and modeling centers stems mainly from their different philosophies, emphases, formats, architectures, and purposes. One can obtain similar data sets from various nationally funded data archives as well as from VxOs. The existence of duplicative capabilities, each with significantly different purpose and implementation philosophy, provides greater, more flexible access at the cost of generating confusion about which path to follow to the data. National and international agencies have not identified a common goal, nor have they adopted a standard approach for funding and implementing data facilities and archives.

Current modeling centers, such as the CCMC, have multiple sponsors and allow researchers to run simulations using community-provided models that cover vastly different domains, such as the solar corona, the solar wind, the radiation environment in the heliosphere and Earth's radiation belts, and the magnetic and electric field environments of the magnetosphere and ionosphere. Although some space weather modeling groups have developed end-to-end models, often the component modules employ controversial techniques and are based on assumptions with inherent strengths and weaknesses. Only a small fraction of all models can be run interactively, and even fewer can be coupled. This makes it difficult to validate different models and to model interesting space weather events.

Future Goals and Directions

Heliophysics is poised to make a natural transition from being driven predominantly by the pursuit of basic scientific understanding of physical processes toward one that must also address more operational, application-specific needs, much like terrestrial weather forecasting. This transition requires (1) instant unfettered access to a wide array of data sets from distributed sources in a uniform, standardized format, (2) incorporation of the results of community-developed models, and (3) the ability to perform simulations interactively and to couple different models to track ongoing space weather events.

NASA has already taken the important first step in integrating many of these data sets and tools to form the HPDE. The main objective of the HPDE is to implement a distributed, integrated, flexible data environment. HPDE modeling centers should serve as a sound foundation for a future, fully integrated heliophysics data and modeling center.

The key ingredients necessary for any successful centralized data and modeling environment are (1) full involvement of data providers, (2) rapid, open access to scientifically validated data, (3) peer-reviewed data systems driven by community needs and standards, (4) coordinated, user-friendly analysis tools, (5) reliable high-performance computing facilities and data storage, (6) uniform terminology and adequate documentation describing data products and sources, (7) flexible, interoperable, and interconnected data archives, modeling centers, and VxOs, and (8) effective communication among data providers, national and international partners, and data users.

The tremendous quantity of heliophysics data that will become available in the next decade will strain the financial, personnel, hardware, and software resources available to individual scientists, teams, and even national agencies. The dramatic advances in computing and data storage technology over the past decade are likely to continue, so the cost of future data systems and modeling centers will be dominated by personnel and software development rather than securing ultrafast computing or data storage. To achieve these goals efficiently, the national agencies will need to develop a common approach for funding data facilities, archives, modeling centers, and VxOs and coordinate the development of data systems infrastructure, including the development of data systems software, data analysis tools, and training for personnel.

Opportunities in New Data Systems

Community Input to and Control of the Integrated Data Environment

A number of virtual observatory and other data identification and access tools have appeared or are under development. These efforts could be strengthened, better focused, and more efficiently managed if more user feedback were incorporated into their governance, perhaps by formalizing community oversight of such emerging, integrated data systems in an ad hoc group such as the NASA Heliophysics Data and Computing Working Group. Interagency coordination of the data environment as a whole would benefit researchers whose efforts are funded by multiple agencies.

Emerging Technologies

The information technology industry continues to generate novel technologies and capabilities faster than any federally funded, competitively sourced research program can hope to match. Agencies must be agile enough to exploit emerging technologies without investing in their original development. The best approach is to (1) focus on commercially viable technologies for which there is a demonstrated need, such as high-performance computing clusters, and (2) otherwise invest modestly in the evaluation of emerging commercial technologies through existing mission and small-scale data center activities.

Virtual Observatories

NASA has funded virtual observatories and related "middleware" development. Some of these have led to useful targeted data identification and access technologies, and some are still under development. Mature capabilities should not continue to compete with research proposals for funding. A more effective approach would be for NASA and its agency partners to establish a heliophysics-wide data infrastructure, selecting the most useful efforts for stable funding and bringing other efforts to a close. Future developments can be managed through the supplemental funding mechanisms discussed in the sections "Emerging Technologies" or "Community-Based Software Tools."

Community-Based Software Tools

In a few subdisciplines, such as solar physics, the availability of integrated open-source data reduction and analysis tools makes a significant difference in the ability of researchers to access and manipulate data. In areas where such tools are not available, immediate agency investment in community-based development would be highly productive. Where tools are already available, support to maintain and evolve them as new data sets and capabilities emerge should continue. Capabilities should expand to include data mining and assimilation in order to enable full exploitation of the large new heliophysics data sets.

Semantic Technologies

The astrophysics and geophysics communities have taken the lead in adopting modern, "semantic" technologies, where machines "understand" the context and meaning of data, to enable cross-discipline data access. Promoting the development of semantic technology would enable the emerging data access capability in heliophysics to share data and knowledge with other fields.

A National Approach to Data Policies

The heliophysics data policies of the funding agencies differ or are in some cases lacking. The NSF, for instance, now requires a data management plan in all research proposals, but geosciences does not yet have a uniform data access and preservation policy. NASA Heliophysics has a well-developed data management policy, but long-term preservation of data is in a state of flux. It would be wise for the agencies to formulate a national policy for curation of data from taxpayer-funded scientific research. For heliophysics, the Committee on Space Weather could review and monitor agency data policies.

SPACECRAFT TECHNOLOGIES AND POLICIES

Space technology has matured over the past five decades, enabling reliable access to both near-Earth space and beyond. Nonetheless, continued progress in heliophysics, carried out with robotic spacecraft, requires infusion of new technology to advance the scientific program affordably. Meeting the survey's science goals and maintaining leadership in heliophysics requires improved spacecraft technologies, as well as appropriate new sensors and data analysis tools.

The most significant advances in heliophysics over the next decade and beyond are most likely to derive from new observational techniques in new locations. Such techniques require a synergistic combination of spacecraft capabilities, sensors, and data-processing capability. Innovation is most likely to occur in an environment that allows ready access to advanced technology in space. Here "access" refers both to the number of launch opportunities and to appropriate risk policies. Available financial resources ultimately limit all NASA robotic scientific missions, and a very significant driver is launch vehicle cost. Thus a major motivation for technology investment is to provide more scientific capability with fewer spacecraft resources. While mass is typically the primary resource, it is also intimately connected with power, propulsion, and data return capability.

Heliophysics will benefit from developments in spacecraft technologies and policies in six broad areas, based on a review of community white papers and panel reports:

1. Constellations of small (<~20 kg) spacecraft
2. Spacecraft propulsion systems
   a. Solar sails
   b. High-drag environment
3. Communication systems
4. Spacecraft power systems
5. Access to advanced fabrication
6. Policy—International Traffic in Arms Regulations (ITAR), risk management, and radio frequency spectrum allocation.

Items 1 through 4 have been mentioned in surveys and NASA roadmaps for the past decade or longer. Both heliophysics and planetary exploration have interest in items 3 and 4. While some commonality exists in principle for item 2, the implementations differ; for example, aerocapture and aerobraking are more useful for planetary exploration, whereas solar sails, a potentially enabling technology for a variety of heliophysics missions, are less significant for planetary exploration.

Constellations

The study of the heliophysics system requires multipoint observations to develop understanding of the coupling between disparate regions—solar wind, magnetosphere, ionosphere, and thermosphere, and mesosphere—on a planetary scale and to resolve temporal and spatial ambiguities that limit scientific understanding. Most AIMI and SWMI missions require multiple spacecraft. Approximately 25 community white papers are associated with this topic, and past NASA Heliophysics roadmap concept-missions suggest constellations of 20 to 90 spacecraft.

Small satellites in the 1- to 20-kg range enable the possibility of large constellations. The utility of multipoint observations in heliophysics has been demonstrated by NASA's Time History of Events and Macroscale Interactions during Substorms (THEMIS) mission, a constellation of five 100-kg probes. The Space Technology 5 (ST5) spacecraft, at ~25 kg each, were flown under the New Millennium program before it was terminated.

To enable future missions, it would be wise to accelerate the development of spacecraft technologies for supporting small satellites, including constellation operations and inter-spacecraft coordination. Also useful would be investigating system engineering tradeoffs for designing a large constellation of small, scientific satellites, including balancing the risk of using modern, low-power electronics in space versus spacecraft lifetime.

Propulsion

Propulsion—Solar Sails

Heliophysics can benefit from observations of the Sun, Earth, and heliosphere from orbits requiring continuous propulsive activity to maintain or reach in a timely fashion. Vantage points above Earth's poles, at sub-L1 locations, and at high ecliptic latitudes have unique properties that enable important observations. Six community white papers advocated science enabled by this technology, and previous strategic studies have advocated solar sails. Past heliophysics mission concepts include a solar polar imager, a stationary Earth polar observatory, upstream solar wind monitoring at a sub-L1 location, an L5 mission, Solar Sentinels, and Interstellar Probe.

Significant investments have already been made in the United States and abroad. The Japan Aerospace Exploration Agency (JAXA) and NASA carried out dedicated tests of solar sailing for primary thrust using the IKAROS spacecraft and NanoSail-D2. Without an appropriate demonstration mission that goes beyond the small-scale, heavy-sail tests to date, the possibility of using this technology in the future will remain in doubt. NASA's Office of the Chief Technologist has suggested a technology demonstration line that could cover 75 percent of the cost of a $200 million solar sail demonstration mission. The committee strongly urges that this potential opportunity be pursued to demonstrate a 25 g m^{-2}, ~40-m solar sail.[1]

Propulsion—High Drag

Heliophysics system science requires an understanding of plasma-neutral coupling, global composition, and structure of Earth's upper atmosphere (< 300 km altitude). Long-term in situ observations in this relatively high-drag region are required to develop scientific understanding of this coupling region between space and Earth's upper atmosphere. Reaching this altitude requires increased performance propulsion systems, innovative ways to reduce dependence on expendables, and use of aerodynamic effects to enable satellite operations in high-drag regions. The concept of using "dipper" satellites that maneuver in and out of high-drag regions is the conventional approach, but it does not significantly increase the observational time. Capable but "disposable" (very small, very-low-cost) satellites for the exploration of high-drag regions is an approach that has yet to be studied in detail. Such observational platforms could be deployed from the ISS or another host spacecraft if a propulsion system were developed to enable the deployment and maintenance of small satellite constellations in and above these high-drag environments.

[1] See Section 2.4.4.4 of NASA, *In-Space Transportation Capability Portfolio*, 2005.

Communications

Study of the heliophysics system requires data-intensive observations from distant vantage points or from small, resource-constrained spacecraft. As for planetary missions, optical communications could enable large data rates. It would be prudent to start development of space and ground-station communications for a swarm of small, low-power, Earth-orbiting satellites and for distant platforms at L5, a solar polar orbiter at high ecliptic latitude, or ultimately an interstellar probe.

Spacecraft Power

In situ study of the outer heliosphere requires operations past the orbit of Jupiter. At such large heliocentric distances, solar power is impractical. Other spacecraft power systems are needed. This applies both to heliophysics and planetary exploration missions. Advanced Stirling Radioisotope Generators are a potential solution. There should be a sufficient supply of the radioactive isotope plutonium-238 for use in advanced spacecraft power systems, regardless of the power conversion technology employed.

Advanced Fabrication

The heliophysics community needs access to advanced design and fabrication techniques for new sensing elements, new instrument techniques, and the application of greater computing power to enable scientific progress throughout the field. An agency-supported center could provide valuable assistance to spacecraft teams, instrument designers, and computing groups, serving as a consultant, provider of services, or broker for government or industrial technologies useful in aerospace applications. Rapid and cost-effective creation of custom hardware for the implementation of computational algorithms is needed for advanced sensor systems and for advanced heliophysics modeling. Custom hardware for numerical simulations can exceed by orders of magnitude the speed of general computer implementations.

The experimental community must be able to design and fabricate custom analog, digital, mixed-signal, and microelectromechanical systems (MEMS) devices rapidly and cost-effectively. Even complex current technologies such as field-programmable gate arrays (FPGAs) continue to drive costs and deliveries. Broad use of these techniques requires access to both design and fabrication methodologies at reasonable cost. Access to advanced fabrication has the potential to revolutionize heliophysics sensor and spacecraft systems.

Policy Issues—ITAR, Risk Management, Frequency Spectrum

International Traffic in Arms Regulations

The United States seeks to protect its security and foreign-policy interests, in part, by actively controlling the export of goods, technologies, and services that are or may be useful for military development in other nations. "Export" is defined not simply as the sending abroad of hardware but also as the communication of related technology and know-how to foreigners in the United States and overseas.[2]

The International Traffic in Arms Regulations (ITAR), which controls defense trade, includes the U.S. Munitions List (USML), which specifies categories of defense articles and services covered by the regulations. In 1999, space satellites were added to the USML. However, in 2002, ITAR was amended to exempt

[2]National Research Council (NRC), *Space Science and the International Traffic in Arms Regulations: Summary of a Workshop*, The National Academies Press, Washington, D.C., 2008.

U.S. universities from having to obtain ITAR licenses when performing fundamental research involving foreign countries and/or persons. Because universities often collaborate with foreign partners in research and teach or employ foreign graduate students and other researchers, ITAR has a substantial effect on university activities in the space sector. Many university activities are considered to be fundamental research and thus are excluded from ITAR control; however, academic regulatory-compliance administrators and researchers alike still encounter problems with space-related activities because of the narrow and somewhat ambiguous conditions that enable research to be considered "fundamental" and therefore excluded from licensing under ITAR.[3]

The use of technology developed for commercial purposes in spacecraft systems and science instrumentation has enormous potential for advancing science. However, space-qualified electronics are regulated under ITAR, so commercial developers avoid ITAR restrictions by avoiding dual-use recognition. ITAR "deemed export" rules limit the exchange of technical information on instruments and spacecraft technologies, thus impeding the development of scientific tools. Uncertainty stifles collaboration between domestic industry and research universities and discourages opportunities for progress involving U.S. and foreign scientists and students. Obstacles to partnership have driven Europe, Japan, and other nations to push ahead with their own small-size missions that foster training of non-U.S. engineers and scientists.

Programmatic Risk Management

Policies for assessing, managing, mitigating, or even futilely attempting to eliminate risk should be regularly updated so as not to impede technological development and scientific advancement. Continual evaluation of risk polices is needed, including those in NASA Procedural Requirement (NPR) 7120.5x, and appropriate application must be made for large, high-profile versus small, low-cost programs. Major drivers of program risk, such as uncertain funding profiles or delays in launch services, are often external to a project.

Spectrum

Frequency allocation policies for satellites and the congestion of current bands available for space research favor high-profile missions. This can disadvantage typically smaller heliophysics science missions and those that make use of secondary launch opportunities. Current frequency licensing and allocation policies also require knowledge of orbital parameters long before launch; this limits the opportunistic pairing of small satellites in containerized deployment systems with launch opportunities when orbital parameters are not known in advance. It would be helpful if NASA would engage proactively with the International Telecommunication Union.

[3]NRC, *Space Science and the International Traffic in Arms Regulations*, 2008.

C

Toward a Diversified, Distributed Sensor Deployment Strategy

Breakthrough research in heliophysics is enabled by research platforms of a variety of sizes and with a range of functions: billion-dollar-scale strategic missions with payloads up to 100 kg can enable a new set of measurements otherwise not accessible; however, as described in Part I of this report, there are opportunities for leading-edge research with $375 million medium-size Explorers; $150 million small Explorers with payloads of 20 kg; and CubeSats, at $1 million scale, with payloads of 1 kg. In addition to these space-based platforms, suborbital programs provide unique, relatively low-cost opportunities for research and technology demonstration. For example, sounding rockets provide the only means for in situ sampling in regions inaccessible to aircraft, balloons, or satellite platforms.[1] Ground-based facilities offer an entirely different, but no less necessary, "platform" for solar and space physics research and long-term observations. Finally, there are unique opportunities for solar and space physics research to be carried out by instruments hosted on commercial and government space platforms that carry payloads for other purposes.

A diversity of approaches is required for achieving the top-level objective in solar and space physics—to create system-wide understanding—but the data from these platforms need to be integrated into distributed yet coordinated approaches that create the best system-wide understanding from the data, which may very well be collected by a variety of platforms.

This appendix reviews the platforms that are currently available to pursue research in solar and space physics and examines prospects for the coming decade. Its review is not comprehensive (and discussions of NASA spacecraft are left to Part I of this report); however, it does provide illustrative examples of the scientific utility of selected platforms. The appendix also describes examples of data integration in constellations and so-called heterogeneous facilities—a set of distributed measurements from a variety of measurement vantage points, integrated into a greater whole.

---

[1] In particular, direct in situ sampling in the important region of the lower ionosphere/thermosphere and mesosphere below 120 km altitude is not possible with aircraft and balloons, which operate well below this height, or with satellites, which to avoid atmospheric drag must be placed in orbits at higher altitude. See, NASA, "NASA Sounding Rocket Science," at http://rscience.gsfc.nasa.gov/srrov.html.

GROUND-BASED SOLAR MEASUREMENTS

Significant science results of the past decade were accomplished at ground-based facilities—probing solar processes from the interior through the solar surface to the chromosphere and corona. Existing ground-based optical platforms include the Mauna Loa Solar Observatory (operated by HAO), the Sacramento Peak Observatory, Kitt Peak Observatory, GONG, and SOLIS (operated by the National Solar Observatory), the Mees Solar Observatory (University of Hawaii), the Big Bear Solar Observatory (New Jersey Institute of Technology), the San Fernando Observatory (CSUN), and the Mt. Wilson 60-foot tower (USC) and Mt. Wilson 150-foot tower (UCLA). Existing ground-based radio platforms include the Owens Valley Radio Observatory (OVSA; operated by NJIT), the Long-Wavelength Array (New Mexico State University), and the Haystack Mountain Observatory (MIT).

The Advanced Technology Solar Telescope (ATST) will provide U.S. leadership in large-aperture, high-resolution, ground-based solar observations and will be a unique complement to space-borne and existing ground-based observations. Full-Sun measurements by existing synoptic facilities (e.g., GONG, SOLIS, and ISOON), and new initiatives such as the Coronal Solar Magnetism Observatory (COSMO) and the Frequency-Agile Solar Radiotelescope (FASR), have the potential to balance the narrow field of view captured by ATST and are essential for the study of transient phenomena.

A major science goal is to continue comparison of highly resolved observations with numerical models in order to critically define the physical nature of photospheric features such as sunspots, faculae, and cool molecular clouds. Another goal is to better understand the physical behavior of magnetic fields in the chromosphere and corona—both on small and large scales—in order to study the flow, storage, and eruption of energy and mass in these poorly understood regions. The recent, mostly unexpected, behavior of the solar cycle and solar scientists' inability to predict its future course makes synoptic studies of the magnetic field and surface and interior mass flows that are related to the solar dynamo a high-priority goal. Ground-based observations are increasingly used in near-real-time data-driven models of the heliosphere and space weather. A goal is to improve the quality of these measurements and to extend them upward into the chromosphere and corona.

Compared with space missions, ground-based facilities can be far larger, more flexible and exploratory, and longer-lived. Accordingly, emphasis is on achieving high spatial resolution (e.g., NST, ATST), making unique measurements of physical processes at long wavelengths (e.g., OVSA, FASR), and collecting sufficient light flux to make high-time-resolution, high-precision measurements of magnetic and velocity fields, long-term synoptic observations, and novel frontier observations of various kinds (e.g., COSMO).

GROUND-BASED ATMOSPHERE-IONOSPHERE MEASUREMENTS

Ionospheric modification is an incisive tool for probing the upper atmosphere from the ground and offers a means of performing repeatable experiments and obtaining reproducible results. Ionospheric modifications use powerful high-frequency transmitters to induce phenomena in ionospheric plasmas. Some of these phenomena give insights into complicated plasma physics processes that may occur elsewhere in nature but that are difficult or impossible to explore in the laboratory or numerically. Other processes provide diagnostics of naturally occurring ionospheric phenomena and of natural rate constants that are otherwise hard to quantify. Ionospheric modification experiments affect the propagation of radio signals passing through the modified volume, which is how the phenomenon was first discovered (i.e,. the radio Luxembourg effect). These experiments generate airglow and radio emissions, which can be observed from the ground; create field-aligned plasma density irregularities that can be interrogated by small coherent scatter radars; generate low-frequency radio signals, which have practical societal utility; accelerate electrons, mimicking auroral processes; and modify plasma density and electron and ion temperatures and enhance the plasma and ion lines observed by incoherent scatter. Because of this, heaters are most productive when located close to incoherent scatter radars.

The Department of Defense (DOD) operates and maintains the world's largest ionospheric modification facility, the High Frequency Active Auroral Research Program (HAARP), near Gakona, Alaska. HAARP, which became fully operational during the past decade, is powerful, modular, and flexible and is especially well suited for studying ionospheric modification phenomena under different beam-pointing, emission frequency, and modulation conditions. HAARP is not co-located with an incoherent scatter radar, however, and its potential has therefore not been fully realized, since the phenomena it creates cannot be fully diagnosed.

Another ionospheric modification facility is currently under construction at the site of the Arecibo Radio Observatory. While this facility will be modest in power compared to HAARP, its co-location with Arecibo, the world's most sensitive incoherent scatter radar, raises the prospect of discovery science in the areas of artificial and naturally occurring ionospheric phenomena. The Arecibo heater came about through close collaboration between DOD and the National Science Foundation (NSF). The collaboration included community support and involvement from the beginning that will continue through the planning and execution of heating campaigns. The committee regards this kind of interagency cooperation as a model to be followed for the utilization of existing ionospheric modification facilities as well as the planning and development of new ones.

Another recent, important development is the emergence of Advanced Modular Incoherent Scatter Radar (AMISR)-class incoherent scatter radars, which were supported by the 2003 National Research Council (NRC) decadal survey, *The Sun to the Earth and Beyond*.[2] These are ultrahigh-frequency phased-array radars that can be electrically steered from pulse to pulse. AMISR-class incoherent scatter radars are currently deployed near Poker Flat, Alaska, and Resolute Bay, Canada. The latter of these includes two full radar faces and came about through international collaboration with Canada. Plans are being developed to deploy at least one additional radar in Antarctica. These facilities represent the current state of the art in high-power, large-aperture radars used for aeronomy and space physics.

An important hallmark of AMISR-class radars is their portability. These radars have been designed to be disassembled, shipped, and reassembled. The permanent infrastructure required for an AMISR is modest compared to the relocatable components, and shipping costs are expected to be modest compared to production costs. The objective of relocation is to enable discovery science through temporary deployments in geophysically interesting or under-instrumented regions. Additional benefits can accrue when relocation brings an AMISR-class radar into collaborative arrangements with other scientific assets, such as optical instruments, rocket ranges, or other radars. For example, current plans call for the redeployment of the radar near Poker Flat, Alaska, to La Plata, Argentina, where it could support an investigation into magnetic conjugacy effects with Arecibo pertaining to natural and heater-induced ionospheric phenomena. This particular relocation would also entail extensive collaborations with Argentine universities, faculty, and students, which are very welcome.

NSF also sponsored an NRC study on a distributed array of small instruments (DASI),[3] which could have major impact for future measurements. Due to the near omnipresence of wireless or phone connectivity, sensors can be distributed and worked as an integrated whole. The instruments in question could include ground-based imagers, optical interferometers, and spectrometers, magnetometers, radio beacon

---

[2]National Research Council, *The Sun to the Earth and Beyond—A Decadal Research Strategy in Solar and Space Physics*, The National Academies Press, Washington, D.C., 2003.

[3]National Research Council, *Distributed Arrays of Small Instruments for Solar-Terrestrial Research: Report of a Workshop*, The National Academies Press, Washington, D.C., 2006.

and Global Positioning System (GPS) receivers, ionosondes, coherent scatter radars, and other small, relatively inexpensive devices. Broadly interpreted, the instruments could also include small satellites, sounding rockets, and balloons. The arrays could be assembled on an ad hoc basis for campaigns of short or intermediate duration. Instruments could be rotated in and out of service, possibly in a rotation in the EarthScope model. Instruments could possibly be checked in and out of central reserves in the way that seismometers are currently within the Earth sciences. Arrays of instruments could furthermore be anchored by large, class-I facilities (incoherent scatter radars, wind-temperature lidars, and so on, as needed).

Among such measurements, there has been much progress in lidar measurements. A giant leap forward in understanding of meteorological influences and neutral-plasma interactions on the atmosphere-ionosphere-magnetosphere (AIM) system can be achieved if neutral wind, temperature, and mass density measurements can be obtained simultaneously at subscale-height altitude resolution from the lower atmosphere to the mid-thermosphere (~200 km altitude). Beyond providing the first continuous measurements of Earth's thermal and wind structure from the lower atmosphere to the mid-thermosphere, this measurement capability would enable new discoveries in understanding wave-mean flow interactions, gravity wave propagation into the thermosphere, mesosphere turbulence, secondary wave generation in the thermosphere, eddy diffusivity, wave fluxes of momentum, heat, and constituents, and heating and cooling processes. In addition, these areas of study would significantly advance models, as many of these physical processes are presently parameterized in numerical simulations. For example, eddy diffusivity in the mesosphere is a parameter that is poorly characterized and parameterized in models but is absolutely crucial to understanding and modeling the chemistry and thermal balance of the mesosphere and thermosphere.

Combining such a neutral gas measurement capability with existing altitude-resolved plasma measurements provided by the incoherent scatter radar technique, along with other complementary instrumentation, would further broaden the potential for scientific discovery in areas of ion-neutral thermal and momentum exchange, Hall and Pedersen conductivity behavior, neutral wind effects on current flow and dissipation, neutral wind dynamo processes, and ion-neutral chemical interactions.

HOSTED PAYLOADS ON COMMERCIAL AND GOVERNMENT PLATFORMS

There are opportunities to leverage distributed systems for commercial and government use by adding sensors and finding secondary uses of data from these systems, in the context of breakthrough science investigations of solar and space physics.

As a prime example for secondary data use, the International GNSS Service (IGS) has been very successful in creating a leveraged global network of GPS/GNSS (Global Navigation Satellite Systems) receivers that provides excellent science data to the geophysics community without unduly taxing the resources of any single country or agency. The AIM community has been leveraging this global network to great success. The AIM community would benefit from adopting a proactive approach to the evolving and growing global GNSS network. The community should develop a leadership role in coordinating GNSS networks for AIM science, following the IGS model of leveraging an international effort across multiple institutions. Activities could include developing standards for data access; facilitating institutional hosting of GNSS receivers for AIM science; and providing scientific leadership in the use of global GNSS networks for AIM science.

An example of a successfully hosted payload is the Two Wide-Angle Imaging Neutral-Atom Spectrometer (TWINS) instrument, which provides stereoscopic imaging of the magnetosphere from twin platforms hosted on DOD spacecraft. TWINS is in a Molniya orbit with 63.4 degree inclination and 7.2 Earth radius (R_E) apogee, which is ideal for its imaging objectives. Also, the Los Alamos National Laboratory geosynchronous Earth orbit satellites have been providing important particle measurements for decades.

A very cost-effective approach to creating a robust, long-term data set to meet AIM system science objectives is to host GPS remote sensing instruments on research and operational satellites deployed by NASA and the National Oceanic and Atmospheric Administration. GPS deployed on commercial satellites and constellations is another viable option. GPS is relatively simple to host because (1) it is passive (low electromagnetic emissions); (2) it operates autonomously; and (3) it is self-calibrating. The committee supports coordination with the Technology and Innovation Working Group to develop a new generation of space-borne GPS receivers that can perform high-quality ionospheric measurements with low mass and power (~1-2 kg, 1-2 W). Another technology development option is technology transfer to commercial providers of navigation receivers to add an ionospheric measurement capability. Such receivers can acquire ionospheric measurements of reasonable quality on a non-interference basis because they perform the operational navigation function.

Science objectives relevant for NASA require other orbits, but hosted opportunities are currently and increasingly available in a variety of orbits and even provide an opportunity for carrying out constellation missions that would be prohibitive in cost if dedicated spacecraft were utilized. For example, the Iridium NEXT constellation will replace the current Iridium communications satellites and will consist of 72 low-Earth-orbit (LEO) satellites (66 + 6 on-orbit spares). The orbits are nearly polar (86.4 degree inclination) at an altitude of 780 km, which is ideal for studies of particle precipitation or total electron content maps, for example. Each satellite can support up to 50 kg, 50 W average (200 W peak), and a 1-Mbps peak data rate. Several studies are underway to investigate the feasibility of placing science payloads on some or all of the Iridium NEXT satellites. For example, scientists at Johns Hopkins University's Applied Physics Laboratory have organized a grassroots effort for the GEOscan project, which would place several different types of instruments on the satellites and would also provide the infrastructure for some CubeSats. The constellation would provide continuous coverage over the entire Earth, allowing for high-time-resolution studies on a global scale. Despite the fact that this constellation provides an unprecedented opportunity that will possibly never happen again, it is not clear whether there is sufficient time for GEOscan or something similar to materialize, since the first Iridium launch will occur in 2015. This illustrates the importance of NASA or NSF developing a capability to respond quickly to similar kinds of opportunities.

In addition to LEO opportunities, there are also many geostationary Earth orbit (GEO)-hosted opportunities. GEO communications satellites, such as StarBus, are launched 2 to 3 times a year and typically have excess capacity for 100 kg, up to 300 W, and downlink rates up to 70 Mbps. They can provide pointing control of 0.1 degrees and have greater than 5-year lifetimes. This is particularly important because it is difficult to get an orbit slot in GEO. GEO sits at the outer edge of the radiation belts and the inner edge of the plasma sheet and is thus perfectly situated for investigations of the link between substorm injections and the energization of particles in the inner magnetosphere. It is also the location of many commercial and government assets. Thus the ability to understand the space weather environment is very important.

ROCKETS

NASA's Sounding Rocket Program provides regular, inexpensive access to near-Earth space for a broad range of space science disciplines, including solar, geospace, and astrophysics research missions. The program has been extremely successful throughout its history, consistently providing high science return for the funding invested. Scientific payloads launched on sounding rockets provide the only means to gather data along nearly vertical profiles and are the only platforms capable of in situ sampling of the mesosphere and lower ionosphere regions (40-150 km altitude), which constitute Earth's critical interface region between the atmosphere and space. The obtained data rates can sometimes exceed those of satellites by orders of magnitude. The rocket program is also used to provide timely calibration and correlative data for several of NASA's larger satellite scientific missions and enables new instrument concepts to be developed and tested in space. Another critical attribute of the Sounding Rocket Program is that it provides training for scores of university graduate students, many of whom are now among the nation's leaders in the field of space research. The program can uniquely carry out this training because other programs are too risk averse or span too long a period of time for a graduate student to be involved from start to finish.

Although new instruments have been developed within the rocket program, the emphasis for mission selection in the past 10 to 15 years continues to be governed primarily by science and the promise of closure of critical science questions. It is conceivable that a percentage of rocket flights could be dedicated to technology development as their main objective. A new NASA-developed capability, the High Altitude Sounding Rocket with apogees of ~3,000 km, providing approximately 40 minutes of observing time above 100 km, which is significantly longer than the 5- to 10-minute typical mission-duration periods of apogees of 200-1000 km, would have a huge impact on the utility of rockets for science and technology development. Such rockets would also include approximately 1-meter-diameter experiment sections, which are significantly larger than current payload diameters (40-50 cm). These new platforms would enable significantly longer observing times for solar missions that track developing features on the solar disk, as well as enable direct penetration of the cusp and auroral acceleration regions and also the inner radiation belt, by geospace missions.

BALLOONS

High-altitude balloon experiments have a rich history in solar and space physics. Balloons continue to offer a unique science platform, and, like sounding rockets, they provide opportunities for instrument development and program management that are essential in the training of the next generation of scientists and engineers. Investigations using balloons are contributing to fundamental research advances across the discipline areas that constitute solar and space physics, with observations that range from gamma-ray solar flares to particle precipitation to large-scale magnetospheric electric fields.

In solar physics, balloons offer a low-cost method for carrying heavy payloads high above the atmosphere. Such missions have led to scientific discoveries and are an ideal platform for developing and testing new spacecraft instrumentation. For example, in the 1980s, hard-X-ray microflares and superhot flare plasmas were discovered during missions using balloon-based X-ray instruments. Balloon missions (e.g., HEIDI, HIREGS) were also essential for the development of the RHESSI (Ramaty High Energy Solar Spectroscopic Imager) small Explorer that has operated for nearly a decade.

Balloons have also made important contributions to understanding both auroral and radiation belt particle precipitation. For example, electron microbursts were discovered with balloons,[4] and it is now recognized that microbursts are an important loss mechanism for the radiation belts and may be an ideal test case for studying nonlinear wave-particle interactions. Balloons also offer a unique platform for precipitation studies that is complementary to spacecraft measurements. The BARREL (Balloon Array for RBSP Relativistic Electron Precipitation) project will fly 40 small (~20 kg) balloon payloads during two Antarctic campaigns in 2013 and 2014 to provide a global view of electron precipitation during the RBSP (renamed Van Allen Probes) mission and be able to distinguish complex temporal and spatial variations.

The Ultra Long Duration Balloon (ULDB) program remains critical for solar research; X-ray, gamma-ray, and neutron instruments are generally very heavy due to the amount of power required to stop high-energy photons and the long observing window required to catch rare large gamma-ray flares.

[4]K.A. Anderson and D.W. Milton, Balloon observations of X rays in the auroral zone 3: High time resolution studies, *Journal of Geophysical Research* 69(21):4457-4479, doi:10.1029/JZ069i021p04457, 1964.

Balloons remain the cheapest and most effective method for carrying large (>1,000 kg) payloads to the edge of space. The increased availability of the ULDB program provides a critical enhancement of capability for a variety of science disciplines.

VERY SMALL SPACECRAFT

Through March 2011 55 CubeSats had been launched worldwide. CubeSats are distinguished by their conformance to a size standard that allows them to be carried into orbit within a generic dispenser/carrier, the so-called P-POD. The P-POD fully encapsulates the CubeSats, providing a high level of protection for the launch vehicle and any primary payload. Owing to its simplicity and ease of adaptability to a wide variety of launch vehicles, the P-POD has been qualified for flight and utilized on a number of launch vehicles. To date, all CubeSats launched have conformed to size requirements of the standard 3-U P-POD, which accommodates up to three CubeSats measuring 10 × 10 × 10 cm.

The scientific utility of very small satellites is being demonstrated through NSF's CubeSat-based Science Missions for Space Weather and Atmospheric Research Program. Seven scientific CubeSat missions have been competitively selected (through July 2011) from many more proposals submitted to the NSF program through its first three solicitations. The missions (FireFly, RAX, FIREBIRD, CINEMA, CSSWE, DICE, and CADRE) will pursue science goals, including studies of lightning-induced terrestrial gamma rays; electromagnetic emissions near ionospheric radar beams; mechanisms responsible for relativistic electron microbursts; energetic radiation belt electrons; electrons, ions, and neutral atoms in the ionosphere; ion-neutral coupling in the ionosphere; and the atmospheric density response in the thermosphere to extreme forcing. The NSF CubeSat program is substantially oversubscribed.

Technological gains in miniature, low-power, highly integrated electronics, microelectromechanical system devices, and other nanoscale manufacturing techniques have enabled revolutionary approaches to experimental space science that have not widely been put into practice by the heliophysics community. Small, low-cost satellites provide opportunities not typically available to traditional, large spacecraft. Low-cost "expendable" spacecraft may be deployed into regions where satellite lifetimes are constrained but where important, yet not well characterized, science linkages take place. The lower ionosphere well below 300 km and reaching down into the thermosphere at even lower altitudes is an important coupling region between the upper atmosphere and the ionosphere, yet atmospheric drag severely limits satellite lifetime against orbital decay at these altitudes. The solar chromosphere is another poorly characterized coupling region where harsh environments limit satellite lifetime. The science return per dollar to perform in situ measurements in these types of regimes, to conduct even limited lifetime measurements, is enhanced by low-cost platforms. Constellation missions, described elsewhere in this report, composed of dozens of science platforms are enabled by low-cost satellites that can be launched in large numbers from a single launch vehicle or from a number of launch vehicles as secondary or rideshare payloads at very low launch costs. Very small instrument-carrying satellites, e.g., with total mass-to-orbit in the range of 1 to 20 kg, can accommodate instruments of great utility for heliophysics. Their utility for collecting unique science measurements is greatly enhanced when they are deployed as swarms or into unique environments. Continuing technological developments will lead to even more capable spacecraft with space, weight, and power (SWAP) resources capable of supporting payload complements even more sophisticated than those permitted by the present-day 3-U CubeSats. Already, a number of organizations have developed prototype 6-U deployers and satellites that have significantly enhanced SWAP resources compared to the 3-U standard.

Launch opportunities for conforming CubeSats are becoming commonplace. The NASA Launch Services Program's Educational Launch of Nanosatellites (ELaNa) program for educational satellites has been established and is well underway, providing launch manifesting for CubeSats. The first ELaNa CubeSats were launched in March 2011 as secondary payloads on the ill-fated Glory launch, and several other ELaNa follow-on missions are in the launch queue. To date 35 satellite projects have been selected for manifesting as secondary payloads under the ELaNa program. A new call for proposals was announced (July 2011) that entertains flight opportunity proposals for 6-U CubeSats, in addition to the traditional 1-U, 2-U, and 3-U form factors. Launch opportunities in the future are expected to increase, especially for certain orbits. Two large spacecraft are under development that will resupply the International Space Station (ISS)—the Dragon spacecraft of Space-X (Space Exploration Technologies Corporation) and the Cygnus spacecraft of Orbital Sciences Corporation. Both Space-X and Orbital have initiated programs to carry CubeSats on these ISS resupply missions. Indeed, the ISS itself could be used to store and launch on demand any number of CubeSats. Such a capability would allow quick responses to geophysically interesting events, or, for example, the measured deployment of CubeSats to form a thermosphere constellation. The feasibility of launching CubeSat-like satellites from the ISS has been demonstrated multiple times through the microelectromechanical systems-based PicoSat Inspector program for the Aerospace Corporation.

Satellites with bigger volumes than CubeSats, but still well below a standard Explorer size, could revolutionize research in AIM and other science disciplines. Because of their substantially smaller cost, such tiny Explorers could be built with a much higher risk-tolerance and could carry a small number of instruments, or novel low-weight sensors, into interesting locations of the space environment. Tiny Explorers would still be carried as secondary payloads and would share some commonalities. However, they would open a new set of opportunities for longer-duration measurements.

CONSTELLATIONS

Constellations of measurement platforms of various sizes have the potential to take advantage of miniaturization technologies, enhanced computational capabilities, and autonomous systems, as well as novel system and network approaches. The value of many breakthrough solar and space physics constellations strongly depends on the number of elements in the system. With the advent of small CubeSats and tiny Explorers, novel means of investigations are enabled that have the potential to provide an unprecedented density of measurements in the upper atmosphere and elsewhere.

Illustrative Examples of Newly Enabled Constellations

Magnetosphere Radiation Belt Constellation

A 6- to 12-month database of continuous ultralow-frequency measurements of the azimuthal mode number spectrum would provide a critical missing piece of information for radiation belt modelers. A constellation of ~30 CubeSats equipped with fluxgate magnetometers in a circular orbit near the geomagnetic equatorial plane (e.g., geosynchronous) would be an ideal platform for such a radiation belt study.

Ionosphere-Magnetosphere Coupling Constellation

A constellation mission utilizing small satellites would radically improve understanding of the dynamics of the coupled thermosphere-ionosphere system. In order to achieve this goal, the following top-level requirements must be met:

1. Each satellite needs to be complemented with instrumentation that is capable of measuring both the neutral and ion state variables, such as density, temperature, and winds. There are a few different instru-

ment packages that are capable of achieving this requirement, with many of them small enough to fit on very small spacecraft.

2. Each satellite needs to pass through the auroral zone, since this is the primary science zone.

3. There must be enough satellites in a single orbital plane that the return time to a location is less than dynamical timescales within the system.

4. Multiple planes of satellites should be utilized to provide information on the longitudinal structure and the dynamics of the system.

HETEROGENEOUS FACILITIES

Measurements from a diverse set of platforms—space missions, suborbital, ground-based, CubeSats, alternatives—can enhance the science of any single-point space missions. Examples of successfully demonstrated integration are THEMIS ground-based, BARREL and RBSP, and the rocket flights that were done during the CRRES mission.

A heterogeneous facility generalizes this integration concept into a comprehensive approach that encompasses all aspects of a mission, but it is applied to a variety of platforms that are integrated through operational means, data analysis, and dissemination to target science objectives in an integrated fashion. Heterogeneous facilities are particularly valuable to system science for understanding mesoscale and global-scale processes. Heterogeneous facilities would combine and coordinate measurement platforms that are developed through a variety of means and financial support.

Illustrative Example of Heterogeneous Facilities: A Storm-Based Campaign

The purpose of a storm-based heterogeneous facility would be to set up and operate a campaign over a given time period focused on a variety of assets, all deployed to understand geomagnetic storms. Global GPS networks have provided a synoptic view of ionospheric storms that emphasizes how different physical processes and regional features work in concert to create the "global ionospheric storm." GPS networks operate continuously so that no special coordination is needed as part of a heterogeneous facility. However, significant questions regarding the physical processes causing the storm-time dynamics requires unraveling coordinated observations. A storm investigation would require that the GPS data be analyzed during an event in which coordinated observations from SuperDarn, NSF radars, and wind measurements are obtained. Mission data would play a role also. These facilities would be operated in a mode that optimizes their data for the particular science objectives being pursued. Mobile facilities could be located where they are most needed to investigate particular mesoscale features that are prominent during storms, such as storm-enhanced density or tongues of ionization. The scientific payoff is understanding the physical drivers of the ionospheric response that is captured by the GPS sensors. This campaign may also involve coordinated operation of some CubeSats and also balloon launches.

D

Education and Workforce Issues in Solar and Space Physics

SUMMARY

Working Group Mission and Challenges

The cross-disciplinary Education and Workforce Working Group assessed the current state of health of the solar and space physics field and provided foundational support and guidance to the survey committee regarding recommendations designed to enhance the community's efforts in education, to support the necessary workforce needed within solar and space physics, and to contribute to the community's broader STEM (science, technology, engineering, and mathematics) education efforts. The assessment was based on the community's first demographic survey and quantitative determination of the state of health of the solar and space physics community.

One of the largest challenges in this undertaking was that the disciplines constituting solar and space physics are spread among many different academic fields (physics, astronomy, the Earth and space sciences, and several engineering disciplines), making the task of assessing a number of basic quantitative measures challenging because of different standards and norms in the various academic environments.

Quantitative Studies of the Health of the Field

To quantitatively answer these and other questions, the Education and Workforce Working Group undertook a series of studies. These included the following: (1) development of a community-wide database of members and programs to determine the size of the community; (2) development of a community demographic survey with the American Institute of Physics (AIP) to determine the age, gender, type of employer, academic background, and participation levels within various NASA and National Science Foundation (NSF) programs, among other questions; (3) development and implementation of graduate student surveys (both online and within focus groups) to determine their pathways into the field; (4) a Ph.D. survey to determine the number of Ph.D.s graduated each year, from which institution, and within what subdiscipline; (5) a job advertisement study to determine the number of positions advertised each year;

and (6) a bibliometric survey of the number of publications in each of the three main subdisciplines (solar and heliospheric, magnetospheric, and ionosphere/thermosphere) in each year over the past decade. This appendix describes each of these studies and provides the results from those that have been completed as well as highlights of some of the other results.

In addition to these studies, the Education and Workforce Working Group received a number of white papers through the decadal survey process and held several town hall meetings and stand-alone working group meetings and teleconferences to receive input and information, including from NASA, NSF, and professional societies (such as the American Geophysical Union (AGU) and the American Physical Society). Key findings are summarized here.

FINDINGS OF THE EDUCATION AND WORKFORCE WORKING GROUP

Graduate Students

Professionals in solar and space physics work in diverse settings, including government labs, industry, research centers, and academic institutions (in colleges and departments that go by many different names), which complicates any demographic study of the field. Also, solar and space physics is not currently listed as a dissertation research area within NSF's Annual Survey of Earned Doctorates.[1] This survey influences other rankings, ratings, and demographic surveys done by the National Research Council and AIP.

Graduate students interviewed at Geospace Environment Modeling (GEM)-Coupling, Energetics, and Dynamics of Atmospheric Regions (CEDAR) and Solar and Heliospheric Influences at Earth (SHINE) meetings often cited a childhood interest in space and astronomy that grew through high school and undergraduate physics courses into an interest in trying out astronomy research as an undergraduate. Once exposed to the solar and space physics field, the reasons they cited for finding space physics more attractive than astronomy or cosmology included the following: (1) the domain of physical reality being studied is not so remote and can be accessed by humans; (2) the research content is easier to explain to family and friends; (3) the research has more societal relevance; and (4) there is a more compelling opportunity to constrain theory with in situ data.

Graduate students surveyed about how their graduate program could be better at preparing them cited a desire for more networking, a sense of community, and access to in-depth academic courses that are specific to space physics (e.g., If someone developed a great course on complexity theory and space physics, how could it be more broadly disseminated in the community?). Solar and space physics is most often taught or introduced in summer schools at the graduate level, so opportunities to become aware of the discipline have been more limited at the undergraduate and pre-college levels. This raises the issue of what kinds of activities captured the attention of current students in the field.

Approximately 50 percent of the graduate students surveyed at NSF-supported GEM-CEDAR and SHINE meetings during the summer of 2011 reported an undergraduate research experience in the solar and space physics field (e.g., via the NSF Research Experiences for Undergraduates (REU) program). This highlights the importance of such programs for recruiting students in the future. As a consequence, it seems prudent to develop a community of solar and space physics REU and Research Experiences for Teachers sites and programs—to facilitate the sharing of resources, collaboration through distance-learning technologies, training of mentors and program staff regarding issues of diversity, and utilization of existing summer schools—to augment REU programs and enhance the opportunities for attracting a diverse and talented

[1] See National Science Foundation, Survey of Earned Doctorates, available at http://www.nsf.gov/statistics/srvydoctorates/.

graduate student population. The Los Alamos National Laboratory's post-baccalaureate program,[2] which provides recent college graduates the opportunity to explore research experiences within solar and space physics, is another model for recruiting graduate students.

Challenges of Educational Institutions

The studies described in detail below in this appendix show that while the Ph.D. production rate for solar and space physics has increased over the past decade, the number of advertised positions in the field has decreased. Indeed, the number of advertised faculty positions reached a decadal low in the last year surveyed, 2010. Although, historically, many solar and space physics graduates find jobs in areas other than academic research, the trend—an increasing number of students being trained versus a decreasing number of faculty hires to train them—indicates the continued importance of the NSF Faculty Development in Space Science program. This program grew out of a recommendation by the 2003 decadal survey[3] and led to the creation of eight tenure-track faculty positions, most of whose awardees have already become tenured. The program is widely viewed, along with CAREER awards, as an exemplary means of sustaining space physics within the university and promoting the science and engineering workforce. In order to increase the reach of this program and the diversity of students exposed to opportunities in solar and space physics, eligibility for these awards could be expanded to include 4-year institutions, not just Ph.D.-granting research universities.

Regardless of whether a student remains in the field, there are skills all students need in order to become successful professionals. These skills include interpersonal and communication skills, career awareness, leadership, grantsmanship, and laboratory management skills. The formal discipline training provided in graduate school does not usually provide adequate mentoring and training opportunities for developing such skills.

NASA ended the 30-year-old NASA Graduate Student Research Program (GSRP) in 2012. Providing a research stipend and a set amount of fees, tuition, and travel, the program supported solar and space physics graduate students working closely with NASA scientists as research mentors. It was cost-effective because universities generally helped to subsidize the cost of tuition and stipend, which the GSRP did not cover fully, thus enabling the NASA research dollar to go further. While another program, NASA's Earth and Space Science Fellowship (NESSF) program, will provide funding for graduate students, there was concern expressed that the formal link between the student and a NASA center will disappear, and there was a strong desire to see NESSF support for solar and space physics maintained at levels as high as those the GSRP historically provided.

Vibrant university-based solar and space physics education and research programs that extensively involve experimental science and engineering undergraduate and graduate students focused on instrument development and space systems are vital to maintaining the health of the field. Hurdles to this include lack of low-cost access to space, long development times for spacecraft missions, funding limitations, and the inhibiting qualities of International Traffic in Arms Regulations.[4]

---

[2]See Los Alamos National Laboratory, About the LANL Undergraduate Student Program: Program Overview, available at http://www.lanl.gov/education/undergrad/about.shtml.

[3]National Research Council, *The Sun to the Earth—and Beyond: A Decadal Research Strategy in Solar and Space Physics,* The National Academies Press, Washington, D.C., 2003.

[4]See Department of State, International Traffic in Arms Regulations, available at http://www.pmddtc.state.gov/regulations_laws/itar_official.html.

Education and Public Outreach and K-12 Issues

Since the time of the 2003 decadal study, the community of NASA-supported, discipline-focused education and public outreach (EPO) professionals (e.g., the Heliophysics Education and Public Outreach Forum[5]) has grown in size and capacity to connect the scientists, missions, and results of the solar and space physics research community with potent dissemination partners in the world of K-12 education (e.g., science teacher professional societies and teacher networks in physics and Earth science) and informal education (i.e., science centers and planetariums). Scientific professional societies like AGU now host more than 20 education sessions at their meetings, and the AGU Space Physics and Aeronomy (SPA) Division has a very strong EPO committee. Thus EPO professionals have become close and vital allies in the cultivation of a healthy scientific workforce and should be considered members of that workforce in solar and space physics.

Many members of the community felt that NASA and NSF should consider programs and workshops to support the EPO skills and partnership of solar and space physics EPO professionals and solar and space physics researchers. EPO professionals and physics educators are essential members of the solar and space physics workforce who are working collaboratively with scientists to develop the solar and space physics workforce as well as public support and interest in solar and space physics.

Of particular note is the existence of a growing national standards movement that will have an increasing impact in science and mathematics education nationwide. These standards (the Common Core[6] and the Next Generation Science Standards[7]) are being adopted by many states and will guide instruction through the next decade. The NASA Heliophysics Education Forum will play a critical role in providing bridges between the scientific research community and the standards movement.

Background for Findings and Priorities and Description of Research Projects Undertaken by the Education and Workforce Working Group

The cross-discipline Education and Workforce Working Group's mission as defined by the decadal survey committee was to identify trends, strengths, weaknesses, and opportunities, and to give strategic guidance related to education and workforce issues in solar and space physics. The working group identified four broad themes (workforce, the community, university programs, and K-12 education) around which to organize its findings and conduct surveys. Six studies were initiated by the working group, as detailed in the following sections of this appendix.

COMPILATION OF CURRENT SUMMER SCHOOLS

In recent years there has been an increase in the popularity of "summer schools" in which specialized solar and space physics content is taught, with graduate students being the largest (but not only) audience. The summer school established by the Center for Integrated Space Weather Modeling (CISM) has been seen as particularly successful.[8] Table D.1 is a list of summer schools identified by the Education and Workforce Working Group.

---

[5] See NASA, SMD EPO Forums, available at http://smdepo.org/node/546.

[6] See Common Core State Standards Initiative website, available at http://www.corestandards.org.

[7] See Next Generations Science Standards website, available at http://www.nextgenscience.org.

[8] R.E. Lopez and N.A. Gross, Active learning for advanced students: The Center for Integrated Space Weather Modeling Graduate Summer School, *Advances in Space Research* 42:1864-1868, doi:10.1016/j.asr.2007.06.056, 2008; S. Simpson, A Sun-to-mud education in two weeks, *Space Weather Quarterly* 2(7):S07002, doi:10.1029/2004SW000092, 2004.

TABLE D.1 2011 Summer Schools in Solar and Space Physics

Name	Sponsorship Agency, Program	Intended Audience
Heliophysics Summer School	NASA, Living With a Star	Career scientists, postdoctoral researchers, graduate students, and physics undergraduates in astrophysics, geophysics, plasma physics, or space physics.
2011 Research Experiences for Undergraduates Program in Solar and Space Physics	National Science Foundation (NSF)	Current sophomore and junior undergraduates with an interest in solar and space physics.
Center for Integrated Space Weather Modeling	NSF	Students who are about to enter, or are in their first year of, graduate school; also for anyone new to space weather who would like to enhance their understanding.
University of Arizona and National Solar Observatory 2008 Solar Physics Summer School	Supported in part by a grant from NSF	Graduate to advanced undergraduates with an intense interest in solar physics, space physics, or related fields.
American Astronomical Society, Solar Physics Division, High Energy Solar Physics	NASA, Living With a Star, Targeted Research and Technology program	Specifically incoming graduate students but also graduate students at any level and recent Ph.D.s.
Polar Aeronomy and Radio Science (PARS)	NSF; University of Alaska Fairbanks, Geophysical Institute	Provides upper-atmosphere/ionosphere instruction and hands-on experimental experience for students and their graduate advisors.
AMISR summer school	NSF; MIT Haystack Observatory	Graduate and advanced undergraduate students, as well as scientists new to the incoherent scatter radar technique.
CEDAR-GEM and SHINE meeting summer schools	NSF	Graduate and advanced undergraduate students.
Los Alamos Space Weather	Institute of Geophysics and Planetary Physics	Designed for graduate students enrolled in a graduate program in the United States.
Joint Space Weather Summer School	University of Huntsville and German Aerospace Center (DLR)	Graduate and undergraduate students.

GRADUATE STUDENT SURVEY

The co-chairs of the working group led the development of a graduate student interview protocol to determine the different pathways that graduate students took into solar and space physics. This effort was inspired by an appeal from graduate students who attended a town hall meeting convened by the decadal survey in early 2011. The protocol was reviewed and revised by working group members and AIP partners, and vetted through the Institutional Review Board (IRB) process of both Georgia State University (Morrow) and the University of Michigan (Moldwin). Interviews with graduate students (two or four groups per day, with four to six graduate students per group) attending NSF-supported summer meetings (e.g., CEDAR-GEM, SHINE) were conducted. The group interviews combined individual written responses to a short questionnaire and the oral discussion related to the written responses, which involved all group members. Some highlights of these interviews are included in the findings.

DEMOGRAPHIC SURVEY

The Education and Workforce Working Group, with support from an NSF grant, collaborated with AIP to conduct the first-ever comprehensive demographic study of solar and space physics and related fields. This included scientists who investigate (1) the Sun, (2) the nature of the region of space affected by the Sun, and (3) the response of Earth and the rest of our solar system to this space environment. Such scientists study the predictability of changes in our space environment that can impact life and society on Earth. They also explore ways that investigating the Sun and the space environment can teach us about important physical processes that occur throughout the universe. The names of the disciplinary areas in which these scientists operate include solar physics, heliophysics, space physics, aeronomy, and upper atmospheric physics.

As part of the demographic survey, a database of individuals and their email addresses was compiled. This database drew on four sources: the AGU's Solar and Space Physics member directory, the APS Division of Solar Physics member directory, the NSF AGS list of research grants and proposals, and the NASA Heliophysics SR&T grant and proposal databases.

The objectives of the survey were to:

1. Establish a baseline of statistical facts about the pipeline in solar, space, and upper atmospheric physics.
2. Identify factors that affect career paths and career development, and in particular determine how people find their way into solar, space, and upper atmospheric physics, and how they are distributed among universities, government labs, and industry.
3. Examine the community's current perceptions about the health and vitality of the field.
4. Investigate participation in programs and projects run by NSF, NASA, or other funding agencies, and also try to determine what fraction of the community is dependent on "soft money."

Email announcements describing the survey were sent out to the community on October 6, 2011. These were followed up on October 11, 2011, with an email with the URL for the survey that went to 2,560 unique email addresses (from the AGU Space Physics and Aeronomy Section, American Astronomical Society (AAS) Solar Physics Division (SPD), Space Weather Week attendee lists, and NSF principal investigator lists). There were 1,305 responses (51 percent) of which 1,171 were from individuals working in the areas of solar, space, and upper atmospheric physics and who work and live in the United States. The survey generated 125 pages of single-spaced responses to a number of open-ended questions, such as respondent comments concerning factors leading to career success and barriers to success. Some highlights from the survey are presented here.

Of the respondents, 83 percent were men and 17 percent women. Most were white (81 percent), while 13 percent were Asian or Asian American, with 6 percent other. The median age of the respondents was 51 years, with a symmetric distribution (the middle 50 percent were between 40 and 62 years old). Physics was by far the most common undergraduate degree (62 percent). For those earning their Ph.D.s in 1999 or earlier, physics was the most common degree field (40 percent), while the most common degree for those receiving Ph.D.s since 2000 is space physics (36 percent, with physics having dropped to 27 percent). Almost three-quarters of graduate student support and two-thirds of undergraduate student research support are from NASA or NSF.

The number of women in solar and space physics is greater than for physics, but women and (especially) minorities remain underrepresented, with African Americans and Hispanics constituting only 3 percent of the community. This disparity highlights the need for more efforts to increase diversity in solar and space physics. The activities of CISM have shown that focusing on diversity can result in significantly improved recruitment from historically underrepresented groups,[9] so solar and space physics as a community could benefit from programs that specifically target enhancing diversity within solar and space physics, such as the NSF Opportunities for Enhancing Diversity in the Geosciences program.[10]

Nearly three-quarters of recent Ph.D. recipients (receiving their degree since 2000) participated in some form of undergraduate research. About half of the respondents reported that they were involved at some level in K-12 education and/or public outreach. This highlights the importance of education and public outreach and the role that the Heliophysics Education and Public Outreach Forum plays in connecting the scientific community with education resources.

The survey asked a number of Likert-scale questions, giving respondents the opportunity to elaborate on their answers—two of which are highlighted here. One question asked if they strongly agree, agree, disagree, strongly disagree, or "don't know" regarding the following statement: "The next generation of scientific leadership is emerging in my field, and I am confident that they will be able to answer the scientific questions of the next decade." Two-thirds of the respondents agreed or agreed strongly with the statement, while one-quarter disagreed or disagreed strongly. The open-ended comments were generally optimistic about the abilities of the next generation but pessimistic about reduction in funding and NASA missions. Another question asked, "What have been the barriers to your career up until this point?" and the responses were divided between men and women. More than 31 percent (48 of 154) of the written responses from women indicated some form of gender discrimination or lack of family-friendly policies as barriers.

PH.D. PRODUCTION IN SOLAR AND SPACE PHYSICS

To assess the state of health of the field of solar and space physics and the discipline areas it comprises, an analysis of the number of Ph.D.s produced each year was done for the decade 2001-2010. The University of Michigan Ph.D. Dissertation Archive (Proquest) was queried for solar and space physics dissertations. The discipline areas generated between 30 to 40 Ph.D.s per year from 2001 to 2006 but then saw the number increase to 56 to 71 per year for the rest of the decade. Most of the increase was due to the doubling of ionosphere-thermosphere-meosphere (ITM) Ph.D.s (from an annual average of 12.4 Ph.D.s in the first half of the decade to 24.4) and solar Ph.D.s (from an average of 5.6 to 13). Magnetospheric Ph.D.s also increased, but by a smaller number (from an annual average of 12 to 16.5).

The Proquest dissertation database was queried for dissertations published between 2001 and 2010 for Ph.D.s in the broad field of solar and space physics. The keywords used to identify dissertations were the following: space physics, solar physics, magnetosphere, ionosphere, thermosphere, aurora, auroral, aurorae, heliosphere, space weather, aeronomy, solar corona, solar wind, solar chromosphere, helioseismology, solar flare, and solar active region. These terms were looked for in the dissertation keywords, title, and abstract. Once identified, the dissertation titles and/or abstracts were then analyzed to eliminate dissertations more focused on astrophysics, space engineering, or other nonrelated disciplines (like medicine). Only Ph.D. dissertations were included. The Proquest database contains most of the dissertations produced in the United States and Canada, but only a few from other universities around the world; therefore, the results include only dissertations produced in North America. The dissertations were divided

---

[9]D.E Chubin, E. Derrick, I. Feller, and P. Phartiyal, *AAAS Review of the NSF Science and Technology Centers Integrative Partnerships (STC) Program, 2000-2009, Final Report,* December 17, 2010, American Association for the Advancement of Science, Washington, D.C., 2010.

[10]See National Science Foundation, Opportunities for Enhancing Diversity in the Geosciences (OEDG), available at http://www.nsf.gov/funding/pgm_summ.jsp?pims_id=12726.

into five subdisciplines (solar, heliosphere, magnetosphere, ionosphere/thermosphere/upper atmospheric electricity, and space plasmas).

A total of 475 Ph.D.s were identifed. Figure D.1 shows the solar and space physics Ph.D. production as a function of year. Note the significant increase in Ph.D. production in 2007 that persisted through 2010. Figure D.2 shows the Ph.D. production as a function of year for each of the five subdisciplines. The increase in Ph.D. production in the second half of the decade is primarily due to ITM and, to a lesser extent, solar Ph.D.s.

The 475 Ph.D.s were produced at 76 different institutions. The top 10 institutions produced 238 (or 50 percent) of the total. Thirty of the 76 institutions produced only 1 solar and space physics Ph.D. during the decade. Table D.2 shows the top universities for solar and space physics production (all universities that averaged at least 1 Ph.D. per year). It should be noted that most of these universities have multiple departments that produce Ph.D.s in multiple subdisciplines. The overal trend of increasing Ph.D. production in the latter half of the decade is mirrored in these top departments (data not shown). It is difficult to assess the state of health of individual departments using one decade's worth of Ph.D.-production numbers because, even in the top Ph.D.-producing departments, the year-to-year fluctuation was large and the average time to Ph.D. is 5 to 6 years.

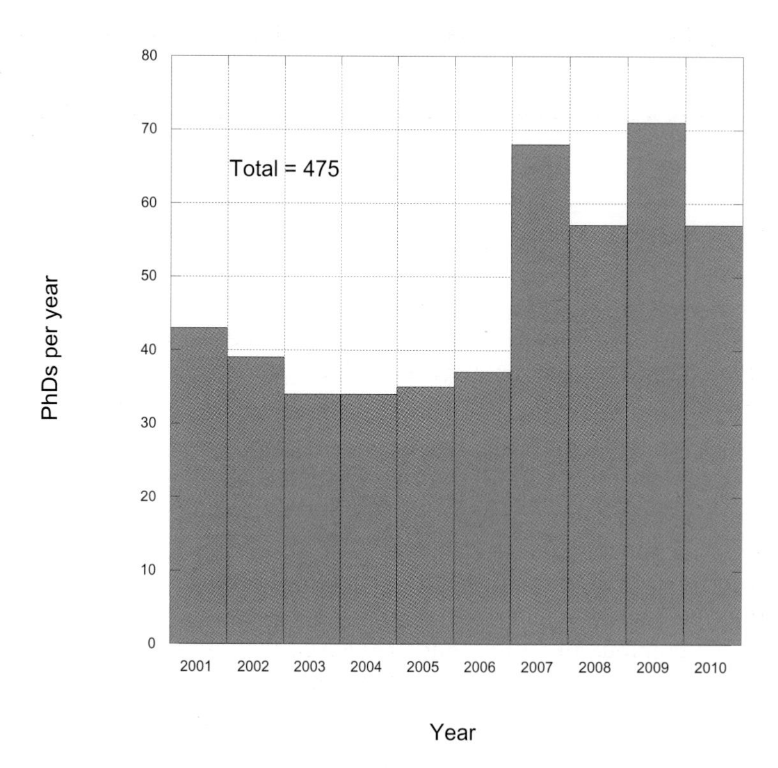

FIGURE D.1 U.S. and Canadian Ph.D.s in solar and space physics produced in the past decade. SOURCE: Courtesy of Mark Moldwin, Department of Atmospheric, Oceanic and Space Sciences, University of Michigan, for the Education and Workforce Working Group of the Decadal Strategy for Solar and Space Physics (Heliophysics).

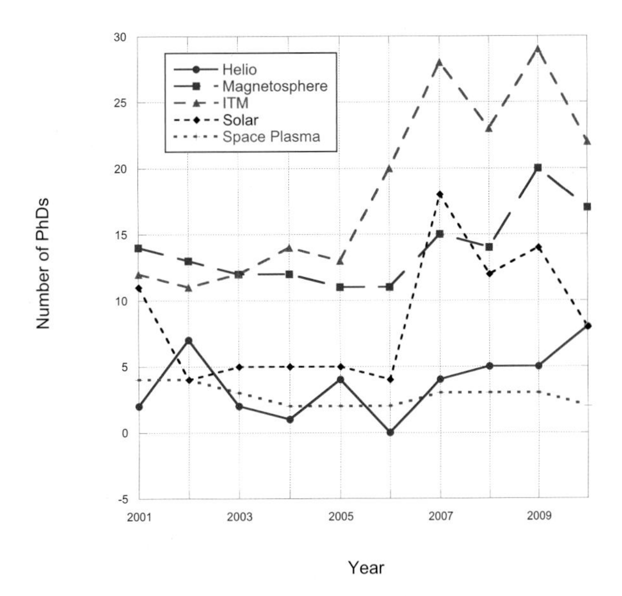

FIGURE D.2 Ph.D. production as a function of year in each of the five subdisciplines of solar and space physics. SOURCE: Courtesy of Mark Moldwin, Department of Atmospheric, Oceanic and Space Sciences, University of Michigan, for the Education and Workforce Working Group of the Decadal Strategy for Solar and Space Physics (Heliophysics).

EMPLOYMENT OPPORTUNITIES STUDY

To assess the state of health of the field of solar and space physics, an analysis of the number of job postings produced each year was done for the decade 2001-2010. A compilation of the job advertisements listed in the AAS's SPD and the AGU's SPA electronic newsletters was done for the period 2001 to 2010. The positions were sorted into four types (faculty, postdoctoral researcher, scientist, and staff), institution type (academic, government laboratory, or industry), and whether the position was located inside or outside the United States. The scientist position includes any non-faculty or postdoctoral position that required a Ph.D. and included civil service and soft-money research positions. Support staff positions, such as an observatory telescope operator or computer programmer, often did not require a Ph.D. One finding from the job advertisement survey was that there was very little (about 20 percent) overlap in the job postings (i.e., jobs advertised in the SPD newsletter were generally not advertised in the SPA newsletter, and vice versa), indicating that the communities are distinct. However, this was not true for faculty advertisements. Most faculty positions were advertised in both the SPD and SPA newsletters. Worldwide, 949 solar and space physics positions were advertised over the decade, with 52 percent of the jobs located outside the United States.

Figure D.3 shows unique positions advertised for the United States (SPD + SPA) over the decade by job type (faculty, scientist, and postdoctoral researcher). Jobs advertised in both SPA and SPD were not double counted. A disconcerting trend is a decline in job advertisements in 2010 for all types of positions. The total number advertised for each type in 2010 was the lowest number in the decade. For both com-

TABLE D.2 Top Solar and Space Physics Ph.D.-Producing Departments in the United States and Canada

Institution	Number of Total Ph.D.s, 2001-2010
University of Colorado, Boulder	37
University of California, Los Angeles	34
University of Michigan	31
Stanford University	27
Rice University	25
Utah State University	22
Boston University	18
University of New Hampshire	16
University of Washington	14
Dartmouth College	14
Cornell University	14
New Jersey Institute of Technology	13
University of Texas, Dallas	11
University of California, Berkeley	11
University of Maryland	10
University of Alaska, Fairbanks	10

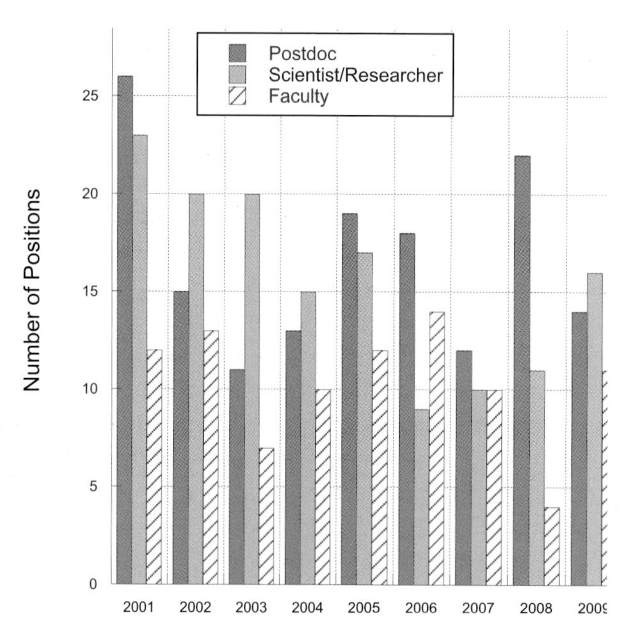

FIGURE D.3 Number of advertised positions in solar and space physics. SOURCE: Courtesy of Mark Moldwin, Department of Atmospheric, Oceanic and Space Sciences, University of Michigan, for the Education and Workforce Working Group of the Decadal Strategy for Solar and Space Physics (Heliophysics).

munities within the United States, the total job advertisements reached their lowest levels in the decade (14), approximately half the decadal average number of job advertisements.

Figure D.4 shows the number of U.S. faculty jobs advertised in SPD and SPA and the total number of non-duplicative faculty advertisements. Note that most, if not all, faculty jobs advertised in SPD were advertised in SPA. Significant year-to-year variation is seen. Figure D.5 shows the total number of unique faculty jobs advertised worldwide in solar and space physics. The average for the decade is about 18 per year.

Figures D.6 and D.7 are similar to Figures D.4 and D.5, except for postdoctoral positions. Note that there are more unique field-specific postdoctoral positions that are advertised only in SPA or SPD, compared to faculty positions. Note the negative trend of postdoctoral positions as well as the imbalance since 2008 in the total number of worldwide postdoctoral position advertisements and the total number of U.S. and Canadian Ph.D.s produced (shown in Figure D.1). Also note that the majority of postdoctoral position advertisements are from non-U.S. institutions. The number of postdoctoral positions advertised in any given year does not count the large number of postdoctoral positions that are filled in the community without advertising. Figure D.8 shows the worldwide total of scientist/researcher positions (consisting of civil service, leadership, and soft-money positions). Again, the 2010 numbers are significantly less than the average during the decade.

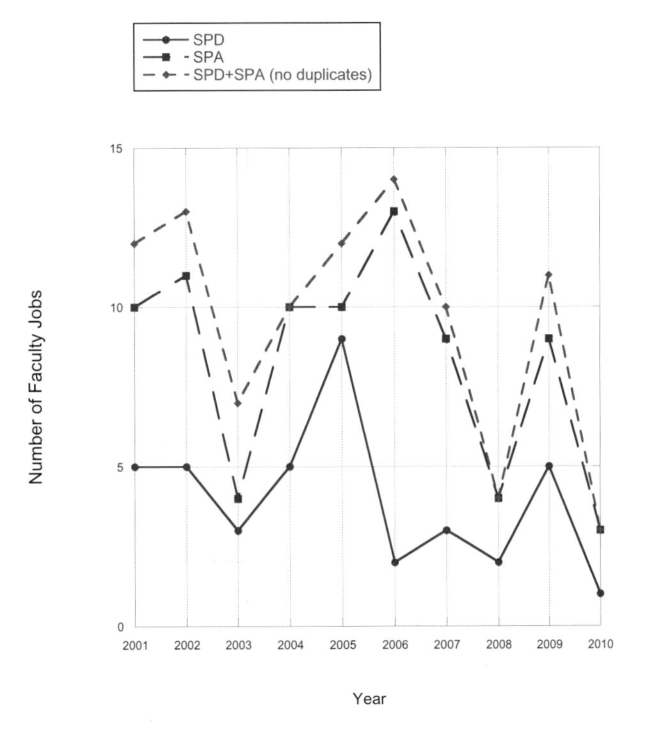

FIGURE D.4 Number of advertised U.S. faculty positions in solar and space physics. SOURCE: Courtesy of Mark Moldwin, Department of Atmospheric, Oceanic and Space Sciences, University of Michigan, for the Education and Workforce Working Group of the Decadal Strategy for Solar and Space Physics (Heliophysics).

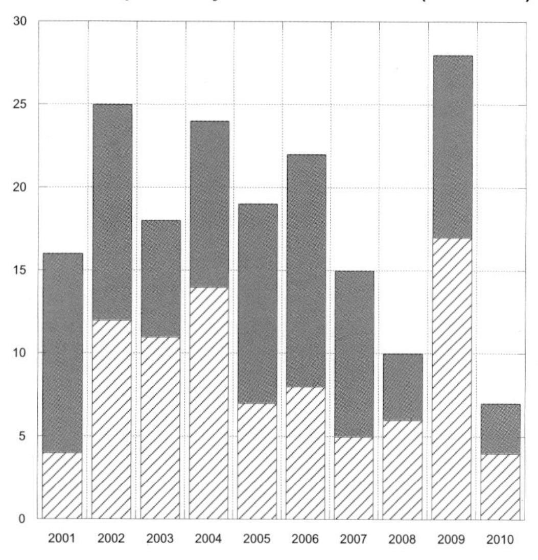

FIGURE D.5 Number of advertised faculty positions in solar and space physics, worldwide. SOURCE: Courtesy of Mark Moldwin, Department of Atmospheric, Oceanic and Space Sciences, University of Michigan, for the Education and Workforce Working Group of the Decadal Strategy for Solar and Space Physics (Heliophysics).

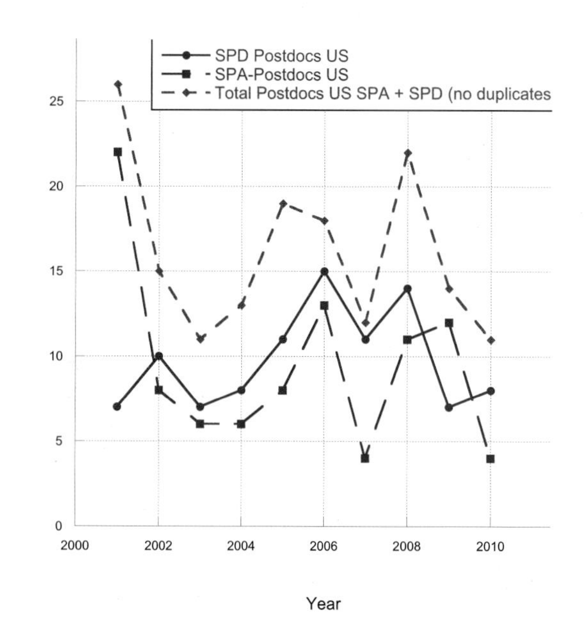

FIGURE D.6 Number of advertised U.S. postdoctoral positions in solar and space physics. SOURCE: Courtesy of Mark Moldwin, Department of Atmospheric, Oceanic and Space Sciences, University of Michigan, for the Education and Workforce Working Group of the Decadal Strategy for Solar and Space Physics (Heliophysics).

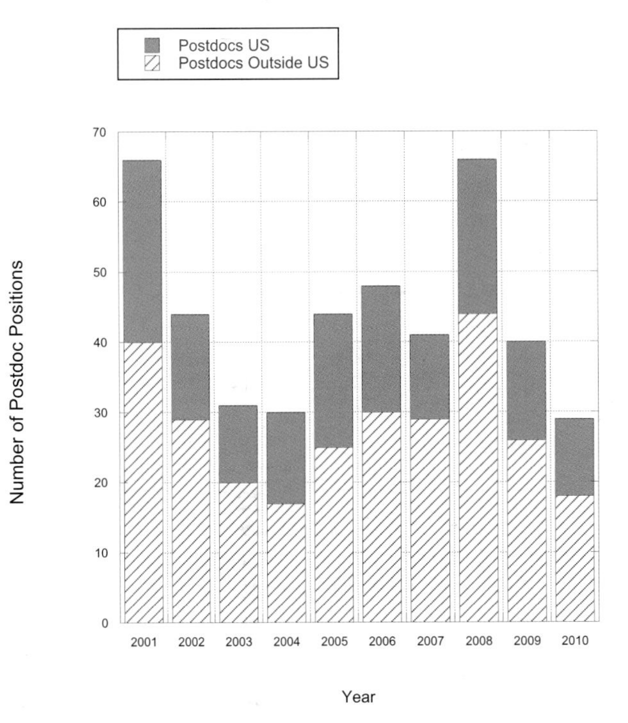

FIGURE D.7 Number of advertised postdoctoral positions in solar and space physics, worldwide. SOURCE: Courtesy of Mark Moldwin, Department of Atmospheric, Oceanic and Space Sciences, University of Michigan, for the Education and Workforce Working Group of the Decadal Strategy for Solar and Space Physics (Heliophysics).

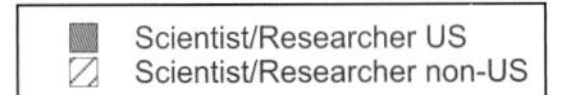

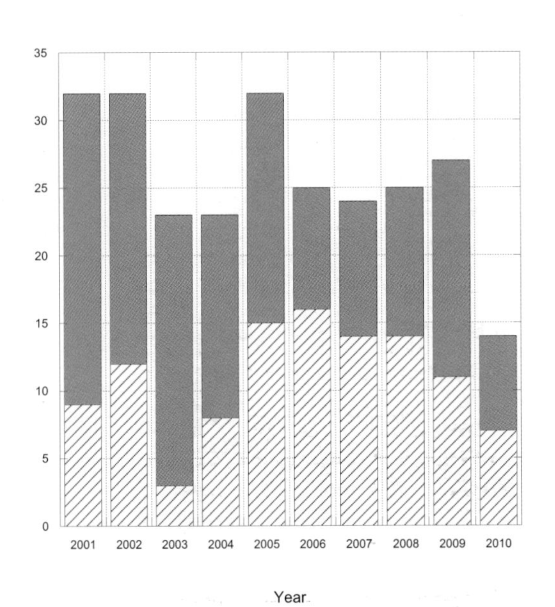

FIGURE D.8 Number of advertised scientist positions in solar and space physics, worldwide. SOURCE: Courtesy of Mark Moldwin, Department of Atmospheric, Oceanic and Space Sciences, University of Michigan, for the Education and Workforce Working Group of the Decadal Strategy for Solar and Space Physics (Heliophysics).

BIBLIOMETRIC SURVEY

A bibliometric survey of the fields of solar and space physics was conducted to provide a measure of the health of the field. Using the Thomsen-ISI (Institute for Scientific Information) Web of Science bibliometric database, the number of publications in the three broad areas of solar and space physics (solar and heliospheric physics, magnetospheric physics, and upper atmosphere and ionospheric physics) were examined for the period 2001 to 2009. The results shown in Figure D.9 indicate that the fields have experienced significant growth over the decade, with productivity increasing overall in the latter half of the decade. The share of U.S. publications relative to the rest of the world over this time interval has remained constant, with about 50 percent of the total papers published by U.S. investigators. The subdiscipline with the most variability year-to-year is magnetospheric physics, and it shows a 7 percent decline from 2008 to 2009.

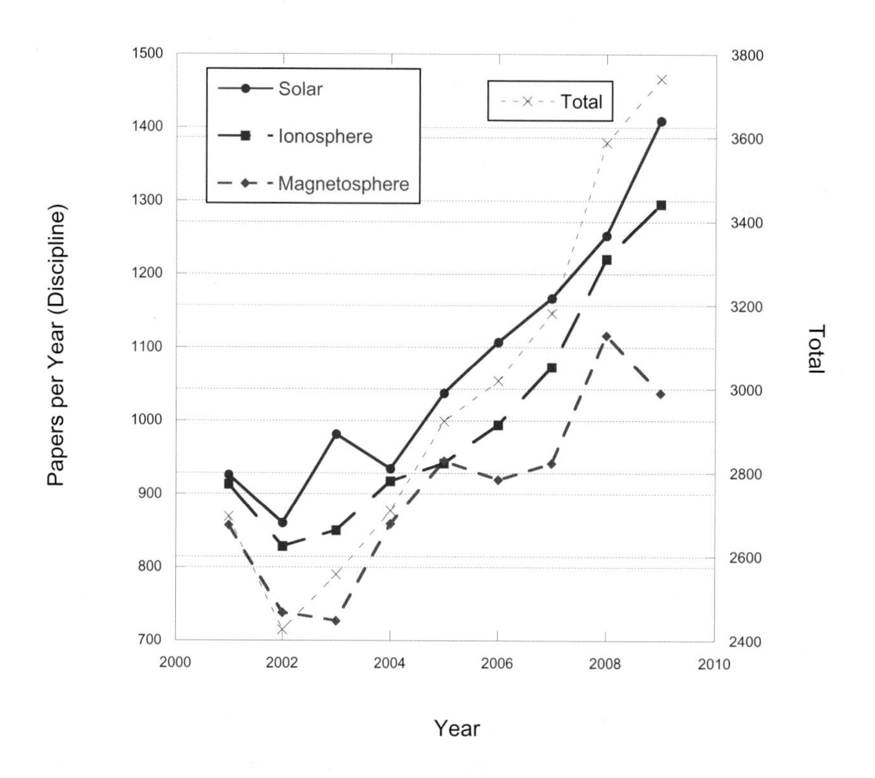

FIGURE D.9 Number of papers published in solar and space physics, worldwide. SOURCE: Courtesy of Mark Moldwin, Department of Atmospheric, Oceanic and Space Sciences, University of Michigan, for the Education and Workforce Working Group of the Decadal Strategy for Solar and Space Physics (Heliophysics).

E

Mission Development and Assessment Process

INTRODUCTION

This report follows the astronomy and astrophysics and the planetary science decadal survey reports as the third survey employing the cost and technical evaluation (CATE) process implemented by the Aerospace Corporation. CATE is an independent analytical approach for realistically assessing the expected cost and risk related to recommended initiatives that are at an early stage of formulation (typically pre-Phase-A). A unique attribute of the process is the ability to employ technical methods in combination with statistical methods to develop mission concepts based on their assessed complexity, risk, and lien factors. An essential driver to the CATE process is the requirement that all missions be treated using systematic and consistent methods in order to facilitate a uniform comparison of the relative cost, schedule, and risk of projects proposed for implementation. Secondary, but also essential, to the success of the process is the incorporation of iterative technical feedback from stakeholders, combined with a proven method for independent validation of the results. The successful implementation of these elements as an integrated process was fundamental to achieving confidence in the CATE process and consequent confidence in the assessment results, implementation assumptions, and decision rules applied to the recommended program.

The CATE process is the third leg of a science investigation triad following science mission implementation recommendations and an intermediate stage consisting of mission conceptual design activities. The flow of activities began with white paper submissions by the science community in November 2010, followed by concept evaluation and a mission design effort culminating with initial mission selection in April 2011. The subsequent iterative CATE process began in April and ended in September 2011.

As part of the CATE process, mission descoping options were introduced in recognition of the budget pressures NASA is likely to be facing over the next several years. A unique strength of the heliophysics discipline is the long experience of its investigators in doing exceptional science in a principal-investigator (PI)-led mode using small and medium Explorers. A key finding of the CATE process used in this study was that all of the proposed concepts were technically feasible at a reasonable cost. Furthermore, the descoped missions were of a scope, complexity, and cost that are consistent with successful mid-size PI-led missions executed for NASA's Discovery program.

MISSION DESIGN ACTIVITY

The initial science assessment and prioritization effort yielded 12 mission concepts, with 4 concepts put forward by each of the three science discipline panels. Each of the concepts were then organized and systematized using a standard design questionnaire intended to provide a relatively uniform set of design parameters as an input to the mission design process. The 12 mission designs were then produced over a 5-week period in February and March of 2011 at the Concept Design Center (CDC) of the Aerospace Corporation in El Segundo, California. The CDC team consisted of experts who were "firewalled" from the CATE process. Each mission was supported by a group of both shared and dedicated CDC team members, one or more members of the survey committee, and a "champion" designated by the science panels to represent each investigation concept.

The cornerstone assumption for the mission design process was that cost-effective mission concepts demand a requirements-focused design combined with overall system simplicity. The general system design baseline resulting from these assumptions was a single-string (with selective redundancy in some cases) architecture using commercially available spacecraft bus designs similar in size to a medium Explorer. The bus structure and specific capabilities such as power and pointing were sized to accommodate the required science instrumentation. Mission-unique features were also incorporated to meet requirements for orbit and downlink, radiation hardness, magnetic cleanliness, deployable subsystems, dispensing functions, and de-orbit capability (if determined to be necessary). To constrain the system design and control cost, the system architecture was also scaled such that all missions were both sized and intended for orbits so as to be compatible with a launch on a medium-size launch vehicle or smaller. The largest launch vehicle required for any mission concept was the Falcon 9.

The design process was intense but also effective at developing a realistic set of concepts consistent with the above requirements and constraints. The effort was well supported by all stakeholders in the process, with detailed standardized reports prepared for each of the resulting system designs. The quality and consistent nature of the reports allowed for a straightforward comparison of missions leading into the subsequent down-select process. The reports were also designed to be compatible with the input parameters required by the CATE process.

Following distribution of the mission design reports, the science panels performed an initial review and made recommendations to the committee for the CATE down-select from 12 to 6 missions. The recommendations were based on science objectives, complexity factors, and relative cost assumptions. The following recommended concepts were presented by the science panels, endorsed by the committee, and moved forward into the first stage of the CATE process:

1. Dynamical Neutral Atmosphere-Ionosphere Coupling (DYNAMIC) mission,
2. Interstellar Mapping and Acceleration Probe (IMAP),
3. Solar Eruptive Events (SEE),
4. Magnetosphere-Ionosphere Source Term Energetics (MISTE),
5. Magnetosphere Energetics, Dynamics, and Ionospheric Coupling (MEDICI) mission, and
6. Geospace Dynamics Constellation (GDC).

CATE PROCESS

The CATE process has been discussed at length in previous reports and is based on the following principles, which have remained consistent between recent past surveys and this survey. The CATE flow chart provided as Figure E.1 diagrams the end-to-end effort as described by these principles and executed for this report.

- Use multiple methods and databases regarding past space systems so that no one model or database biases the results.
- Use analogy-based estimating that anchors cost and schedule estimates to NASA systems that have already been built with a known cost and schedule.
- Use system-level estimates as well as a build-up to system-level estimates by appropriately summing subsystem data so as not to underestimate system cost and complexity.
- Use cross-checking tools, such as Complexity Based Risk Assessment (CoBRA), to cross-check cost and schedule estimates for internal consistency and risk assessment.
- In an integrated fashion, quantify the total threats to costs from schedule growth, the costs of maturing technology, and the threat to costs owing to mass growth resulting in the need for a larger, more costly launch vehicle.

The general CATE methodology for cost assessment is presented in Figure E.2. As with most cost assessment approaches, the CATE process uses mission analogies, growth metrics, and other relevant criteria applied to each work breakdown structure element that are derived from measured cost and schedule performance on past NASA (and selected Department of Defense) missions. Probabilistic methods are then employed using a triangle distribution to calculate the reserves needed to move the sum of the most

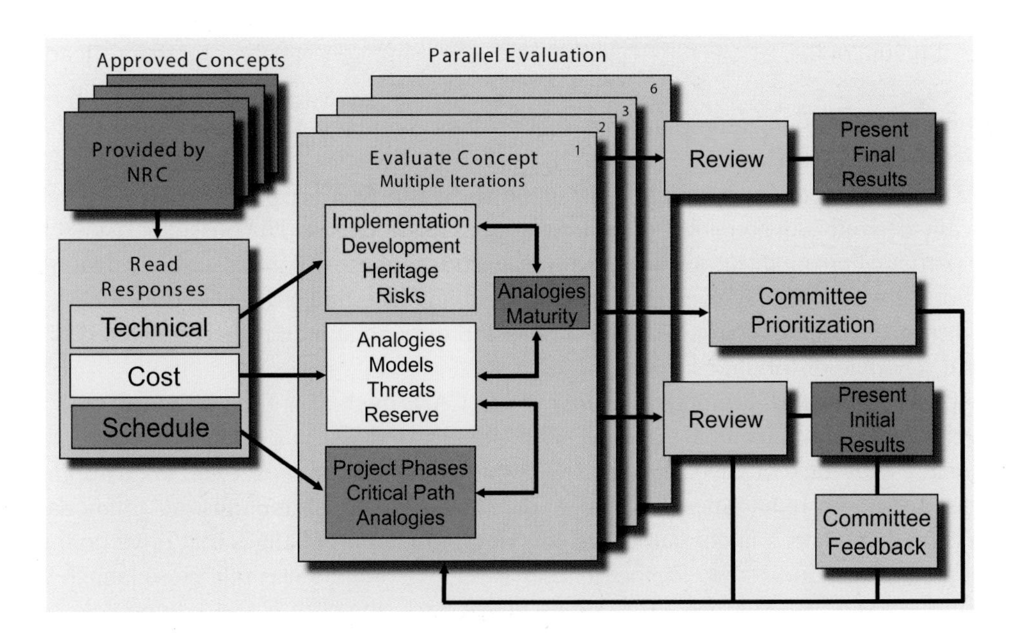

FIGURE E.1 Flow chart for the CATE process. SOURCE: Provided by the Aerospace Corporation under contract with the National Research Council.

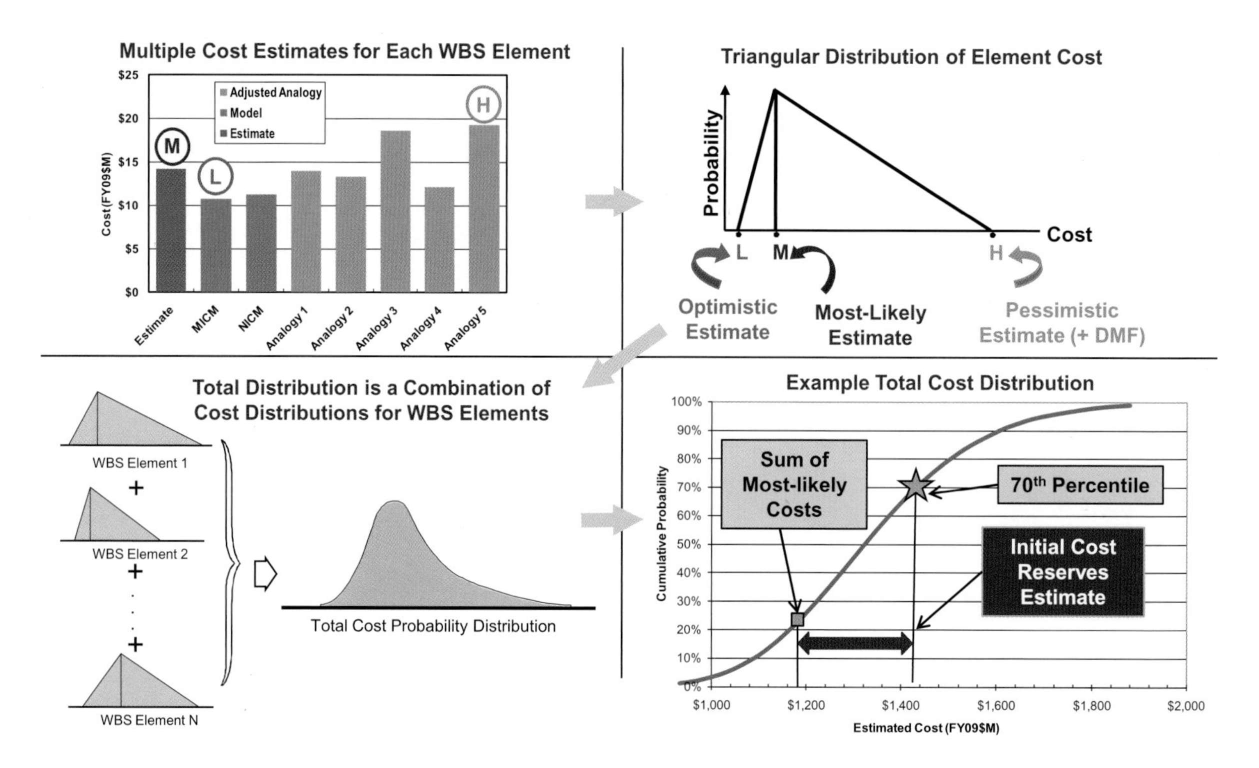

FIGURE E.2 CATE cost estimating methodology. SOURCE: Provided by the Aerospace Corporation under contract with the National Research Council.

likely cases up to a 70th percentile point[1] on the resulting S-curve describing the statistical distribution for mission cost.

The CATE methodology for schedule assessment shown in Figure E.3 is also analogy based, with a probabilistically derived 70th percentile output. Analogous missions based on mission class, technical similarities, and participating organizations are used with historical phase durations for key mission milestones. These durations are used to develop triangular distributions of estimated phase duration. A cumulative probability distribution of the total schedule from Phase-B through launch is then generated by combining the individual schedule distributions.

As previously mentioned, a unique attribute of the CATE method is the ability to "grow forward" a program from a very early conceptual stage, where the threats and liens are often much larger than for missions that are at a later stage of development. The dominant factors are most often the natural growth that occurs in mass and power requirements on a mission as its system design matures. Accommodating growth in these areas often requires a larger launch vehicle or other mission changes that drive up the final cost.

A major area of cost uncertainty for the heliophysics CATE process was related to launch vehicle cost and availability. Launch vehicles are furnished by NASA and subject to NASA competitive procurement practices. Consequently, specific vehicle costs are unknown or could not be provided except in a broad cost range enveloping numerous vehicle types. As discussed above, smaller vehicles, beginning with the

[1] The 70th percentile point represents the total cost (including reserves) within which there is a 70 percent likelihood that the mission can be accomplished.

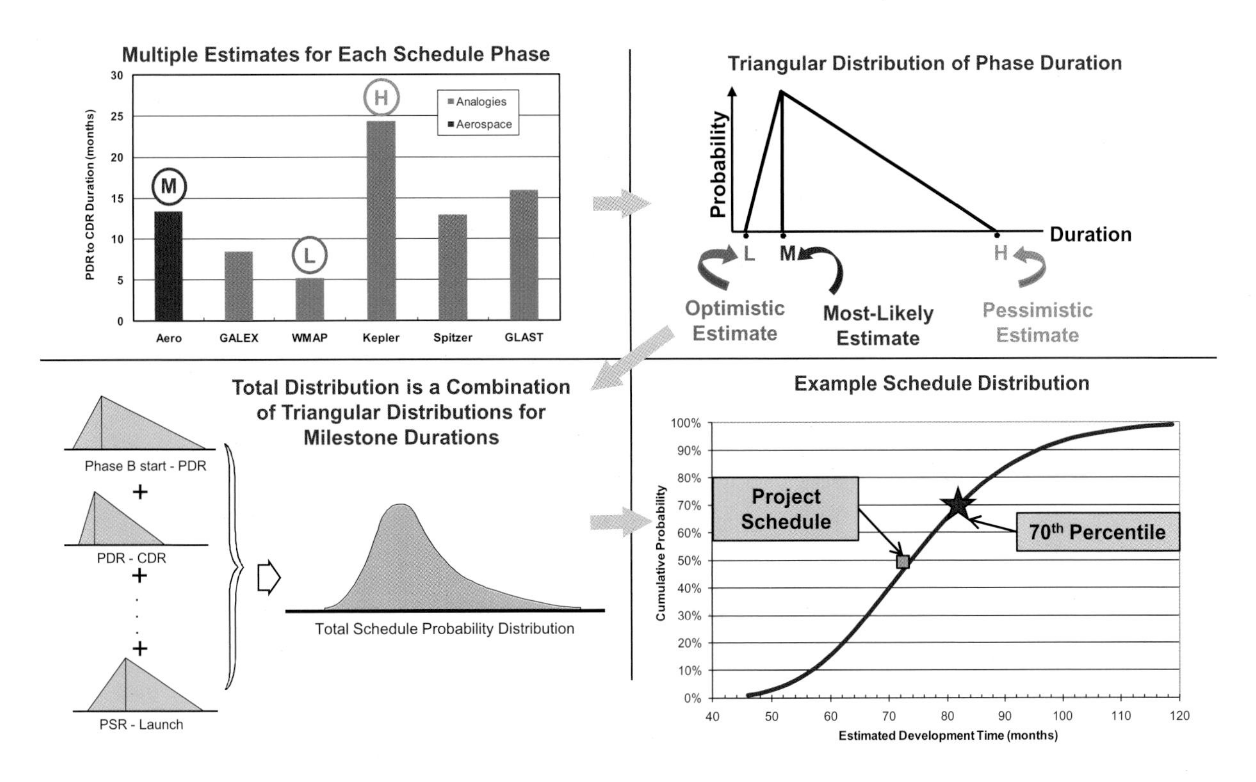

FIGURE E.3 CATE schedule estimating methodology. SOURCE: Provided by the Aerospace Corporation under contract with the National Research Council.

Taurus-XL 2XXX, Taurus-XL 3XXX, Taurus-II, and Falcon 9, were baselined and are capable of meeting all of the mission design cases that were studied. The Falcon 1 and Athena were treated as variants of the Taurus-class vehicle and are considered cost-neutral substitutes for the majority of the applications. The Delta-II was not considered to be available at the time of the mission design activity but is also a viable substitute with respect to launch capability, although at an unknown cost. It is also unclear whether the Delta-II will retain launch capabilities at the Eastern Test Range launch facility.

All vehicles (or comparable vehicles at comparable cost) were considered to be available later in the decade when the first missions would approach a state of launch readiness. The CATE process assumed the launch vehicle cost to be a pass-through based on an analysis using multiple publicly available sources. The cost range used in the process started at $60 million for the Taurus-XL 2XXX class and extended up to $125 million for the Falcon 9 in 2012 dollars. For the vehicles baselined in the design studies, the potential for cost growth and associated cost risk is considered low, although there is some risk related to availability of a particular vehicle type.

MISSION COST RESULTS

The cost comparison between the six mission concepts and associated descoped missions is shown in Figure E.4. For each mission there is a base cost resulting from the probabilistic cost analysis plus the analyzed cost for mission-specific threats and liens. The corresponding schedule results are shown in Figure E.5.

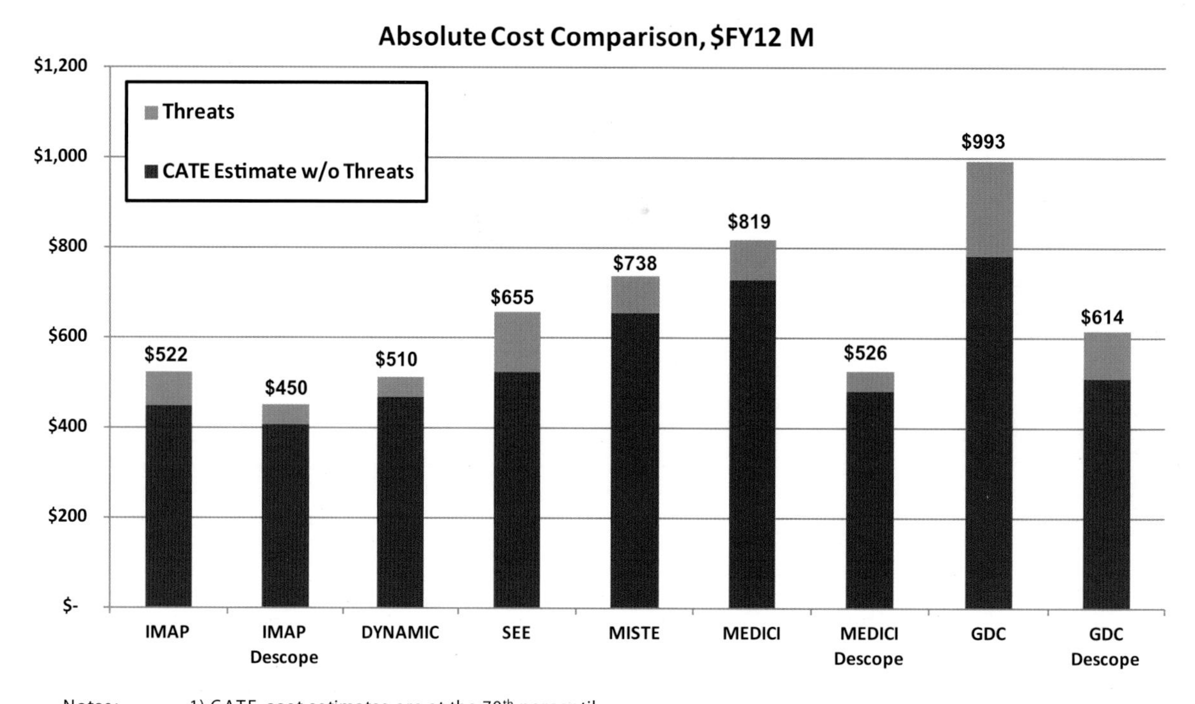

FIGURE E.4 CATE 70th percentile cost results for missions, including descoped alternatives. Base cost plus threats and liens shown (in fiscal year (FY) 2012 $ million). SOURCE: Provided by the Aerospace Corporation under contract with the National Research Council.

Specific to this survey is the recognition that several large missions are already through formulation or in the late stages of formulation such that they will become operational in stages that envelop most of the coming decade. Given the evident funding challenge early in this decade, a frequent mission cadence via the implementation of small- and medium-class missions over the entire decade is considered more valuable to the community than the implementation of, at most, one additional large mission.

This "simplified mission strategy" biased toward targeted executable investigations becomes clearer when examining Figure E.4 and Figure E.6, which represent the CATE assessment of cost and technical risk, respectively, for each of the six baseline missions plus the three descope options. It can be seen that the system approaches chosen for the missions have been successful at limiting the technical risk to the medium-low and medium categories, with a resulting payback in both affordability and cadence. The risk rankings are particularly illuminating when compared to previous surveys, in which most missions were ranked in the medium-high and high categories.

A combined assessment of cost and technical risk leads to the conclusion that simplified mission concepts similar to IMAP, DYNAMIC, and the descoped MEDICI recommended by the survey committee fall into the range of PI-led missions that can be successfully executed within the scope of the Solar-Terrestrial Probes (STP) mission line. These missions, along with GDC as the recommended Living With a Star (LWS) mission, have their key parameters summarized in Boxes E.1 through E.4 (provided by the Aerospace Corporation under contract with the National Research Council).

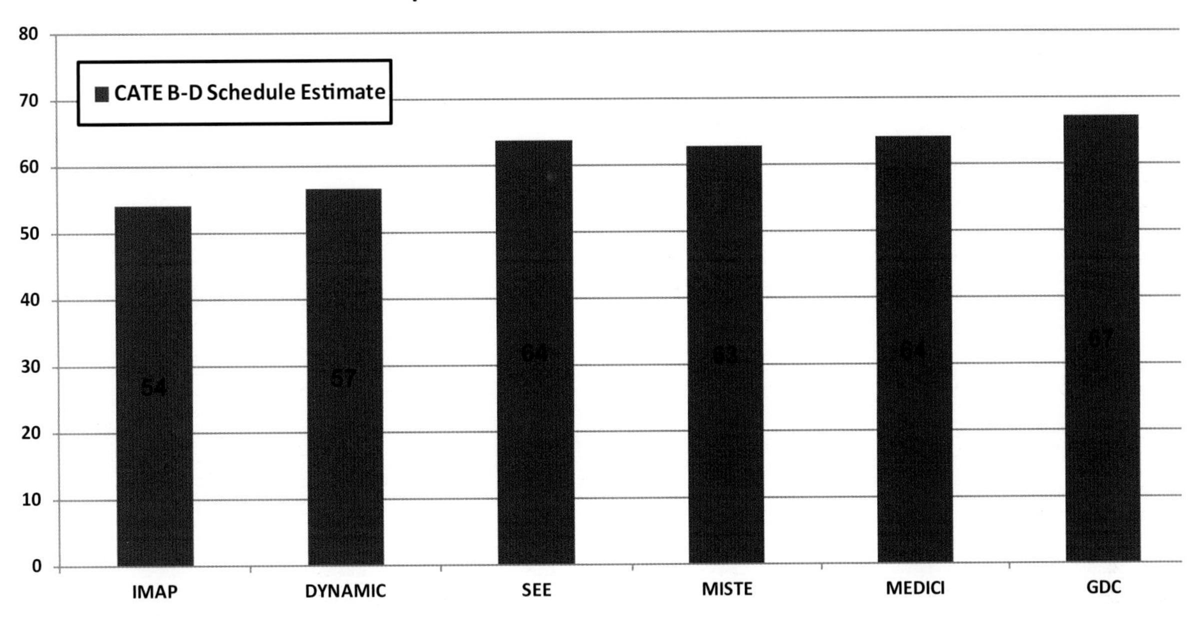

Note: GDC descope reduces CATE B-D by 2 months. IMAP and MEDICI unchanged by descopes.

FIGURE E.5 CATE 70th percentile Phase-B through Phase-D schedule results. GDC descope reduces schedule by 2 months. IMAP and MEDICI descopes do not affect schedule. SOURCE: Provided by the Aerospace Corporation under contract with the National Research Council.

VALIDATION APPROACH

The final step of the CATE process consisted of a series of validation reviews intended to ensure that the assumptions and corresponding cost and schedule results are based on defensible assumptions, are fair in their representation of the assessed missions, and are consistent with past missions of similar content and complexity. The reviews consisted of a mix of both internal and external reviews conducted by the Aerospace Corporation and by the committee, respectively. The CATE validation tool used for this survey and all past surveys is the CoBRA tool developed by the Aerospace Corporation. Figures E.7 and E.8 map the mission candidates onto cost and schedule grids for comparison with successful missions of varying complexity plus "anchor missions" considered to be similar in size and complexity.

The results show excellent correlation between CATE results for mission concepts compared to successful missions and anchor missions. Therefore, the CATE results are considered compatible with experience, with respect to both cost and schedule. As one would expect, the CATE missions representing a 70th-percentile distribution are generally above the mean. One interesting result, discussed in the next section, is that the heliospheric anchor missions are generally below the mean trend line.

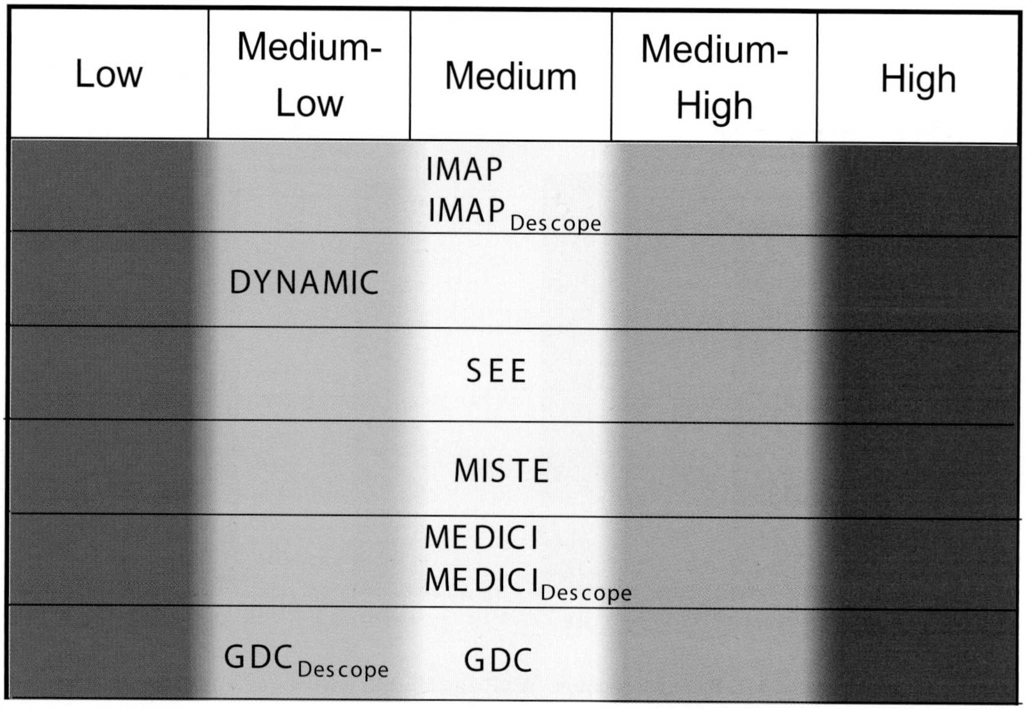

Low	Medium-Low	Medium	Medium-High	High
		IMAP $IMAP_{Descope}$		
	DYNAMIC			
		SEE		
		MISTE		
		MEDICI $MEDICI_{Descope}$		
	$GDC_{Descope}$	GDC		

FIGURE E.6 CATE technical risk estimate. This estimate is provided to represent the level of technical risk resulting from the CATE assessment. Note that in the case of the IMAP and MEDICI investigations the descope resulted in a cost reduction but no change in the technical risk posture. An advantage of using a standardized CATE process is that the metrics for evaluating technical risk are based on technical criteria that are consistent across programs within a single survey as well as across surveys from other disciplines. Note: Red indicates that the concept is likely to be technically infeasible as proposed and/or that required technology development prevents achievement of science/mission objectives. Blue indicates that the concept has adequate margin and is consistent with a heritage heliophysics mission. SOURCE: Provided by the Aerospace Corporation under contract with the National Research Council.

PI-MODE IMPLEMENTATION

As described in the 2006 National Research Council report *Principal-Investigator-Led Missions in the Space Sciences*,[2] NASA introduced the PI-led mode of implementation with the Discovery program and has since extended it to the Explorer, New Frontiers, and Mars Scout programs. Recent CoBRA analysis of 18 successful PI-led missions demonstrates that such missions with complexity indices of greater than or equal to 40 percent have costs that are typically approximately 30 percent less than non-PI-led missions having the same complexity. Figure E.9 shows the cost versus complexity for the set of successful missions from Figure E.7, with the PI-led missions indicated. The costs of only three of the PI-led missions with complexity greater than or equal to 40 percent exceeded the complexity-dependent median cost for all missions.

The distributions of individual mission costs relative to the overall median cost are shown in Figure E.10 for PI-led missions and separately for the other missions with 40 to 80 percent complexity. The two cost

[2] National Research Council, *Principal-Investigator-Led Missions in the Space Sciences*, The National Academies Press, Washington, D.C., 2006.

BOX E.1 CATE OUTPUT PRODUCT FOR THE INTERSTELLAR MAPPING AND ACCELERATION PROBE MISSION

IMAP — Interstellar Mapping and Acceleration Probe

Spinner Spacecraft

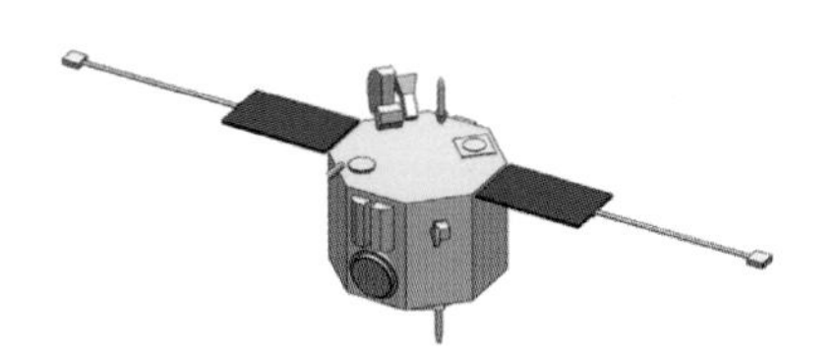

Key Challenges

- **Payload Suite**
  - *Design uncertainty based on significant differences between instrument concepts and heritage systems as well as long elapsed time since heritage missions*
- **System Mass**
  - *Low mass growth allowance*
  - *Risk of moving to larger launch vehicle*
- **System Power**
  - *Low power growth allowance*
  - *Non-rechargeable battery*
  - *Liens against power subsystem design*

Scientific Objectives

- **Observations/measurements**
  - *High-Resolution & Time Resolved Mapping of Heliosphere Boundary*
  - *Physical Properties and Composition of the Interstellar Medium and the Outer Heliosheath*
  - *Seed Populations for Energetic Particles and Their Time Variation*
- **Key science themes cited:**
  - *What is the nature of Earth's home in the solar system?*
  - *How does variability in Earth's space environment affect our lives?*
  - *What does our space environment teach us about universal, fundamental physical processes?*

CATE Cost Estimate

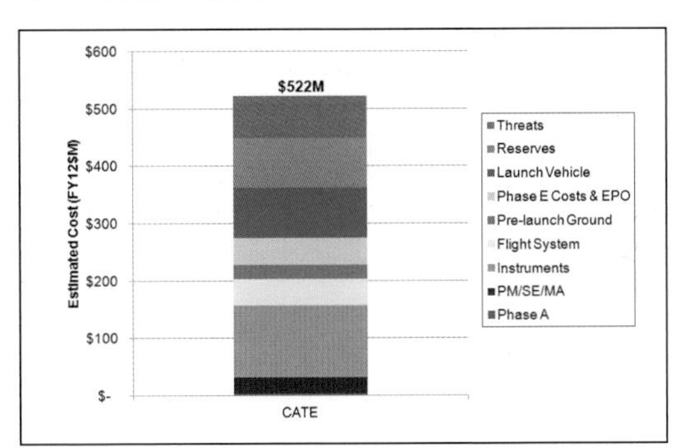

Key Parameters

- **Payload (10 instruments)**
  - *High Sensitivity, High Resolution ENAs: Solar Wind and PUI ENAs, High Energy ENAs*
  - *Samples of Interstellar Matter: Interstellar Neutrals, Pickup Ions, Interstellar Dust, Ly-α Photometry*
  - *Real Time Solar Wind*
  - *Energetic Particle Injection, Acceleration and Propagation*
- **Power: 1 m^2 GaAs array, 270 W End of Life**
- **Launch Mass: 336 kg**
- **Launch Date: 2018+ (on Taurus XL 3213 w/ 5th stage)**
- **Orbit: Sun-Earth L1 Lissajous Orbit**
- **Mission Duration: 2.5 years**

Cost Risk Analysis S-Curve

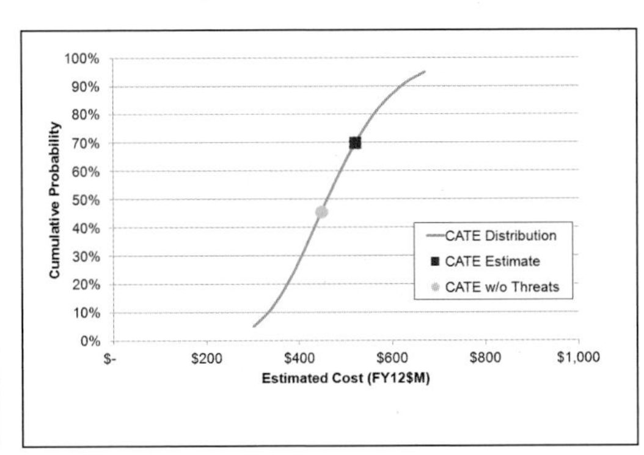

BOX E.2 CATE OUTPUT PRODUCT FOR THE DYNAMICAL NEUTRAL ATMOSPHERE-IONOSPHERE COUPLING MISSION

DYNAMIC — Dynamical Neutral Atmosphere-Ionosphere Coupling

DYNAMIC Two Spacecraft Constellation

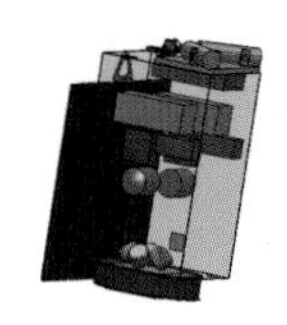

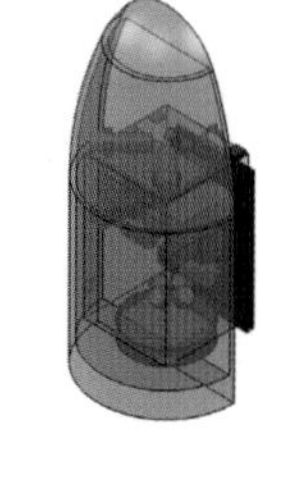

Key Challenges

- **Launch Schedule and Spacecraft Reliability**
  - *Need to closely schedule 2 launches*
  - *Low constellation reliability for 2 out of 2 satellites over full mission duration*
- **WIND Instrument Definition**
  - *Preferred design approach not yet selected and options being considered are very different*
- **System Power**
  - *Power growth allowance is low*

Scientific Objectives

- **Observations/measurements**
  - *Combines in situ and remotely sensed information from two s/c to resolve spatial and temporal structure in the upper atmosphere*
  - *Neutral and ion density, auroral inputs, vector wind fields, ionospheric irregularities*
- **Key science themes cited:**
  - *How does the lower atmosphere drive geospace?*
  - *Do we understand how this driving occurs and what the important spatial and temporal scales are?*
  - *How does behavior of the neutral lower atmosphere couple into plasma behavior in the upper atmosphere?*

CATE Cost Estimate

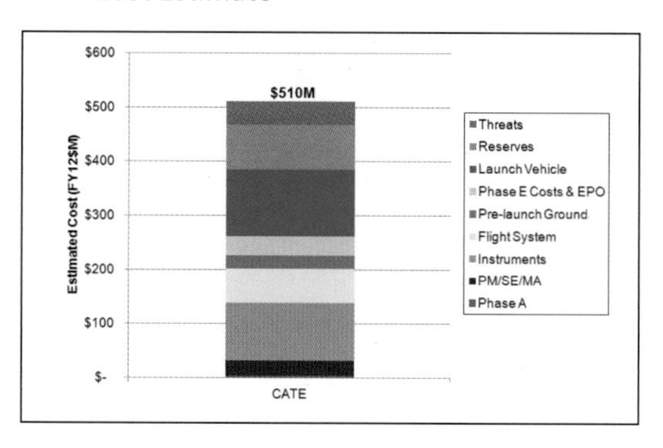

Key Parameters

- **Payload**
  - *WIND: Along track vector wind fields, density, temp*
  - *FUV: Cross track scanner*
  - *NWM: Neutral wind meter*
  - *IVM: Ion velocity meter*
  - *INMS: Neutral/ion mass spectrometer*
- **Power: 1.5 m^2 GaAs two-axis array, 416 W End of Life**
- **Launch Mass: 286 kg**
- **Launch Date: 2015+ (on two Taurus 2110 vehicles)**
- **Orbit: 600 km circular, 80 deg inclination**
- **Mission Duration: 2 years (dual spacecraft)**

Cost Risk Analysis S-Curve

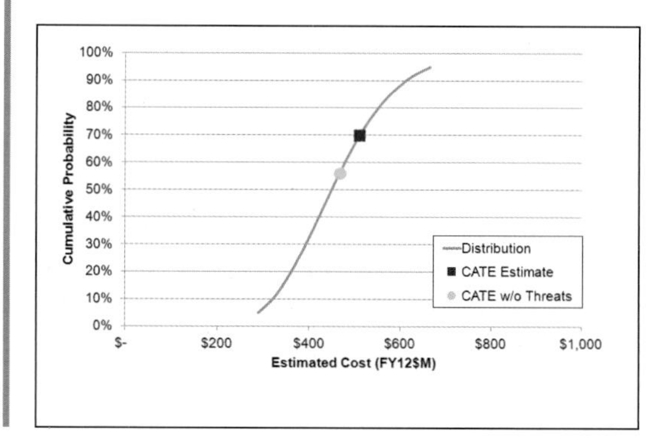

BOX E.3 CATE OUTPUT PRODUCT FOR THE DESCOPED MAGNETOSPHERE ENERGETICS, DYNAMICS, AND IONOSPHERIC COUPLING MISSION

MEDICI — Magnetosphere Energetics, Dynamics, and Ionospheric Coupling Descope

Two Spacecraft Constellation

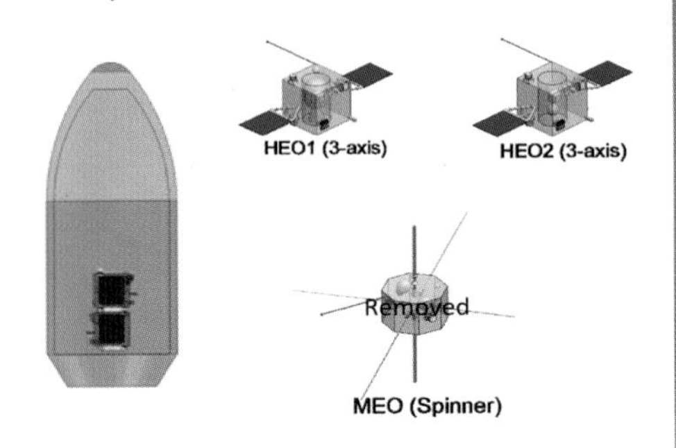

Key Challenges

- **Constellation Reliability**
  - *Loss of science objectives with less than 2 out of 2 spacecraft*
  - *Potential liens against system redundancy*
- **System Mass**
  - *Low mass growth allowance*
  - *Mass lien against HEO1 solid rocket motor*
- **System Power**
  - *Low power growth allowance*

Scientific Objectives

- **Observations/measurements**
  - *Energetic Neutral Atom magnetospheric images*
  - *EUV plasmasphere images*
  - *FUV auroral and ionospheric images*
  - *Plasma and fields in situ at high and mid altitudes*
- **Key science themes cited:**
  - *Global solar wind-magnetosphere-ionosphere coupling*
  - *Ionospheric particle outflow and acceleration*
  - *Magnetospheric energy input to ionosphere/atmosphere*

CATE Cost Estimate

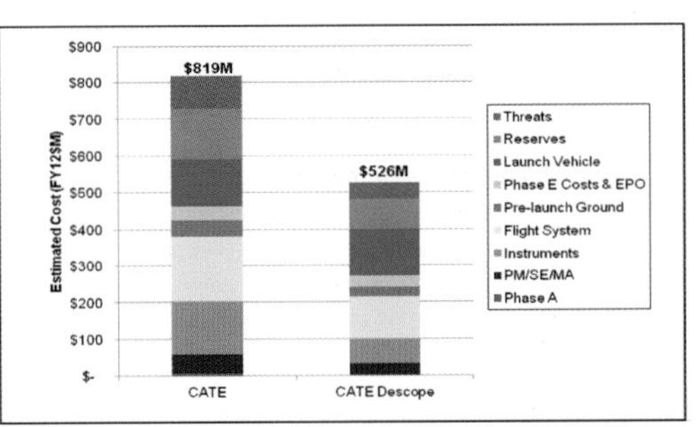

Key Parameters

- **Payload**
  - *HEO (4 or 5 instruments): ENA and EUV imaging; in situ plasma composition and magnetometer; FUV imaging on one HEO only*
  - *Removed Items: MEO Spacecraft (with 5 instruments); Spectroscopic FUV imager from both HEOs; FUV LBH imager from one HEO*
- **Power:**
  - *HEO: 2.5 m^2 GaAs array, 194 W End of Life*
- **Launch Mass: 1057 kg**
- **Launch Date: 2018+ (on Falcon 9)**
- **Orbit:**
  - *8 Re circular, 90 deg inclination*
- **Mission Duration: 2 years**

Cost Risk Analysis S-Curve

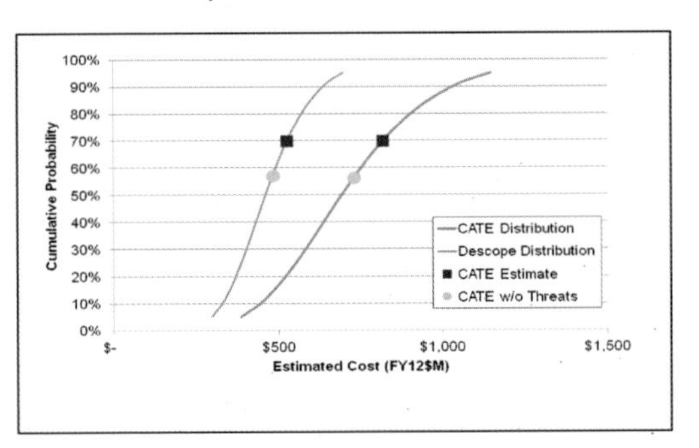

NOTE: The version of the MEDICI concept that included FUV LBH (Lyman-Birge-Hopfield) long- and short-wavelength cameras on both HEO spacecraft—a configuration strongly advocated by the SWMI panel—was not evaluated; however, the cost of adding a duplicate far-ultraviolet LBH imager to an otherwise identical spacecraft should not add significantly to the overall mission cost, which is estimated to be $526 million. The proposed launch vehicle for MEDICI is a Falcon 9, which has lift capability that easily accommodates the additional imager.

BOX E.4 CATE OUTPUT PRODUCT FOR THE GEOSPACE DYNAMICS CONSTELLATION MISSION

GDC — Geospace Dynamics Constellation

Six Spacecraft Constellation

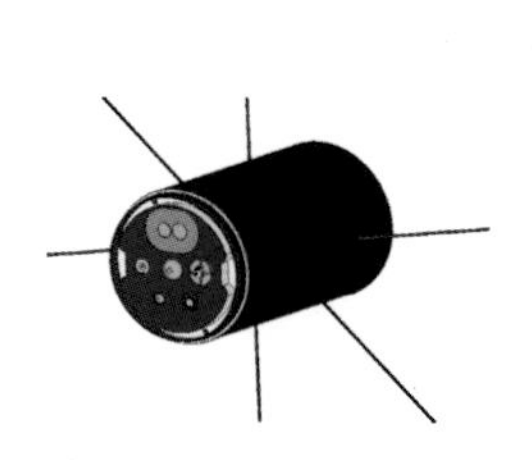

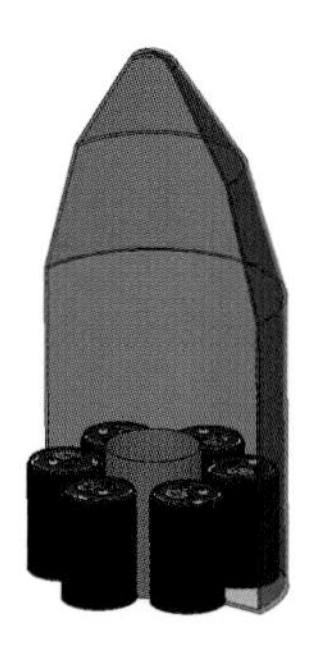

Key Challenges

- **System Reliability**
  - *Single-string spacecraft contribute to risk of constellation degradation before deployment is complete and during full capability mission*
- **Magnetic Cleanliness**
  - *Several instruments are sensitive to SC and PL generated magnetic fields*
- **System Mass**
  - *Mass growth allowance is low*
- **System Power**
  - *Power growth allowance is low*

Scientific Objectives

- **Observations/measurements**
  - *Neutral density, temperature, winds*
  - *Ion density, drifts, temperatures, rough composition*
  - *Magnetometer*
  - *Energetic electrons*
- **Key science themes cited:**
  - *Understand the global-scale electrodynamic response of the AIM system to solar wind driving*
  - *Discover how the AIM system responds globally to magnetic storms*
  - *Explore the self-consistent response of the global thermospheric winds to forcing from the lower atmosphere*

CATE Cost Estimate

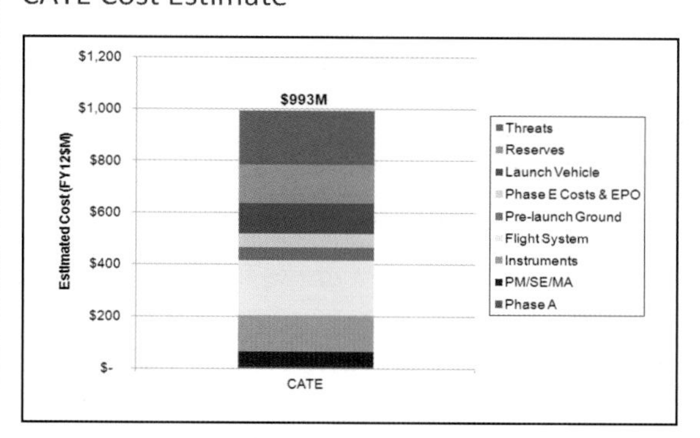

Key Parameters

- **Payload**
  - *Primary: Langmuir Probe, Ion Velocity Meter, Ionization Gauge, Neutral Wind Meter, Magnetometer, Electron & Ion Spectrometer*
  - *Secondary: Temperature & Wind Mass Spectrometer, Vector Electric & Fields Investigation*
- **Power: 6.5 m^2 GaAs body mounted array, 308 W End of Life**
- **Launch Mass: 2716 kg (including six spacecraft)**
- **Launch Date: 2018+ (on Taurus II Enhanced)**
- **Orbit: 320-450 km, 80 deg inclination**
- **Mission Duration: 1 year deployment + 2 years full operation capability**

Cost Risk Analysis S-Curve

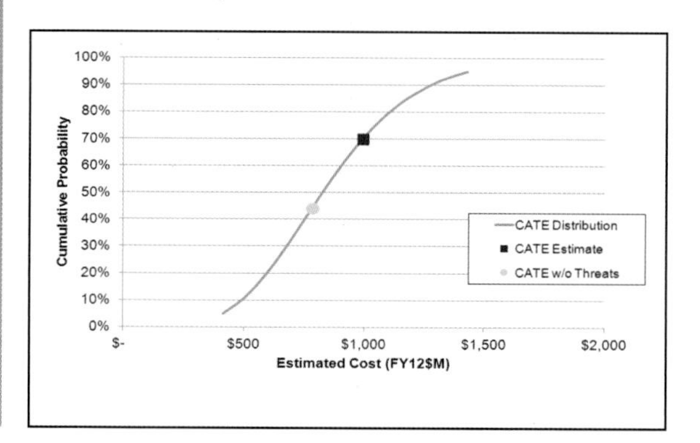

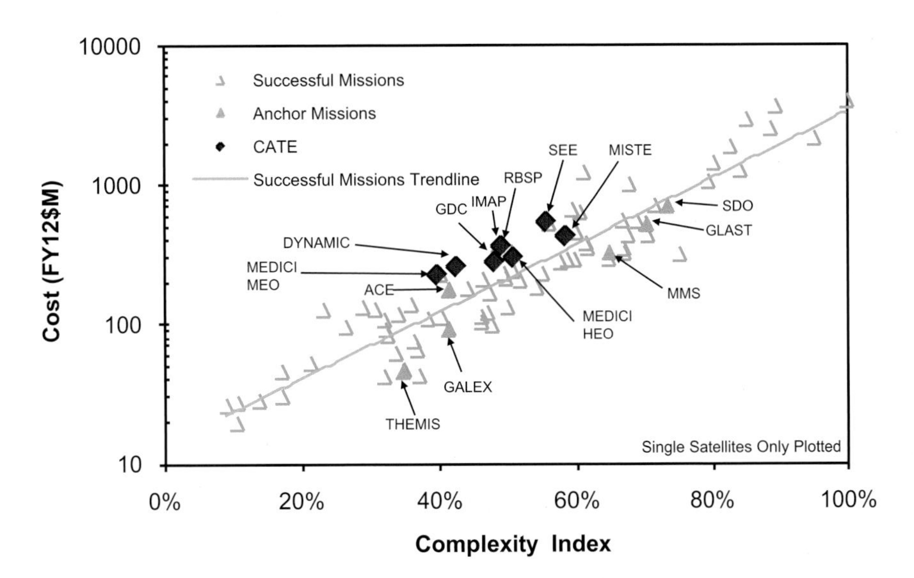

FIGURE E.7 CATE missions on Complexity Based Risk Assessment (CoBRA) cost versus complexity plot. Missions subjected to the CATE process mapped on to a standard CoBRA plot showing both successful missions and "anchor missions" used for correlation. The trendline represents the mean 50th percentile case. The CATE missions are statistically representative of a 70th percentile case. Results are for a single satellite in cases where multiple satellites are used. MEDICI is a special case since it has different satellites for medium Earth orbit and high Earth orbit. SOURCE: Provided by the Aerospace Corporation under contract with the National Research Council.

distributions are significantly different, with the median cost of PI-led missions 0.7 times that of the other missions. In addition, the cost risk of PI-led missions is also reduced, as indicated by the nearly symmetric distribution of PI-led mission costs about their median, with a small tail of three missions that exceeded the median cost of the other missions. In comparison, the cost distribution for other missions is distinctly asymmetric about its 50th percentile, with a larger fractional variance extending to higher costs.

The attributes of the PI-led mode that contributed to the success of these programs were summarized in the *Principal-Investigator-Led Missions* report and included:

- Direct involvement of the PI in shaping the decisions and the mission approach to realizing the science objectives;
- Two-step competitive selection: initial selection of two to four missions that receive funding for preparing mission concept studies prior to final selection of one or two missions for implementation;
- Standard evaluation process of technical, management, cost, and other factors prior to final selection; and
- Capped mission cost with a termination review and threat of cancellation if the mission is projected to exceed its cost cap.

The cost estimates from the CATE process do not take into account the potential cost reduction associated with PI-led mode implementation. However, the reduced cost and reduced cost risk associated with PI-led missions offer an important opportunity for optimizing science return in an era of stringent budget constraints.

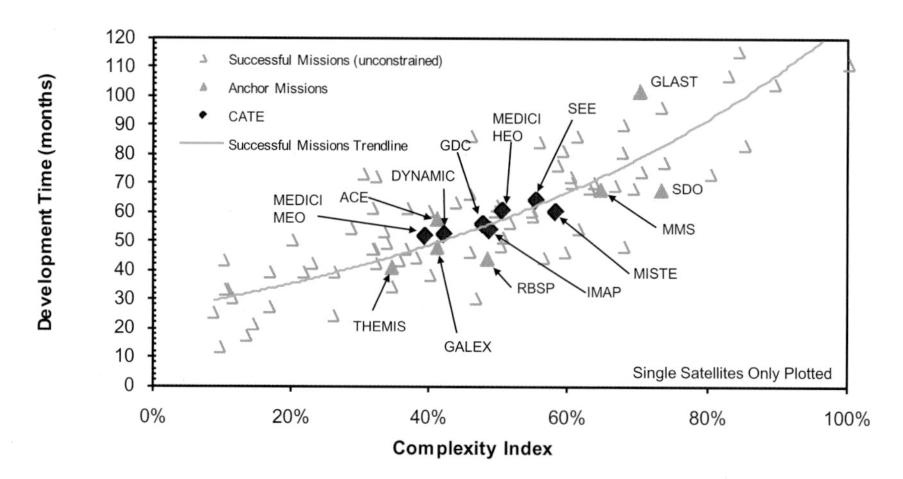

FIGURE E.8 CATE missions on Complexity Based Risk Assessment (CoBRA) schedule versus complexity plot. Missions subjected to the CATE process mapped on to a standard CoBRA plot showing both successful missions and "anchor missions" used for correlation. Results are for a single satellite in cases where multiple satellites are used. MEDICI is a special case since it has different satellites for medium Earth orbit and high Earth orbit. SOURCE: Provided by the Aerospace Corporation under contract with the National Research Council.

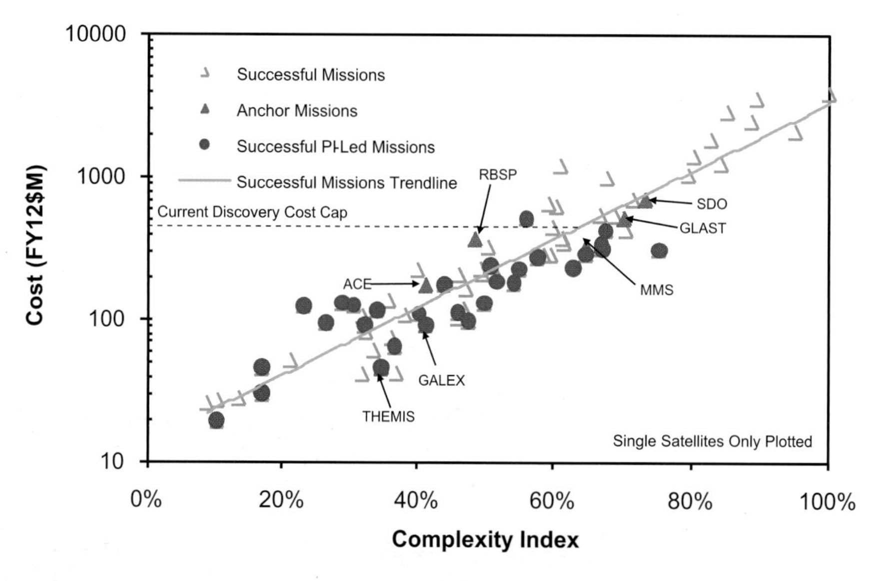

FIGURE E.9 CATE missions on Complexity Based Risk Assessment cost versus complexity plot. The green trendline is the median of the complexity-dependent costs of all successful missions. The costs of principal-investigator (PI)-led missions with >40 percent complexity are mostly below the median cost line. SOURCE: Provided by the Aerospace Corporation under contract with the National Research Council.

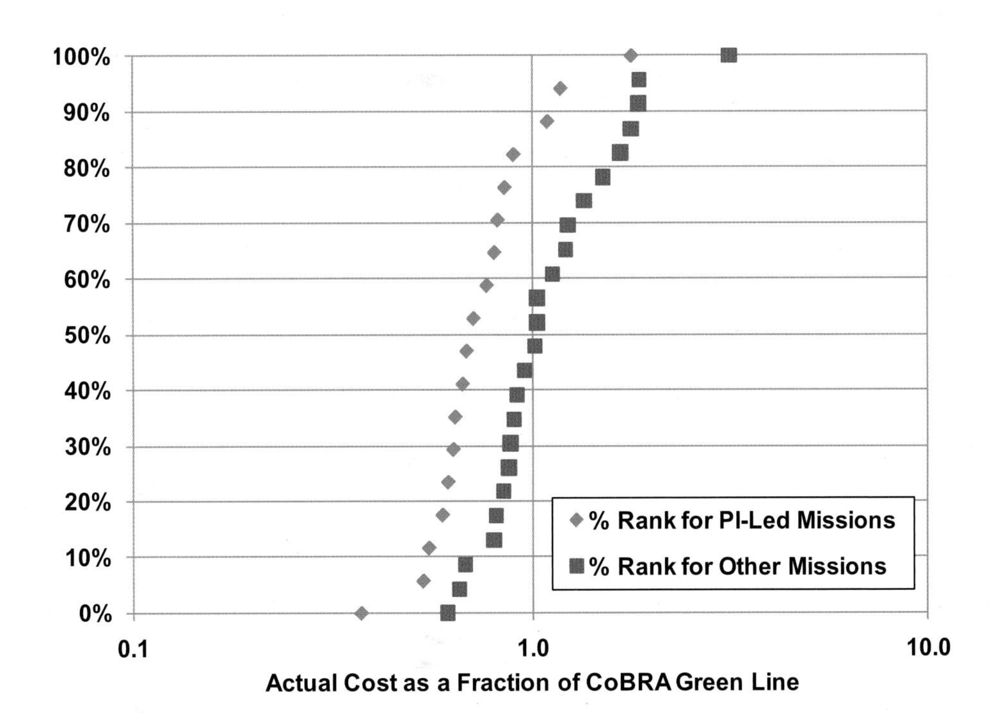

FIGURE E.10 Distribution of principal-investigator (PI)-led mission costs (blue) and non-PI-led mission costs (red) as a fraction of the median cost of all missions with 40 to 80 percent complexity (green trendline in Figure E.9). The median cost of PI-led missions is 0.7 times that of the other missions, with only 16 percent of the PI-led mission costs exceeding the median of the costs of other missions. The PI-led missions also have a smaller fractional cost variance. SOURCE: Provided by the Aerospace Corporation under contract with the National Research Council.

F

Committee, Panels, and Staff Biographical Information

COMMITTEE

DANIEL N. BAKER, *Chair*, is director of the Laboratory for Atmospheric and Space Physics (LASP) at the University of Colorado, Boulder, where he also holds appointments as a professor of astrophysical and planetary sciences and as a professor of physics. His primary research interest is the study of plasma physical and energetic particle phenomena in planetary magnetospheres and in Earth's vicinity. He conducts research in space instrument design, space physics data analysis, and magnetospheric modeling. He currently is an investigator on several NASA space missions, including the MESSENGER mission to Mercury, the Magnetospheric Multiscale mission, the Radiation Belt Storm Probes mission, and the Canadian ORBITALS mission. Dr. Baker has published more than 700 papers in the refereed literature and has edited six books on topics in space physics. In 2010, Dr. Baker was elected to the National Academy of Engineering (NAE) for leadership in studies, measurements, and predictive tools for Earth's radiation environment and its impact on U.S. security. He is a fellow of the American Geophysical Union (AGU), the International Academy of Astronautics (IAA), and the American Association for the Advancement of Science (AAAS). Among his other awards are the 2007 University of Colorado's Robert L. Stearns Award for outstanding research, service, and teaching; the 2010 American Institute of Aeronautics and Astronautics' (AIAA) James A. Van Allen Space Environments Award for excellence and leadership in space research; and his selection in 2004 as a national associate of the National Academy of Sciences (NAS). He earned his Ph.D. in physics from the University of Iowa. Dr. Baker served as president of the Space Physics and Aeronomy section of the AGU (2002-2004), and he currently serves on advisory panels of the U.S. Air Force (USAF) and the National Science Foundation (NSF). He served as chair of the National Research Council (NRC) Committee on Solar and Space Physics and the Committee on the Societal and Economic Impacts of Severe Space Weather Events Workshop and as the co-chair of the Committee on Assessment of Impediments to Interagency Cooperation on Space and Earth Science Missions. He also served as a member of the Space Studies Board (SSB), the NRC's 2003 decadal survey Committee on Solar and Space Physics: A Community Assessment and Strategy for the Future, and the 2006 Committee on an Assessment of Balance in NASA's Science Programs.

THOMAS H. ZURBUCHEN, *Vice Chair*, is a professor of space science and engineering in the Department of Atmospheric, Oceanic, and Space Sciences and the associate dean for entrepreneurship in the College of Engineering at the University of Michigan, where he is leading the Solar and Heliospheric Research Group, which focuses on solar and space physics through novel experiments, data analysis, and theoretical methods. This group has been actively involved in the Advanced Composition Explorer (ACE), WIND, Ulysses, MESSENGER, and Solar Orbiter. His research interests include instruments that measure the composition of plasmas in the heliosphere, new particle detection technologies suitable for future space missions, theoretical concepts and experimental exploration methods of interaction between the heliosphere and local interstellar medium, and developing and analyzing space mission architectures for various exploration and commercial applications. Dr. Zurbuchen is a recipient of a Presidential Early Career for Scientists and Engineers Award. He earned his Ph.D. in physics from the University of Bern, Switzerland. Dr. Zurbuchen served on the NRC Panel on the Sun and Heliospheric Physics, the Plasma Science Committee, and the Workshop Organizing Committee on Solar Systems Radiation Environment and NASA's Vision for Space Exploration. Dr. Zurbuchen served as vice chair of the Committee on Solar and Space Physics.

BRIAN J. ANDERSON is a physicist at the Johns Hopkins University (JHU), Applied Physics Laboratory (APL). Dr. Anderson has management experience with a number of missions, including serving as instrument scientist for the NEAR Magnetometer and for the MESSENGER Magnetometer, MESSENGER advance science planning lead/deputy project scientist, and SRP magnetic fields section supervisor. He has extensive experience in space magnetometry, spacecraft magnetics, and basic space plasma physics, with concentrations in pulsations, currents, wave-particle interactions, and geomagnetic storms. Dr. Anderson has conducted data analysis of AMPTE/CCE magnetic field data and data validation and was the archiving and data processing and analysis lead for UARS magnetic field data. He holds a Ph.D. in physics from the University of Minnesota. Dr. Anderson served on the NRC's Panel on Solar Wind and Magnetospheric Interactions for the 2003 decadal survey.

STEVEN J. BATTEL is president of Battel Engineering, providing engineering, development, and review services to NASA, the Department of Defense (DOD), and university and industrial clients. Prior to becoming president of Battel Engineering, he worked as an engineer, researcher, and manager at the University of Michigan, the Lockheed Palo Alto Research Laboratory, University of California (UC) Berkeley, and the University of Arizona Lunar and Planetary Laboratory. His areas of specialization include program management, systems engineering, advanced technology development, spacecraft avionics, power systems, high-voltage systems, precision electronics, and scientific instrument design. Mr. Battel was a member of the Hubble Space Telescope External Readiness Review Team for SM-2, SM3A, and SM3B, the AXAF/Chandra Independent Assessment Team, the TDRS-H/I/J Independent Review Team, the Mars Polar Lander Failure Review Board, and the Jet Propulsion Laboratory (JPL) Genesis Failure Review Board. Mr. Battel received a B.S. in electrical engineering from the University of Michigan. He has extensive NRC membership service, including on the SSB, the Committee for a Decadal Survey of Astronomy and Astrophysics 2010, the Committee on Earth Studies, and the Committee on Assessment of Options for Extending the Life of the Hubble Space Telescope.

JAMES F. DRAKE, JR., is a professor of physics at the University of Maryland, College Park. After completing his doctorate, Dr. Drake became a postdoctoral scholar at the University of California, Los Angeles (UCLA), and then moved to the University of Maryland, first as a postdoctoral scholar and then as a member of the teaching faculty in the Department of Physics and the Institute for Physical Science and Technology. He has worked on a very broad range of topics in the general area of theoretical plasma physics, using both analytical and numerical techniques. His work has applications spanning a variety of physical systems, including the solar corona, Earth's magnetosphere and ionosphere, magnetically confined plasma, and the interaction of intense lasers with plasma. His present focus is on magnetic reconnection with space physics applications and on turbulence and transport with applications to the magnetic fusion program. In recognition of his contributions to the field of plasma physics, he was made a fellow of the American Physical Society (APS) and was awarded a Humboldt Senior Scientist Research Award. Dr. Drake is also a national associate of the NAS. He received his Ph.D. in theoretical physics from UCLA. Dr. Drake has served on numerous NRC studies and has been a member of the Board on Physics and Astronomy (BPA) and of the Panel on Theory, Computation, and Data Exploration for the 2003 Decadal Survey on Solar and Space Physics.

LENNARD A. FISK is the Thomas M. Donahue Distinguished University Professor of Space Science in the Department of Atmospheric, Oceanic, and Space Sciences at the University of Michigan. Dr. Fisk was previously the associate administrator for Space Science and Applications and chief scientist at NASA. He has served as a professor of physics and as vice president for research and financial affairs at the University of New Hampshire. He is an active researcher in both theoretical and experimental studies of the solar atmosphere and its expansion into space to form the heliosphere. He is a member of the NAS. Dr. Fisk is a member of the board of directors of Orbital Sciences Corporation and co-founder of Michigan Aerospace Corporation. He received his Ph.D. in applied physics from the University of California, San Diego. Dr. Fisk was a member of the NRC Committee on Earth Science and Applications from Space. His prior service also includes chair of the SSB and membership on the Committee on Scientific Communication and National Security, the Committee on Fusion Science Assessment, the Committee on International Space Programs, the Air Force Physics Research Committee, and the Committee on Solar and Space Physics.

MARVIN A. GELLER is a professor of atmospheric sciences at the School of Marine and Atmospheric Sciences at Stony Brook University. His research deals with atmospheric dynamics, the middle and upper atmosphere, climate variability, and aeronomy. He became the fourth Stony Brook professor sharing the 2007 Nobel Peace Prize with Al Gore for his participation in the Inter-Governmental Panel on Climate Change (IPCC). Dr. Geller received the congratulatory letter from the United Nations Environment Programme on January 22, 2008, for his contribution in the assessment of stratospheric ozone depletion and climate change that led to the Montreal Protocol. Dr. Geller has served on many national and international advisory committees on atmospheric science, the upper atmosphere, and near-space environment and is currently president of the Scientific Committee on Solar-Terrestrial Physics (SCOSTEP); he has served as co-chair of the World Climate Research Programme's SPARC (Stratospheric Processes and Their Role in Climate) project, president of the AGU's Atmospheric Sciences section, chair of NASULGC's Board on the Oceans and Atmosphere, and president of ICSU's SCOSTEP (Scientific Committee on Solar-Terrestrial Physics). He is a fellow of the American Meteorological Society (AMS) and the AGU and past president of the AGU's Atmospheric Sciences Section. He earned his Ph.D. in meteorology from the Massachusetts Institute of Technology (MIT). He has served on numerous NRC panels and committees, including the Committee on Solar-Terrestrial Research (chair) and as a member of the Board on Atmospheric Sciences and Climate (BASC) and the Board on International Scientific Organizations.

SARAH GIBSON is currently a scientist at the High Altitude Observatory (HAO) at the National Center for Atmospheric Research (NCAR) in Boulder, Colorado. Dr. Gibson's positions prior to her arrival at HAO included a 1-year visit to Cambridge University in England as a NATO/NSF postdoctoral fellow and nearly 4 years at NASA's Goddard Space Flight Center (GSFC)—first as an NRC associate—as well as a research assistant professor at the Catholic University of America. Her primary interest is in the magnetic structure and dynamic evolution of coronal mass ejections (CMEs), and she uses theoretical CME models to explain a wide variety of space- and ground-based observations of CMEs from pre-eruption, through initiation and eruption, to their post-eruption state. A particular focus is observations and models of coronal prominence cavities, which represent dynamic equilibrium states that store magnetic energy, and Dr. Gibson leads an ISSI international working group to study coronal cavities. She is also a leader of the Whole Sun Month and Whole Heliosphere Interval international coordinated observing and modeling efforts to characterize the three-dimensional, interconnected solar-heliospheric-planetary system. Dr. Gibson was the recipient of the AAS-SPD 2005 Karen Harvey Prize. She obtained her Ph.D. in astrophysics from the University of Colorado, Boulder. She is a scientific editor for the *Astrophysical Journal* and serves on the Heliophysics Subcommittee of the NASA Advisory Council, on the AURA Solar Observatory Council, and as a member of the ATST Science Working Group. She has served on the NRC's Committee on Solar and Space Physics, the Committee on Distributed Arrays of Small Instruments for Research and Monitoring in Solar-Terrestrial Physics: A Workshop, and the Astro2010 Panel on Radio, Millimeter, and Submillimeter from the Ground.

MICHAEL HESSE is an astrophysicist and director of the Heliophysics Science Division at NASA GSFC. Dr. Hesse has also served as director of the Community Coordinated Modeling Center at the Laboratory for Solar and Space Physics, as acting branch head for the Geospace Physics Branch, and as the project scientist for theory and modeling for NASA's Living With a Star Program. Prior to his work at GSFC, Dr. Hesse was a principal scientist at Hughes System Corporation and a postdoctoral researcher at Los Alamos National Laboratory (LANL). His professional interests include research into fundamental physical processes in space plasmas, particularly studies of magnetospheric, solar physical, and astrophysical problems. He has been a recipient of a NASA Group Achievement Award for the Community Coordinated Modeling Center and of eleven GSFC performance awards. He was a participant in the NASA Sun-Earth Connection Roadmap. Dr. Hesse earned his Ph.D. in theoretical physics from Ruhr University, Bochum, Germany. He served on the NRC Panel on Solar Wind and Magnetospheric Interactions for the 2003 decadal survey.

J. TODD HOEKSEMA is a senior research scientist in the W.W. Hansen Experimental Physics Laboratory at Stanford University. His professional experience includes research administration, system and scientific programming, and the design, construction, and operation of instruments to measure solar magnetic and velocity fields from both the ground and space. He is co-investigator and magnetic team lead for the Helioseismic and Magnetic Imager on NASA's Solar Dynamics Observatory (SDO) and the instrument scientist for the Michelson Doppler Imager instrument on the Solar and Heliospheric Observatory that was launched by NASA and the European Space Agency (ESA). He has been associated with the Wilcox Solar Observatory at Stanford for three sunspot cycles. His primary scientific interests include the physics of the Sun and the interplanetary medium, solar-terrestrial relations, the large-scale solar and coronal magnetic fields, solar velocity fields and rotation, helioseismology, and education and public outreach. Dr. Hoeksema was chair of the Solar Physics Division of the American Astronomical Society (AAS) and has served on the heliophysics subcommittee of the NASA Advisory Council Science Committee. He served for 4 years as a solar physics discipline scientist at NASA. Dr. Hoeksema led NASA's 2005 Heliophysics Roadmap Team. He has been awarded the NASA distinguished public service medal and is a member of the AAS, AGU, International Astronomical Union (IAU), American Scientific Affiliation, and AAAS. He earned his Ph.D. in applied physics from Stanford University. For several years, Dr. Hoeksema was the vice chair of Commission E.2 of the Committee on Space Research. He served on the NRC's Astro2010 Panel on Optical and Infrared Astronomy from the Ground.

MARY K. HUDSON is the Eleanor and A. Kelvin Smith Distinguished Professor of Physics and served for 8 years as chair of physics and astronomy at Dartmouth College. Dr. Hudson has served as one of the principal investigators with the NSF-funded Center for Integrated Space Weather Modeling (CISM), where researchers study the weather patterns that originate from a solar eruption, following the energy and mass transfer through the interplanetary medium, all the way to Earth's ionosphere. Current areas of investigation include the evolution of the radiation belts; how the ionized particle outflow known as the solar wind and the magnetic field of the Sun interact with the magnetic field of Earth, producing electrical currents in the ionosphere; and the effects of solar cosmic rays on radio communications near Earth's poles. Dr. Hudson is also funded by NASA's Supporting Research and Technology program, studying related effects of Earth's space radiation environment that can affect both astronaut safety and satellite systems. Along with her students and postdoctoral research staff, she is modeling sudden changes in relativistic electron fluxes and solar cosmic rays at and inside the 24-hour orbital period of many communication and navigation satellites, and effects of global oscillations of Earth's magnetic field, associated with changes in solar wind conditions that have their origins at the Sun. Dr. Hudson received her Ph.D. in physics from UCLA. She has served as chair of the NSF Geospace Environment Modeling program and is funded for research in that program on geomagnetic storms. Dr. Hudson was the vice chair of the NRC's 2003 Panel on Atmosphere-Ionosphere-Magnetosphere Interactions and a member of the Committee on Solar and Space Physics and the Plasma Science Committee.

DAVID L. HYSELL is a professor in the Department of Earth and Atmospheric Sciences at Cornell University. As a graduate student there, he worked as a postdoctoral researcher in space plasma physics. He has also worked at Clemson University as an associate professor in the physics department. Dr. Hysell's research interests are in the area of upper atmospheric physics, space plasmas, and radar remote sensing. His research also focuses on theoretical and experimental investigations of space plasmas in Earth's ionosphere between 80 and 1500 km altitude. Dr. Hysell has designed and built a number of small, portable coherent scatter radars for studying plasma instabilities and irregularities in Earth's ionosphere at low, middle, and high latitudes. He uses these radars and radar interferometry and imaging techniques similar to those applied in radio astronomy and medicine, to observe the growth, propagation, and decay of ionospheric plasma irregularities in three spatial dimensions and in time. He received his Ph.D. in physics from Cornell University. His NRC committee experience includes serving on the U.S. National Committee for the International Union of Radio Science.

THOMAS J. IMMEL is an associate research physicist and senior fellow at the Space Sciences Laboratory at UC Berkeley. His expertise lies in interpretation of remote-sensing data and modeling of physical processes in the upper atmosphere and ionosphere. Dr. Immel's work has included ultraviolet imaging observations from four NASA missions: Dynamics Explorer, Polar, Imager for Magnetopause-to-Aurora Global Exploration (IMAGE), and Thermosphere-Ionosphere-Mesosphere Energetics and Dynamics (TIMED). His research efforts have extended out from the upper atmosphere and ionosphere to understand auroral and inner magnetospheric dynamical processes and have also focused on the upward coupling of energy and momentum from the lower atmosphere and the subsequent modification of conditions in the space environment. Dr. Immel received his Ph.D. in physics from the University of Alaska, Fairbanks. He served on the NASA 2009 Heliophysics Roadmap Team and is currently serving on the NSF CEDAR Science Steering Committee and the NASA Geospace Mission Operations Working Group.

JUSTIN KASPER is an astrophysicist in the Solar and Stellar X-Ray Group in the High Energy Astrophysics Division of the Harvard-Smithsonian Center for Astrophysics and is a lecturer in the Department of Astronomy at Harvard University. He is also a visiting scholar at Boston University, Center for Space Physics. Dr. Kasper has worked on the development, construction, and analysis of instrumentation for the in situ and remote measurement of particles and fields, including space-based plasma probes and particle telescopes such as the Faraday Cups on Wind, and ground-based radio telescopes including the Mileura Wide-Field Array Low Frequency Demonstrator (MWA-LFD). Currently, he is leading the design and operation of the Faraday Rotation Subsystem for MWA-LFD and participating in the radio transients, sky survey, and ionospheric calibration efforts. Dr. Kasper studies the flow of energy in astrophysical plasmas, including the solar corona, the solar wind, and planetary magnetospheres. His research focuses on the role of non-thermal velocity distribution functions, plasma micro-instabilities, magnetic reconnection, turbulence, and dissipation in the physical processes of heating, bulk acceleration, collisionless shocks, energetic particle acceleration, and radio emission. Dr. Kasper received his Ph.D. in physics from MIT. He was a member of the U.S. organizing and instrumentation committees for the 2007 International Heliophysical Year and the project scientist for the Cosmic Ray Telescope for the Effects of Radiation (CRaTER), on the Lunar Reconnaissance Orbiter.

JUDITH L. LEAN is a senior scientist for Sun-Earth system research in the Space Science Division of the U.S. Naval Research Laboratory (NRL). After completing her Ph.D. she worked at the Cooperative Institute for Research in Environmental Sciences at the University of Colorado, Boulder, from 1980 to 1986, joining NRL in 1988. Dr. Lean is the recipient of a number of NASA research grants, in collaboration with other SSDs and U.S scientists, and is currently a co-investigator on the SORCE, TIMED/SEE, SDO/EVE, and GLORY/TIM space missions. The focus of her research is to understand the Sun's variability using measurements and models, and to determine the impact of this variability on the Earth system, including climate change, the ozone layer, and space weather. Dr. Lean has published more than 100 papers in journals and books and delivered over 250 presentations documenting her research. She is a member of the AGU, AAS Solar Physics Division (SPD), and AMS International Association of Geomagnetism and Aeronomy. Dr. Lean was elected a fellow of the AGU in 2002 and a member of NAS in 2003. She has a Ph.D. in atmospheric physics from the University of Adelaide, Australia. Dr. Lean has served on a variety of NASA, NSF, and National Oceanic and Atmospheric Administration (NOAA) advisory committees. Her prior NRC service includes serving as a member of the Panel on Options to Ensure the Climate Record from the NPOESS and GOES-R Spacecraft and as a member of BASC.

RAMON E. LOPEZ is a professor of physics at the University of Texas, Arlington. He is the co-director for diversity of the Center for Integrated Space Weather Modeling, a science and technology center funded by NSF. His current research focuses on solar wind-magnetosphere coupling, magnetospheric storms and substorms, and space weather prediction. Dr. Lopez is also active in education research involving student perception and interpretation of images and visualizations. He is a fellow of the APS and was awarded the Nicholson Medal for Humanitarian Service. Dr. Lopez received his Ph.D. in space physics from Rice University. In 2003, he was elected vice chair of the APS Forum on Education and served as chair in 2005. Dr. Lopez has also served on various education-related committees of the AGU and as a member of the board of directors of the Society for Advancement of Chicanos/Latinos and Native Americans in Science. His previous NRC service includes membership on the Committee on Solar and Space Physics, the Committee on Strategic Guidance for NSF's Support of the Atmospheric Sciences, the Committee on Undergraduate Science Education, and the Steering Committee on Criteria and Benchmarks for Increased Learning in Undergraduate Science, Technology, Engineering and Mathematics (STEM).

HOWARD J. SINGER is chief scientist at the NOAA Space Weather Prediction Center. Previously, he served as the chief of the research and development division of the Space Environment Center (SEC) and as the project leader for the current and future NOAA Space Environment Monitor instruments on the GOES spacecraft and the responsible scientist for the GOES spacecraft magnetometers. Prior to joining SEC, Dr. Singer was with the Air Force Geophysics Laboratory, where he was the principal experimenter for the fluxgate magnetometer on the joint USAF-NASA Combined Release and Radiation Effects satellite. His research is in the area of solar-terrestrial interactions, ultralow-frequency waves, geomagnetic disturbances, storms, and substorms. He has received awards from the Air Force, NASA, and NOAA, including the prestigious Department of Commerce Gold Medal for Leadership, and he is the recipient of the Antarctica Service Medal for spending more than 1 year at South Pole Station, Antarctica, where a geographic feature is named for him. Dr. Singer received his Ph.D. in space physics and geophysics from UCLA. He is currently on the NSF Geospace Environment Modeling Steering Committee and is Editor's Choice Editor of *Space Weather: The International Journal of Research and Applications*. Dr. Singer has served on various NASA and NSF committees, including service on the NASA Living with a Star Geospace Mission Definition Team. He served on the NRC's 2003 Solar and Space Physics Survey Panel on Atmosphere-Ionosphere-Magnetosphere Interactions and on the Committee on Solar and Space Physics.

HARLAN E. SPENCE is a professor of physics and director of the Institute for the Study of Earth, Oceans and Space at the University of New Hampshire. Prior to that, Dr. Spence was a professor and department chair of the Department of Astronomy and a member of the Center for Space Physics at Boston University. He has also served as a senior member of the technical staff at the Aerospace Corporation. Dr. Spence's research interests include theoretical and experimental space plasma physics, cosmic rays and radiation belt processes, and the physics of the heliosphere, planetary magnetospheres, and the aurora. He received his Ph.D. in geophysics and space physics from UCLA. Dr. Spence has served on multiple NASA and NSF advisory panels, including the NASA Living with a Star Management Operations Working Group and the NASA Earth-Sun System Subcommittee. He previously served on the NRC's Panel on Solar Wind and Magnetospheric Interactions, Panel on Space Sciences, and Committee on Solar and Space Physics.

EDWARD C. STONE is the David Morrisroe Professor of Physics at the California Institute of Technology (Caltech) and vice provost for special projects. He is a former director of JPL. Dr. Stone served as the chair of Caltech's Division of Physics, Mathematics, and Astronomy and oversaw the development of the Keck Observatory as vice president for astronomical facilities and chairman of the California Association for Research in Astronomy. Since 1972, Dr. Stone has been the project scientist for the Voyager mission at JPL, coordinating the scientific study of Jupiter, Saturn, Uranus, and Neptune and Voyager's continuing exploration of the outer heliosphere and search for the edge of interstellar space. Following his first instrument on a Discoverer satellite in 1961, Dr. Stone has been a principal investigator on nine NASA spacecraft and a co-investigator on five other NASA missions for which he developed instruments for studying galactic cosmic rays, solar energetic particles, and planetary magnetospheres. Dr. Stone is a member of the NAS and the American Philosophical Society, president of the IAA, and a vice president of COSPAR. Among his awards and honors, he has received the National Medal of Science from President Bush, the Magellanic Premium from the American Philosophical Society, and Distinguished Service Medals from NASA. In 1996, an asteroid (5841) was named after him. Dr. Stone received his Ph.D. in physics from the University of Chicago. He has served on the NRC Planning Committee for a Workshop on "Sharing the Adventure with the Public"—Communicating the Value and Excitement of "Grand Questions" in Space Science and Exploration, the Committee on the Scientific Context for Space Exploration, and the SSB.

PANEL ON ATMOSPHERE-IONOSPHERE-MAGNETOSPHERE INTERACTIONS

JEFFREY M. FORBES, *Chair*, is a professor and the Glenn Murphy Endowed Chair of the Department of Aerospace Engineering Sciences at the University of Colorado, Boulder. His research interests include the upper-atmosphere environments of Earth, Mars, and other planets and the coupling of these environments to lower altitudes and to solar variability; geomagnetic storm effects on satellite drag variability; the vertical propagation of tides and planetary waves in planetary atmospheres and their electrodynamic and chemical effects; utilization of accelerometer, satellite drag, and satellite remote sensing data to elucidate atmospheric variability; and testing, validating and developing upper-atmosphere models. He also conducts numerical simulations of the above phenomena. Dr. Forbes is also the principal investigator of the Multidisciplinary University Research Initiative on Atmospheric Density Prediction of the Air Force Office of Scientific Research and the chair of the Academic Affairs Committee for the AIAA. He is a fellow of the AGU and an associate fellow of the AIAA and received the 2010 AIAA Robert M. Losey Atmospheric Sciences Award. Dr. Forbes received his Ph.D. in applied physics from Harvard University. His previous NRC service includes the Committee for the Study "Review of the Next Decadal Mars Architecture," the Committee on Solar-Terrestrial Research, and the Committee on Solar and Space Physics.

JAMES H. CLEMMONS, *Vice Chair*, is the principal director of the Space Science Applications Laboratory at the Aerospace Corporation. In his 13 years at Aerospace, Dr. Clemmons has led development of approximately 20 scientific instruments, flown on sounding rockets and satellites, to investigate a variety of phenomena in Earth's magnetosphere as well as its ionosphere-thermosphere-mesosphere system. Before joining Aerospace, he worked at NASA Goddard Space Flight Center (GSFC), the Swedish Institute for Space Physics, and the Max Planck Institute for Extraterrestrial Physics on related research. Dr. Clemmons is the author of numerous publications, including studies of observations conducted with the Freja satellite and other missions characterizing electric, magnetic, and plasma phenomena in the space environment. He is a member of the AGU, APS, and the American Chemical Society. Dr. Clemmons has participated in several NASA advisory groups and is the recipient of several awards by NASA and the Aerospace Corporation. He was a Fulbright Scholar and a resident associate of the NRC. Dr. Clemmons received B.S. degrees in physics and chemistry from the University of Illinois, Urbana-Champaign, and an M.S. and a Ph.D. in physics from UC Berkeley.

ODILE de la BEAUJARDIERE is the principal investigator of the Communication/Navigation Outage Forecasting System at the Air Force Research Laboratory's Space Vehicles Directorate. She previously worked at SRI International. Dr. de la Beaujardiere was a member of the NSF Coupling, Energetics and Dynamics of Atmospheric Regions (CEDAR) Science Steering Committee. She previously served as a member of the NRC Committee on Solar-Terrestrial Research.

JOHN V. EVANS is a retired chief technical officer of COMSAT Corporation, a position he took after serving as president and director of COMSAT Laboratories, the largest research center devoted entirely to satellite communications research. Prior to joining COMSAT, Dr. Evans was assistant director of the MIT Lincoln Laboratory. His current technical interest lies in satellite communications technology, including both space and ground segments. Dr. Evans is the co-editor of *Radar Astronomy* and has published more than 100 papers on the topics of radar reflection and high-power radar studies of the upper atmosphere and ionosphere. He is a member of the IAU, AGU, and AIAA. Dr. Evans is a fellow of the Institute of Electrical and Electronics Engineers and a member of the NAE. In 1975, he was awarded the Appleton Prize by the Council of the Royal Society of London for his contributions to ionospheric physics. Dr. Evans earned his Ph.D. in physics from Manchester University. He served as chair on the NRC Committee on Solar-Terrestrial Research and was a member of the Committee on Earth Studies and the Panel for Electronics and Electrical Engineering.

RODERICK A. HEELIS is the Cecil and Ida Green Honors Professor of Physics and director of the William B. Hanson Center for Space Sciences at the University of Texas, Dallas. His research specialization covers planetary atmospheres, ionospheres, and magnetospheres, and the physical phenomena coupling these regions. Dr. Heelis has published more than 130 papers in the refereed literature and presented numerous invited papers at national and international meetings. He serves as principal investigator for grant and contract research sponsored by DOD and NSF. Dr. Heelis graduated from the University of Sheffield with a Ph.D. in applied and computational mathematics. In addition to his research activities, he also serves on a number of advisory committees and working groups, including being a member of the NASA Sun-Earth Connections Advisory subcommittee. Dr. Heelis was a co-chair on the NRC Committee on Heliophysics Performance Assessment and served as a member of the Committee on the Assessment of the Role of Solar and Space Physics in NASA's Space Exploration Initiative.

THOMAS J. IMMEL, *see committee entry above.*

JANET U. KOZYRA is the George R. Carignan Collegiate Research Professor at the Space Physics Research Laboratory at the University of Michigan. She has also served as a summer faculty member at the Air Force Geophysics Laboratory at the Hanscom Air Force Base in Massachusetts. Dr. Kozyra is active in space plasma physics and aeronomy, concentrating on processes that couple the atmosphere and ionosphere with near-Earth space. Her research emphasis has been on the development of theoretical models of geophysical regions and the comparison of model results with satellite observations. Dr. Kozyra was the first demonstrator of the importance of high-energy oxygen ions in producing stable auroral red arcs through collisions with thermal electrons at high altitudes. She is currently a co-investigator on the TIDE instrument onboard the POLAR spacecraft and has also been selected as an interdisciplinary scientist on the proposed TIMED mission. Dr. Kozyra is an elected fellow of the AGU and is a former associate editor for the *Journal of Geophysical Research* and *Geophysical Research Letters.* She received her Ph.D. in space physics and aeronomy from the University of Michigan. She served as a member of the NRC AFOSR Atmospheric Sciences Review Panel.

WILLIAM LOTKO is a professor of engineering at the Thayer School of Engineering at Dartmouth College. He has also held the positions of interim dean and senior associate dean of the Thayer School. Before moving to Dartmouth, he was a research physicist at the Space Sciences Laboratory at UC Berkeley. Dr. Lotko is currently investigating and developing simulation models for ionospheric outflows into the magnetosphere, electron precipitation into the high-latitude ionosphere and thermosphere, plasma kinetics that enable superfluent ion outflows and electron precipitation, and the effects of all of these processes on global geospace dynamics. He is the principal investigator for Dartmouth's Heliophysics Theory Project sponsored by NASA, co-investigator for the NSF-sponsored Center for Integrated Space Weather Modeling, and team leader for one of NASA's Living With a Star projects that focuses on geospace dynamics during storms. He is the editor-in-chief of the *Journal of Atmospheric and Solar-Terrestrial Physics* and is an elected fellow of the AGU. He holds a Ph.D. in physics from UCLA. Dr. Lotko served on the NRC Panel on Theory, Computation, and Data Exploration.

GANG LU is a senior scientist in the Terrestrial Impacts of Solar Output section of the HAO at NCAR. Her primary research covers high-latitude ionospheric electrodynamics, solar wind-magnetosphere-ionosphere-thermosphere coupling, and space weather. She is the associate editor for *JGR*, was elected as the secretary for the aeronomy section of the AGU Space Physics and Aeronomy (SPA) section, and was awarded editor citation for excellence in refereeing for *JGR-Space Physics*. Dr. Lu received her Ph.D. in space physics from Rice University. She serves as the scientific discipline representative to SCOSTEP. Dr. Lu is a member of NSF's Geospace Environment Modeling (GEM) Steering Committee and a member of the Auroral Plasma Physics Working Group at the International Space Science Institute. Her NRC experience includes service on the Committee on the Assessment of the Role of Solar and Space Physics in NASA's Space Exploration Initiative and the Committee on Exploration of the Outer Heliosphere: A Workshop.

KRISTINA A. LYNCH is an associate professor of physics and astronomy at Dartmouth College. Prior to arriving at Dartmouth, Dr. Lynch was a research assistant professor at the University of New Hampshire. Her research revolves around auroral space plasma physics; ionospheric and mesospheric sounding rocket experiments, instrumentation, and data analysis; and wave-particle interactions in the auroral ionosphere. She has a Ph.D. in physics from the University of New Hampshire. Dr. Lynch was a member of the Committee on Plasma 2010: An Assessment of and Outlook for Plasma and Fusion Science.

JENS OBERHEIDE is an associate professor of physics in the Department of Physics and Astronomy at Clemson University. Previously, he was an associate professor in atmospheric physics at the University of Wuppertal, Germany. Dr. Oberheide is a specialist in satellite data analysis and conducts empirical modeling of global-scale wave dynamics in Earth's upper atmosphere. His research interests include the dynamics of Earth's mesosphere-thermosphere-ionosphere system; the forcing and vertical propagation of tides, planetary waves, and gravity waves, including their effects on chemistry and electrodynamics; geospace environment coupling to the atmosphere below and to solar activity; and utilization of satellite and ground-based remote sensing data to resolve variability and vertical coupling processes in the atmosphere. Dr. Oberheide is a recipient of the NASA Group Achievement Award to the TIMED team. He is an associate editor for the *Journal of Geophysical Research-Atmospheres*. He received his Ph.D. in physics from the University of Wuppertal. He served on the NASA Senior Review panel of the 2009-2012 Mission Operations and Data Analysis Program for the Heliophysics Operating Missions. Currently, Dr. Oberheide serves on the Steering Committee of SCOSTEP's Climate and Weather of the Sun-Earth System program and leads one of its working groups, investigating the geospace response to variable waves from the lower atmosphere.

LARRY J. PAXTON is a staff scientist and head of the Atmospheric and Ionospheric Remote Sensing Group at JHU/APL. He is the co-principal investigator for the global ultraviolet imager on NASA's TIMED and the principal investigator on the Defense Meteorological Satellites Program's (DMSP) special sensor ultraviolet spectrographic imager (SSUSI). His research focuses on the atmospheres and the ionospheres of the terrestrial planets, in particular the aeronomy of Earth's upper atmosphere and the role of solar cycle and anthropogenic change in creating variability in the dynamics, energetics, and composition of the upper atmosphere. Dr. Paxton was APL's chief scientist for the Ultraviolet and Visible Imagers and Spectrograph Imagers (UVISI) on the Midcourse Space Experiment (MSX). He has been involved in more than a dozen satellite, shuttle, and sounding rocket experiments. Dr. Paxton has published nearly 200 papers on planetary and space science, instruments, remote sensing techniques, and space mission design. He earned his Ph.D. in astrophysical, planetary, and atmospheric sciences from the University of Colorado, Boulder. He has served on several NASA and NSF committees, panels, and working groups and currently chairs the

International Academy of Astronautics Commission 4 on Space Systems Utilization and Operations. He was a member of the NRC Committee on the Effects of Solar Variability on Earth's Climate.

ROBERT F. PFAFF is a space scientist in the Laboratory for Extraterrestrial Physics at NASA's GSFC. Prior to arriving at Goddard, he was a research support specialist at Cornell University's School of Electrical Engineering. He is the principal investigator for the Communications/Navigation Outage Forecasting System (C/NOFS) Vector Electric Field Instrument and co-investigator for the San Marco Satellite Electric Field Experiment, the Polar Electric Field Investigation, and the Cluster Electric Field Investigation. In addition to these projects, Dr. Pfaff is also either principal or co-investigator of numerous sounding rockets. He received his Ph.D. in physics from Cornell University.

JOSHUA SEMETER is an associate professor in the Department of Electrical and Computer Engineering and director of the Center for Space Physics at Boston University. He was previously a senior research engineer at SRI and a scientist at the Max Planck Institute for Extraterrestrial Physics. His research interests revolve around ionospheric and space plasma physics, spectroscopy of atmospheric airglow and the aurora borealis, image processing, and radar systems and radar signal processing. Dr. Semeter was associate editor of the *Journal of Geophysical Research* from 2004 to 2006. He won the 2009 Boston University Electrical and Computer Engineering Faculty Teaching Award and the 2006 NSF CAREER Award. Dr. Semeter graduated from Boston University with a Ph.D. in electrical engineering.

JEFFREY P. THAYER is an associate professor of remote sensing in Earth and space science in the Aerospace Engineering Sciences Department at the University of Colorado, Boulder, where he has led the design, manufacturing, and testing of lidar systems for lower- and upper-atmosphere research with deployments in remote locations, such as Greenland. He was director of the NSF Sondrestrom Upper Atmosphere Research Facility in Greenland and performed experiments using incoherent scatter radar. His research interests include remote sensing instrumentation for atmospheric and space science, optical engineering for lidar system design and deployment, geophysical fluid dynamics, ionospheric electrodynamics, and thermosphere dynamics and composition. Dr. Thayer is a recipient of the Stanford Research Institute (SRI) Presidential Achievement Award and the University of Michigan Alumni Merit Award. He earned his Ph.D. in atmospheric and space physics from the University of Michigan. Dr. Thayer served as chair of the NSF Coupling, Energetics, and Dynamics of Atmospheric Regions (CEDAR) program for the past 3 years. He has also served on the NASA Geospace Mission and Operations Working Group, the NASA Sun-Earth Connections Roadmap Team, the NASA Science and Technology Definition Team for the Solar-Terrestrial Probe Geospace Electrodynamics Connections Mission, and the NSF CEDAR Science Steering Committee.

PANEL ON SOLAR WIND-MAGNETOSPHERE INTERACTIONS

MICHELLE F. THOMSEN, *Chair*, is a fellow of LANL. Having worked as a staff scientist from 1981 until her retirement in 2009, Dr. Thomsen now works on contract with the laboratory. Her primary research activities have involved the analysis and interpretation of spacecraft data, especially plasma data from the ISEE satellites, the Cassini spacecraft to Saturn, and the Los Alamos geosynchronous satellites. Previously, she served as the principal investigator for the plasma instruments on the geosynchronous satellites, as well as the chief scientist for space environment in the LANL High Altitude Space Monitoring program. Dr. Thomsen has also served as the acting director of the LANL Center for Space Science and Exploration. In addition, she served for 2 weeks as a Regents Lecturer at UCLA. Dr. Thomsen is the author or co-author of over 360 publications. She has received an outstanding alumni award from the University of Iowa and an honorary doctorate in science from Colorado College, and has been elected a fellow of the AGU. Dr. Thomsen received her Ph.D. in physics from the University of Iowa. She has served on a number of committees and advisory and review panels for NASA, NSF, and the AGU. Dr. Thomsen served a term as secretary for AGU's Space Physics and Aeronomy Section (magnetosphere), has twice served as associate editor for *Geophysical Research Letters,* and is currently on the editorial board of *Space Science Reviews.* She is a member of NASA's Heliophysics Advisory Subcommittee. Dr. Thomsen served a term as chair of NASA's former Earth-Sun Systems Subcommittee. She has served on the NRC's Committee on Solar and Space Physics and the 2003 Panel on Solar Wind and Magnetospheric Interactions.

MICHAEL WILTBERGER, *Vice Chair,* is a scientist at NCAR in the Earth and Sun Systems Laboratory of the High Altitude Laboratory. A component of his work involves modeling of ionospheric outflows with data from the Solar Extreme-ultraviolet Experiment (SEE) on the NASA TIMED satellite used as input to the NCAR Thermosphere-Ionosphere-Electrodynamics General Circulation Model (TIE-GCM) for a simulation spanning the declining phase of solar cycle 23 from 2001 to 2007. Model simulations of the neutral thermospheric density at ~400 km were compared with density measurements obtained from observations of the atmospheric drag on five spherical satellites in low Earth orbit. Previously, Dr. Wiltberger served as a research assistant professor in the Department of Astronomy at Dartmouth College. He has been published in multiple scientific journals. NASA awarded him the group achievement award and the AGU awarded him an outstanding student paper award in 1998. Dr. Wiltberger earned his B.S. in physics from Clarkson University and his M.S. in physics and his Ph.D. in space plasma physics from the University of Maryland.

JOSEPH BOROVSKY is a scientist with the Space Science and Applications Group of LANL. A 19-year veteran of LANL, he works primarily on NASA-funded research in space physics, plasma physics, and surface physics. He served as a guest investigator at the Max Planck Institute for Aeronautics in Germany. His recent efforts have been focused on turbulence, the aurora, and Earth's magnetosphere. Dr. Borovsky has contributed to numerous articles and professional journals. He is a member of and past editor for the AGU. Dr. Borovsky earned his Ph.D. from the University of Iowa.

JOSEPH F. FENNELL is a distinguished scientist in the Space Science Applications Laboratory of the Physical Sciences Laboratory at the Aerospace Corporation. Dr. Fennell has been heavily involved in the development, fabrication, testing, and flight of many different particle instruments. These have covered the range from auroral and magnetospheric plasma instruments to medium- and high-energy electron and ion sensors. The most recent instrumentation efforts have been with the energetic particles and energetic ion composition measurements on the SCATHA, VIKING, CRRES, POLAR, Cluster II, Radiation Belt Storm Probe (RBSP), and operational Department of Defense satellites. His professional activity has included studies of artificially injected and natural geomagnetically trapped particles, solar cosmic rays, particle access to the magnetosphere, particle transport within the magnetosphere, energetic particle composition, ring current development, space radiation effects, and spacecraft charging and the impact the charging has on satellite systems. Recent studies have included observations of radiation belt particle transport and losses; studies of magnetic storm and substorm processes; high-altitude plasma sheet, cusp, plasma mantle, and ring current composition studies; and studies of the plasma sheet boundary. He is a fellow of the AGU and received the Aerospace Presidents Award for his work in satellite charging and its effects. Dr. Fennell earned his Ph.D. in physics from Saint Louis University, Missouri. He has served on the NRC Committee on Solar-Terrestrial Research, the Committee on Solar and Space Physics, the Panel on Solar Wind and Magnetosphere Interactions for the 2003 solar and space physics decadal survey, and on the Committee on the Societal and Economic Impacts of Severe Space Weather Events Workshop.

JERRY GOLDSTEIN is a staff scientist at the Southwest Research Institute (SwRI) and an adjunct professor of physics at the University of Texas at San Antonio (UTSA). Dr. Goldstein specializes in the study of the inner magnetospheres of Earth and Saturn. His primary research involves data analysis and modeling of data from the IMAGE mission, the Two Wide-angle Imaging Neutral-atom Spectrometers (TWINS) mission, and the Cassini mission. Dr. Goldstein teaches graduate-level courses at UTSA and is the Science Operations Center lead for TWINS. He has authored or co-authored dozens of articles and has been cited hundreds of times in multiple journals. Dr. Goldstein is a fellow of the AGU and recipient of the 2006 Macelwane Medal in recognition of his plasmaspheric research. In 2006, *Popular Science Magazine* named him one of its annual "Brilliant 10" young scientists, and the *San Antonio Business Journal* included him in its "Forty Under 40." In 2009, Dr. Goldstein received the Young Alumnus Award from Brooklyn College. He earned his Ph.D. in space physics from Dartmouth College.

JANET C. GREEN is a physicist at the NOAA Space Weather Prediction Center. She acts as the instrument scientist for the particle detectors on the NOAA GOES and POES satellites. Her main area of expertise is the physics of Earth's radiation belts and their effects on satellite electronics and performance. Dr. Green guides the design and implementation of new NOAA particle instruments, monitors current data and instrument performance, and transitions into operations new algorithms and products that rely on the data. She also collaborates with the research community to improve understanding and modeling of Earth's radiation belts and works with the user community to understand and mitigate satellite anomalies. Dr. Green received her Ph.D. from UCLA.

DONALD A. GURNETT is a professor at the University of Iowa. His primary research interests are in the area of magnetospheric radio and plasma wave research, and he has more than 450 scientific publications. Dr. Gurnett has participated in more than 25 spacecraft projects, including Voyager 1 and 2, Galileo, and Cassini. He is a member of the NAS and a fellow of the AAAS. Among his numerous research awards are the John Howard Dellinger Gold Medal from the International Scientific Radio Union, the John Adam Fleming Medal from the AGU, the Excellence in Plasma Physics Award from APS, and the Hannes Alfven Medal from the European Geosciences Union. He has also received several teaching awards, including the Iowa Board of Regents award for faculty excellence. Dr. Gurnett received his Ph.D. in physics from the University of Iowa. He served on the NRC Panel on Space Sciences and the Committee on Planetary and Lunar Exploration.

LYNN M. KISTLER is a professor of physics in the Department of Physics and in the Space Science Center at the University of New Hampshire. Her major research interests are in the impact of heavy ions on dynamics of the magnetosphere, particularly the ring current and the magnetotail. Dr. Kistler is also interested in space instrumentation to measure ion composition and has been involved in developing instruments for the CLUSTER, FAST, Equator-S, ACE, and STEREO missions. She earned her Ph.D. in physics from the University of Maryland, College Park. She was involved in the NASA Sun-Earth Connections Roadmap Committee in 1999, the NASA Heliophysics Lunar Science Subpanel in 2006, and the NASA Heliospheric Mission Planning Working Group (Roadmap) in 2008. Dr. Kistler was the AGU Space Physics and Aeronomy-Magnetospheric Physics Secretary from 2008 to 2010.

MICHAEL W. LIEMOHN is an associate professor of space science and engineering in the Department of Atmospheric, Oceanic, and Space Sciences at the University of Michigan, where he has led the development of several numerical models for energetic particle transport and the use of these models for the interpretation of ground-based and spacecraft measurements. His current research activities include investigations of the storm-time inner magnetosphere (electrons, ring current, and plasmasphere) and understanding both the large-scale and small-scale processes of relevance, including magnetosphere-ionosphere coupling, ionospheric conductance influences, and magnetospheric plasma sources. He is also involved in data analysis and modeling of energetic electrons and ions around Mars, especially using the Mars Global Surveyor and Mars Express data sets. Dr. Liemohn completed an NRC-sponsored postdoctoral position at NASA Marshall Space Flight Center (MSFC) before returning to the University of Michigan in 1998. He earned his Ph.D. in atmospheric and space science from the University of Michigan. Dr. Liemohn has served as chair of the NASA Geospace Management and Operations Working Group and chair of the NSF Geospace Environment Modeling steering committee, and he has served on various other NASA, NSF, and LANL advisory committees.

ROBYN MILLAN is an assistant professor of physics and astronomy at Dartmouth College. Her research includes the use of high-altitude scientific balloon experiments to study Earth's radiation belts, specifically the loss of relativistic electrons from the outer radiation belts into Earth's atmosphere. Dr. Millan is principal investigator for the Balloon Array for RBSP Relativistic Electron Losses (BARREL) project, which is being planned for flight in association with the RBSP mission. Her prior positions include research appointments at Dartmouth and at UC Berkeley. She received her Ph.D. in physics at UC Berkeley in 2002. Dr. Millan served on the NRC Committee on the Role and Scope of Mission-Enabling Activities in NASA's Space and Earth Science Missions.

DONALD G. MITCHELL is the Cassini spacecraft instrument scientist and IBEX co-investigator at JHU/APL. Dr. Mitchell has been with JHU/APL since 1976. He was the lead investigator for the High Energy Neutral Atom (HENA) imager for the IMAGE mission. Dr. Mitchell is currently the instrumentation scientist for the Magnetospheric Imaging Instrument for the Cassini Saturn mission and the Radiation Belt Science of Protons and Ion Composition Experiment. He has many publications in Earth magnetospheric, solar wind, and outer-planets magnetospheric physics. Dr. Mitchell is a member of the AGU, AAAS, and IAA. He earned his Ph.D. in physics from the University of New Hampshire in 1975. Dr. Mitchell has served on the NRC Committee on International Space Programs and the Committee on Solar and Space Physics.

TAI D. PHAN is a senior fellow at the Space Sciences Laboratory of UC Berkeley. He has worked as a visiting postdoctoral scientist with the Max Planck Institut für extraterrestrische Physik in Germany and as a research associate for the Herzberg Institute of Astrophysics at the National Research Council in Canada. Dr. Phan is a co-investigator of NASA's Time History of Events and Macroscale Interactions during Substorms (THEMIS) mission to study the cause of magnetospheric substorms. He leads an interdisciplinary science team of the Magnetospheric Multiscale Mission to study the microphysics of magnetic reconnection. His research interests include solar wind interaction with Earth's magnetosphere, and magnetic reconnection in the solar wind, magnetosheath, and magnetosphere. He earned his Ph.D. in engineering from Dartmouth College.

MICHAEL SHAY is an associate professor in the Department of Physics and Astronomy at the University of Delaware. Dr. Shay studies plasma physics using analytical theory and massively parallel computer simulations. He has extensively studied one multiscale process called magnetic reconnection, in which a large amount of magnetic energy is explosively released in the form of energetic particle acceleration, heating, and plasma flows. Dr. Shay is also studying novel simulation techniques that may provide a means to directly simulate multiscale phenomena. He has received a Faculty Early Career Development Award from NSF. Dr. Shay earned his Ph.D. in physics from the University of Maryland.

HARLAN E. SPENCE, *see committee entry above.*

RICHARD M. THORNE is a professor of atmospheric physics in the Department of Atmospheric and Oceanic Sciences at UCLA. He was a member of the Galileo Energetic Particle Detector team and chair of the GEM working group on energetic electron variability. He is currently chair of the GEM focus group on diffuse auroral precipitation, co-investigator and chair of the Radiation Working Group on the NASA New Frontiers JUNO mission, and co-investigator and lead theorist on the NASA Living With a Star RBSP-ECT and EMFISIS teams. His principal research involves theoretical studies of the interactions between waves and particles in geophysical plasmas, including the origin of many different classes of plasma waves found in the highly tenuous solar system plasmas, the role of wave-particle scattering on the dynamics of the energetic radiation belts, and the effects of particle precipitation on the upper atmosphere. Dr. Thorne is a fellow of the AGU. He earned his Ph.D. in physics from MIT. Dr. Thorne has served on numerous NASA panels, including the Geosciences Mission Definition Team for NASA's Living With a Star Program, the CASSINI Extended Mission Senior Review Board, and the JUNO project Radiation Advisory Board.

PANEL ON SOLAR AND HELIOSPHERIC PHYSICS

RICHARD A. MEWALDT, *Chair,* is a senior research associate in the Space Radiation Laboratory at Caltech. Dr. Mewaldt's research interests cover spacecraft and balloon-borne measurements of energetic nuclei and electrons accelerated in solar energetic particle events, galactic cosmic rays, the heliosphere, and Earth's magnetosphere. His work has focused specifically on studies of elemental and isotopic composition and the implications of these measurements for energetic particle origin, acceleration, and transport; on solar particle and cosmic-ray impacts on space weather; and on the development of high-resolution instrumentation to extend these measurements. Dr. Mewaldt has been a co-investigator on the NASA missions IMP-7, IMP-8, and ISEE-3 and a guest investigator on HEAO-3, and he is currently a co-investigator on the NASA Solar, Anomalous, and Magnetospheric Explorer (SAMPEX) and on STEREO, and mission scientist for the Advanced Composition Explorer. He earned his Ph.D. in physics from Washington University. Dr. Mewaldt's NRC membership includes the Committee on Cosmic-Ray Physics and the Panel on Particle, Nuclear, and Gravitational-wave Astrophysics.

SPIRO K. ANTIOCHOS, *Vice Chair,* is a research astrophysicist in the Heliophysics Division of NASA GSFC. Dr. Antiochos is also an adjunct professor in the Department of Atmospheric, Oceanic, and Space Sciences at the University of Michigan. His fields of expertise include theoretical solar physics and plasma physics. Dr. Antiochos's work consists primarily of developing theoretical models to explain observations from NASA space missions. During his career he has worked on a number of problems related to the Sun and heliosphere, in particular, the physics of magnetic-driven activity and the structure of the Sun's corona. Dr. Antiochos previously served as the head of the Solar Theory Section in the Space Science Division at NRL. He also served as a postdoctoral fellow at the NCAR and as a research associate at Stanford University. Dr. Antiochos is a co-investigator on NASA's STEREO mission, part of the Solar-Terrestrial Probes program. He served as chair of the Solar Physics Division for AAS. Dr. Antiochos has authored or coauthored more than 100 refereed papers in archival journals. He is a fellow of the AGU, a recipient of the AAS George Ellery Hale Prize for outstanding contributions to the field of solar astronomy, and a recipient of the National Research Laboratory's E.O. Hulburt Award, the NRL's highest honor for scientific achievement. Dr. Antiochos received his Ph.D. in applied physics from Stanford University. He served previously on the NRC's Panel on Theory, Computation, and Data Exploration and the Committee on an Assessment of Balance in NASA's Science Programs.

TIMOTHY S. BASTIAN is assistant director of the Office of Science and Academic Affairs at the National Radio Astronomy Observatory, where he has been an astronomer since 1987. He also is an adjunct faculty member in the Astronomy Department at the University of Virginia. Dr. Bastian's research interests include solar and stellar radiophysics. He is currently the principal investigator for the Solar Radio Burst Spectrometer project and served on the faculty of the NCAR Summer School on Heliophysics. Dr. Bastian served as scientific editor of the *Astrophysical Journal*. He received a B.S. in mathematics from the University of Chicago and a Ph.D. in astrophysics from the University of Colorado. Dr. Bastian served on the 2003 NRC Panel on the Sun and Heliospheric Physics.

JOE GIACALONE is an associate professor of planetary sciences at the University of Arizona. Prior to coming to the University of Arizona, Dr. Giacalone was a research associate at Queen Mary and Westfield College, London. His core research interests include understanding the origin, acceleration, and propagation of cosmic rays and other charged-particle species in the magnetic fields of space, and general topics in space plasma physics and astrophysics. He develops physics-based theoretical and computational models that are used to interpret in situ spacecraft observations. Dr. Giacalone is a recipient of the NSF's Early CAREER award. He currently serves as a member of NASA's Living With a Star TR&T Steering Committee and as secretary for the SPA/SH subdivision of the AGU. Dr. Giacalone earned his Ph.D. in physics from the University of Kansas. He has also served on a NASA senior review panel for NASA Data and Modeling Centers and on the Steering Committee for NSF's SHINE program.

GEORGE M. GLOECKLER is distinguished university professor, emeritus, University of Maryland, and research professor in the Atmospheric, Oceanic and Space Sciences Department at the University of Michigan. Dr. Gloeckler's research focuses on space plasma physics, particularly the properties of the local interstellar medium, such as its magnetic field, the density and composition of its gas, and its interaction with the solar system. He is known for developing a new experimental measurement technique based on observations of interstellar pickup ions and for pioneering discoveries and the invention of instruments carried on satellites and deep-space probes, including the two Voyagers, Ulysses and Cassini. Elected to the NAS in 1997, Dr. Gloeckler is also a fellow of the AGU and APS and a recipient of the COSPAR Space Science Award. He earned his Ph.D. in physics from the University of Chicago. Dr. Gloeckler has served on the NRC's Committee on Fusion Science Assessment, the Committee for a Review of Scientific Aspects of the NASA Triana Mission, and the Committee on Solar and Space Physics.

JOHN W. (JACK) HARVEY is an astronomer at the National Solar Observatory (NSO), where he studies solar magnetic and velocity fields and helioseismology. Dr. Harvey's major efforts have been in the design and development of instrumentation for community use in these research areas. His more recent research has focused on unambiguous observations of permanent magnetic field changes associated with solar flares, discovery that the quiet solar photosphere has a ubiquitous, rapidly changing, mainly horizontal magnetic field, and solar chromospheric magnetic field structure associated with coronal holes and prominences. Dr. Harvey is a member of the NSO Scientific Personnel Committee, instrument scientist for the GONG project, and project scientist for the SOLIS project. He received his Ph.D. from the University of Colorado, Boulder. In the outside community, Dr. Harvey serves on NASA and NSF review panels and is a past co-editor of the journal *Solar Physics*. He chaired recent reviews of solar programs in Japan and Switzerland. Dr. Harvey has served on the Committee on Solar and Space Physics, as well as other NRC panels and projects.

RUSSELL A. HOWARD is an astrophysicist at NRL. Dr. Howard's research has centered on understanding the physics of the solar corona and the coronal mass ejection phenomenon—its initiation, propagation, and eventual interplanetary effects. He is currently the principal investigator for the operating experiments SOHO/LASCO and STEREO/SECCHI and two experiments under development, the Solar Orbiter/SoloHI and the Solar Probe Plus WISPR. Dr. Howard developed the CCDs and CCD cameras for LASCO and EIT, for which he received an NRL Royalty Award, and he is currently developing the APS/CMOS sensor for SoloHI and WISPR. He was the project scientist for the development of the Solwind and LASCO coronagraphs and led the development of the LASCO/EIT flight software and ground system. Dr. Howard has been a co-investigator on numerous NASA projects, including an XUV CCD detector development program. He has more than 200 papers in the refereed literature. He received the E.O. Hulburt Science Award, which is the highest award that NRL gives to a scientist, and the NASA Exceptional Scientific Achievement Award. Dr. Howard earned a B.S. in mathematics and a Ph.D. in chemical physics, both from the University of Maryland.

JUSTIN KASPER, *see committee entry above.*

ROBERT P. LIN is a professor in the Department of Physics at UC Berkeley.[1] Dr. Lin is a world-renowned experimentalist in space science. Through numerous innovative instruments that have flown on NASA missions, he has revealed the behavior of electrons and ions accelerated by the Sun and has detected the accompanying X-ray and gamma-ray emissions. As an astrophysicist, his primary interest is in how particles are accelerated to high energies in nature. To study these processes, Dr. Lin has developed instruments to directly measure the plasma, fields, and energetic particles and flown them on spacecraft into regions where acceleration is occurring. He is particularly interested in the Sun as the most powerful accelerator in our solar system, accelerating particles to the highest energies. Dr. Lin conducts imaging and spectroscopy of the X rays and gamma rays emitted by energetic particles at the Sun, as well as directly detecting the accelerated particles that escape to the interplanetary medium. He studies the acceleration that occurs in transient events that involve the phenomena of magnetic reconnection or collisionless shock waves. Dr. Lin is the principal investigator on the Gamma Ray Imager/Polarimeter for Solar Flares experiment, which will utilize high-altitude balloons but is not funded by the NASA suborbital program but by the Supporting Research and Technology program. He is a member of the NAS. Dr. Lin received his Ph.D. from UC Berkeley in 1967. He is a member of the NRC's SSB and has served on many committees, including the Committee on NASA's Suborbital Research Capabilities and the Panel on Solar and Space Physics.

GLENN M. MASON is on the senior professional staff at JHU/APL, where he is currently an investigator for the Remote Analysis Site for the Ultra Low Energy Isotope Spectrometer particle instrument in the ACE mission. Dr. Mason was a professor in the Department of Physics and at the Institute for Physical Science and Technology at the University of Maryland, College Park. He has worked on the development of novel instrumentation that allows determination of the mass composition of solar and interplanetary particles in previously unexplored energy ranges. His research has included work on galactic cosmic rays, solar energetic particles, and the acceleration and transport of particles both in the solar atmosphere and in the interplanetary medium. Dr. Mason is principal investigator on the NASA SAMPEX spacecraft mission and is co-investigator on energetic particle instruments for the NASA Wind spacecraft and the ACE spacecraft. He received his A.B. in physics from Harvard College and his Ph.D. in physics from the University of Chicago. Dr. Mason was former chair of the NASA Sun-Earth Connections Advisory Subcommittee, the NASA Space Science Advisory Committee, and the Steering Committee of the Space Science Working

[1] The survey committee notes with regret that Dr. Lin died on November 17, 2012.

Group of the Association of American Universities. He previously served on the NRC Committee on Solar and Space Physics.

EBERHARD MOEBIUS is a professor of physics at the University of New Hampshire. He worked as a research scientist at the Max Planck Institut für extraterrestrische Physik in Germany, and has been on the physics faculty at University of New Hampshire. His research interests include the acceleration of ions in Earth's magnetosphere, in interplanetary space, and in solar flares; the interaction of interstellar gas with the solar wind; and the study of the local interstellar medium. Dr. Moebius's group is finishing the PLASTIC instrument to measure the solar wind and suprathermal ion composition for NASA's STEREO mission and is involved in several studies for future missions to Earth's magnetosphere and the heliosphere. He earned his Ph.D. in laboratory plasma physics at Ruhr University in Bochum, Germany.

MERAV OPHER is an assistant professor of astronomy at Boston University. Prior to that, Dr. Opher was an associate professor at George Mason University and was a research scientist at JPL, where she conducted research on the interaction between the solar and interstellar winds found at the edge of the solar system. As a postdoctoral associate with the Plasma Group in the Department of Physics and Astronomy at UCLA, she investigated the effects of electromagnetic fluctuations on nuclear reaction rates and how these plasma effects can influence stellar evolution and early universe calculations. Her research interests focus on plasma effects in space physics and astrophysics. Dr. Opher received a Ph.D. in plasma astrophysics from the University of São Paulo, Brazil. She served on the NRC Committee on Solar and Space Physics.

JESPER SCHOU is currently a senior research scientist at Stanford University. Dr. Schou is the instrument scientist and co-investigator for the Helioseismic and Magnetic Imager on SDO. His research interests include solar variability, solar magnetic activity, and helioseismology. He has written over 70 refereed papers. Dr. Schou holds a Ph.D. in astronomy from the University of Aarhus. He has been chairman of the GONG Data Management and Analysis Center Users Committee, a member of the NASA Solar and Heliospheric Management Operations Working Group, and a member of the scientific organizing committees for SOHO14/GONG 2004, SDO 2008, and SOHO 24.

NATHAN A. SCHWADRON is an associate professor of astronomy at Boston University and the science operations lead for the Interstellar Boundary Explorer Mission. Dr. Schwadron's previous experience includes positions as a senior research scientist, a principal scientist and a staff scientist at the SwRI in San Antonio, Texas, an assistant research scientist at the University of Michigan, a senior research scientist at the International Space Science Institute in Bern, Switzerland, and a postdoctoral scholar in the University of Michigan's Atmospheric, Oceanic and Space Science Department. His research interests include heliospheric phenomena related to the solar wind, the heliospheric magnetic field, pickup ions, cometary X rays, energetic particles, and cosmic rays. He received a B.A. with honors in physics from Oberlin College and a Ph.D. in physics from the University of Michigan. Dr. Schwadron served on the NRC Committee on Priorities for Space Science Enabled by Nuclear Power and Propulsion and is a member of the Committee on Solar and Space Physics.

AMY R. WINEBARGER is an astrophysicist at NASA MSFC. She previously worked as an assistant professor at Alabama A&M University, a research scientist at NRL, and an astrophysicist at the Smithsonian Astrophysical Observatory. Her research interests include solar coronal heating, solar flare heating, energy growth and release in coronal mass ejections, comparisons between simulation results and observables, analysis of spectroscopic and filter data, development and testing of filter response functions, and hydrodynamic code validation and verification. She is the recipient of an NSF CAREER Award. Dr. Winebarger received an M.S. and a Ph.D. in physics from the University of Alabama, Huntsville.

DANIEL WINTERHALTER is a principal scientist with JPL, California Institute of Technology. His research interests include the spatial evolution of the solar wind into the outer reaches of the heliosphere, as well as its interaction with, and influence on, planetary environments. He has published articles in refereed journals and edited two books on this subject. Most recently he has been interested in the low-frequency radio emissions from the (presumed) magnetospheres of extrasolar planets, for which his team has carried out observations with the world's largest radio telescopes. As a member of several flight teams over the years, Dr. Winterhalter is and has been intimately involved with the planning, launching, and operating of complex spacecraft and space science missions. He received a NASA Special Recognition Certificate for his work on Mars Observer. Dr. Winterhalter is the experiment representative for the Mars Global Surveyor magnetometer team, and until recently was the investigation scientist for the Cassini Radio Science Experiment. He was the study scientist for the space science Mercury Orbiter effort in 1996 and the pre-project scientist for the Mars Science and Telecom Orbiter. Dr. Winterhalter has received achievement awards for his participation on the Voyagers 1 and 2, Pioneer 11, and Mars Observer, Mars Global Surveyor, and Cassini interplanetary probes. He received an M.S. and a Ph.D. in geophysics and space physics from UCLA. Dr. Winterhalter previously served on the NRC's NASA Technology Roadmaps: Instruments and Computing Panel.

THOMAS N. WOODS is the associate director for Technical Divisions at LASP at the University of Colorado, Boulder. He previously held research scientist positions at LASP and HAO. His research is focused primarily on solar ultraviolet irradiance and its effects on Earth's atmosphere. Dr. Woods is the principal investigator for numerous experiments, including the EUV Variability Experiment on the NASA SDO; X-Ray Sensor and EUV Sensor on NOAA GOES-R; Solar Radiation and Climate Experiment as a mission for NASA's Earth Observing System; and the Solar EUV Experiment on the Thermosphere, Ionosphere, Mesosphere Energetics and Dynamics mission. He received an M.A. and a Ph.D. in physics from Johns Hopkins University.

Staff

ARTHUR A. CHARO, *Study Director*, joined the SSB as a senior program officer in 1995. He has directed studies that have resulted in some 30 reports, notably the first NRC decadal survey in solar and space physics (2003) and in Earth science and applications from space (2007). Dr. Charo received his Ph.D. in physics from Duke University in 1981 and was a postdoctoral fellow in chemical physics at Harvard University from 1982 to 1985. He then pursued his interests in national security and arms control at Harvard University's Center for Science and International Affairs, where he was a research fellow from 1985 to 1988. From 1988 to 1995, he worked as a senior analyst and study director in the International Security and Space Program in the U.S. Congress's Office of Technology Assessment. Dr. Charo is a recipient of a MacArthur Foundation Fellowship in International Security (1985-1987) and a Harvard-Sloan Foundation Fellowship (1987-1988). He was the 1988-1989 American Institute of Physics AAAS Congressional Science Fellow. In addition to NRC reports, he is the author of research papers in molecular spectroscopy, reports on arms control and space policy, and the monograph "Continental Air Defense: A Neglected Dimension of Strategic Defense" (University Press of America, 1990).

ABIGAIL A. SHEFFER joined the SSB in fall 2009 as a Christine Mirzayan Science and Technology Policy Graduate Fellow to work on the report *Visions and Voyages for Planetary Science in the Decade 2013-2022*. She continued with the SSB to become an associate program officer. Dr. Sheffer earned her Ph.D. in planetary science from the University of Arizona and her A.B. in geosciences from Princeton University. Since coming to the SSB, she has worked on several studies, including *Defending Planet Earth: Near-Earth Object Surveys and Hazard Mitigation Strategies, Assessment of Impediments to Interagency Collaboration on Space and Earth Science Missions,* and *The Effects of Solar Variability on Earth's Climate: A Workshop Report.*

MAUREEN MELLODY has been a program officer with the Aeronautics and Space Engineering Board (ASEB) since 2002, where she has worked on studies related to NASA's aeronautics research and development program, servicing options for the Hubble Space Telescope, and many other projects in space and aeronautics. Previously, she served as the 2001-2002 AIP Congressional Science Fellow in the Office of Representative Howard L. Berman (D-Calif.), focusing on intellectual property and technology transfer. Dr. Mellody also worked as a postdoctoral research scientist at the University of Michigan in 2001. She received her Ph.D. and M.S. degrees in applied physics from the University of Michigan and her B.S. in physics from Virginia Polytechnic Institute and State University.

LEWIS B. GROSWALD, research associate, joined the SSB as the Autumn 2008 Lloyd V. Berkner Space Policy Intern. Mr. Groswald is a graduate of George Washington University, where he received a master's degree in international science and technology policy and a bachelor's degree in international affairs, with a double concentration in conflict and security and Europe and Eurasia. Following his work with the National Space Society during his senior year as an undergraduate, Mr. Groswald decided to pursue a career in space policy, with a focus on educating the public on space issues and formulating policy. He has worked on NRC reports covering a wide range of topics, including near-Earth objects, orbital debris, life and physical sciences in space, and planetary science.

CATHERINE A. GRUBER, editor, joined the SSB as a senior program assistant in 1995. Ms. Gruber first came to the NRC in 1988 as a senior secretary for the Computer Science and Telecommunications Board and also worked as an outreach assistant for the National Science Resources Center. She was a research assistant (chemist) in the National Institute of Mental Health's Laboratory of Cell Biology for 2 years. She has a B.A. in natural science from St. Mary's College of Maryland.

DANIELLE PISKORZ, a Fall 2011 Lloyd V. Berkner Space Policy Intern in the SSB, graduated from MIT with a degree in physics and a minor in applied international studies. She worked on various research projects at L'Institut d'Astrophysique de Paris, LANL, and JPL and spent her junior year studying at the University of Cambridge. Ms. Piskorz began graduate studies in Fall 2012 in geophysics.

LINDA M. WALKER, a senior project assistant, has been with the NRC since 2007. Before starting with the SSB, she was on assignment with the National Academies Press. Prior to working at the NRC, she was with the Association for Healthcare Philanthropy in Falls Church, Virginia. Ms. Walker has 28 years of administrative experience.

TERRI BAKER was a senior program assistant with the SSB until April 2012. She came to the SSB from the National Academies' Center for Education. Mrs. Baker has held numerous managerial, administrative, and coordinative positions.

BRUNO SÁNCHEZ-ANDRADE NUÑO was a National Academies Christine Mirzayan Science and Technology Policy Graduate Fellow for the SSB. Dr. Sánchez is currently the director of science and technology at the Global Adaptation Institute (GAIN), where he is responsible for model development, data mining, methodological development, and visualization techniques to support the GAIN Index. He previously worked for 2 years as a space and rocket scientist at NRL. His work has focused on solar data analysis and project planning for NASA rockets and satellites. He was also on the faculty of George Mason University. In 2008, Dr. Sánchez obtained his Ph.D. in astrophysics at the Max Planck Institute for Solar System Research in Germany.

HEATHER D. SMITH was a Christine Mirzayan Science and Technology Policy Graduate Fellow for the SSB. Dr. Smith is currently a Planetary Science Division management postdoctoral fellow at NASA Headquarters. Her research area is astrobiology instrumentation and life in extreme environments, in particular microbial physical habitats. Dr. Smith completed her doctorate in biological engineering at Utah State University, where her dissertation was on the design of a native fluorescence life detection instrument for soils. Prior to this, she earned a master of science in space studies from International Space University analyzing the flight checkout data for the Surface Science Package on board the Huygens probe, worked for several years at NASA Ames Research Center as a research assistant for the SETI Institute, earned an undergraduate degree in physics from Evergreen State College, and earned a bachelor's degree in psychology from the University of North Texas.

MICHAEL H. MOLONEY is the director of the SSB and the ASEB at the NRC. Since joining the NRC in 2001, Dr. Moloney has served as a study director at the National Materials Advisory Board, BPA, the Board on Manufacturing and Engineering Design, and the Center for Economic, Governance, and International Studies. Before joining the SSB and ASEB in April 2010, he was associate director of the BPA and study director for the Astro2010 decadal survey for astronomy and astrophysics. In addition to his professional experience at the NRC, Dr. Moloney has more than 7 years' experience as a foreign-service officer for the Irish government and served in that capacity at the Embassy of Ireland in Washington, D.C., the Mission of Ireland to the United Nations in New York, and the Department of Foreign Affairs in Dublin, Ireland. A physicist, Dr. Moloney did his graduate Ph.D. work at Trinity College Dublin in Ireland. He received his undergraduate degree in experimental physics at University College Dublin, where he was awarded the Nevin Medal for Physics.

G

Acronyms

AAS	American Astronomical Society
ACE	Advanced Composition Explorer
ACR	anomalous cosmic ray
AFOSR	Air Force Office of Scientific Research
AFRL	Air Force Research Laboratory
AFWA	Air Force Weather Agency
AGS	Atmospheric and Geospace Sciences Division
AGU	American Geophysical Union
AIA	Atmospheric Imaging Assembly
AIM	atmosphere-ionosphere-magnetosphere (system); Aeronomy of Ice in the Mesosphere (mission)
AIMI	atmosphere-ionosphere-magnetosphere interactions
AIP	American Institute of Physics
AMISR	Advanced Modular Incoherent Scatter Radar
AMPERE	Active Magnetosphere and Planetary Electrodynamics Response Experiment
AMPTE	Active Magnetospheric Particle Tracer Explorers
AO	Announcement of Opportunity
APS	American Physical Society; active pixel sensor
AST	Division of Astronomical Sciences
ATST	Advanced Technology Solar Telescope
AU	astronomical unit
AURA	Association of Universities for Research in Astronomy
AXAF	Advanced X-ray Astrophysics Facility
BARREL	Balloon Array for RBSP Relativistic Electron Losses
BBSO	Big Bear Solar Observatory

CADRE	CubeSat Investigating Atmospheric Density Response to Extreme Driving
CAREER	Faculty Early Career Development
CASSIOPE	Cascade, Small Satellite, and Ionospheric Polar Explorer
CATE	cost and technical evaluation
CAWSES	Climate and Weather of the Sun-Earth System
CCD	charge-coupled device
CCE	Charge Composition Explorer
CCMC	Community Coordinated Modeling Center
CDC	Concept Design Center
CEDAR	Coupling, Energetics, and Dynamics of Atmospheric Regions
CHAMP	Challenging Mini-Satellite Payload
CINDI	Coupled Ion Neutral Dynamic Investigation
CINEMA	CubeSat for Ions, Neutrals, Electrons, Magnetic Fields
CIR	corotating interaction region
CISM	Center for Integrated Space Weather Modeling
CME	coronal mass ejection
CMOS	Complementary Metal Oxide Semiconductor
C/No	carrier signal-to-noise strength
C/NOFS	Communications/Navigation Outage Forecasting System
COBE	Cosmic Background Explorer
CoBRA	Complexity Based Risk Assessment
COSMIC	Constellation Observing System for Meteorology, Ionosphere, and Climate
COSMO	Coronal Solar Magnetism Observatory
COTS	commercial off-the-shelf
CPU	central processing unit
CRRES	Combined Release and Radiation Effects Satellite
CSSWE	Colorado Student Space Weather Experiment
DASI	distributed array of small instruments
DC	direct current
DICE	Dynamic Ionosphere Cubesat Experiment
DLR	German Aerospace Center
DMSP	Defense Meteorological Satellites Program
DOD	Department of Defense
DOE	Department of Energy
DRIVE	Diversify, Realize, Integrate, Venture, Educate
DSCOVR	Deep Space Climate Observatory
DYNAMIC	Dynamical Neutral Atmosphere-Ionosphere Coupling (mission)
EASCO	Earth-Affecting Solar Causes Observatory
ECT	Energetic Particle, Composition, and Thermal Plasma Suite
EIA	equatorial ionization anomaly
EISCAT	European Incoherent Scatter Scientific Association
ELaNa	Educational Launch of Nanosatellites
ENA	energetic neutral atom
ENASI	Energetic Neutral Atom Spectroscopic Imager

EPO	education and public outreach
e-POP	Enhanced Polar Outflow Probe
EPP	energetic particle precipitation
ESA	European Space Agency
ESCAPE	Energetics, Sources and Couplings of Atmosphere-Plasma Escape
ETA	equatorial temperature anomaly
EUV	extreme ultraviolet
EUVI	extreme ultraviolet imager
EVE	Extreme Ultraviolet Variability Experiment
FAA	Federal Aviation Administration
FASR	Frequency-Agile Solar Radiotelescope
FAST	Fast Aurora Snapshot
FDSS	Faculty Development in Space Science
FIREBIRD	Focused Investigations of Relativistic Electron Burst Intensity, Range, and Dynamics
FOXSI	Focusing Optics X-ray Spectroscopic Imager
FUV	far ultraviolet
FWHM	full-width, half-maximum
FY	fiscal year
GCR	galactic cosmic ray
GDC	Geospace Dynamics Constellation
GEM	Geospace Environment Modeling
GEO	geostationary Earth orbit
GI	guest investigator
GIC	geomagnetically induced current; ground-induced current
GNSS	Global Navigation Satellite Systems
GOCE	Gravity Field and Steady-State Ocean Circulation Explorer
GOES	Geostationary Operational Environmental Satellite
GONG	Global Oscillation Network Group
GPS	Global Positioning System
GRACE	Gravity Recovery and Climate Experiment
GRIPS	Gamma-Ray Imager/Polarimeter for Solar Flares
GRIS	Gamma-Ray Imaging Spectrometer
GSFC	Goddard Space Flight Center
GSRP	Graduate Student Research Program
GUVI	Global Ultraviolet Imager
GW	gravity waves
HAARP	High Frequency Active Auroral Research Program
HAO	High Altitude Observatory
HEAO	High Energy Astronomy Observatory
HEIDI	High Energy Imaging Device
HEO	high Earth orbit (above GEO)
HF	high frequency
HI	Heliospheric Imager

HIREGS	High Resolution Gamma-Ray and Hard X-Ray Spectrometer
HITDP	Heliophysics Instrument and Technology Development Program
HPDE	Heliophysics Data Environment
HSC	heliophysics science center
HSO	Heliophysics Systems Observatory
HSS	high-speed solar wind streams
HSW	halo solar wind
HTP	Heliophysics Theory Program
HXR	hard X-ray
IBEX	Interstellar Boundary Explorer
IGS	International GNSS Service
IGY	International Geophysical Year
IHY	International Heliophysical Year
IKAROS	Interplanetary Kite-craft Accelerated by Radiation of the Sun
IMAGE	Imager for Magnetopause-to-Aurora Global Exploration
IMAP	Interstellar Mapping and Acceleration Probe
IMF	interplanetary magnetic field
IMP	Interplanetary Monitoring Platform
IMPACT	In Situ Measurements of Particles and CME Transients
INCA	Imaging Neutral Atom Camera
INMS	Ion Neutral Mass Spectrograph
IR	infrared
IRIS	Interface Region Imaging Spectrograph
ISCCP	International Satellite Cloud Climatology Project
ISEE	International Sun-Earth Explorer
ISI	Institute for Scientific Information
ISM	interstellar medium
ISMF	interstellar magnetic field
ISOON	Improved Solar Observing Optical Network
ISR	incoherent scatter radar
ISS	International Space Station
IT	ionosphere-thermosphere
ITAR	International Traffic in Arms Regulations
ITM	ionosphere-thermosphere-mesosphere
IVM	Ion Velocity Meter
JAXA	Japan Aerospace Exploration Agency
JPL	Jet Propulsion Laboratory
L1	Sun-Earth Lagrangian Point 1
L5	Sun-Earth Lagrangian Point 5
LANL	Los Alamos National Laboratory
LBH	Lyman-Birge-Hopfield
LCAS	Low-Cost Access to Space
LECP	low-energy charged-particle

LEO	low Earth orbit
LET	Low Energy Telescope
LISM	Local Interstellar Medium
LPV	Localizer Performance with Vertical Guidance
LT	local time
LWS	Living With a Star
MAC	Magnetospheric-Atmosphere Coupling (mission)
MagCat	Magnetospheric Constellation and Tomography (mission)
MagCon	Magnetospheric Constellation (mission)
MEDICI	Magnetosphere Energetics, Dynamics, and Ionospheric Coupling Investigation
MEMS	microelectromechanical system
MESSENGER	Mercury Surface, Space Environment, Geochemistry, and Ranging
MHD	magnetohydrodynamic
MI	magnetosphere-ionosphere
MIDEX	Mid-size Explorer
MISTE	Magnetosphere-Ionosphere Source Term Energetics (mission)
MIT	magnetosphere-ionosphere-thermosphere
MLSO	Mauna Loa Solar Observatory
MLT	mesosphere and lower thermosphere
MMS	Magnetospheric Multiscale (mission)
MO&DA	mission operations and data analysis
MODIS	Moderate Resolution Imaging Spectroradiometer
MPS	Mathematical and Physical Sciences
MREFC	Major Research Equipment and Facilities Construction (program)
MRF	Medium-Scale Research Facility
MSO	Mees Solar Observatory
MWO	Mt. Wilson Observatory
NASA	National Aeronautics and Space Administration
NCAR	National Center for Atmospheric Research
NESDIS	National Environmental Satellite, Data, and Information Service
NESSF	NASA's Earth and Space Science Fellowship (program)
NGDC	National Geophysical Data Center
NIST	National Institute of Standards and Technology
NJIT	New Jersey Institute of Technology
NOAA	National Oceanic and Atmospheric Administration
NRC	National Research Council
NSF	National Science Foundation
NSO	National Solar Observatory
NST	New Solar Telescope
NSWP	National Space Weather Program
NuSTAR	Nuclear Spectroscopic Telescope Array
NWM	Neutral Wind Meter
NWS	National Weather Service

OCT	Office of the Chief Technologist
OEDG	Opportunities for Enhancing Diversity in the Geosciences (program)
ORBITALS	Outer Radiation Belt Injection, Transport, Acceleration and Loss Satellite
OVSA	Owens Valley Solar Array
PARS	Polar Aeronomy and Radio Science
PFISR	Poker Flat Incoherent Scatter Radar
PI	principal investigator
POES	Polar Operational Environmental Satellite
P-POD	Poly Picosatellite Orbital Deployer
PUI	pickup ion
PW	planetary waves
QBO	quasi-biennial oscillation
R_E	Earth radius
R_S	Sun radius
R&A	research and analysis
RAX	Radio Aurora Explorer
RBSP	Radiation Belt Storm Probe; renamed Van Allen Probes
REPT	Relativistic Electron-Proton Telescope
REU	Research Experiences for Undergraduates
RFI	request for information
RHESSI	Ramaty High-Energy Solar Spectroscopic Imager
RISR	Resolute Bay Incoherent Scatter Radar
ROSES	Research Opportunities in Space and Earth Sciences
RPA	Ion Retarding Potential Analyzer
SABER	Sounding of the Atmosphere using Broadband Emission Radiometry
SALMON	Stand Alone Missions of Opportunities Notice
SAMPEX	Solar Anomalous and Magnetospheric Particle Explorer
SAPS	subauroral polarization streams
SCOSTEP	Scientific Committee on Solar-Terrestrial Physics
SDO	Solar Dynamics Observatory
SEE	Solar Eruptive Events (mission)
SEP	solar energetic particle
SFO	San Fernando Observatory
SHINE	Solar, Heliosphere and Interplanetary Environment
SHP	solar and heliospheric physics
SKR	Saturnian Kilometric Radio
SM	servicing mission
SMD	Science Mission Directorate
SMEI	Solar Mass Ejection Imager
SMEX	Small Explorer
SNOE	Student Nitric Oxide Explorer
SNR	signal-to-noise ratio

SO	Solar Orbiter
SOHO	Solar and Heliospheric Observatory
SOLIS	Synoptic Optical Long-term Investigations of the Sun
SOON	Solar Observing Optical Network
SORCE	Solar Radiation and Climate Experiment
SPA	Space Physics and Aeronomy
SPARC	Stratospheric Processes and Their Role in Climate
SPASE	Space Physics Archive Search and Extract
SPD	Solar Physics Division
SPI	Solar Polar Imager
SPP	Solar Probe Plus
SRAG	Space Radiation Analysis Group
SRP	solar radiation pressure
SR&T	supporting research and technology
SSP	solar sail propulsion
SSPIS	Solar and Space Physics Information System
STE	suprathermal electron
STEM	science, technology, engineering, and mathematics
STEREO	Solar-Terrestrial Relations Observatory
STP	Solar-Terrestrial Probes (program)
SuperDARN	Super Dual Auroral Radar Network
SUVIT	Solar UV-Visible-IR Telescope
SWaC	space weather and climatology
SWAP	space, weight, and power
SWICS	Solar Wind Ion Composition Spectrometer
SWL	Space Weather Laboratory
SWMI	solar wind-magnetosphere interactions
SWPC	Space Weather Prediction Center
SXT	Soft X-Ray Telescope
TDRSS	Tracking and Data Relay Satellite System
TEC	total electron content
TEX	tiny Explorers
THEMIS	Time History of Events and Macroscale Interactions during Substorms
TIM	Total Irradiance Monitor
TIMED	Thermosphere-Ionosphere-Mesosphere Energetics and Dynamics
TR&T	Targeted Research and Technology
TRACE	Transition Region and Coronal Explorer
TRL	technology readiness level
TS	termination shock
TSI	total solar irradiance
TWINS	Two Wide-Angle Imaging Neutral-Atom Spectrometers
UARS	Upper Atmosphere Research Satellite
UHURU	Small Astronomical Satellite 1, Swahili for *Freedom*
ULDB	Ultra Long Duration Balloon

USAF	U.S. Air Force
USGS	U.S. Geological Survey
USML	U.S. Munitions List
UT	universal time
UV	ultraviolet
UVCS	Ultraviolet Coronagraph Spectrometer
V1	Voyager 1
V2	Voyager 2
VxO	(domain-specific) virtual observatory
WAAS	Wide Area Augmentation System
WACCM	Whole Atmosphere Community Climate Model
WBS	work breakdown structure
WHI	Whole Heliosphere Interval
WLC	White Light Coronagraph
WPI	wave-particle interaction
WSA-Enlil	Wang-Sheely-Arge-Enlil
WSO	Wilcox Solar Observatory

H

Request for Information from the Community

The request for information from the solar and space physics community is reprinted below.

To: Members of the Solar and Space Physics Community
From: Daniel Baker and Thomas Zurbuchen, Chair and Vice Chair of the 2013-2022 NRC Decadal Survey in Solar and Space Physics
Date: September 10, 2010

Background: As you may know, the Space Studies Board of the National Research Council (NRC) is beginning to organize the next "decadal survey" in solar and space physics (heliophysics). This community-based effort will develop a comprehensive science and mission strategy for the field that updates and extends the first-ever NRC survey, which was completed in 2002. An open website has been created to describe the survey and to provide an opportunity for community input throughout the study process: http://tinyurl.com/2b4e4sh.

The statement of task, which is posted on the website, directs the survey committee to:

1. Provide an overview of the science and a broad survey of the current state of knowledge in the field, including a discussion of the relationship between space- and ground-based science research and its connection to other scientific areas;
2. Identify the most compelling science challenges that have arisen from recent advances and accomplishments;
3. Identify—having considered scientific value, urgency, cost category and risk, and technical readiness—the highest priority scientific targets for the interval 2013-2022, recommending science objectives and measurement requirements for each target rather than specific mission or project design/implementation concepts; and
4. Develop an integrated research strategy that will present means to address these targets.

The scope of the study will include:

- The structure of the Sun and the properties of its outer layers in their static and active states;
- The characteristics and physics of the interplanetary medium from the surface of the Sun to interstellar space beyond the boundary of the heliosphere; and
- The consequences of solar variability on the atmospheres and surfaces of other bodies in solar system, and the physics associated with the magnetospheres, ionospheres, thermospheres, mesospheres, and upper atmospheres of the Earth and other solar system bodies.

The section of the report that provides recommendations to NASA will reflect NASA's responsibility for flight mission investigations. (However, the survey will consider the space-based activities and programs of other agencies; e.g., NSF's CubeSat program.)

Recommendations related to ground-based implementations (e.g., ground-based solar observatories) will be directed to the NSF. As noted above, the survey committee is charged with establishing science targets and then matching these targets to notional mission designs. In a change from the previous survey, we will employ an independent contractor to assist in the cost estimation and technical analysis of the notional missions.

We anticipate that the survey committee will recommend a mix of small ($250-$500 million), medium ($500-$750 million), and large ($750-$1000 million) strategic flight investigations appropriate to goals of the Solar-Terrestrial Probes program and the Living With a Star program. Flagship (> $1 billion) flight investigations will be considered if permitted by the budget guidelines provided to the survey committee by NASA. Candidates for individual Explorer and lower-cost flight projects, including sub-orbital programs, will not be prioritized; they will be solicited and selected as PI-led missions on the basis of science merit via NASA's standard AO processes. However, the strategic value of these smaller mission programs to the overall science agenda will be appraised and described in the report from the study.

Request to the Community: ***We invite you to write a concept paper (e.g., about a mission or extended mission, observation, theory, or modeling activity) that promises to advance an existing or new scientific objective, contribute to fundamental understanding of the Sun-Earth system, and/or facilitate the connection between science and societal needs (e.g., improvements in space weather prediction).***

Note that missions that had been prescribed in the 2003 survey (*The Sun to the Earth—and Beyond*) or other recent NRC reports but have not yet been formally confirmed for implementation will need to be reprioritized for inclusion in the present survey (with the exception, per NASA, of Solar Orbiter and Solar Probe Plus).

The ideas and concepts received will be reviewed by one or more of the survey's study panels, which are organized to address the following themes:

- Atmosphere-Ionosphere-Magnetosphere Interactions
- Solar Wind-Magnetosphere Interactions
- Solar and Heliospheric Physics

Concept papers can range from a specific instrument or mission proposal to a white paper that examines an issue of broad concern to the community. They may be directed for action by NASA or the NSF; indeed, our audience includes all federal agencies responsible for the conduct of solar and space physics research and operations. We encourage concept papers that, for example, examine the potential of novel instruments or measurement techniques, including the use of orbital or suborbital vehicles of various sizes. In addition, we welcome concept papers on the use of existing or new ground-based facilities, as well as the potential of laboratory experiments. In all cases, *concept papers should include the identification of a science and/or societal objective and motivate the importance of achieving this objective in the context of the state of the field.*

Initiatives and concepts that are cross-cutting among these themes are particularly encouraged. We anticipate that one or more of the concepts will be selected by each panel for a detailed technical and cost assessment. Panel recommendations and the technical and cost assessments will be provided to the decadal survey steering committee, which is charged with writing a final report that includes targeted recommendations for the agencies. All responses will be considered non-proprietary public information for distribution with attribution. The concept papers should be no longer than seven single-spaced pages in length including figures and tables (but not including footnotes), and should provide the following information, if possible (see next page for more formatting requirements):

1. For proposed observations, technology innovations, or missions, a summary of the science concept, including the observational variable(s) to be measured, the characteristics of the measurement if known (accuracy, spatial and temporal resolution), and domain of the observation. For proposed theory, modeling, or cross-cutting science themes, a summary of the science concept, key opportunities to be addressed, and approach used.
2. A description of how the proposed science concept will help advance solar and space physics science or provide a needed operational capability for the next decade and beyond.

3. For proposed observations, technology innovations and missions, a rough estimate of the total cost (FY10 dollars), including launch, of a proposed mission over ten years, if it can be estimated. For an "operational" mission, the costs should include one-time costs associated with building the instrument and launch and ongoing operational costs. Please describe the supporting rationale for the total mission cost estimate to understand the confidence that the mission considered could fall within the cost bins identified (i.e. small, medium or large). For proposed theory, modeling, and cross-cutting science themes, an estimated total cost, including all key phases of the proposed project.

4. A description of how the proposed concept meets one or more of the following criteria, which will be used to evaluate the candidate proposals:

a. Is identified as a high priority or requirement in previous studies or roadmaps (for example, the earlier NRC decadal survey);
b. Makes a significant contribution to more than one of the Panel themes;
c. Contributes to important scientific questions facing solar and space physics today;
d. Contributes to applications and/or policy making (operations, applications, societal benefits);
e. Complements other observational systems or programs available;
f. Is affordable (cost-benefit);
g. Has an appropriate degree of readiness (technical, resources, people);
h. Fits with other national and international plans and activities.

The description should provide enough detail that the potential value and feasibility can be evaluated by an independent group of experts. We note that the National Research Council's ongoing Astronomy and Astrophysical Decadal Survey solicited Science White papers for their "Astro2010" activity. Examples of community input to their request can be found at http://sites.nationalacademies.org/BPA/BPA_050603.

White Paper Format Requirements

- One-inch margins (8.5" x 11" page)
- No more than 7 pages including figures and tables (but not including footnotes)
- 11-12 point font

Submissions to the survey, which will be posted on a public website, can be made by following instructions on the survey website (http://tinyurl.com/2b4e4sh).

For full consideration, please submit the concept paper *no later than* 5:00pm Friday, November 12, 2010.

Questions about the RFI may be directed to the study director, Art Charo (acharo@nas.edu, by telephone at (202) 334-3477, or by fax at (202) 334-3701), or to the survey's Chair and Vice Chair, Dan Baker (Daniel.Baker@lasp.colorado.edu) and Thomas Zurbuchen, (thomasz@umich.edu), respectively.

I

List of Responses to Request for Information

Listed in Table I.1 are the responses received by the Committee on a Decadal Strategy for Solar and Space Physics (Heliophysics) in response to its request for information (RFI) sent in September 2010 to the solar and space physics community (see Appendix H). The full-text versions of the RFI responses are included in the compact disk that contains this report and are also available online through links at the survey's website at http://sites.nationalacademies.org/SSB/CurrentProjects/SSB_056864.

TABLE I.1 Responses to Request for Information

RFI Response Number	First Author	Response Title	Summary Description
1	Ali, Nancy A., et al.	Recommendations for Education/Public Outreach (EPO) Programs: A White Paper Submitted for Consideration to the NRC Decadal Survey in Solar and Space Physics	Addresses EPO as a major contributing factor to workforce development in solar and space physics as well as in creating a scientifically literate U.S. public.
2	Araujo-Pradere, Eduardo A.	Research to Operations (R2O) Activities, a Natural Conclusion of Research	Details transitions of academic models to operations, which require an organizational structure and a clear financial commitment that barely exists today.
3	Ayres, Thomas R., and D. Longcope	Ground-Based Solar Physics in the Era of Space Astronomy	Provides a synopsis of a 2009 report commissioned by AURA in advance of the Astro2010 decadal survey concerning the future of ground-based solar physics.
4	Bach, Bernhard, et al.	The Use of a Z-pinch Facility as a Platform for Laboratory Solar and Heliophysics	Explores the utilization of Z-pinch facilities to create and investigate the physics of high-energy-density plasmas.

TABLE I.1 Continued

RFI Response Number	First Author	Response Title	Summary Description
5	Bailey, Scott M., et al.	A Mission to Study the Coupling of Atmospheric Regions by Precipitating Energetic Particles	Proposes a mission to understand the atmospheric response to energetic particles, specifically the coupling of atmospheric regions, as the particle energy is redistributed via dynamical, chemical, and radiative processes.
6	Baker, Joseph B.H., et al.	The Importance of Distributed Measurements of the Ionospheric Electric Field for Advancement of Geospace System Science and Improved Space Weather Situational Awareness	Explains the value of networks of ionospheric radars to the space physics community because they provide spatially distributed electric field measurements and should be further developed.
7	Bala, Ramkumar	Space Weather Forecasting through Association	Advocates for stronger community participation.
8	Balch, Christopher C.	The Next Step in Heliospheric Modeling—Increasing the Interplanetary Observing Network	Proposes an enhanced interplanetary observing network consisting of up to 1,000 CubeSats and recommends use of known technologies and modeling techniques to make order of magnitude improvements in the accuracy of physics-based models for the solar wind and interplanetary coronal mass ejections (CMEs).
9	Bandler, Simon R., et al.	High Spectral Resolution, High Cadence, Imaging X-ray Microcalorimeters for Solar Physics	Describes a solar-optimized X-ray microcalorimeter that provides high-resolution spectra at arcsecond scales to enable a wide range of studies, such as the detection of microheating in active regions, ion-resolved velocity flows, and the presence of non-thermal electrons in hot plasmas.
10	Bellan, Paul M.	Using Laboratory Experiments to Study Solar Corona Physics	Recommends that the next decade of heliospheric research include advanced laboratory plasma experiments designed to tackle specific, outstanding coronal issues.
11	Bernasconi, Pietro N., and N.-E. Raouafi	Solar Magnetized Regions Tomograph (SMART) Mission	Presents a mission to map the solar vector magnetic fields at high spatial resolution at several heights in the solar atmosphere from the photosphere to the chromosphere across the magnetic transition region.
12	Bhattacharjee, Amitava, et al.	Advanced Computational Capabilities for Exploration in Heliophysical Science (ACCEHS)—A Virtual Space Mission	Recommends that NASA, perhaps in partnership with National Science Foundation (NSF) and other agencies, lead by establishing a new peer-reviewed program in which critical-mass groups of heliophysicists, computational scientists, and applied mathematicians are brought together to address transformational science quests.
13	Bishop, Rebecca L., et al.	Understanding Tropospheric Influences on the Mesosphere/Thermosphere/ Ionosphere Region	Presents specific science goals and observational platforms required to perform investigations into tropospheric and thermospheric/ ionospheric coupling.
14	Bishop, Rebecca L., and J. Roeder	The International Space Station: Platform for Future Upper Atmospheric Investigations	Presents potential upper atmospheric investigations, sensors, and customers utilizing the International Space Station.

continued

TABLE I.1 Continued

RFI Response Number	First Author	Response Title	Summary Description
15	Bookbinder, Jay, et al.	The Solar Spectroscopy Explorer Mission	Presents the Solar Spectroscopy Explorer, a small strategic mission built around an X-ray microcalorimeter and a high spatial resolution extreme ultraviolet (EUV) imager.
16	Bortnik, Jacob, and Y. Nishimura	MMAP: A Magnetic-Field Mapping Mission Concept	Outlines a novel mission that aims to observationally map the geomagnetic field from geosynchronous Earth orbit to the ionosphere in near-real time using the recently described link between pulsating aurora and chorus waves.
17	Bortnik, Jacob	The Critical Role of Theory and Modeling in the Dynamic Variability of the Radiation Belts and Ring Current	Highlights the critical role that was played by theoretical and modeling projects in the preceding decade and urges the decadal committee to support further modeling efforts dealing with wave-particle interactions in controlling the structure and dynamics of the radiation belts.
18	Brown, Benjamin P., et al.	An Experimental Plasma Dynamo Program for Investigations of Fundamental Processes in Heliophysics	Advocates for community-scale laboratory plasma experiments that offer unique opportunities to probe heliophysically relevant phenomena.
19	Brown, Michael R., et al.	Intermediate-Scale MHD Wind Tunnel for Turbulence and Reconnection Studies	Proposes an intermediate scale magnetohydrodynamic (MHD) wind tunnel for turbulence studies in order to illuminate MHD turbulence processes such as observed in the solar wind.
20	Budzien, Scott A., et al.	Evolved Tiny Ionospheric Photometer (ETIP): A Sensor for Ionospheric Specification	Addresses the requirements for space weather sensors with adequate flexibility for accommodation on a range of future flight opportunities, including microsatellite constellations.
21	Budzien, Scott A., et al.	The Volumetric Imaging System for the Ionosphere (VISION)	Describes a mission for volumetric characterization of the ionosphere using optical tomography.
22	Budzien, Scott A., et al.	Heterogeneous Measurements for Advances in Space Science and Space Weather Forecasting	Emphasizes that space weather forecasting with new, full-physics models requires heterogeneous datasets with complementary characteristics—not merely a higher volume of any single data type.
23	Burch, James L., et al.	Magnetospheric Causes of Saturn's Pulsar-Like Behavior	Proposes a three-spacecraft mission to identify the cause of Saturn's periodicity.
24	Burger, Matthew H., et al.	Understanding Mercury's Space Environment-Magnetosphere-Exosphere System: A Unified Strategy for Observational, Theoretical, and Laboratory Research	Recommends a strong program to combine ground-based and spacecraft observations, laboratory measurements, and numerical modeling to maximize the science return from these missions.
25	Burkepile, Joan R., et al.	The Importance of Ground-Based Observations of the Solar Corona	Proposes the use of a new K-coronagraph that will provide dramatically better data of the very low corona.

TABLE I.1 Continued

RFI Response Number	First Author	Response Title	Summary Description
26	Carpenter, Kenneth G., et al.	Stellar Imager (SI): Developing and Testing a Predictive Dynamo Model for the Sun by Imaging Other Stars	Proposes a mission to resolve surface magnetic activity and subsurface structure and flows of a population of Sun-like stars in order to accelerate the development and validation of a predictive dynamo model for the Sun and enable accurate long-term forecasting of solar/stellar magnetic activity.
27	Cassak, Paul, et al.	The Development of a Quantitative, Predictive Understanding of Solar Wind-Magnetospheric Coupling	Proposes a multi-pronged and interdisciplinary effort to understand observationally and theoretically what controls solar wind-magnetospheric coupling and how to predict it.
28	Chakrabarti, Supriya, et al.	Domestication of Scientific Satellites	Proposes the use of a flexible and scalable satellite system design in order to fill the wide gap between CubeSats and small Explorer missions and make space accessible to a new generation of Explorers.
29	Chandran, Benjamin D.G., et al.	Theoretical Research on Solar Wind Turbulence	Describes several areas in which future research on theory of solar wind turbulence holds particular promise and offers brief policy recommendations.
30	Chau, Jorge L., et al.	An Ionospheric Modification Facility for the Magnetic Equator	Proposes the deployment of an ionospheric modification facility, also called ionospheric heater, near the geomagnetic equator.
31	Chi, Peter J., et al.	A National Ground Magnetometer Program for Heliophysics Research	Recommends the establishment of a national ground magnetometer program to help coordinate, maintain, and enhance the magnetometer networks in North America.
32	Chollet, Eileen E., et al.	Career Development for Postdoctoral and Early Career Scientists	Discusses some career development issues early career scientists face and recommends some community changes that will help the field retain young talent.
33	Christe, Steven D., et al.	The Focusing Optics X-ray Solar Imager (FOXSI)	Proposes a mission to learn how and where electrons are accelerated, along which field lines they travel away from the acceleration site, where they are stopped, and how some electrons escape into interplanetary space, using the Focusing Optics X-ray Solar Imager.
34	Christian, Eric R., et al.	Heliophysics Instrument and Technology Development Program (HITDP)	Describes a program to reinvigorate hardware development, provide a pathway for new technology to be infused into missions, develop the next generation of instrument scientists, and ensure a healthy science mission program.
35	Christensen, Andrew B., et al.	The International Space Station as a Space Physics Observation Platform V2	Demonstrates that the Exposed Facility on the Japanese Experiment Module is suitable for siting optical instrumentation and conducting scientific experiments.

continued

TABLE I.1 Continued

RFI Response Number	First Author	Response Title	Summary Description
36	Chu, Xinzhao, et al.	Space Lidar Mission to Study Middle and Upper Atmosphere Dynamics and Chemistry	Proposes a mission to make high-resolution temperature, wind, and Na-density measurements in the mesosphere and lower thermosphere region and to study the upper atmosphere chemistry, structure, and dynamics, especially the impact of gravity waves, using space Na Doppler lidar.
37	Chu, Xinzhao, and J. Thayer	Whole Atmosphere LIDAR for Whole Atmosphere Study	Proposes a mission to profile wind and temperature through the whole atmosphere from ground to 120 km with superlative accuracy and with whole atmosphere lidar.
38	Chua, Damien H., et al.	Geospace Dynamics Imager: A Mission Concept for Heliospheric and Magnetospheric Imaging and Space Weather Forecasting	Proposes a mission to provide the first direct, global images of the solar wind-magnetosphere system with the Geospace Dynamics Imager.
39	Chutjian, Ara, et al.	Laboratory Solar Physics from Molecular to Highly-Charged Ions: Meeting Future Space Observations of the Solar Plasma and Solar Wind	Describes an addition to the Jet Propulsion Laboratory facility to provide a compact storage ring with the new, required measurement capabilities.
40	Cirtain, Jonathan W., et al.	The High-Latitude Solar-C International Collaboration: Observing the Polar Regions of the Sun and Heliosphere	Proposes a mission to fly a focused suite of instruments designed to study the solar interior flows (by helioseismology), surface magnetic fields, transition region, and extended corona from an orbit inclined at least 40 degrees to the ecliptic plane.
41	Clarke, John T., et al.	White Paper on Comparative Planetary Exospheres	Recommends the observation of planetary and satellite exospheres by enhanced ground-based and new Earth-orbiting telescopic instruments.
42	Claudepierre, Seth G., et al.	A CubeSat Constellation to Study Magnetospheric Ultra-Low Frequency Pulsations	Proposes a mission to constrain the azimuthal mode number spectrum of magnetospheric ultralow-frequency pulsations.
43	CoBabe-Ammann, Emily, et al.	The Importance of Student Instrument Programs in the Workforce Development in Solar and Space Physics	Details how student instrument programs attract, retain, and move students through the higher education pipeline into graduate studies and the scientific and engineering workforce.
44	Codrescu, Mihail	Data Assimilation for the Thermosphere and Ionosphere	Develops global data assimilation schemes using coupled thermosphere ionosphere models and large amounts of diverse data.
45	Cohen, Christina M.S., et al.	Protecting Science Mission Investment: Balancing the Funding Profile for Data Analysis Programs	Addresses the necessary balance between data analysis funds and support for guest investigator programs.
46	Colgate, Stirling A.	Experiments to Demonstrate Solar and Astrophysical Dynamos	Explains how experiments have shown that turbulence leads primarily to enhanced resistive diffusion and not a dynamo.
47	Conde, Mark G.	Constructively Growing the Sounding Rocket Program: A Technology Development Line of Sounding Rocket Launches	Proposes that a competitive line of sounding rocket launches be added to the existing scientifically competed program.

TABLE I.1 Continued

RFI Response Number	First Author	Response Title	Summary Description
48	Cooper, John F., et al.	Space Weathering Impact on Solar System Surfaces and Mission Science	Explains how surfaces and atmospheres directly exposed to space environments of planetary magnetospheres, the heliosphere, and the local interstellar environment are eroded and chemically modified.
49	Coster, Anthea J., et al.	Investigations of Global Space Weather with GPS	Proposes improvements in the global distribution of ionospheric sensors.
50	Cranmer, Steven R., et al.	Ultraviolet Coronagraph Spectroscopy: A Key Capability for Understanding the Physics of Solar Wind Acceleration	Describes how ultraviolet coronagraph spectroscopy enables measurements of the collisionless processes responsible for producing the solar wind.
51	Davila, Joseph M., et al.	The International Space Weather Initiative (ISWI)	Details the deployment of 100 new instruments in Africa and around the world, including GPS, magnetometers, particle detectors, H-alpha telescopes, and radio spectrographs.
52	Davila, Joseph M., et al.	Understanding Magnetic Storage, Reconnection, and CME Initiation	Recommends the tracing of magnetic fields with a high-resolution coronagraph.
53	de la Beaujardiere, Odile, and C. Fesen	Global scintillation prediction	Proposes a systems-approach mission to predict scintillations from ultra-high frequency to L-band at all latitudes.
54	de la Beaujardiere, Odile, and D. Ober	Long-Term Changes in the Ionosphere/ Magnetosphere System and Reliable Platform for Innovations in Space Sensors	Recommends that the Defense Weather Satellite System accommodate instruments that provide the observations required for long-term trends in the magnetosphere ionosphere system as well as provide a reliable "home" for flight opportunity to test new instruments.
55	Denig, William F., et al.	On the Utility of Operational Satellite Data to Solar and Space Physics Research	Recommends increased interaction between the operational and research communities.
56	Desai, Mihir I., et al.	Particle Acceleration and Transport in the Heliosphere (PATH)	Proposes a mission to determine the mechanisms responsible for the acceleration and propagation of SEPs through the inner heliosphere.
57	Donovan, Eric	The Great Geospace Observatory and Simultaneous Missions of Opportunity	Presents a novel concept of a cost-effective multiagency initiative to fly the "Great Geospace Observatory" and provide a revolutionary three-dimensional view of Earth's plasma environment.
58	Dorelli, John C., et al.	A Proposal for a Computational Heliophysics Innovation Program (CHIP)	Proposes that NASA create a program to ensure that the heliophysics community keeps up with the rapidly advancing high performance computing frontier over the next decade.
59	Doschek, George A.	A Concept White Paper for a New Solar Flare Instrument Designed to Determine the Plasma Parameters in the Reconnection Region of Solar Flares at Flare Onset	Describes a possible Bragg crystal spectrometer experiment that could provide spectroscopic plasma diagnostics of the reconnection region of solar flares, such as electron temperature, turbulence, flows, and polarization.

continued

TABLE I.1 Continued

RFI Response Number	First Author	Response Title	Summary Description
60	Doschek, George A., et al.	The High Resolution Solar-C International Collaboration	Proposes a collaboration to observe simultaneously the photosphere, chromosphere, transition region, and corona at high spatial, spectral, and temporal resolution with proposed high-resolution Solar-C mission.
61	Dyrud, Lars, et al.	A Crucial Space Weather Effect: Meteors and Meteoroids	Describes the importance of the interplanetary meteoroid and dust environment to support studies of solar system evolution, solar wind, upper atmospheric physics, planetary atmospheres and ionospheres, planetary geology, and manned and unmanned spacecraft.
62	Dyrud, Lars, et al.	Commercial Access to Space for Scientific Discovery and Operations	Describes the observations required to achieve the greatest scientific advances with arrays of scientific sensors distributed throughout the system gathering data.
63	Eastes, Richard W.	Far Ultraviolet Imaging of the Earth's Thermosphere and Ionosphere	Proposes a mission to provide full disk images of atmospheric temperature and composition during the daytime and electron densities in the F2 region of the ionosphere at night using far ultraviolet (FUV) spectral imaging from geostationary orbit.
64	Ebbets, Dennis, C., et al.	Flight Opportunities for Hosted Payloads on the Iridium NEXT Satellites	Invites ideas for flying sensors as hosted payloads on the NEXT constellation of commercial communications satellites being developed by Iridium Satellite.
65	Elkington, Scot R., and X. Li	MORE/ORBITALS: An International Mission to Advance Radiation Belt Science	Proposes support for the MORE/ORBITALS mission, an international collaboration to build a spacecraft to study the dynamical evolution of the radiation belts.
66	Emmert, John T., et al.	Geospace Climate Present and Future	Describes the importance of the systematic response of geospace to natural and anthropogenic forcing for societal utilization of this environment.
67	England, Scott, L et al.	Concept Paper: An Investigation of the Coupling of the Earth's Atmosphere to Its Plasma Environment	Proposes a mission concept that addresses ion-neutral coupling.
68	Englert, Christoph, R., et al.	Spatial Heterodyne Spectroscopy: An Emerging Optical Technique for Heliophysics and Beyond	Proposes the use of the Spatial Heterodyne Spectroscopy for NASA missions.
69	Eparvier, Francis G.	The Need for Consistent Funding of Facilities Required for NASA Missions	Explores how to avoid the risk of the haphazard, precarious, and inconsistent funding of calibration and test facilities, which are necessary for the success of NASA missions such as NIST SURF Beamline-2.

TABLE I.1 Continued

RFI Response Number	First Author	Response Title	Summary Description
70	Erickson, Philip J., et al.	Investigations of Plasmasphere Boundary Layer Processes in the Coupled Earth-Sun Geospace System	Proposes a focused attack on advancing knowledge of ionospheric structuring and space weather effects driven by magnetosphere/ionosphere coupling in the critically important subauroral plasmasphere boundary layer, through use of multipoint measurement networks in both the American and Australian sectors.
71	Fennell, J.F., et al.	The Magnetospheric Constellation Mission	Reconsiders the original Magnetospheric Constellation Mission concept for implementation in the next decade.
72	Fennell, J.F., and P.T. O'Brien	Mission to Understand Electron Pitch Angle Diffusion and Characterize Precipitation Bands and Spikes	Proposes a mission to understand the processes and answer the questions raised by observations of precipitation bands.
73	Fennell, J.F.	Enhancement of POES Instruments to Provide Better Space Weather Electron Data	Encourages the National Oceanic and Atmospheric Administration (NOAA) to seriously consider flying an electron sensor that measures the precipitating and trapped electron fluxes in the 40-2000 keV energy range on a continuous basis to fulfill operational and science needs discussed.
74	Fennell, J.F., et al.	Transition Region Exploration (TREx) Mission	Describes the science need for a mission that spans the L* region from 4.5 to 8.5 in the equatorial plane.
75	Fleishman, Gregory D., et al.	Uncovering Mechanisms of Coronal Magnetism via Advanced 3D Modeling of Flares and Active Regions	Recommends capitalizing on new (or soon to be available) facilities such as Solar Dynamic Observatory (SDO), the Advanced Technology Solar Telescope (ATST), and the Frequency-Agile Solar Radiotelescope and the challenges they present.
76	Florinski, Vladimir, et al.	The Outer Heliosphere—Solar System's Final Frontier	Discusses the need for a dedicated theoretical program to study the physics of the outer heliosphere and identifies four main thrust areas: global structure, pickup ions and anomalous cosmic rays, galactic cosmic rays, and physics of the termination shock and heliopause.
77	Foster, John	DASI: Distributed Arrays of Scientific Instruments for Geospace and Space Weather Research	Demonstrates the importance of a larger-perspective point of view to appreciate how the individual features of geospace come together.
78	Frahm, Rudy A.	Interaction of the Solar Wind with a Partially Magnetized Planet	Proposes a spacecraft experiment that deals with our nearest neighbor planet, Mars, and its interaction with the space environment.
79	Frazier, Jr., Jesse R.	Alternate Magnetic Thermodynamics	Presents the author's unconventional views on magnetism and energy.
80	Fritts, Dave, et al.	Solar Forcing of the Thermosphere and Ionosphere from Below: Coupling via Neutral Wave Dynamics	Describes the motivations for a mission addressing neutral atmosphere-ionosphere coupling via neutral waves propagating into the thermosphere and ionosphere from the lower atmosphere and the auroral zone.

continued

TABLE I.1 Continued

RFI Response Number	First Author	Response Title	Summary Description
81	Fritz, Theodore, A., and B.M. Walsh	Particle Acceleration and Entry of Solar Wind Energy into the Magnetosphere	Recommends the investigation of energetic particle acceleration in the cusp and the contributions of such a population to the plasma sheet and radiation belt.
82	Fry, Dan J., et al.	Solar Proton Event Risk Modeling for Variable Duration Human Spaceflight	Proposes a strategy that will utilize agency science to feed the development of needed near-term probabilistic models that assess solar proton event risk for long-duration human exploration.
83	Fuller-Rowell, Tim, et al.	Forecasting Ionospheric Irregularities	Discusses the pressing need to develop a capability to forecast the likely occurrence of ionospheric irregularities and their detrimental impact on communications and navigation.
84	Fung, Shing F., et al.	Magnetosphere-Ionosphere Connector (MAGIC): Investigation of Magnetosphere-Ionosphere Coupling from High-to-Low Latitudes	Proposes a mission with a high-altitude satellite for auroral and plasmaspheric imaging and multiple lower-orbiting spacecraft for simultaneous in situ and radio sounding measurements.
85	Fuselier, Stephen A., et al.	Stereo Magnetospheric Imaging (SMI) Mission	Proposes energetic neutral atom (ENA) imaging mission using two spacecraft at the lunar L4 and L5 Lagrange points to investigate plasma processes at the bow shock and magnetopause, and in the cusps and magnetotail, lunar interactions, and heliospheric and interstellar phenomena.
86	Gary, Dale E., et al.	The Frequency-Agile Solar Radiotelescope (FASR)	Proposes that a wide range of science goals can be addressed with a solar-dedicated radio telescope in a high state of readiness, with superior imaging capability and broad frequency coverage.
87	Gary, Dale E., et al.	Particle Acceleration and Transport on the Sun	Describes the comprehensive observations required to understand particle acceleration and particle transport on the Sun.
88	Gentile, L.C., et al.	Scintillation and Energy Input for Space Situational Awareness and Monitoring the Environment (SESSAME)	Provides a new generation of space environmental monitoring instruments to measure high-latitude energy input and scintillation at both high and equatorial latitudes.
89	Gentile, L.C., et al.	Constellation for Heliospheric and Ionospheric Equatorial Forecasting of Scintillation (CHIEFS)	Proposes to advance understanding of ionospheric effects on communication and navigation systems with a constellation of small, dedicated satellites orbiting the equator, combined with data from ground-based instruments.
90	Gentile, L.C., et al.	Polar and Equatorial Communication Outage Satellites (PECOS)	Gives a low-cost solution that meets many current space weather objectives with three paired constellations of small satellites flying in multiple orbits.

TABLE I.1 Continued

RFI Response Number	First Author	Response Title	Summary Description
91	Giampapa, Mark S., et al.	Asteroseismology: The Next Frontier in Solar-Stellar Physics	Recommends comparative studies of the influence of parameters such as rotation and convection zone structure on dynamo-related magnetic activity at all relevant timescales.
92	Giampapa, Mark, S., et al.	Causes of Solar Activity	Emphasizes the critical importance of a sustained program of long-term, high-continuity observations of the solar magnetic field by a network of ground-based synoptic-type instruments.
93	Gilbert, Jason A.	What Composition Measurements Could Have Done for Solar Probe Plus	Examines the science benefits that composition data would have brought to the Solar Probe Plus missions, which explore the source regions of the solar wind and of inner-source pickup ions.
94	Gjerloev, Jesper W., et al.	Auroral Forms and Their Role in the Dungey Convection Cycle	Urges the committee to include the fundamental science objectives: What is the role of meso-scale auroral forms in the Dungey global convection cycle?
95	Gjerloev, Jesper W., et al.	SuperMAG: The Global Ground Based Magnetometer Initiative	Urges the decadal survey committee to acknowledge the strength of the ground-based magnetometer data set and the need for global collaborations such as SuperMAG.
96	Golub, Leon, et al.	RAM: The Reconnection and Microscale Mission	Outlines a new approach to understanding the dynamic activity of hot, magnetized plasmas using the best example available, the solar corona, with the goals of determining the configurations that lead to energy release and locating sources of high-energy particles.
97	Goncharenko, Larisa P., et al.	Coupling Through Planetary Waves: From the Stratosphere to Ionospheric Irregularities	Proposes an observational strategy that would investigate potential effects of planetary waves on irregularities.
98	Goode, Philip	The 1.6 m Clear Aperture Optical Solar Telescope in Big Bear—The NST	Describes how the largest aperture (1.6 m) solar telescope will provide an essential complement to SDO, Hinode, and other satellite data, especially as a probe of the space weather.
99	Gopalswamy, Nat, et al.	Earth-Affecting Solar Causes Observatory (EASCO): A New View from Sun-Earth L5	Outlines the concept of a mission that will make remote sensing and in situ measurements from the Sun-Earth Lagrange point L5 to understand the origin and evolution of large-scale solar disturbances such as coronal mass ejections and corotating interaction regions.
100	Gross, Nicholas, and J. Sojka	Value and Need of Helio and Space Physics Summer Schools	Highlights the value of space physics summer schools and encourages the continued funding of similar efforts.
101	Gross, Nicholas	Value of Enhanced Mentoring in Space and Helio Physics	Outlines the value of enhancing mentoring that a center provides through exposure of students to a broad range of mentors and activities.
102	Grotheer, Emmanuel B., et al.	Determination of Optical Spectra and G-values for Negative Ions of Low-Mass Atoms and Molecules	Recommends research to determine negative ions' emission spectra and g-values and discusses interactions of solar wind with Mercury's magnetosphere and surface.

continued

TABLE I.1 Continued

RFI Response Number	First Author	Response Title	Summary Description
103	Habbal, Shadia R., et al.	Exploring the Physics of the Corona with Total Solar Eclipse Observations	Supports total solar eclipse observations in the visible and near infrared wavelength range to explore the physics of the corona, in particular on August 21, 2017.
104	Heelis, Rod	Magnetosphere Atmosphere Coupling Mission (MACM)	Proposes a mission to discover the spatial and temporal scales over which the ionosphere and thermosphere respond to magnetospheric energy inputs and determine how magnetospheric sources and wind dynamos contribute to the electric field in the ionosphere and thermosphere.
105	Heelis, Rod	Space-Atmosphere Boundary Layer Electrodynamics (SABLE)	Proposes a mission to uncover the pathways through which energy from the magnetosphere and solar wind is redistributed in Earth's atmosphere.
106	Hess, Sebastien L.G., et al.	Exploration of the Uranus Magnetosphere	Proposes a middle-size mission to explore the Uranus magnetosphere with a minimal set of instruments that are necessary to address the most compelling questions about the Uranus magnetosphere and to improve understanding of the solar wind-magnetosphere interactions in general.
107	Hill, Frank, et al.	Helioseismology	Supports ground-based multi-wavelength observations and space-based multi-viewpoint measurements in order to further understanding of space physics.
108	Hill, Frank, et al.	The Need for Synoptic Optical Solar Observations from the Ground	Discusses the value of long-term observations in understanding the Sun and its activity cycle and the importance of providing sufficient resources to obtain and improve the measurements.
109	Holzworth, Robert H.	Lightning Influence on Ionosphere and Magnetosphere Plasma	Discusses the influence of lightning on ionoshere and magnetosphere processes, gives history behind current knowledge, and suggests that lightning-generated plasma waves may be much more important to magnetospheric and ionospheric physics than is realized.
110	Horanyi, Mihaly, et al.	iDUST: Interstellar and Interplanetary Dust Near Earth: A Mission Concept for "Dust Tomography" of the Heliosphere	Proposes a mission to observe the inward transport of interstellar dust and the outflow of near-solar dust and explore dusty plasma processes throughout the heliosphere.
111	Huang, Cheryl	A Satellite Mission Concept to Study Thermosphere-Ionosphere Coupling	Presents a satellite mission concept to study ion-neutral coupling with an improvement in satellite drag modelling.
112	Huba, Joseph D., et al.	A Comprehensive, First-Principles Model of Equatorial Ionospheric Irregularities and Turbulence	Discusses the development of a new modeling capability that describes the onset and development of equatorial ionospheric irregularities covering a spatial range of tens of centimeters to thousands of kilometers.

TABLE I.1 Continued

RFI Response Number	First Author	Response Title	Summary Description
113	Hudson, Hugh S., et al.	Solar Flares and the Chromosphere	Emphasizes the need to put new programs in place to follow up the Hinode and SDO successes and take advantage of modern modeling prowess and ground-based data to understand this complicated but physically fundamental domain.
114	Hughes, W. Jeffrey	The Future of Modeling the Space Environment	Details the need for a cohesive team of space and computational scientists and software engineers to develop a modern space environment model.
115	Intrator, Thomas P., et al.	Fundamental Heliophysics Processes: Unsteady Wandering Magnetic Field Lines, Turbulence, Magnetic Reconnection, and Flux Ropes	Suggests the use of complementing Earth-based experimental collaborations and observations, simulations, and theory to understand unsteady, three-dimensional, MHD-like energy conversion.
116	Israel, Martin H., et al.	The Effect of the Heliosphere on Galactic and Anomalous Cosmic Rays	Recommends improving observations of galactic cosmic rays and anomalous cosmic rays to better understand how the interplanetary magnetic field modulates both galactic and anomalous cosmic rays in the inner solar system.
117	Jackson, Bernard V., et al.	SWIRES, a Solar Wind Instrument for Remote Sensing	Proposes a visible-light imager that provides solar wind bulk density measurements from an 840 km Sun-synchronous terminator polar orbit.
118	Jackson, Bernard V., et al.	PERSEUS, A Pegasus Explorer for Remote Sensing and In-Situ Space Science	Proposes PERSEUS instruments to provide all-sky coverage to enable mapping and three-dimensional reconstruction of the heliosphere.
119	Jensen, Elizabeth A., et al.	Campaign Observations of the Heliosphere During the STEREO Superior Conjunction	Recommends the simultaneous measurement of the magnetic, velocity, and density fields of the heliosphere using the radio signal from natural sources and spacecraft in superior conjunction.
120	Ji, Hantao, et al.	Next Generation Experiments for Laboratory Investigations of Magnetic Reconnection Relevant to Heliophysics	Describes the scientific opportunity for next-generation laboratory experiments to study magnetic reconnection in regimes directly relevant to space and solar plasmas.
121	Ji, Hantao, and S. Prager	Strengthening Heliophysics Through Coordinated Plasma Astrophysics Programs with Laboratory Plasma Physics and Astrophysics	Introduces scientific opportunities articulated by the Workshop on Opportunities in Plasma Astrophysics and recommends close coordination with laboratory plasma physics and astrophysics to strengthen heliophysics programs.
122	Johnson, Les, et al.	Solar Sail Propulsion: Enabling New Capabilities for Heliophysics	Reports on a sampling of missions enabled by solar sails, the current state of the technology, and what funding is required to advance the current state of technology such that solar sails can enable these missions.
123	Johnston, Janet C., and D.F. Webb	Detecting and Tracking Solar Ejecta with Next-Generation Heliospheric Imaging Systems	Discusses the need for a low-risk L5 imager, and/or an L1 or LEO Sun-Earth line imager, to bank on knowledge gained from the Solar Mass Ejection Imager and STEREO.

continued

TABLE I.1 Continued

RFI Response Number	First Author	Response Title	Summary Description
124	Jones, Andrew R., and F. Eparvier	The Importance of Fundamental Laboratory Measurements to NASA Heliophysics	Stresses the importance of fundamental laboratory measurements of quantities such as cross sections, atomic scattering factors, and reaction rates that are vital to interpreting data from existing and future missions, as well as essential for instrument design.
125	Judge, Philip G.	Measuring Magnetic Free Energy in the Solar Atmosphere	Offers a credible method for measuring free magnetic energy in the solar atmosphere with infrared imaging technology.
126	Kanekal, Shrikanth G., et al.	Heliospheric Particle Explorer: Advancing Our Understanding of Magnetospheric, Solar Energetic Particle, and Cosmic Ray Physics	Proposes a low-Earth-orbiting satellite with an instrument payload that measures energetic particles over a wide range of energies and species.
127	Kashyap, Vinay L., et al.	The Sun as a Star	Argues for increased focus on studies that target solar-stellar connections.
128	Keeley, Helena	Using KEEL Technology for Vehicle Prognostics and Diagnostics, and for Other Space Applications	Addresses how Compsim's KEEL (Knowledge Enhanced Electronic Logic) technology can be applied horizontally in the heliophysics realm (theory and modeling; innovations: technology, instruments, and data systems) and describes how KEEL can satisfy NASA's future space needs, which would otherwise cost billions of dollars.
129	Keesee, Amy M., et al.	A Campaign to Understand Mechanisms Responsible for Ion Heating in Magnetic Reconnection	Proposes a multidisciplinary campaign to address the mechanisms of ion heating in magnetic reconnection.
130	Keil, Stephen L., et al.	Science and Operation of the Advanced Technology Solar Telescope	Outlines the science goals for ATST and expresses its ability to impact understanding of the Sun.
131	Keil, Stephen L., et al.	Generation, Evolution, and Destruction of Solar Magnetic Fields	Proposes a project to measure the Sun's magnetic fields on their natural physical scales with a large-aperture solar telescope, namely ATST.
132	Keiling, Andreas	Science and Mission Concept of a Holistic Ionosphere-Auroral Zone-Magnetosphere Investigation	Outlines a holistic ground-spacecraft mission with four-point conjunctions along magnetic flux tubes connecting the ionosphere, the auroral acceleration region, and the outer magnetosphere.
133	Kepko, Larry, et al.	A NASA-funded CubeSat Program	Argues for a small augmentation to the Suborbital and Special Orbital Projects program to allow for CubeSats as an available science and technology platform.
134	Kepko, Larry, and G. Le	Magnetospheric Constellation	Describes a mission to trace the transport of mass and energy across the boundaries of and within Earth's magnetosphere using a constellation of up to 36 small satellites.

TABLE I.1 Continued

RFI Response Number	First Author	Response Title	Summary Description
135	Klimchuk, James A.	Maximizing NASA's Science Productivity	Argues that the science productivity of NASA's Heliophysics Science Division is not maximized by the current program balance and recommends that research and analysis (R&A) funding be gradually increased by 10% of the Heliophysics Science Division budget to an eventual target of 20-25% of the total budget.
136	Klumpar, David M., et al.	The Technological Case for TinySats (CubeSats and Nanosatellites) in Support of Heliophysics Research and the National Space Weather Program	Establishs that rapid developments in electronics miniaturization, new manufacturing techniques, and new materials and the development of a cadre of commercial suppliers of small satellite subsystems fortify the technical readiness of CubeSats for heliophysics research.
137	Ko, Yuan-Kuen et al.	Breakthrough Toward Understanding the Solar Wind Origin	Argues that a breakthrough in the next decade toward understanding the origin of the solar wind will require the continued collection and analysis of solar spectroscopic and in situ solar wind ion composition data, and future mission designs that address the solar wind origins need to take into account the coexistence and coordination of these two types of instruments.
138	Ko, Yuan-Kuen, and G.A. Doschek	Systematic Science for Future Missions	Examines strategies for future fleets of space missions to achieve systematic, optimal science, identifies key measurements that should not be sacrificed, and proposes coordination of the locations and timing of data availability.
139	Komjathy, Attila, et al.	Detecting Tsunami Generated Ionospheric Perturbations Using GPS Measurements	Argues for a concept of ionospheric sounding that would provide a method of tsunami confirmation using NASA's global network of real-time GPS receivers.
140	Korendyke, Clarence M., et al.	Fine-Scale Advanced Coronal and Transition Region Spectrometer (FACTS) Mission: An Imaging Spectroscopy Mission to Observe Physical Processes of the Solar Chromosphere, Transition Region, and Corona	Proposes a mission to determine and characterize the dominant physical processes responsible for the structure, dynamics, and evolution of the upper solar atmosphere.
141	Kosovichev, Alexander, et al.	Solar Dynamo	Proposes that a significant breakthrough can be made to advance understanding of the physical mechanisms of magnetic field generation and formation of magnetic structures on the Sun with the help of targeted funding to support coordinated interdisciplinary groups of observers, theorists, and modelers, working together on solar dynamo as a single complex problem.
142	Krall, Jonathan, and J.D. Huba	Physics-Based Modeling of the Plasmasphere	Argues the case that a key priority of U.S. space physics enterprise should be to develop a physics-based numerical model of the coupled magnetosphere-ionosphere system that will describe the plasmasphere and its interactions with the ionosphere, magnetosphere, and ring current.

continued

TABLE I.1 Continued

RFI Response Number	First Author	Response Title	Summary Description
143	Kucharek, Harald, et al.	Multi-Scale Investigations of Fundamental Physical Processes	Proposes research to understand the role of coupling between different scales in particle acceleration, energy dissipation, and plasma transport in shocks, reconnection, and turbulence.
144	Laming, J. Martin, et al.	Science Objectives for an X-Ray Microcalorimeter Observing the Sun	Presents the science case for a broadband X-ray imager with high-resolution spectroscopy, including simulations of X-ray spectral diagnostics of both active regions and solar flares.
145	Laming, J. Martin, et al.	Understanding the Coronal Abundance Anomalies of the Sun	Emphasizes the importance of the first ionization potential abundance anomaly in regions of the solar corona and wind, especially for the insight it provides into wave-particle interactions in the solar atmosphere and how these might inform models of coronal heating.
146	Larsen, Miguel F., and G. Lemacher	Diffusion and Transport Near the Turbopause	Addresses the poorly understood turbulent diffusion and transport processes in the lower thermosphere and the need for more extensive in situ observations of the neutral dynamics in the region.
147	Lawrence, David J., et al.	Using Solar Neutrons to Understand Solar Acceleration Processes	Explains the need for robust neutron measurements in concert with coordinated observations of gamma rays, energetic ions, and electrons, extreme ultraviolet, and radio waves to fully understand solar acceleration mechanisms.
148	Lazio, Joseph, et al.	Magnetospheric Emission from Extrasolar Planets	Describes the effort needed to detect and use magnetospheric emissions from extrasolar planets to help with understanding the nature of planets and magnetospheres.
149	Lehmacher, Gerald	Small-Scale Neutral-Ion Coupling in the Mesosphere	Addresses the fundamental science questions, What governs the coupling of neutral and ionized species in the mesosphere? and What is responsible for the variability of the ionization layers in the mesosphere?, and urges mesospheric investigations for the revitalization of the sounding rocket program.
150	Lemon, Colby L., et al.	The Importance of Ion Composition and Charge State Measurements for Magnetospheric Physics	Advocates for more ion composition instruments in the magnetosphere on future missions in order to resolve outstanding questions in magnetospheric physics.
151	Lepri, Susan T., et al.	Solar Wind and Suprathermal Ion Composition Measurements: An Essential Element of Current and Future Space Missions	Discusses the vital role of solar wind and suprathermal composition measurements in resolving major outstanding science questions regarding reconnection, particle acceleration, and improving space weather predictions.
152	Lessard, Marc R., et al.	The Importance of Ground-Observations and the Role of Distributed Arrays in Polar Regions	Describes the importance of high-latitude ground-based observations, emphasizing the importance of instrument development, distributed arrays of instruments, and multi-instrument observations.

TABLE I.1 Continued

RFI Response Number	First Author	Response Title	Summary Description
153	Lewis, Laurel M.	The Determination of the Effects of Major Impacts on Global Geophysical and Geological Parameters	Proposes the study of global geologic and geophysical parameters in response to solar/galactic fields in order to better understand the past influence impact has had on Earth history as well as to better determine the probability of future events.
154	Li, Gang, et al.	A "Swarm" Mission to Study Particle Acceleration at Interplanetary Shocks	Proposes a mission to study particle acceleration at interplanetary shocks via a swarm of spacecraft.
155	Li, Xinlin, et al.	Energetic Particles from a Highly Inclined Constellation (EPIC)	Addresses solar flares and solar energetic particles reaching at Earth, the loss rate of Earth's radiation belt electrons, and the effect of these energetic particles on the chemistry and dynamics of Earth's middle and upper atmosphere.
156	Liewer, Paulett C., et al.	Solar Polar Imager: Observing Solar Activity from a New Perspective	Proposes a mission to target the unexplored polar regions by enabling crucial observations not possible from lower latitudes with a 0.48-AU orbit with an inclination of 75° and a solar sail.
157	Lin, Chin S., and F.A. Marcos	Predict Neutral Density	Proposes research to improve predictions of satellite drag with physics-based atmospheric density models.
158	Lin, Chin S., and F.A. Marcos	Research-to-Operation of Predicting Neutral Density	Proposes research to use physics-based atmospheric density models to improve research-to-operation of predicting satellite.
159	Lin, Chin S., and F.A. Marcos	CubeSat Orbital Drag Experiment	Proposes an experiment to exploit the CubeSat opportunity, which provides the unprecedented capability of long-term, routine, high-accuracy measurements of thermospheric variability by accelerometers.
160	Lin, Robert P., et al.	Expansion of the Heliophysics Explorer Program	Points out that Heliophysics Explorer missions have the best success record in all respects of any space missions.
161	Lin, Robert P., et al.	The Multi-Spacecraft Inner Heliosphere Explorer (HELIX)	Proposes a five-spacecraft mission to study large-scale solar transients in the inner heliosphere and their acceleration of particles to high energies.
162	Lin, Robert P., et al.	Solar Eruptive Events (SEE) 2020 Mission Concept	Proposes a complement of advanced new instruments that focus on the coronal energy release and particle acceleration sites of major solar eruptive events.
163	Lind, Frank D.	Next Generation Space Science with the Geospace Array	Addresses science topics from the lower atmosphere through the ionosphere and heliosphere and to the surface of the Sun and beyond with a globally deployed geospace array.
164	Livi, Stefano A., et al.	Solar Wind Ion Composition Measurements	Proposes an instrument to establish physical links between the outward transport of solar energy and the solar wind by providing direct measurements of solar eruption products in coronal mass ejections.

continued

TABLE I.1 Continued

RFI Response Number	First Author	Response Title	Summary Description
165	Love, Jeffrey J.	Long-term Coordinated Ground and Satellite Monitoring of the Ring Current	Proposes improved, long-term, low-latitude ground- and space-based magnetometer monitoring for at least an entire solar cycle.
166	Luhmann, Janet G., et al.	Guest Investigator and Participating Scientist Programs	Expresses that Guest Investigator and Participating Scientist programs and mission modeling and theory teams provide major enhancements to the science potential of NASA's missions.
167	Luhmann, Janet G., et al.	Extended Missions: Engines of Heliophysics System Science	Summarizes arguments for supporting the Heliophysics Systems Observatory, the engine of heliospheric systems science.
168	Lyons, Larry R.	Conceptual Framework for Space Weather Dynamics: An Interplay of Large and Mesoscale Structure within the Nightside Magnetosphere-Ionosphere-Thermosphere System	Describes a concept to allow for unprecedented comprehensive interdisciplinary study of the coupled magnetosphere-ionosphere-thermosphere system with new facilities and model development and to coordinate the use of these capabilities for transformational understanding of structure, dynamics, and disturbances.
169	Lystrup, Makenzie, et al.	A Multi-Spacecraft Jupiter Space Plasma Explorer	Proposes a multi-spacecraft Jupiter explorer mission to measure the jovispace plasma environment in key locations simultaneously—within and without the magnetosphere, in the plasma disc, in boundary regions, in the dawn and dusk flanks—all while monitoring solar wind and auroral energy output.
170	Mabie, Justin J.	A Comprehensive and Continuous Record of Ground Based Space Weather Observations	Recommends steps to modernize a comprehensive climatology of ground-based space weather observations collected from magnetometers, ionosondes, and other methods.
171	MacDonald, Elizabeth A., et al.	A Science Mission Concept to Actively Probe Magnetosphere-Ionosphere Coupling	Describes how directly mapping magnetic field lines from a magnetospheric satellite to their ionospheric footpoints using an on-board electron emitter and ground-imaging techniques can answer long-standing fundamental questions of magnetosphere-ionosphere coupling.
172	MacDowall, Robert J., et al.	A Radio Observatory on the Lunar Surface for Solar Studies (ROLSS)	Proposes an observatory to image solar radio bursts at frequencies <10 MHz with a lunar-based radio telescope.
173	Makela, Jonathan J., et al.	A North American Thermosphere Ionosphere Observation Network	Recommends a network of ground-based multi-instrument sites to understand fundamental spatio-temporal processes in Earth's ionosphere/thermosphere/mesosphere system.
174	Mannucci, Anthony J.	Global Ionospheric Storms	Discusses global ionospheric storms as an important subfield of study within solar and space physics.
175	Mannucci, Anthony J.	Research to Operations: Continuous Improvement	Suggests a path forward that, over time, will lead to steadily improving operational capabilities in space weather.

TABLE I.1 Continued

RFI Response Number	First Author	Response Title	Summary Description
176	Mannucci, Anthony J., et al.	GNSS Geospace Constellation (GGC): A CubeSat Space Weather Mission Concept	Recommends technology investment in miniaturized GPS receivers that can be deployed on CubeSats for ionospheric remote sensing.
177	Mannucci, Anthony J., et al.	Estimating the Forces That Drive Ionosphere and Thermosphere Variability: Continuous Data and Assimilative Modeling	Advocates for developing a model-based approach to retrieving the driving forces from measurements of electron density structure and dynamics.
178	Marshall, Robert A.	Ionospheric Forcing from Below: Effects of Lightning	Demonstrates that extension of single-discharge studies to global effects are required to quantify the coupling in ionosphere forcing.
179	Marshall, Robert A.	Ground-based Space Weather Instrument Suites	Proposes that the geoscience community foster a class of instrumentation that involves multi-instrument, ground-based suites similar to satellite instrumentation.
180	Martinis, Carlos	Neutral Winds in the Upper Atmosphere	Discusses the need to measure neutral winds in the upper atmosphere in a global spatial and temporal scale to understand many processes involving electro-dynamics and ion-neutral coupling.
181	Matthaeus, William H., et al.	Turbulence and Nonlinear Dynamics and Its Many Effects in Solar and Heliospheric Physics	Calls attention to the broad implications of nonlinearity and turbulence within the complex, coupled solar and heliospheric system.
182	Mauel, Michael E., et al.	Development and Validation of Space Weather Models Using Laboratory Dipole Experiments	Describes the value of laboratory dipole experiments to develop and validate space weather models of magnetospheric dynamics.
183	Mazur, Joseph E.	Ultra-Heavy Nuclei in Solar Flare: The Rarest Elements in the Sun	Discusses the measurement of ultra-heavy energetic particles to understand nucleosynthesis and further processing of matter in flare acceleration, interplanetary acceleration, and transport.
184	Mazur, Joseph E.	Low-Impact Space Environment Sensors Required on Every NASA Space Vehicle	Recommends that space vehicles be required to include low-impact sensors and that the data be collected and synthesized in a centralized repository.
185	Mazur, Joseph E.	Need to Measure Solar Energetic Particle Ionization States from ~1 to Above 100 MeV/nucleon	Proposes one or more charged particle sensors in low-Earth polar orbit to infer the ionization state using the geomagnetic cutoff technique.
186	Mazur, Joseph E.	Unintended Effects of Increasing Reliance on Science Requirements	Details the unintended consequences of the practice of tracking NASA science missions with project management systems that establish and monitor requirements for science.
187	Mazur, Joseph E.	Unexploited Heliophysics Data Sets	Explains the value of the particle and plasma data sets collected from highly inclined and low-Earth-orbit research into the sources and dynamics of the near-Earth trapped and precipitating particle environments.
188	McComas, David J., et al.	Interstellar Mapping Probe (IMAP) Mission Concept: Illuminating the Dark Boundaries at the Edge of Our Solar System	Proposes the IMAP mission concept for the discovery of the detailed processes of the heliosphere/local interstellar medium interaction.

continued

TABLE I.1 Continued

RFI Response Number	First Author	Response Title	Summary Description
189	McConnell, Mark L., et al.	X-Ray and Gamma-Ray Polarimetry of Solar Flares	Reviews the study of polarization at X-ray and gamma-ray energies for a greater understanding of particle acceleration in solar flares.
190	McCormack, John P., and S. Eckermann	High Altitude Data Assimilation: Characterizing the Effects of Solar Variability from the Ground to the Thermosphere	Proposes a high-altitude data assimilation system capable of exploiting satellite- and ground-based observations using state-of-the-art assimilation techniques to improve the observational characterization of the atmospheric response to solar variations.
191	McDonald, Sarah E., et al.	The Importance of Thermospheric Winds for Ionospheric Modeling	Presents specific examples where neutral wind measurements are needed to enable reliable ionospheric modeling due to variations in the neutral wind, which drive a complex system of ionospheric currents and electric fields, profoundly influencing the structure and composition of the ionosphere.
192	McHarg, Matthew G., and D. Knipp	Measuring Energy Inputs and ITM Response Using a Constellation of Small Satellites	Proposes a constellation of inexpensive small satellites to provide measurements for understanding the fundamental physics that will enable ionosphere-thermosphere-mesosphere forecasting.
193	McIntosh, Scott W., et al.	The Solar Chromosphere: The Inner Frontier of the Heliospheric System	Discusses the status of chromospheric physics and the frontiers that are opening up following recent observational discoveries.
194	McIntosh, Scott W., et al.	ChroMag: The Community Synoptic Chromospheric Magnetograph	Proposes a chromospheric magnetometry mission to provide a comprehensive, synoptic spectro-polarimetric observational data set from spectral lines formed at multiple "heights" of the chromosphere and the very base of the corona.
195	McNutt, Ralph L., Jr., et al.	Interstellar Probe	Proposes an interstellar probe that can be launched during the coming decade.
196	Merkin, Viacheslav G., et al.	Synergy Between Large Data Sets, First-Principles and Empirical Models of the Magnetosphere	Advocates for the need of support for programs that build on synergy between currently available and future large data sets of ionospheric and magnetospheric measurements and physics-based models of the ionosphere-thermosphere-magnetosphere system.
197	Mertens, Christopher J., et al.	Nowcast of Atmospheric Ionizing Radiation for Aviation Safety	Describes the Nowcast of Atmospheric Ionizing Radiation for Aviation Safety, a prototype operational model for predicting commercial aircraft radiation exposure from galactic and solar cosmic rays.
198	Mertens, Christopher J., et al.	Ionospheric E-Region Chemistry and Energetics	Proposes an Earth-observing, multisatellite science mission to explore the last remaining frontier in upper atmospheric research—the ionospheric E-region.

TABLE I.1 Continued

RFI Response Number	First Author	Response Title	Summary Description
199	Miesch, Mark S., et al.	The Importance of Polar Observations in Understanding the Solar Dynamo	Supports an out-of-ecliptic heliophysics mission that focuses on observations of the magnetic structure, dynamics, and solar-cycle evolution of the polar regions of the Sun that will provide data of critical importance for understanding the solar dynamo mechanism and the cyclic nature of solar activity.
200	Millan, Robyn M., et al.	NASA's Balloon Program: Providing World-Class Science, Technology Development, and Vital Training of the Next Generation of Space Physicists	Summarizes examples of balloon-based science accomplishments and future opportunities and advocates continued support of the Ultra Long Duration Balloon Program, increased support for small and mid-sized payloads and balloon flotillas, and appropriately scaled funding for development of new experiments.
201	Miller, Ethan S.	Initiation of Irregularities in the Equatorial F-Region Ionosphere	Explores several theories and a variety of space- and ground-based measurements that will be useful to test theories on the initiation of irregularities in the equatorial F region ionosphere.
202	Mitchell, Donald G., et al.	Geospace Magnetosphere-Ionosphere-Neutral Interaction (GEMINI)	Describes GEMINI, which uses two identical spacecraft in an 8 R_E circular polar orbit for global, continuous imaging of the ring current, plasma sphere, atmospheric ultraviolet, and auroral emissions.
203	Mitchell, Elizabeth J.	Center for Magnetosphere and Ionosphere Decoupling Investigations	Outlines an initiative to correct the erroneous assumption of polar cap symmetry in ring current and radiation belt models.
204	Mlynczak, Martin G., et al.	Spectral Signatures of Geospace Climate Change	Proposes continued measurement of infrared spectral signatures of the energy balance of the geospace environment in order to identify and attribute causes and consequences of geospace climate change.
205	Moebius, Eberhard, et al.	NASA's Explorer Program as a Vital Element to Further Heliophysics Research	Details the importance of Explorers to a mix of large and small mission opportunities for heliophysics in maintaining innovative research and a diverse infrastructure and in training the future workforce.
206	Moore, Thomas E., et al.	Mechanisms of Energetic Mass Ejection (MEME)	Proposes MEME to achieve the overarching objective of the 2009 Heliophysics Roadmap: Origins of Near Earth Plasmas—to understand the transport of terrestrial gas and plasma from its atmospheric source into the Magnetosphere and downstream Solar Wind.
207	Moore, Thomas E.	Laboratory for Active Space Experiment Research (LASER) Program	Presents a case for the resumption of active space experimentation in heliophysics.
208	Moses, J.D.	Magnetic Properties of the Solar Atmosphere (SolmeX Cosmic Vision Mission)	Summarizes the Solar Magnetism Explorer (SolmeX) mission proposal being submitted to the 2010 ESA Cosmic Vision call, which proposes to overcome the observational gap in the measurements of the coronal magnetic field.

continued

TABLE I.1 Continued

RFI Response Number	First Author	Response Title	Summary Description
209	Nossal, Susan M.	Long Term Observations for Trend Studies	Addresses the importance of long-term observations for understanding the chemical and physical processes affecting the whole atmosphere system and recommends observations to enable long-term trend studies into the future.
210	Oberheide, Jens, et al.	Short-term Variability of the IT System	Proposes a comprehensive observation program to untangle the complex web of interacting processes and wave coupling that causes day-to-day variability in the ionosphere-thermosphere system.
211	Oberoi, Divya, et al.	Heliospheric Science at Low Radio Frequencies	Describes how the new generation of low-radio-frequency telescopes provides effective means to exploit the electromagnetic propagation effects to probe the heliosphere and presents the possibility of characterizing it in unprecedented detail.
212	Oberoi, Divya, et al.	High-Fidelity Coronal Imaging at Low Radio Frequencies	Describes how spectroscopic imaging of the Sun at low radio frequencies with next-generation radio interferometers can play a crucial role in addressing many long-standing puzzles related to magnetic fields and heating mechanisms in the solar corona.
213	O'Brien, Thomas P.	Long-term Monitoring of the Global Space Environment	Proposes a long-term inner magnetosphere monitor in a geosynchronous transfer orbit, NASA-to-NOAA (research to operations) transfer of future NASA explorers after the end of their science mission.
214	Onsager, Terrance G.	Need for Explicit Basic and Applied Research Funding	Advocates for distinct lines of funding by the U.S. civilian agencies for basic space physics research and the development of space weather applications, maintaining distinct requirements for both.
215	Osten, Rachel A.	Deepening the Solar/Stellar Connection for a Better Understanding of Solar and Stellar Variability	Encourages deepening of the solar/stellar connections, by recognizing a mutual relationship from which both solar physicists and stellar astronomers can benefit.
216	Oza, Nikunj C., et al.	Data Mining for Heliophysics	Recommends new data mining techniques that are needed to support breakthroughs in heliophysics.
217	Papadopoulos, Dennis	Active Experiments in Space—Ionospheric Heaters	Describes a program that uses ionospheric heaters with ground and space diagnostics to study "cause and effect" space plasma processes of key importance to space plasma physics and geophysics.
218	Pevtsov, Alexei A.	Current and Future State of Ground-Based Solar Physics in the U.S.	Draws attention to need for development of a comprehensive long-term plan for ground-based solar physics.

TABLE I.1 Continued

RFI Response Number	First Author	Response Title	Summary Description
219	Pfaff, Robert, et al.	Understanding Geospace on a Grand Scale: The Global Ionosphere/ Thermosphere Constellation	Proposes a Frontier mission consisting of a constellation of observing platforms that would sample the ionosphere and thermosphere at all local times and latitudes and hence provide a revolutionary advance for understanding the processes that define this critical region of geospace.
220	Pfaff, Robert, et al.	Sounding Rockets as Indispensable Research Platforms for Heliophysics Research and Development of a High Altitude Sounding Rocket	Provides an overview of the sounding rocket program's capabilities, which are critical for heliophysics research, and describes the High Altitude Sounding Rocket initiative.
221	Pilewskie, Peter, et al.	The Total and Spectral Solar Irradiance Sensor: Response to the National Academy of Science Decadal Strategy for Solar and Space Physics	Recommends continuity of the measurements of total and spectral solar irradiance from space.
222	Podesta, John, et al.	High-resolution, High-accuracy Plasma, Electric, and Magnetic Field Measurements for Discovery of Kinetic Plasma Structures and Processes in the Evolving Solar Wind	Addresses the need for high-cadence, high-accuracy plasma and field measurements to diagnose kinetic scale processes in the solar wind and interplanetary medium.
223	Rast, Mark P., et al.	Next Steps in Solar Spectral Irradiance Studies	To understand the solar spectral output and its coupling to climate is a challenge in the next decade with full-disk radiometric imaging of the Sun, high-resolution observation and modeling of globally unresolved dynamics, and coupling to models of radiative and dynamical processes at Earth.
224	Reardon, Kevin P., et al.	Approaches to Optimize Scientific Productivity of Ground-based Solar Telescopes	Explains the value in the broader use of high-resolution, ground-based solar observations and the need for university programs to train young scientists in the analysis of such data.
225	Rempel, Matthias, et al.	Modeling of Magnetic Flux Emergence Across Scales	Summarizes recent progress in modeling flux emergence from the base of the convection zone into the solar corona and details the need for modeling capabilities and large coordinated teams of investigators that are currently not supported through available programs.
226	Retterer, John M.	Next Gen IT Modeling Infrastructure for Space Weather Forecasting	Describes new infrastructure required for progress in ionosphere/thermosphere modeling, both for scientific progress and operational utility.
227	Richardson, John D., et al.	The Heliospheric Interaction with the LISM: Observations and Models	Emphasizes the scientific progress to be made by continuation of the Voyager mission with appropriate theoretical and modeling support.
228	Rickard, Lee J., et al.	The Long Wavelength Array (LWA): A Large HF/VHF Array for Solar Physics, Ionospheric Science, and Solar Radar	Provides a new approach for studying the Sun-Earth environment from the surface of the Sun through Earth's ionosphere.
229	Ridley, Aaron J., et al.	A Constellation Mission to Understand the Thermospheric Reaction to Energy Input Across Scales	Proposes an ionosphere/thermosphere constellation mission utilizing micro-satellites to study the dynamics of the upper atmosphere after energy input.

continued

TABLE I.1 Continued

RFI Response Number	First Author	Response Title	Summary Description
230	Roberts, D. Aaron, et al.	The Heliophysics Data Environment as an Enabler of HP Science of the Next Decade	Argues for the continued support of a "heliophysics information system," as recommended in the previous decadal survey, as is being developed by many groups.
231	Roddy, Patrick A., and J.O. Ballenthin	The Atmospheric Density Specification Experiment	Proposes a neutral mass spectrometer mission to feed these models of the neutral thermosphere and improve atmospheric drag forecasts.
232	Roelof, Edmond C., and G.B. Andrews	Telemachus Redux	Describes Telemachus, a technologically ready low-to-medium-cost, dual-mode mission that addresses basic science questions of the dynamics of the subsurface, surface, and coronal dynamics of the polar regions of the Sun and the non-ecliptic solar wind, fields, and energetic particles.
233	Rowland, Douglas E., et al.	The Tropical Coupler Mission	Proposes a mission to provide a complete understanding of the forcing of the ionosphere and lower thermosphere by stratospheric and tropospheric effects at low latitudes and to determine the causes and consequences of internal ionospheric instabilities.
234	Russell, Christopher T., et al.	Determination of How Charged Interplanetary Dust Affects the Flowing Magnetized Solar Wind	Recommends funding opportunities and mission opportunities for research on solar wind turbulence, such as multiscale, multisatellite measurements in the undisturbed solar wind with dust detectors capable of measuring the mass, speed, charge, and elemental/chemical composition of the solar wind entrained grains, as well as sensitive plasma and field instruments.
235	Ryan, James M., et al.	Ground Based Measurements of Galactic and Solar Cosmic Rays	Discusses the merits of measuring and monitoring solar and galactic cosmic rays at ground level.
236	Rymer, Abigail M.	The Case for Exploring Uranus' Magnetosphere	Advocates support from the heliophysics community for a dedicated mission to Uranus, launched in the time frame 2020-2023 and designed to make detailed in situ observations of the unique Uranian magnetosphere.
237	Sanchez, Ennio R., et al.	Magnetic Meridian Ring of Incoherent Scatter Radars: Supporting Science Discovery and Tracing AIM Weather, Climate, and Global Change	Advocates the concept of positioning atmospheric observatories built around incoherent scatter radars at points all around a single geomagnetic meridian to provide global system scientists, modelers, and data consumers with the high-quality data they require.
238	Scherb, Frank	The Abundance of Deuterium and He3 in the Solar Wind	Describes a novel instrument for detecting and measuring solar wind deuterium.
239	Schreiner, William S.	Using Space-Based GNSS Radio Occultation Data for Ionospheric and Space Weather Applications	Addresses the scientific and operational needs for continuing global observations of the ionosphere and lower neutral atmosphere with Global Navigation Satellite System radio occultation data.

TABLE I.1 Continued

RFI Response Number	First Author	Response Title	Summary Description
240	Schrijver, Karel, et al.	The Solar Magnetic Dynamo and Its Role in the Formation and Evolution of the Sun, in the Habitability of Its Planets, and in Space Weather around Earth	Addresses the needs and opportunities for dynamo-related studies in the coming decade(s), including understanding the solar magnetic dynamo by combining numerical studies and theory with observations of the evolving surface field of Sun and stars and of solar and stellar internal flows.
241	Schwadron, N.A., et al.	Research to Operations (Res2Ops)—Opportunities for Center Development	Discusses the development of research-to-operations centers at universities and laboratories to develop highly successful operational tools to fill demonstrated or emerging needs and combines best practices in engineering, physics, computer science, and management.
242	Schwadron, Nathan A., et al.	NESSC Summer School for Undergraduates in Space Physics	Describes the New England Space Science Consortium (NESSC), which can provide both governance and lecturers in a new 1- or 2-week summer school for undergraduate students that provides students with an intellectual background to help in the pursuit of research projects (e.g., an REU gateway) and graduate programs.
243	Semeter, Joshua, et al.	Energy Transfer from the Solar Wind to the Solid Earth	Recommends distributed measurements from ground and from space, coupled through first-principles modeling, to establish the global force balance governing the interaction between the solar wind and the magnetosphere.
244	Shih, Albert Y., et al.	Solar Ion Acceleration and the Flaring Atmosphere	Discusses the open science questions, the remotely observable signatures (e.g., gamma rays, neutrons, and energetic neutral atoms), and several instrument concepts that are associated with understanding solar ion acceleration, as well as the aspects of the flaring atmosphere that are probed by energetic ions.
245	Siskind, David E., et al.	Dynamical Ground-to-Space (G2S) Coupler	Outlines the recent science linking the lower to the upper atmosphere and discusses measurement strategies to understand these couplings.
246	Smith, Charles W., et al.	The Case for Continued, Multi-Point Measurements in Space Science	Argues for preservation of the distributed assets of space science that constitute the heliospheric "Great Observatory" and of this unique creation with all the opportunities it represents.
247	Smith, David M., et al.	The High-Energy Sun at High Sensitivity: A NuSTAR Solar Guest Investigation Program	Describes the extraordinary solar science expected from NuSTAR, a NASA small Explorer satellite to be launched in 2012 primarily for astrophysical observations, and outlines the advantages of a dedicated guest investigator program specifically to support NuSTAR solar observations.

continued

TABLE I.1 Continued

RFI Response Number	First Author	Response Title	Summary Description
248	Smith, Steven M.	Gravity Wave Coupling Processes Between the Lower Atmosphere and the Mesosphere and Lower Thermosphere	Proposes a research strategy to forecast space weather phenomena with a fundamental understanding of the dynamics (e.g., energy and momentum flux budgets, chemistry, etc.) of Earth's atmosphere as a whole system.
249	Smith, Steven M.	Neutral Temperature and Wind Measurements of Earth's Thermosphere	Discusses the need for more accurate neutral temperature and wind climatologies of Earth's lower thermosphere in the altitude range 100-300 km.
250	Snow, Martin A., et al.	The Importance of Solar EUV and FUV Irradiance Measurements for Space Weather and Atmospheric Modeling	Argues that extending the data record of solar extreme ultraviolet (EUV) and far ultraviolet (FUV) irradiances should continue to be a priority of the heliophysics community for understanding both the long-term (climate) and short-term (space weather) influence of the Sun on the atmosphere.
251	Sojka, Jan J.	Adoption of a Paradigm Shift for Space Exploration	Presents a case for collaboration between all space science funding agencies to support synergistic activities that benefit all and provides the links between science and applications that will produce the expected societal benefits.
252	Solomon, Stanley C., and L. Qian	Modeling and Measurement of Upper Atmosphere Climate Change	Proposes a program of model development and space-based measurements for understanding and quantifying the increasing temperatures of the upper atmosphere.
253	Spann, James F., et al.	A Cross-Agency Enabling Effort Focused on Space Weather Observations and Research-to-Operation Transition	Presents an approach to address the lack of a robust national coordinated research-to-application program for space weather and a lack of sufficient relevant space- and ground-based observations by proposing a joint federal agency program involving NASA, NOAA, and NSF.
254	Spann, James F., et al.	A NASA Applied Spaceflight Environments Office Concept	Presents a solution to the NASA problem that there is no coordinated activity to harness the valuable knowledge and products across the field centers related to spaceflight environments, which includes space weather.
255	Spann, James F., et al.	Dynamic Geospace Coupling Mission	Addresses fundamental questions related to plasma processes that determine how energy and momentum from the solar wind propagate downward through geospace to Earth and proposes a multi-spacecraft implementation with imaging and in situ instruments.
256	St. Cyr, O.C., et al.	Space Weather Diamond: A 10x Improvement in Real-Time Forecasting	Promotes an applied heliophysics mission concept to facilitate the connection between science and societal needs (e.g., improvements in space weather prediction) by providing an order of magnitude improvement over present-day L1 monitors.
257	St. Cyr, O.C., et al.	Solar Orbiter: Exploring the Sun-Heliosphere Connection	Describes the ESA/NASA Solar Orbiter mission, including the science and mission design.

TABLE I.1 Continued

RFI Response Number	First Author	Response Title	Summary Description
258	Stephan, Andrew W., et al.	Global Ionosphere-Thermosphere-Mesosphere (ITM) Mapping Across Temporal and Spatial Scales	Provides a robust method for obtaining necessary routine space weather maps of the near-Earth space weather volume for both scientific inquiry and operational users by combining global imaging and low-altitude sensors measuring ultraviolet airglow.
259	Stevens, Michael H., et al.	Direct Observations of Global-Scale Transport in the Lower Thermosphere	Describes a mission to study global-scale dynamics by observing tracers injected by vehicles launched from a satellite in geosynchronous orbit.
260	Stoneback, Russell A.	Effective Aperture Behavior on the Earth and Sun	Proposes continued research on a model to investigate the apparent equivalence between high-latitude currents on both Earth and the Sun and currents used to calculate the diffracted field produced by an aperture.
261	Streltsov, Anatoly V.	Multi-scale Electrodynamics of Magnetosphere-Ionosphere Interactions	Proposes comprehensive, multi-fluid, wave-particle numerical models with predictive capabilities to address observations from satellites, sounding rockets, and radars conducted in the high-latitude magnetosphere and the ionosphere, which frequently measure intense unexplained electromagnetic fields and currents.
262	Strong, Keith T., et al.	4PI: A Global Understanding of the Solar Cycle	Proposes a mission to provide the first continuous view of the solar magnetic field and plasma dynamics mapped from below the surface into the corona, over all solar longitudes with extended polar coverage, and tracks the evolution of the dynamo(s) throughout at least a solar cycle.
263	Swenson, Charles M., et al.	CubeSats in Heliophysics Research	Emphasizes that CubeSat technologies represent a significant opportunity for achieving multipoint observations from within the space environment.
264	Swenson, Gary R.	Remote Sensing the Upper Atmosphere with Lidar from Space	Advocates development of lidar technology for remote sensing of the upper atmosphere/ionosphere.
265	Szabo, Adam, et al.	Solar Wind Kinetic Physics: High Time Resolution Solar Wind Measurements from the DSCOVR Mission	Advocates refurbishment of the DSCOVR spacecraft to obtain unprecedented high time resolution solar wind measurements from the Sun-Earth L1 Lagrange point at minimal NASA cost.
266	Szabo, Adam, et al.	Energetic Particle Propagation and Coronal Mass Ejection Evolution in the Inner Heliosphere: Multi-point In-Situ Solar Sentinels Observations	Proposes the Solar Sentinels mission to provide in situ particle and field observations to answer questions of solar energetic particle transport and interplanetary coronal mass ejection evolution in the inner heliosphere.

continued

TABLE I.1 Continued

RFI Response Number	First Author	Response Title	Summary Description
267	Talaat, Elsayed R., et al.	Electrodynamics Observations with Numerous Satellites	Proposes a suite of small satellites distributed in local time at F-region altitudes with essential instrumentation to address ion-neutral coupling and the roles of the disturbance dynamo, tidal dynamos, and magnetospheric penetration electric fields in determining the global electrodynamics.
268	Tomczyk, Steven, et al.	COSMO—The Coronal Solar Magnetism Observatory	Presents the science justification and technical overview of the Coronal Solar Magnetism Observatory (COSMO), a facility dedicated to the measurement of magnetic fields and plasma properties in the solar corona to advance understanding of the Sun's generation of space weather.
269	Tylka, Allan J.	Heliophysics System Science and Funding for Extended Missions	Offers suggestions and recommendations regarding adequate funding for extended missions concerning all areas of heliophysics science.
270	Valladares, Cesar E.	On Understanding the Origin of Plasma Density Variability Within the Polar Cap	Describes two ionospheric processes that contribute to the high variability of the plasma density within the polar cap: polar cap patches and Sun-aligned arcs.
271	Velli, Marco, et al.	SAFARI: Solar Activity Far Side Investigation	Proposes a mission to explore the origins of solar magnetic activity by carrying out observations of the velocity and magnetic fields at the solar surface from a vantage point widely separated from Earth in longitude.
272	Vial, Jean-Claude	Fast UV Spectro-imagery for Solar Physics	Argues that imaging Fourier transform spectroscopy should be pursued to measure temperature, density, ionization, and abundance, along with magnetic and velocity fields in the solar atmosphere in three dimensions and time, in the high chromosphere, corona, and the transition region.
273	Vourlidas, Angelos, et al.	Mission to the Sun-Earth L5 Lagrangian Point: An Optimal Platform for Heliophysics and Space Weather Research	Argues that a research-to-operations approach is the best strategy to foster the vibrancy of the field in the next decade, spearheaded by a mission to the L5 Lagrangian point.
274	Vourlidas, Angelos, and A. Rymer	A Proposal to Lighten the Burden of International Traffic in Arms Regulations on Heliophysics Research	Describes the problem of the overly restricted regulatory regime on science missions classified as "defense systems" and provides a set of recommendations to improve this situation.
275	Walterscheid, Richard, et al.	Gravity Wave Propagation in the Dissipative and Diffusively Separated Thermosphere	Advocates further understanding of the upward coupling of waves from the lower atmosphere and new observations combined with models of acoustic-gravity waves for thermospheric regions where rapid dissipation and diffusive separation prevail.

TABLE I.1 Continued

RFI Response Number	First Author	Response Title	Summary Description
276	Walterscheid, Richard L., and J.H. Hecht	Effects of Large Amplitude Planetary Waves in the Ionosphere and Thermosphere	Demonstrates clear evidence that tides and planetary waves strongly affect each other, including the particularly interesting case of the rapid amplification of the Southern Hemisphere two-day wave to very large amplitudes via interactions with tides.
277	Walterscheid, Richard L., et al.	The Paired Ionosphere-Thermosphere Orbiters (PITO) Mission: Multipoint Geospace Science in 3D	Proposes a mission to utilize two spacecraft in equal but opposite eccentric orbits, so that when one is at apogee the other is at perigee and located within the field of view of the other, using combined measurements of the best features of remote sensing (coverage) with in situ measurements (detail).
278	White, Stephen M., et al.	Coronal Magnetic Fields	Outlines the difficulties in making advances in coronal magnetism and describes the developments needed to make progress over the next decade.
279	Wilson, Gordon R., and D. Ober	Local Response of the Ionosphere/Thermosphere to High-Latitude Energy Deposition	Designs a mission to monitor the atmospheric response in a local high-latitude region while simultaneously measuring the magnetospheric energy input to that region.
280	Wood, Kent S., et al.	Continuous FUV/EUV Imaging of the Ionosphere from Geosynchronous Orbit	Recommends new imaging systems to generate measurements in two-dimensional formats continuously for large regions with high spatial resolution.
281	Wu, Qian	Global Airglow Interferometer Limb-scanner (GAIL)—A New Thermospheric Wind Instrument	Describes a concept for a high-altitude limb-scan instrument that will measure the thermospheric winds (200 to 300 km) by recording wind-induced Doppler shift in the O 630-nm airglow emission day and night.
282	Yizengaw, Endawoke, et al.	Understanding the Unique Equatorial Electrodynamics in the African Sector	Proposes ground-based scientific instrument arrays in the African sector, a region that has been devoid of ground-based instrumentation for space science, in order to address the physics behind the unique equatorial ionospheric irregularities and bubbles often observed.
283	Zank, G.P., and J.R. Jokipii	A White Paper Advocating a Heliophysics Theory Mission	Recommends a major theory program that has the status of a mission.
284	Zank, G.P.	Role of the National Science Foundation ATM/GEO Directorate in Promoting and Supporting Space Physics	Recommends that the NSF Division of Atmospheric Sciences (ATM) Geosciences Directorate (GEO) not limit funding of heliospheric research to within 1 AU.
285	Zhang, Shunrong, et al.	Understanding Upper Atmospheric Climate and Change	Details the effects that the changes in Earth's upper atmosphere, thermospheric density, and ionospheric electron density will have on human activities.
286	Zhang, Yongliang, and L.J. Paxton	Partition and Variability of the Magnetospheric Energy Input into the Polar Ionosphere	Addresses questions on partition and variability of the magnetospheric energy input to the polar ionosphere, its dissipation, and its relation to the solar wind condition with multisatellite and ground-chain measurements.

continued

TABLE I.1 Continued

RFI Response Number	First Author	Response Title	Summary Description
287	Zhou, Xiaoyan, et al.	Dayside Aurora and Auroral Conjugacy	Discusses the scientific significance of dayside and conjugate auroras that have less ambiguity in the connection to their causes.
288	Zhu, Ping	Meso Scale Transients in Magnetotail and Their Roles in Substorm Dynamics	Proposes to systematically investigate meso-scale transients in the magnetotail, which may play key roles in mediating and regulating the transition process from the late substorm growth phase to the beginning of onset expansion.